Primatology Monographs

Series Editors

Tetsuro Matsuzawa
Inuyama, Japan

Juichi Yamagiwa
Kyoto, Japan

For further volumes:
http://www.springer.com/series/8796

Juichi Yamagiwa • Leszek Karczmarski
Editors

Primates and Cetaceans

Field Research and Conservation of Complex Mammalian Societies

Editors
Juichi Yamagiwa
Professor
Laboratory of Human Evolution Studies
Graduate School of Science
Kyoto University
Sakyo, Kyoto 606-8502, Japan
yamagiwa@jinrui.zool.kyoto-u.ac.jp

Leszek Karczmarski
Associate Professor
The Swire Institute of Marine Science
School of Biological Sciences
The University of Hong Kong
Cape d'Aguilar, Hong Kong
leszek@hku.hk

ISSN 2190-5967 ISSN 2190-5975 (electronic)
ISBN 978-4-431-54522-4 ISBN 978-4-431-54523-1 (eBook)
DOI 10.1007/978-4-431-54523-1
Springer Tokyo Heidelberg New York Dordrecht London

Library of Congress Control Number: 2013953884

Cover image: Front cover: *Upper*: A silverback male gorilla in Kahuzi-Biega National Park, Democratic Republic of Congo. Photo by Juichi Yamagiwa. *Center left*: A dusky dolphin performs a noisy leap in Admiralty Bay, New Zealand. Photo by Chris Pearson. *Center middle*: Flipper-to-body rubbing in Indo-Pacific bottlenose dolphins off Mikura Island, Japan. Photo by Mai Sakai. *Center right*: A pair of white-handed gibbons in a fig tree in Khao Yai National Park, Thailand. Photo by Kulpat Saralamba. Back cover: *Upper*: Male and female juvenile bottlenose dolphins pet each other in Shark Bay, Australia. Photo by Ewa Krzyszczyk. *Center left*: A male Japanese macaque in Yakushima Island, Japan. Photo by Juichi Yamagiwa. *Center middle*: Grooming chain of three chimpanzee females in Mahale Mountain National Park, Tanzania. Photo by Michio Nakamura. *Center right*: Indo-Pacific humpback dolphin mother and newborn calf in Hong Kong waters. Photo by Stephen Chan.

Printed on acid-free paper

Springer is part of Springer Science+Business Media (www.springer.com)

Foreword: From the World of Cetaceans

Cetaceans were long valued for their body mass, and modern cetacean biology relied greatly on whaling operations for biological materials and ecological information from whales. This situation is still true in some communities such as Japan, where more than 10,000 individuals of 15 species of cetaceans are killed annually through fishery operations and scientific activities and are subsequently consumed. Many of these studies aimed at the safe management of whale stocks, but resulted in such an ironical consequence: that we increased our knowledge about whales with an accompanying shift of whaling operations from one area to another and from larger species to smaller ones when the whale stocks thus became depleted.

Examples of such management failures are numerous in the history of Antarctic whaling. The total population of ordinary blue whales migrating into the Antarctic Ocean declined from 250,000 in 1904, when modern whaling entered the ocean, to about 1,000 in the mid-1960s when the hunting was stopped for population recovery; the current number is still only around 2,000. Three other great baleen whales and the largest of the toothed whales (the sperm whale) had experienced a more or less similar fate in the Antarctic by the early 1980s. Such management failure also occurred with large whales in other oceans, and smaller toothed whales (dolphins and porpoises) were not totally exempt, as seen with striped dolphins off Japan.

Our failures in managing cetacean populations were related, at least in some degree, to the biology of the species such as their aquatic habitat and low reproductive rates, and to the nature of the human community wherein decision making and enforcement were often difficult. Information on biology and abundance was hard to obtain or was not available for management, and whaling industries were powerful enough to exert control over their governments for their short-term benefit by sacrificing management. Additionally, whaling was often so important to the economy of fishing nations that governments ignored the management advice of scientists or attempted to control their scientists.

All cetacean species are believed to have a high capacity for vocal communication and a range of communication that extends, particularly among baleen whales, beyond the range of our visual perception: this has limited our understanding of their biology and social interactions. For example, we do not know how baleen whales locate the swarms of plankton on which they feed in waters with limited visibility, or how individual baleen whales cooperate in the activity. Observation of some dolphin species in captivity has revealed that they are able to identify images of themselves in a mirror, a suggestion that their level of intelligence is comparable to or greater than that of apes. Toothed whales in general are apparently more social than baleen whales, and they live in groups of various degrees of stability. In a community of such species, contributions of individuals to the group must vary by sex, age, and amount of accumulated experience. Baleen whales cannot be excluded from this generalization in view of the current insufficient level of our knowledge of their biology. Management of cetacean populations is likely to fail if these elements are ignored, as has happened in the past. It is difficult for the traditional methodology of cetacean biology to meet such research needs.

A new type of cetacean research emerged in the 1970s for killer whales, using photographs as a tool for individual recognition, and expanded to bottlenose dolphins, humpback whales, and some other cetacean species in coastal and riverine habitats. In association with genetic, acoustic, and radio technologies, this new approach has produced valuable information on the social structure, life history, and movement of cetaceans. However, these new methodologies still have limitations. Cetaceans spend most of their life underwater and show themselves to scientists only for a small fraction of their time, usually in association with breathing, so scientists are now forced to descend underwater to overcome this limitation. Because cetaceans lifespans cover from 25 to more than 100 years, it takes almost the entire life of a biologist before we can obtain a picture of the whole life of one cetacean species, and during that time human activities can significantly alter the aquatic environment and thus the life of cetaceans as well. A recent promising finding is an aging technique using fatty acid composition in biopsied blubber samples, although we still need to calibrate the data using animals whose ages are known. Under such difficult circumstances, combining the results of recent nonlethal behavior studies with those of the earlier fishery-derived carcass studies has revealed fascinating aspects of the social structure of some toothed whales and the variability among species.

It is our current understanding that most baleen whales and some toothed whales (e.g., harbor porpoise, Dall's porpoise, finless porpoises, and perhaps some river dolphins) have mother-and-calf associations lasting only several months or a year as the most stable social units in their communities, whereas other toothed whales have evolved longer mother–calf associations. Many members of Delphinidae such as bottlenose and striped dolphins live in fission–fusion communities and have some degree of matrilineal tendencies, wherein weaned juveniles, particularly males, leave their mothers' group to spend the prepubertal period with other individuals at the same growth stage. More cohesive matrilineal groups are known in some Delphinidae (killer whale, two species of pilot whales, and perhaps false killer whale) and Physeteridae (sperm whale), where the mother–daughter association is believed to last for life, although patterns of mother–son association are not the same among the species. These species are probably at the top of the specialization of the social system achieved by recent toothed whales.

In the last-mentioned group of toothed whales, juveniles are known to suckle occasionally to the age of 10 years or more, and some females are found to lactate for a similar duration after their last parturition, where communal nursing remains a possibility to be confirmed. Females of this group have extended lives after cessation of ovulation or conception. For example, about 25 % of sexually mature females of the short-finned pilot whales off Japan are over the age of oldest conception (36 years) and live to an age of 62 years. Such information invites speculation that an old female is contributing to the life of her kin's offspring rather than bearing her own offspring, using her accumulated experience through her extended life. So-called social sex, identified in the short-finned pilot whale based on the finding of spermatozoa in the uteri of non-estrous (including postreproductive) females, is interpreted as contributing to the stability of the community. This observation bears similarity to the behavior of some primates, for example, bonobos and humans.

The social structure of Ziphiidae, a group of middle-sized, poorly known toothed whales of 21 pelagic species, merits particular attention. The age structure of a few species suggests that, in contrast to other toothed whales, males of Ziphiidae live longer than females. For example, males of Baird's beaked whales off Japan have a higher survival rate and live about 30 years longer than females, which live to 54 years and are reproductive for life. More information has to be obtained before we understand their social structure and its diversity within the family Ziphiidae.

Evidence of culture in whales and dolphins is still limited by technical difficulties in detection but is known for both baleen and toothed whales. Individual cetaceans have the ability to accumulate information and behavior patterns during their extended lifetime, and the social structure provides them with opportunities to transmit the acquired information and behavior to other members of their community. This is the formation of culture, which I define as information and behavior traits retained in a community by learning. Culture allows the community to adapt quickly to a new or a changing environment and thus enables the community to better utilize resources in an unpredictable environment. If a species contains communities with diverse cultures, it allows the species to utilize a broad spectrum of habitats and thus increases the species opportunity to survive. Therefore, it is important for social species to conserve both social structure and cultural diversity. Driving a whole pilot whale school to shore for subsequent slaughter, which is practiced in Japan and the Faroe Islands, is hard to accept for the conservation and management of the species from this point of view.

Cetaceans entered into the aquatic environment in the early Eocene period, and through a history of more than 50 million years in that environment they have acquired a morphology which is quite different from that of terrestrial mammals.

Their life history and social structure must have also undergone modifications during this long period and must have achieved broad radiation. Although our knowledge of this part of cetacean biology is still in its infancy, I believe it is interesting to see some similarities between the radiation pattern of the social structure of cetaceans and that of primates which evolved in the terrestrial environment. Comparison of social structure and its diversity between the two animal groups will help us to understand the evolution of mammal communities.

Toshio Kasuya

Preface

Cetaceans and primates have developed comparable cognitive abilities in different environments. Their social systems vary from dynamic fission–fusion to long-term stable societies; from male-bonded to bisexually bonded to matrilineal groups. Despite obvious differences in morphology and eco-physiology, there are many cases of comparable, sometimes strikingly similar, patterns of sociobehavioral complexity. Recent studies suggest that many of these similarities and differences are influenced by the ecological factors of their natural environments. A number of long-term field studies have accumulated a substantial amount of data on the life history of various taxa, their foraging ecology, social and sexual relationships, demography, and various patterns of behavior. We can now attempt to view primates and cetaceans in a comparative perspective: such comparisons between social animals that are evolutionarily distant but live in comparable complex, sociocognitive environments boost our appreciation of their sophisticated mammalian systems and may advance our understanding of the ecological factors that have shaped their social evolution.

Cetologists and primatologists, however, rarely get together to explore common interests of their research. To facilitate such an exchange of ideas, we initiated a sharing of knowledge and a discussion of the topics common to the studies of both taxa in a symposium at the 9th International Mammalogical Congress (IMC), held in 2005 in Sapporo, Japan. Many important topics were discussed, involving mating strategies, social behavior, social networks, foraging strategies, communication, social learning, culture, economics of behavior, behavioral plasticity, social evolution, and application of new research tools in field studies. Each topic was presented by a speaker and commenter from the two fields. The Congress was preceded by a symposium on the social ecology of primates and cetaceans held two days earlier in Kyoto. Another joint meeting took place in Cape Town, South Africa, during the 17th Biennial Conference of the Society for Marine Mammalogy in 2007, where we were joined by colleagues working on other group-living mammals including carnivores, ungulates, and pinnipeds. This edited volume was prepared as a collection of selected presentations from these three symposia and additional invited contributions. All chapters were reviewed by two or more external referees. The publication took a long time owing to the difficulties of topic selection, logistics of communication with multiple authors, and occasional incompatibility of thoughts between authors addressing similar topics. We believe, however, that the time spent in compiling this volume increases the value of its content. We sincerely thank the authors for their contributions and fruitful discussions during the editing process.

This volume consists of four sections. The first section presents topics on social ecology of cetaceans and primates. Recent review of the theory of socioecology points out a need for multiple formulas rather than one comprehensive model to explain social relationships. Six chapters present cases of social and behavioral plasticity in primates and delphinids and offer views that in some cases might go beyond earlier interpretations. The second section, with five chapters, provides cases of long-term studies that address topics of social evolution and life history strategies. Because of the slow life histories of primates and cetaceans, longitudinal studies are necessary to gain an understanding of how ecology and natural history influence their behavior and social evolution. The third section presents five chapters with subjects ranging from behavior to demography, to population genetics and eco-toxicology, all directly related to current issues in conservation. The authors point out the susceptibility of primates and cetaceans to anthropogenic pressures and the importance of sound ecological research in addressing the challenges of conservation and management. The fourth section presents five selected topics on comparative studies of behavior: three of them resulted from the fruitful discussions at IMC 9. This comparative look at primates and cetaceans may at times bring us to new points of view that go beyond previous perceptions, facilitating a better understanding of the day-to-day challenges these animals face in the human-dominated world, which in turn may improve our capacity and capabilities of promoting conservation.

In the early days of our work on this volume, we requested that Prof. Toshio Kasuya and Prof. Toshisada Nishida write the Foreword and Afterword, respectively, because both these distinguished colleagues had dedicated their lifelong work to studying cetaceans and primates and advocating their conservation. They accepted our request, and Prof. Kasuya assumed also the role of convener of the symposium at IMC 9. The Foreword is now accompanying this volume; however, the Afterword was not completed as a result of the untimely death of Prof. Nishida in 2011. He had continuously encouraged us to edit this book and was patiently looking forward to its publication. We express our deepest sympathy to his loved ones and dedicate this book to him.

As mentioned earlier, the idea of creating this book emerged from a symposium at IMC 9 held in Sapporo in 2005. Our editing process was funded by the IMC 9 Memorial Project and funding from the Texas Institute of Oceanography at Texas A&M University, Galveston. We wish to thank Prof. Noriyuki Ohtaishi, Prof. Koichi Kaji, Prof. Takashi Saito, Prof. Keisuke Nakata, the members of the IMC 9 Organizing Committee, and Dr. Tammy Holiday of Texas A&M University for their support. The symposium in Kyoto was sponsored by Kyoto University's 21 COE Program "Formation of a Strategic Base for the Multidisciplinary Study of Biodiversity" and by the Japan Society for the Promotion of Science (JSPS)

Core-to-Core Program HOPE "Primate Origins of Human Evolution." The Comparative Ecology Workshop in Cape Town was cosponsored by the University of Pretoria's Mammal Research Institute and the Iziko South African Museum of Cape Town. We thank the participants at the symposia, as without their stimulating talks and the discussions with them, this book would not have been realized. We further express our gratitude to the Laboratory of Human Evolution Studies at Kyoto University, the Texas Institute of Oceanography at Texas A&M University in Galveston, the Mammal Research Institute at University of Pretoria, and the Swire Institute of Marine Science at the University of Hong Kong for hosting and facilitating our editorial efforts. We also thank Aiko Hiraguchi and Yoshiko Shikano, of Springer, for their advice and assistance.

Kyoto, Japan — Juichi Yamagiwa
Cape d'Aguilar, Hong Kong — Leszek Karczmarski

Contents

Part I
Social Ecology

Chapter 1
How Ecological Conditions Affect the Abundance and Social Organization of Folivorous Monkeys

Colin A. Chapman, Tamaini V. Snaith, and Jan F. Gogarten

C.A. Chapman (✉)
Department of Anthropology and McGill School of Environment, McGill University, Montreal, QC, Canada H3A 2T7

Wildlife Conservation Society, 185th Street and Southern Boulevard, Bronx, NY 10460, USA
e-mail: Colin.Chapman@McGill.ca

T.V. Snaith
Department of Anthropology, McGill University, Montreal, QC, Canada H3A 2T7

J.F. Gogarten
Department of Biology, McGill University, Montreal, QC, Canada H3A 2T7

J. Yamagiwa and L. Karczmarski (eds.), *Primates and Cetaceans: Field Research and Conservation of Complex Mammalian Societies*, Primatology Monographs, DOI 10.1007/978-4-431-54523-1_1, © Springer Japan 2014

Abstract A fundamental issue in ecology is identifying factors influencing animal density, and this issue has taken on new significance with the need to develop informed conservation plans for threatened species. With primates, this issue is critical, because tropical forests are undergoing rapid transformation. Similarly, a fundamental question in behavioral ecology is understanding how ecological conditions shape the social organizations of animals. During the past two decades, our research group has been investigating the ecological variables influencing the abundance and social structure of two folivorous monkey species, the red colobus (*Procolobus rufomitratus*) and the black-and-white colobus (*Colobus guereza*) in Kibale National Park, Uganda. We have documented much variation in the abundance of these colobus monkeys over very small spatial scales. This variation is partially caused by variation in quality of the food resources, particularly the availability of high-protein resources relative to their fiber content. However, this is not the whole story, and minerals, disease, and the interaction between disease and stress also appear to play important roles. Further, despite examining all these factors over multiple decades, our understanding is too limited to explain observed changes in colobine abundance over the past 40 years. Emerging from our studies of determinants of primate abundance were investigations of feeding competition. Our findings suggest that, counter to previous claims, feeding competition is occurring in these folivores, and if food competition proves to be biologically significant for folivores, our interpretations of primate behavior will need to be refined, and current theoretical models of primate social organization may need revision.

Keywords Colobus • Competition • Conservation • Determinant of abundance • Folivore • Kibale National Park • Minerals • Nutrition • Parasites

1.1 Introduction

Animal distribution is primarily determined by the nature of available habitat, and identifying the ecological factors that underlie this relationship is a key component of population ecology (Boutin 1990; Chapman and Chapman 2000b). Ecological factors are also fundamental in shaping animal social organization, and understanding the nature of this relationship is the central focus of behavioral ecology (Eisenberg et al. 1972; Wursig and Wursig 1979; Isbell 1991; Sterck et al. 1997). These issues have taken on new significance with the need to develop informed management plans for threatened and endangered species. For primates, this is especially critical because their tropical forest habitats are undergoing rapid transformations that have resulted in severe population reductions for many species. In 2010, the Food and Agriculture Organization of the United Nations (FAO) estimated that 16 million ha of forest was lost globally each year in the 1990s (Camm et al. 2002), and approximately 12.5 million ha/year was lost in countries with primate populations, an area just smaller than Greece (Chapman and Peres 2001;

Kristan 2007; Chapman and Gogarten 2012). In contrast, in the past decade, the global rate of deforestation has decreased to approximately 13 million ha/year, and reforestation and natural expansion of forests in some countries significantly reduced the net loss of forest (Camm et al. 2002). Although the rate of deforestation has decreased, many areas are still being dramatically affected, and to help manage this situation, academics need to develop a predictive understanding of the interactions between ecological variables, primate behavior, and population persistence.

During the past two decades, our research group has been investigating the ecological variables influencing the abundance and social structure of two folivorous monkey species, the red colobus (*Procolobus rufomitratus*) and the black-and-white colobus (*Colobus guereza*), in Kibale National Park, Uganda. In this chapter, we provide an overview of this work and our interpretation of the results relative to current theoretical discussions and practical conservation issues. We first describe our motivation for establishing this research program and our initial investigation of how food resources could determine folivore abundance. Unsatisfying initial results led us to incorporate nutritional considerations and to investigate the potentially synergistic interactions between nutrition and disease, and finally to testing our understanding experimentally and through a long-term contrast of population and ecological change. Although the initial focus of these studies was to understand the relationship between ecology and population density, some surprising results led us to question the assumptions and conclusions of current theoretical models that seek to explain variation in primate social organization. Throughout the chapter, we attempt to illustrate gaps in our understanding and point to ways forward for researchers interested in the population and behavioral ecology and conservation biology of tropical primates.

1.2 Ecological Determinants of Folivore Abundance

Kibale National Park protects 795 km^2 of moist evergreen forest near the foothills of the Rwenzori Mountains in western Uganda (Struhsaker 1997; Chapman and Lambert 2000). Our first studies were started in 1989, and the research continues to this day. As we came to know Kibale, we became intrigued by the apparent variation in primate abundance and realized that, if these patterns were real, this variation offered an excellent opportunity to investigate the ecological determinants of animal abundance. Previous studies of primate abundance had almost universally showed contrasts between sites that are geographically widely separated (McKey et al. 1981; Oates et al. 1990) and were based on the premise that distant sites would have sufficient variation in ecological conditions to permit detection of differences in response variables (e.g., primate density). However, if significant differences in ecological conditions occur over shorter distances, as they do in Kibale, small-scale studies may provide more sensitive tests of primate responses to variation in ecological variables, because small-scale analyses provide controls for some methodological problems associated with large-scale studies. For example, unmeasured ecological parameters (e.g., composition of the predator community) are less likely to differ among neighboring populations than would be the case among widely separated populations (Butynski 1990; Chapman and Fedigan 1990), and because comparisons are made within species, or even within breeding populations, analyses are not confounded by phylogenetic non-independence (Nunn 2011).

To determine the extent of variation in red colobus density, we conducted intensive line transect surveys at six sites within Kibale, typically every second week, for 2 years (Chapman and Chapman 1999). To establish which food resources were important, we collected more than 1,000 h of feeding observations. We then determined the abundance of the major food resources at each of the six sites to evaluate whether red colobus abundance was related to food availability. We found that red colobus numbers were fairly high at most sites, even in disturbed areas. However, a surprisingly low population density was found at Dura River, a relatively undisturbed riverine site in the middle of the park. As we predicted, red colobus density was significantly related to the cumulative size of important food trees, but only if the Dura River site was excluded (Chapman and Chapman 1999). It was this anomalous site that provided the impetus for further investigation.

We initially thought there was reason to believe that red colobus monkeys were below carrying capacity at Dura River. A small number of censuses conducted in 1970 and 1971 (Struhsaker 1975) estimated the red colobus group density to be 2.7 times greater than we recorded in 1996–1997, and an epidemic reportedly killed a number of male red colobus monkeys in Kibale in the early 1980s (T. Struhsaker, personal communication). This epidemic might have involved one of a number of novel viruses that have been identified in red colobus (Goldberg et al. 2008, 2009). If such epidemics are common and are restricted to one or a few sites, they could periodically reduce populations to below carrying capacity. Based on this evidence, it seemed possible that the low monkey densities at Dura River could be attributable to an epidemic. However, this explanation was unsatisfying, because there was little real evidence to support it, and we were concerned that we were missing something. We began to consider the importance not just of food availability, but also its quality.

In contrast to most primates, colobus monkeys have a specialized alkaline forestomach designed for the digestion of fibrous leaves (Chivers and Hladik 1980). Milton (1979) proposed that the protein-to-fiber ratio of leaves was an important criterion driving leaf selection by small-bodied arboreal primates, with those leaves with a higher protein-to-fiber ratio preferentially selected over those with lower ratios. Fiber is often considered an antifeedant because the energy contained within it requires fermentation by symbiotic microbes to be accessible to a primate, and insoluble fiber (cellulose, hemicellulose, and lignin) is only partially digestible by microbes and is thus largely inaccessible to primates (McNab 2002). Others extended Milton's ideas; for instance, Waterman et al. (1988) suggested that the weighted contributions of the protein-to-fiber ratios of the mature leaves of the most abundant trees in a particular habitat could predict the biomass of folivorous colobines. Subsequently, this index of dietary quality has been used to predict the biomass of small-bodied folivorous monkeys at local (Chapman et al. 2002b;

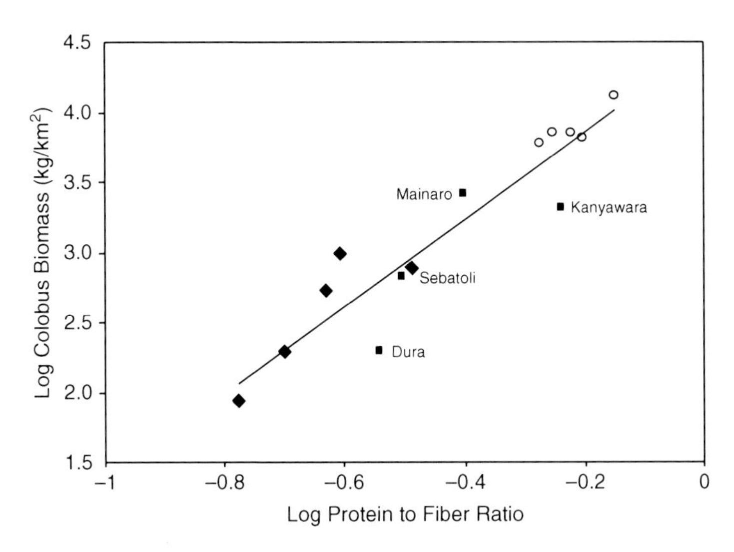

Fig. 1.1 Relationship between mature leaf chemistry and colobine biomass at rainforest sites in Africa and Asia. Chemical values are weighted mean percentages of dry mass, standardized to the species basal area to account for different proportions of the flora being sampled at each site. The weighted values were calculated from $\Sigma (P_i+X_i)/ \Sigma P_i$ where P_i is the proportion of the basal area contributed by species *i* and X_i is the chemical measure for species *i*. This figure is standardized to 100 %. *Diamonds* are sites from around the world (Oates et al. 1990); *squares* are forest sites within Kibale National Park, Uganda and are labeled by location (Chapman et al. 2002a); *open circles* are the forest fragments near Kibale (Chapman et al. 2004)

Ganzhorn 2002) and regional scales (Waterman et al. 1988; Oates et al. 1990; Davies 1994; Chapman et al. 2004; Fashing et al. 2007). The mechanism by which this index of food quality operates to determine folivore biomass is not clearly understood. Davies (1994) suggested that the year-round availability of digestible mature leaves with high protein-to-fiber ratios, which are used by colobus species when other, more preferred foods are unavailable, serves to limit the size of colobine populations (i.e., high protein-to-fiber mature leaves are important fallback foods). However, some colobines rarely eat mature leaves because young leaves are always available in their habitats, yet their biomass is still predicted by this index (Chapman et al. 2004). One possible explanation is that the protein-to fiber ratio of mature leaves in an area is correlated with the protein-to-fiber ratio of foods in general. This idea is supported by the fact that in a sample of leaves from Kibale National Park, we found that the protein-to-fiber ratios of mature and young leaves were strongly correlated ($r=0.837$, $P<0.001$; Chapman et al. 2004). Thus, measuring the protein-to-fiber ratio of mature leaves may be a useful index of the general availability of high-protein, low-fiber foods.

Although our sample size was too small for robust statistical analyses, our results suggested that colobus biomass was positively related to the average protein-to-fiber ratio of mature leaves across sites. Most remarkably, when we accounted for food quality in this manner, the low population density at Dura River was no longer an anomalous outlier (Fig. 1.1). It thus appears that although food is abundant at

Dura River, it is of very low quality, and this is likely the reason that the site does not support a large monkey population (see Struhsaker 2008b for an alternate view).

Although these studies of population density, nutrition, and physiology suggest that the protein-to-fiber ratio of available foods may limit the size of folivore populations, we judged that the data were insufficient to convince managers to use these principles in constructing management plans. To overcome this shortcoming, for the next decade and a half we delved into six issues surrounding the model and then attempted to test these ideas: first, using temporal changes in abundance and ecology, and second, using a natural experiment:

1. Sample size
2. Food selection patterns
3. The importance of energy and minerals
4. Possible synergy between nutrition and disease
5. Temporal changes in abundance and ecology
6. Natural experiment

1.2.1 Sample Size

The most obvious shortcoming of this early research was that our sample size was small and we needed several independent populations to increase the sample. To do this, we turned to a series of forest fragments outside Kibale. These forest fragments vary in size and composition and provide a quasi-experimental setting that allowed us to investigate the influence of the protein and fiber levels of available trees on primate populations. Before making any comparisons across sites, we wanted to establish that each population was stable, a potentially confounding issue that had not previously been examined. If some populations were not at carrying capacity as a result of recent effects of disease, hunting, or deforestation, then correlations between food availability or quality and folivore biomass could be spurious. In 1995, we surveyed the primate communities in 20 of these forest fragments to determine the abundance of black-and-white colobus and the presence of red colobus monkeys. In 2000, we resurveyed these fragments to assess population and forest stability and to compare colobine biomass to the protein-to-fiber ratio of leaves in those fragments that were determined to have stable populations (Chapman et al. 2004).

We discovered that 3 of the 20 fragments inhabited by primates in 1995 had been cleared and that resident primate populations were no longer present. These fragments had remained intact since at least the 1940s, but recent economic conditions had led to more rapid deforestation. Most fragments had been cleared for charcoal production, gin-brewing, brick-making, or timber extraction. As the road to the region had been improved, all these products could now be sold in the capital city (Chapman et al. 2007). In the remaining fragments, the black-and-white colobus populations had declined by 40 % in just 5 years. Although we had initially hoped that most colobus populations in the fragments would be stable (i.e., approximately the same size as 5 years previously and no sign of deforestation), we found that there were only five stable populations. Although this was alarming from a conservation perspective, these 5 sites increased our sample size sufficiently to conduct a more robust statistical analysis of the protein-to-fiber model. Across these 5 fragments, colobus biomass was correlated with the protein-to-fiber ratio ($r^2=0.730$, $P=0.033$). To more rigorously examine the protein-to-fiber model, we combined the data from the fragments with the 4 sites from within Kibale and five published values from other sites in Africa and Asia. Across all 14 sites, colobine biomass varied from 84 kg/km^2 at Sepilok, Malaysia to 13,160 kg/km^2 in 1 of the fragments outside Kibale (mean biomass across sites=3,405 kg/km^2, $n=14$). The protein-to-fiber ratios reported in these studies showed a similar degree of variation [mean=0.41 (range, 0.17–0.71); $n=14$]. Colobine biomass across all 14 sites could be predicted with a significant level of confidence from the protein-to-fiber ratios of available mature leaves ($r^2=0.869$, $P<0.001$; Fig. 1.1).

1.2.2 Food Selection Patterns

To provide a more in-depth understanding of the nutritional ecology of colobus monkeys and to determine whether individuals behave as if protein and fiber are important, we quantified diet choice with respect to the nutritional value of food items (i.e., protein, fiber, digestibility, and minerals) and a number of secondary compounds that may reduce the value of a food item (i.e., alkaloids, tannins, total phenolics, cyanogenic glycosides, and saponins; Chapman and Chapman 2002). We examined the feeding behavior of red and black-and-white colobus groups in a variety of habitats that likely provided different levels of dietary stress. We sampled two groups of each species in unlogged forest, one group of each species in logged forest, and one group of each species in a forest fragment. We collected more than 6,000 h of observations. Typically, each primate group was observed from dawn until dusk for 1 week per month for 2 years. Every time an animal was observed eating during a scan, we recorded what it was ingesting and the feeding rate. At the end of the week, we collected the food items for chemical analysis, being careful to collect only the specific plant parts that were eaten (e.g., the first 1 cm of the leaf petiole).

We demonstrated that all eight groups consistently fed more on young leaves than on mature leaves. These young leaves had more protein, were more digestible, and had a higher protein-to-fiber ratio than mature leaves. There were few consistent differences among young versus mature leaves with respect to minerals or secondary compounds. Regression analyses predicting foraging effort for each of the eight groups from phytochemical components revealed consistent selection for only one factor: a high protein-to-fiber ratio. We found no evidence that colobus monkeys avoided plants with high levels of secondary compounds. In fact, one of the most preferred trees, *Prunus africana*, was the species with the highest levels of cyanogenic glycosides. The highest saponin levels were found in the young leaves of *Albizia grandibracteata*, which was also frequently eaten.

1.2.3 The Importance of Energy and Minerals

We wanted to test two alternative hypotheses concerning ecological factors that could influence colobus population size: energy availability and mineral content. Based on a study of *Colobus polykomos* on Tiwai Island, Sierra Leone, DaSilva (1992) suggested that energy availability played a critical role in colobine behavior and ecology. In Kibale, we found little evidence that energy was a limiting factor for red or black-and-white colobus populations (Wasserman and Chapman 2003). None of the eight groups studied selected high-energy foods, estimates of energy consumption exceeded expenditure and average daily metabolic needs for all groups, and the average energy content of mature leaves from the 20 most abundant tree species at the four sites was not related to colobine biomass. These studies suggest that energy is not limiting for colobus monkeys; however, these results should be considered with caution because they considered the relationship between gross energy (bomb calorimetry) and colobine abundance (DaSilva 1992; Wasserman and Chapman 2003), but gross energy does not consider the variable digestibility of fiber fractions in leaves. Although cellulose can be used as an energy source for colobines, lignin is completely indigestible (Van Soest 1994; Fearer et al. 2007).

Very little information exists on mineral nutrition of tropical, forest-dwelling species, yet mineral nutrition is critical to growth, reproduction, and survival (McDowell 1992; Robbins 1993). Because plants and animals differ in their mineral requirements, herbivores face the difficult task of identifying appropriate foods and consuming a diet that meets their mineral requirements. For example, although sodium makes up 90 % of total blood cations and is necessary for muscle contraction, nerve impulse transmission, acid–base balance, and water metabolism in animals (Robbins 1993), it is not required by most plants, resulting in very low concentrations of sodium in many plant species (Smith 1976). Thus, mineral deficiencies are common in herbivore populations (McDowell 1992; Robbins 1993; Bell 1995). However, we documented that colobus monkeys in Kibale consumed enough of most nutrients to meet suggested mineral requirements (Rode et al. 2003). The only observable exception was a consistently low sodium intake (Rode et al. 2003). In addition, a number of behaviors pointed to a sodium deficiency, including urine drinking, consumption of high-sodium swamp plants, and use of mud puddles. This apparent sodium deficiency was paradoxical, because only one of eight colobus groups selected foods high in sodium in their everyday diet. Interestingly, however, we documented that foods with high sodium concentrations tended to have low protein content and high concentrations of secondary compounds. This observation suggests that diet choice involves a complex set of decisions and that nutrient availabilities interact to affect selection patterns (Rode et al. 2003). More research is required to understand whether sodium availability influences colobine abundance.

1.2.4 Possible Synergy Between Nutrition and Disease

There is ample evidence that finding single-factor explanations for complex biological phenomena is unlikely. Rather, studies have highlighted the importance of multifactorial explanations. For example, Gulland (1992) studied the interacting effects of nutrition and parasites among wild sheep. She demonstrated that when the population crashed, sheep were emaciated, had high nematode burdens, and showed signs of protein-energy malnutrition. In the field, sheep treated with an anti-helminthic to kill the gastrointestinal parasites had lower mortality rates, whereas experimentally infected sheep with high parasite loads, but fed nutritious diets, showed no sign of malnutrition. Similarly, based on a 68-month study of the effect of the parasitic bot fly (*Alouattamyia baeri*) on howler monkeys (*Alouatta palliata*), Milton (1996) concluded that the annual pattern of howler mortality results from a combination of effects, including age, physical condition, and the larval burden of the parasitized individual, which becomes critical when the population experiences dietary stress.

Disease/parasitism and nutrition often interact to determine the abundance of wildlife populations. Helminthic and protozoan parasites can affect host survival and reproduction directly through pathological effects and indirectly by reducing host condition and increasing predation risk (Hudson et al. 1992; Coop and Holmes 1996; Gogarten et al. 2012a). However, parasites do not necessarily induce negative effects if hosts have adequate energy reserves or nutrient supplies concurrent with infection (Munger and Karasov 1989; Gulland 1992; Milton 1996), suggesting that the outcome of host–parasite associations may be contingent on host nutritional status as well as the severity of infection. Dietary stress may exacerbate the clinical consequences of parasitic infection through immunosuppression (Crompton 1991; Holmes 1995; Milton 1996). If so, then food shortages could result in a higher parasite burden, which in turn could increase nutritional demands on the host and intensify the effects of food shortages.

A limitation of previous studies was the difficulty of monitoring populations over long periods to observe whether changes in nutritional status and parasitism were associated with changes in host population size. We attempted to circumvent this limitation by using the fragments surrounding Kibale to study a series of populations known to differ in nutritional status. Along with our knowledge of the population dynamics in each fragment, this situation offered a replicated quasi-experiment, with each fragment being an independent population (dispersal among fragments was rare) where we could examine whether nutritional status and parasitism could have synergistic effects on red colobus abundance. In addition, because primate population levels may respond very slowly to environmental change, we evaluated how nutritional status and parasite infections influenced fecal cortisol levels, presumably a more immediate measure of individual stress (Chapman et al. 2006). A great deal of research on captive mammals and humans has demonstrated that severe and prolonged elevations of glucocorticoids (cortisol is one type of glucocorticoid) typically reduce survival and reproductive output (Sapolsky 1992;

Wasser et al. 1997; Creel 2001; Bercovitch and Ziegler 2002; Creel et al. 2002; Lee and Rotenberry 2005). Although data on the fitness effects of elevated glucocorticoid levels in the wild are limited, the expectation from captive studies was that fitness would decrease as stress became more severe or more prolonged (Boonstra and Singleton 1993; Creel et al. 2002). If positive associations are found between cortisol and parasite infections or poor nutritional status, it seems reasonable to assume that populations with high cortisol levels are physiologically challenged and over the long term will suffer reduced survival and reproduction.

We focused on red colobus and documented that the populations in these fragments declined an average of 21 % over 5 years; however, population change was highly variable among fragments and ranged from an increase of 25 % to a decline of 57 %. In 2000 and in 2003 we identified and measured every tree in these fragments and quantified changes in food availability. We found that the average cumulative diameter at breast height (DBH) of available food trees declined by 34 %; however, it ranged among fragments from a 2 % gain to a 71 % decline. We collected 634 red colobus fecal samples and described parasite infections. Infections varied dramatically among fragments. For example, nematode prevalence averaged 58 % among fragments, but the range was 29 % to 83 %. We documented that fecal cortisol levels averaged 264 ng/g but ranged from 139 to 445 ng/g. As forest loss increased (decline in cumulative DBH of food trees), population size declined ($r=0.827$, $P=0.006$), and several indices of parasite infection increased (parasite richness: $r=-0.668$, $P=0.035$; nematode prevalence: $r=-0.689$, $P=0.028$; nematode eggs/g: $r=-0.692$, $P=0.029$).

The cortisol findings were not as straightforward, despite our best efforts to control for factors that can cause fecal steroid levels to vary. We found that an increase in deforestation was only marginally related to elevated cortisol levels ($r=-0.599$, $P=0.055$). An increase in cortisol was related to an increase in some of the indices of parasite infection (nematode eggs/g: $r=-0.712$, $P=0.024$), but not others (parasite richness: $r=-0.530$, $P=0.088$; nematode prevalence: $r=-0.414$, $P=0.154$). Also, an increase in cortisol was not associated with a decline in population size ($r=-0.399$, $P=0.164$).

We used a path analysis to further investigate associations among these variables. This technique was very useful because the path coefficients allow the determination of the magnitude of both direct and indirect effects among variables (Kingsolver and Schemske 1991; Chapman et al. 2006). We were particularly interested in the indirect effects of changes in food supply on population change through its impact on susceptibility to parasite infection. We produced simple path analyses that considered each major index of parasite infection separately combined with nutritional status (indexed as the loss in food trees; cm DBH/ha) to predict population change. Each of these analyses indicated that lower nutritional status had a direct effect on population size, leading to decline from either reduced fecundity or increased mortality. In all cases the path coefficient was relatively large (mean $=0.663$; see Fig. 1.2 for one example of such an analysis). Nutritional status also had an indirect effect on population size through its influence on parasite infections. Here nutritional status initially had a fairly strong effect on the indices of parasitism considered

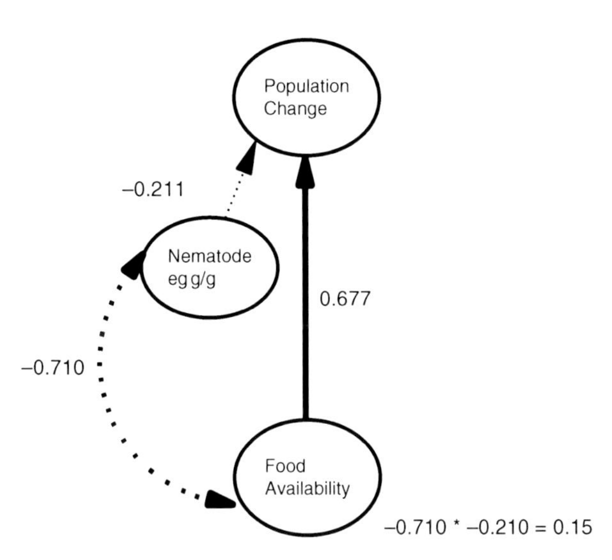

Fig. 1.2 An example path analysis of factors predicted to affect the change in population size of red colobus monkeys (*Procolobus rufomitratus*) in a series of forest fragments neighboring Kibale National Park, Uganda. Positive effects are indicated by *solid lines* and negative effects by *dashed lines*. *Double-headed arrows* indicate positive relationships between predictor variables. The width of the *arrow* indicates the magnitude of the standardized path coefficients

(mean = 0.631), which subsequently had weaker and negative effects on population size (mean = –0.264; Fig. 1.2).

Our findings support suggestions that nutritional status interacts with the host immune response and leads to a synergistic relationship between nutritional status and parasite infection, which can influence population change, but suggest that the direct effect of nutritional status is the strongest factor. This observation raises intriguing questions about what types of anthropogenic disturbance will lead to an increasingly significant role for disease in determining primate population size.

1.2.5 Temporal Changes in Abundance and Ecology

At this point in our research program, we saw that, although we did not understand the mechanism driving the protein-to-fiber ratio model, the correlations were strong and alternative hypotheses (energy, minerals, and disease) did not seem to explain variation in colobine population size. As a result, we sought to test the model. Initially, we quantified if changes in the nature of red colobus food supply predicted temporal changes in their abundance using two to three decades of population and habitat data from Kibale. We calculated primate density (groups/km^2) and encounter rate (groups/km walked) from line transect data and quantified changes in habitat structure (cumulative DBH) and food availability (cumulative DBH of food trees) and food quality (Chapman et al. 2010). We initially reported that while red colobus food availability and quality increased over time in the heavily logged area, their group density did not show a corresponding increase. In the unlogged and lightly logged forestry compartment, a possible decline in red colobus group density and encounter rate was not related to a change in food quality, and in the lightly logged forest the decline in red colobus group density corresponded with an increase in food availability. We subsequently reported that during the past 15 years there had been a general increase in group size of red colobus throughout Kibale (Gogarten et al., in review-b); however, this change does not alter the general population trends we first identified (Chapman et al. 2010). These findings run counter to the general support we had previously been finding for the protein-to-fiber model because changes in the quality of the resources available to the red colobus did not predict changes in their population size or biomass.

1.2.6 Experimental Test

The best way to test the protein-to-fiber model would be through a long-term experiment increasing the proportion of trees with high-protein, low-fiber leaves, while not decreasing tree density. However, for obvious reasons this would be very logistically challenging and ethically suspect. Decreasing the proportion of trees with high-protein, low-fiber leaves would not be as difficult, but it would certainly be unethical considering we would predict a decline in the red colobus, an endangered species (Struhsaker 2008a, 2010). To our luck, a natural experiment was made available by chance. In the 1960s, areas of grassland in Kibale were planted with pine when the area was a forest reserve. When Kibale became a national park, the pines were harvested and the area was left to regenerate (Omeja et al. 2009). We demonstrated that the regenerating areas had many colonizing tree species, the leaves of which had higher concentrations of protein and lower concentrations of fiber than old growth tree species, as predicted from research elsewhere in the tropics (Coley 1983; Gogarten et al. 2012b). Given past interest in energy as a potential determinant of folivorous primate abundance (DaSilva 1992; but see Wasserman and Chapman 2003) and because black-and-white colobus population size in Kibale has been suggested to be limited by the availability of energy (Harris et al. 2010), we also tested if differences in demographic variables might be correlated with indices of energy; total energy, non-protein energy, and nonstructural carbohydrates, which are easily digestible energy sources (Rothman et al. 2012).

We expected groups that had access to this regenerating area to have a greater number of infants per female than a group that did not because of access to higher-quality resources. We also expected that groups with access to regenerating areas would be larger for a number of reasons: there is a delay between birth and dispersal, and thus the number of immature animals will be higher in groups with higher birthrates; second, red colobus males tend to remain in their natal group and a higher birthrate should manifest itself in more males which remain as residents; and last, because regenerating areas should be of higher nutritional quality, home ranges of groups might be smaller (Snaith and Chapman 2007) and there should be less within-group competition over the resources that limit group size (Gogarten et al., in review-a). Although regenerating forests had trees with leaves with high concentrations of protein and low concentrations of fiber, as well as higher total energy (kcal/100 g), there was no corresponding change in the demographic structure of red colobus groups.

1.3 Ecological Determinants of Folivore Social Organization

While attempting to understand the ecological determinants of folivore abundance, we made some interesting observations that appeared to contradict current explanations of folivore social organization and seemed to suggest a solution to a problem which had become known as the "folivore paradox." To investigate further, we launched a parallel research program to examine the ecological determinants of folivore group size and social organization.

Several deductive models have been constructed to explain variation in primate social organization. The nature and intensity of food competition is among the most important variables in these models, and it is considered to be responsible for determining patterns of dispersal and social behavior (Wrangham 1980; van Schaik 1983; Isbell 1991; Sterck et al. 1997; Snaith and Chapman 2007). Scramble competition involves the common depletion of food resources, whereas contest competition involves direct contests over food, and both can occur among group members and between groups (e.g., aggression, displacement, or avoidance (Janson and van Schaik 1988). Scramble-type food competition is thought to relate to group size; larger groups will deplete shared food patches more quickly than smaller groups and will have to visit more food patches each day to feed all group members. The cost of travel between patches is thought to impose a limit on group size (Terborgh and Janson 1986; Chapman 1988; Janson and Goldsmith 1995; Chapman and Chapman 2000b). Interestingly, socioecological models have presumed that, because leaves are highly abundant and evenly dispersed, folivores should experience little or no food competition (Wrangham 1980; Isbell 1991; Janson and Goldsmith 1995; Sterck et al. 1997). Although folivores should therefore be free to form large groups, which are believed to provide protection from predators (van Schaik and van Hooff 1983), many folivores live in surprisingly small groups (Crockett and Janson 2000; Steenbeek and van Schaik 2001; Koenig and Borries 2002), and therein lies the contradiction that has been referred to as the "folivore paradox" (Steenbeek and van Schaik 2001; Koenig and Borries 2002).

These assumptions represented the consensus regarding primate social organization when we started our research and are still widely accepted. However, our research and various pieces of information in the literature led us to question the conventional wisdom (Snaith and Chapman 2005, 2007, 2008). First, our studies of foraging and diet selection demonstrated that red colobus do not simply consume highly abundant and evenly distributed leaf resources. Rather, they preferentially select high-quality young leaves, flowers, and unripe fruits, which are often rare and variable in both quality and availability (Chapman and Chapman 2002). Similar findings had been made earlier, but had been interpreted only with respect to foraging ecology; their implications for folivore social organization had not generally been considered (Oates 1994; Oates and Davies 1994; but see Koenig et al. 1998).

Second, while examining variation in the nutritional value of foods across Kibale to evaluate the protein-to-fiber ratio model, we demonstrated that the average group size of red colobus at these sites was related to the density of the available food resources (Chapman and Chapman 2000a). Although the number of sites is small, this suggests that in areas where food resources are rare, red colobus are constrained from living in large groups by food competition. Around the same time, other studies of colobus monkeys were demonstrating that group size could be predicted by habitat variables such as seasonality, forest size, degree of deciduousness, and degree of disturbance (Struhsaker 2000; Struhsaker et al. 2004). Furthermore, we found that red colobus exhibited fission–fusion behavior (large groups dividing into smaller, temporary subgroups). Other studies had previously reported this behavior in folivores (Skorupa 1988; Oates 1994; Siex and Struhsaker 1999; Struhsaker 2000; Struhsaker et al. 2004; Snaith and Chapman 2008), which may be a short-term response to food competition during periods of low food availability, as has been demonstrated among some frugivores (Chapman 1990; Boesch 1996; van Schaik 1999).

Third, as already described, we demonstrated that colobus biomass can be predicted by the availability of high-quality foods (Chapman and Chapman 2002; Chapman et al. 2004), and previous studies had made similar suggestions (McKey et al. 1981; Waterman et al. 1988; Oates et al. 1990; Davies 1994; Fimbel et al. 2001). These data, along with recent evidence that folivores demonstrate contest competition, both within and between groups (Koenig 2000; Korstjens et al. 2002; Harris 2005), provide further support that food competition can indeed be important for primate populations that rely primarily on leaf resources.

Based on these surprising results, we thought it would be valuable to examine feeding competition in folivores and to reconsider current models of primate social organization. We first examined the relationship between day range and group size in red colobus monkeys. Because the cost of travel is the presumed mechanism by which group size imposes a cost, day journey length has been measured as a behavioral indicator of within-group scramble competition (Isbell 1991; Chapman et al. 1995; Janson and Goldsmith 1995; Wrangham 2000; Isbell and Young 2002). Previous studies had found no relationship between group size and day range or travel costs among folivores (Clutton-Brock and Harvey 1977; Struhsaker and Leland 1987; Isbell 1991; Yeager and Kirkpatrick 1998; Yeager and Kool 2000). However, this evidence is not conclusive because these studies generally did not control for ecological variation among groups or species. Ecological variation can confound correlations between group size and day range, because if large groups only occur in richer habitat (as our data suggest), there may be no need for an increased day range. In our first study, Gillespie and Chapman (2001) found that a large group of red colobus had longer day ranges than a small group, and that day range increased even further in the large group when food availability decreased. Although the sample size was small, this study suggested that inferences drawn on the basis of earlier studies that lacked ecological controls should be reassessed.

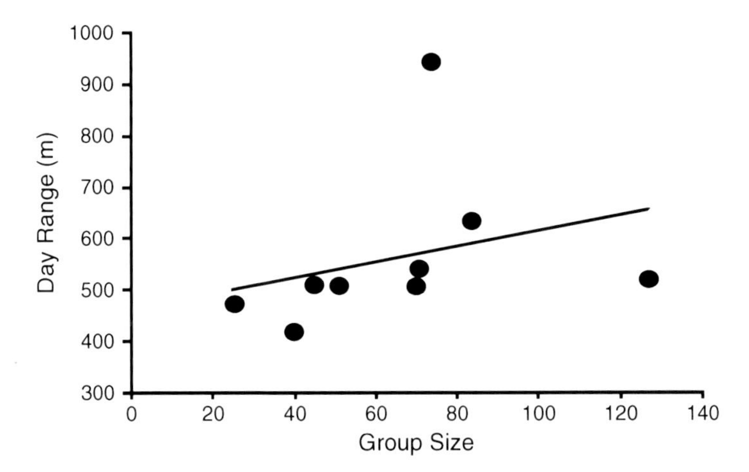

Fig. 1.3 Group size effects on day range in nine groups of red colobus monkeys (*Procolobus rufomitratus*) in Kibale National Park, Uganda (Adapted from Snaith and Chapman 2008)

We followed this up and examined the relationship between day range and group size for nine groups and again found a relationship (Snaith and Chapman 2008 Fig. 1.3). We also examined a single group of red colobus as group size expanded from 50 to more than 100 individuals and documented changes in activity budgets, specifically an increased amount of time spent traveling and a decreased amount of time spent feeding and socializing (Gogarten et al., in review-a). Concurrently, dietary diversity increased, suggesting that there was increased scramble competition at larger group sizes.

Next, we directly addressed a key assumption made in current theoretical models, that scramble competition does not affect folivores because their food resources do not occur in depletable patches (Wrangham 1980; Isbell 1991; Janson and Goldsmith 1995; Sterck et al. 1997). In contrast to what had previously been assumed, we found that red colobus monkeys deplete food patches when feeding on young leaves, as indicated by decreasing gains (intake rate) despite increasing feeding effort (movement while feeding). Furthermore, patch occupancy time was affected by patch size and feeding group size (Snaith and Chapman 2005). This observation provides evidence of a group size effect, where larger groups deplete patches more quickly and will be forced to visit more patches and accrue greater travel costs than smaller groups. These results suggest that red colobus experience within-group scramble competition, and that this type of competition may be an important factor determining group size. We then studied nine red colobus groups and controlled for spatial and temporal variation in food availability. Here we found that larger groups occupied larger home ranges than smaller groups, and that group size was related to increased foraging effort (longer daily travel distance), increased group spread, and reduced female reproductive success (Chapman, unpublished data). We also studied the genetic structure of two differently sized groups and found that average female relatedness was higher in smaller groups, suggesting that a female's decision to disperse may be affected by the degree of scramble competition they face in their natal group (Miyamoto et al., in press). Collectively, these results suggest that the folivorous red colobus do experience within-group scramble competition and possesses a suite of behavioral responses that may mitigate the cost of competition. The results offer an ecological solution to the folivore paradox for this species and present the intriguing notion of plasticity in social systems in response to different environmental conditions.

1.4 Conclusions and Future Directions

Our research in Kibale is ongoing, and we hope that we can make contributions both to conservation biology and to the theoretical fields of population biology and behavioral ecology. In general, our research into ecological determinants of folivore abundance provides support for the notion that the protein-to-fiber ratio is a good predictor of food choice in colobines, but only partially supports the use of this index in predicting colobine biomass. The documentation of the role of parasites adds depth to our understanding of variation in primate abundance. As we have previously demonstrated that human modifications to landscapes can alter interactions between parasites and hosts (Chapman et al. 2005; Gillespie et al. 2005), this raises the intriguing question of what types of anthropogenic disturbances will lead to disease playing a more significant role in determining primate population size. In the future, between-site comparisons should be carefully conducted to explore the effects of specific anthropogenic disturbances (e.g., forest fragmentation with or without elevated rates of human contact), because the focus has now shifted from *whether* anthropogenic habitat change alters primate–disease interactions to *how* anthropogenic change alters primate–disease interactions. Finally, if food competition proves to be biologically significant for folivores, our interpretations of primate behavior will need to be refined, and current theoretical models of primate social organization may need revision (see Snaith and Chapman 2007).

Perhaps the biggest gap in our understanding of folivore abundance and social organization stems from the fact that most socioecological studies conducted to date are based at the group level, and it has not been possible to examine individual strategies (but see Koenig et al. 1998; Koenig 2002). Furthermore, we are missing the critical connection between individual attributes (such as dominance, nutritional status, physiological stress, and parasite burden), and how they affect components of fitness (such as survival probability and reproduction). For example, it would be very useful to assess individual differences in feeding efficiency between groups of different sizes. Although direct measures of reproductive success would be ideal, differences in feeding efficiency may be a sufficient proxy and may help shed light on the costs of grouping in folivores. Following the predictions of the ecological constraints model (Chapman and Chapman 2000b), females in larger groups should have lower caloric intake, consume foods of lower quality (particularly foods with lower protein-to-fiber ratios), have longer day ranges (after controlling for environmental conditions), and be in inferior physical condition relative to females in smaller groups, all else being equal. If the model is correct in linking group size and feeding efficiency, not all females are expected to achieve their full reproductive potential and this may impact dispersal decisions. Our studies suggest that the ecological setting in which a primate population finds itself can strongly influence both its abundance and its social structure. We believe that the most exciting and interesting developments still await us as we come to understand how ecological variables determine individual strategies affecting the behavior and ecology of folivores.

Acknowledgments Funding for this research was provided by the Canada Research Chairs Program, Wildlife Conservation Society (WCS), Natural Science and Engineering Research Council (NSERC) Canada, and the National Science Foundation (NSF), U.S.A. (SBR-990899). J.F.G. was supported by a Graduate Research Fellowship from the National Science Foundation, the Canadian Institutes of Health Research's Systems Biology Training Program, an Explorers Club–Eddie Bauer Youth Grant, and a Quebec Centre for Biodiversity Science Excellence Award. Permission to conduct this research was given by the Office of the President, Uganda, the National Council for Science and Technology, and the Uganda Wildlife Authority. Many people provided assistance or helpful comments on this research, including Lauren Chapman, Karen Bjorndal, Tom Gillespie, Ellis Greiner, Daphne Onderdonk, Mike Huffman, Mike Wasserman, and Toni Zeigler. Ellis Greiner aided in parasite identification.

References

Bell FR (1995) Perception of sodium appetite in farm animals. In: Phillips CJC, Chiy PC (eds) Sodium in agriculture. Chalcombe, Canterbury, pp 82–90

Bercovitch FB, Ziegler TE (2002) Current topics in primate socioendocrinology. Annu Rev Anthropol 31:45–67

Boesch C (1996) Social grouping in Tai chimpanzees. In: McGrew WC, Marchant LF, Nishida T (eds) Great ape societies. Cambridge University Press, Cambridge, pp 101–113

Boonstra R, Singleton GR (1993) Population declines in the snowshoe hare and the role of stress. Gen Comp Endocrinol 91:126–143

Boutin S (1990) Food supplementation experiments with terrestrial vertebrates: patterns, problems, and the future. Can J Zool 68:203–220

Butynski TM (1990) Comparative ecology of blue monkeys (*Cercopithecus mitis*) in high- and low-density sub-populations. Ecol Monogr 60:1–26

Camm JD, Norman SK, Polasky S, Solow AR (2002) Nature reserve site selection to maximize expected species covered. Oper Res 50:946–955

Chapman CA (1988) Patch use and patch depletion by the spider and howling monkeys of Santa Rosa National Park, Costa Rica. Behav Ecol 105:99–116

Chapman CA (1990) Association patterns of spider monkeys: the influence of ecology and sex on social organization. Behav Ecol Sociobiol 26:409–414

Chapman CA, Chapman LJ (1999) Implications of small scale variation in ecological conditions for the diet and density of red colobus monkeys. Primates 40:215–231

Chapman CA, Chapman LJ (2000a) Constraints on group size in red colobus and red-tailed guenons: examining the generality of the ecological constraints model. Int J Primatol 21:565–585

Chapman CA, Chapman LJ (2000b) Determinants of group size in primates: the importance of travel costs. In: Boinski S, Garber PA (eds) On the move: how and why animals travel in groups. University of Chicago Press, Chicago, pp 24–41

Chapman CA, Chapman LJ (2002) Foraging challenges of red colobus monkeys: influence of nutrients and secondary compounds. Comp Biochem Physiol A Physiol 133:861–875
Chapman CA, Fedigan LM (1990) Dietary differences between neighboring cebus monkey groups: local traditions or responses to food availability? Folia Primatol (Basel) 54:177–186
Chapman CA, Gogarten JF (2012) Primate conservation: is the cup half empty or half full? Nature Educ Knowl 3:7
Chapman CA, Lambert JE (2000) Habitat alterations and the conservation of African primates: a case study of Kibale National Park, Uganda. Am J Primatol 50:169–186
Chapman CA, Peres CA (2001) Primate conservation in the new millennium: the role of scientists. Evol Anthropol 10:16–33
Chapman CA, Wrangham RW, Chapman LJ (1995) Ecological constraints on group size: an analysis of spider monkey and chimpanzee subgroups. Behav Ecol Sociobiol 36:59–70
Chapman CA, Chapman LJ, Bjorndal KA, Onderdonk DA (2002a) Application of protein-to-fiber ratios to predict colobine abundance on difference spatial scales. Int J Primatol 23:283–310
Chapman CA, Chapman LJ, Bjorndal KA, Onderdonk DA (2002b) Application of protein-to-fiber ratios to predict colobine abundance on different spatial scales. Int J Primatol 23:283–310
Chapman CA, Chapman LJ, Naughton-Treves L, Lawes MJ, McDowell LR (2004) Predicting folivorous primate abundance: validation of a nutritional model. Am J Primatol 62:55–69
Chapman CA, Speirs ML, Gillespie TR, Holland T, Austad K (2005) Life on the edge: gastrointestinal parasites from forest edge and interior primate groups. Am J Primatol 68:1–12
Chapman CA, Wasserman MD, Gillespie TR, Speirs ML, Lawes MJ, Saj TL, Ziegler TE (2006) Do nutrition, parasitism, and stress have synergistic effects on red colobus populations living in forest fragments? Am J Phys Anthropol 131:525–534
Chapman CA, Naughton-Treves L, Lawes MJ, Wasserman MD, Gillespie TR (2007) The conservation value of forest fragments: explanations for population declines of the colobus of western Uganda. Int J Primatol 23:513–578
Chapman CA, Struhsaker TT, Skorupa JP, Snaith TV, Rothman JM (2010) Understanding long-term primate community dynamics: implications of forest change. Ecol Appl 20:179–191
Chivers DJ, Hladik CM (1980) Morphology of the gastrointestinal tract in primates: comparisons with other mammals in relation to diet. J Morphol 166:337–386
Clutton-Brock TH, Harvey PH (1977) Primate ecology and social organization. J Zool 183:1–39
Coley P (1983) Herbivory and defensive characteristics of tree species in a lowland tropical forest. Ecol Monogr 53:209–233
Coop RL, Holmes PH (1996) Nutrition and parasite interaction. Int J Parasitol 26:951–962
Creel S (2001) Social dominance and stress hormones. Trends Ecol Evol 16:491–497
Creel S, Fox JE, Hardy J, Sands B, Garrot B, Peterson RO (2002) Snowmobile activity and glucocorticoid stress responses in wolves and elk. Conserv Biol 16:809–814
Crockett CM, Janson CH (2000) Infanticide in red howlers: female group size, group composition, and a possible link to folivory. In: van Schaik CP, Janson CH (eds) Infanticide by males and its implications. Cambridge University Press, Cambridge, pp 75–98
Crompton DWT (1991) Nutritional interactions between hosts and parasites. In: Toft CA, Aeschlimann A, Bolis L (eds) Parasite–host associations: coexistence of conflict. Oxford University Press, Oxford, pp 228–257
DaSilva GL (1992) The western black and white colobus as a low energy strategist: activity budget, energy expenditure and energy intake. J Anim Ecol 61:79–91
Davies AG (1994) Colobine populations. In: Davies AG, Oates JF (eds) Colobine monkeys. Their ecology, behaviour and evolution. Cambridge University Press, Cambridge, pp 285–310
Eisenberg JF, Muckenhirn NA, Rudran R (1972) The relation between ecology and social structure in primates. Science 176:863–874
Fashing PJ, Dierenfeld E, Mowry CB (2007) Influence of plant and soil chemistry on food selection, ranging patterns, and biomass of *Colobus guereza* in Kakamega Forest, Kenya. Int J Primatol 28:673–703
Fearer TM, Prisley SP, Stauffer DF, Keyser PD (2007) A method for integrating the Breeding Bird Survey and Forest Inventory and Analysis databases to evaluate forest bird–habitat relationships at multiple spatial scales. For Ecol Manag 243:128–143

Fimbel C, Vedder A, Dierenfeld E, Mulindahabi F (2001) An ecological basis for large group size in *Colobus angolensis* in the Nyungwe Forest, Rwanda. Afr J Ecol 39:83–92
Ganzhorn JU (2002) Distribution of a folivorous lemur in relation to seasonally varying food resources: integrating quantitative and qualitative aspects of food characteristics. Oecologia (Berl) 131:427–435
Gillespie TR, Chapman CA (2001) Determinants of group size in the red colobus monkey (*Procolobus badius*): an evaluation of the generality of the ecological-constraints model. Behav Ecol Sociobiol 50:329–338
Gillespie TR, Chapman CA, Greiner EC (2005) Effects of logging on gastrointestinal parasite infections and infection risk in African primates. J Appl Ecol 42:699–707
Gogarten JF, Bonnell TR, Campenni M, Wasserman MD, Chapman CA (in review-a) Increasing group size alters behavior of a folivorous primate
Gogarten JF, Jacob AL, Ghai RR, Rothman JM, Twinomugisha D, Wasserman MD, Chapman CA (in review-b) Causes and consequences of changing group sizes in a primate community over 15+ years
Gogarten JF, Brown LM, Chapman CA, Marina C, Doran-Sheehy D, Fedigan LM, Grine FE, Perry S, Pusey AE, Sterck EHM, Wich SA, Wright PC (2012a) Seasonal mortality patterns in non-human primates: implications for variation in selection pressures across environments. Evolution 66:3256–3266
Gogarten JF, Guzman M, Chapman CA, Jacob AL, Omeja PA, Rothman JM (2012b) What is the predictive power of the colobine protein-to-fiber model and its conservation value? Trop Conserv Sci 5:381–393
Goldberg TL, Chapman CA, Cameron K, Saj S, Karesh W, Wolfe N, Wong SW, Dubois ME, Slifka MK (2008) Serologic evidence for a novel poxvirus in endangered red colobus monkeys. Emerg Infect Dis 14:801–803
Goldberg TL, Sintasath DM, Chapman CA, Cameron KM, Karesh WB, Tang S, Wolfe ND, Rwego IB, Ting N, Switzer WM (2009) Co-infection of Ugandan red colobus (*Procolobus [Piliocolobus] rufomitratus tephrosceles*) with novel, divergent delta-, lenti-, and spumaretro-viruses. J Virol 83:11318–11329
Gulland FMD (1992) The role of nematode parasites in Soay sheep (*Ovis aries* L.) mortality during a population crash. Parasitol Res 105:493–503
Harris TR (2005) Roaring, intergroup aggression, and feeding competition in black and white colobus monkeys (*Colobus guereza*) at Kanyawara, Kibale National Park, Uganda. Yale University, New Haven
Harris TR, Chapman CA, Monfort SL (2010) Small folivorous primate groups exhibit behavioral and physiological effects of food scarcity. Behav Ecol 21:46–56
Holmes JC (1995) Population regulation: a dynamic complex of interactions. Wildl Res 22:11–20
Hudson PJ, Dobson AP, Newborn D (1992) Do parasites make prey vulnerable to predation: red grouse and parasites. J Anim Ecol 61:681–692
Isbell LA (1991) Contest and scramble competition: patterns of female aggression and ranging behaviour among primates. Behav Ecol 2:143–155
Isbell LA, Young TP (2002) Ecological models of female social relationships in primates: similarities, disparities and some directions for future clarity. Behaviour 139:177–202
Janson CH, Goldsmith ML (1995) Predicting group size in primates: foraging costs and predation risks. Behav Ecol 6:326–336
Janson CH, van Schaik CP (1988) Recognizing the many faces of primate food competition: methods. Behaviour 105:165–186
Kingsolver JG, Schemske DW (1991) Path analysis of selection. Trends Ecol Evol 6:276–280
Koenig A (2000) Competitive regimes in forest-dwelling hanuman langur females (*Semnopithecus entellus*). Behav Ecol Sociobiol 48:93–109
Koenig A (2002) Competition for resources and its behavioral consequences among female primates. Int J Primatol 23:759–783
Koenig A, Borries C (2002) Feeding competition and infanticide constrain group size in wild hanuman langurs. Am J Primatol 57:33–34

Koenig A, Beise J, Chalise MK, Ganzhorn JU (1998) When females should contest for food: testing hypotheses about resource density, distribution, size and quality with hanuman langurs (*Presbytis entellus*). Behav Ecol Sociobiol 42:225–237
Korstjens AH, Sterck EHM, Noe R (2002) How adaptive or phylogenetically inert is primate social behaviour? A test with two sympatric colobines. Behaviour 139:203–225
Kristan WB (2007) Expected effects of correlated habitat variables on habitat quality and bird distribution. Condor 109:505–515
Lee PY, Rotenberry JT (2005) Relationships between bird species and tree species assemblages in forested habitats of eastern North America. J Biogeogr 32:1139–1150
McDowell LR (1992) Minerals in animal and human nutrition. Academic Press, New York
McKey DB, Gartlan JS, Waterman PG, Choo GM (1981) Food selection by black colobus monkeys (*Colobus satanas*) in relation to plant chemistry. Biol J Linn Soc 16:115–146
McNab BK (2002) The physiological ecology of vertebrates: a view from energetics. Cornell University Press, Cornell
Milton K (1979) Factors influencing leaf choice by howler monkeys: a test of some hypotheses of food selection by generalist herbivores. Am Nat 114:363–378
Milton K (1996) Effects of bot fly (*Alouattamyia baeri*) parasitism on a free-ranging howler (*Alouatta palliata*) population in Panama. J Zool Soc Lond 239:39–63
Miyamoto MM, Allen JA, Ting N, Gogarten JF, Chapman CA (in press) Microsatellite DNA demonstrates different levels of genetic structure in two, unequally sized, neighboring groups of red colobus monkeys. Am J Primatol 75:478–490
Munger JC, Karasov WH (1989) Sublethal parasites and host energy budgets: tapeworm infection in white-footed mice. Ecology 70:904–921
Nunn CL (2011) The comparative approach in evolutionary anthropology and biology. University of Chicago Press, Chicago
Oates JF (1994) The natural history of African colobines. In: Davies AG, Oates JF (eds) Colobine monkeys: their ecology, behaviour and evolution. Cambridge University Press, Cambridge
Oates JF, Davies AG (1994) What are the colobines. In: Oates JF, Davies AG (eds) Colobine monkeys: their ecology, behaviour and evolution. Cambridge University Press, Cambridge, pp 1–10
Oates JF, Whitesides GH, Davies AG, Waterman PG, Green SM, DaSilva GL, Mole S (1990) Determinants of variation in tropical forest primate biomass: new evidence from West Africa. Ecology 71:328–343
Omeja PA, Chapman CA, Obua J (2009) Enrichment planting does not promote native tropical tree restoration in a former pine plantation. Afr J Ecol 47:650–657
Robbins CT (1993) Wildlife feeding and nutrition. Academic, New York
Rode KD, Chapman CA, Chapman LJ, McDowell LR (2003) Mineral resource availability and consumption by colobus in Kibale National Park, Uganda. Int J Primatol 24:541–573
Rothman JM, Chapman CA, van Soest PJ (2012) Methods in primate nutritional ecology: a user's guide. Int J Primatol 33:542–566
Sapolsky RM (1992) Neuroendocrinology of the stress-response. In: Becker JB, Breedlove SM, Crews D (eds) Behavioral endocrinology. MIT, Cambridge, pp 287–566
Siex K, Struhsaker TT (1999) Ecology of the Zanzibar red colobus monkey: demography variability and habitat stability. Int J Primatol 20:163–192
Skorupa JP (1988) The effect of selective timber harvesting on rain forest primates in Kibale Forest, Uganda. PhD Thesis, University of California, Davis
Smith WH (1976) Character and significance of forest tree root exudates. Ecology 57:324–331
Snaith TV, Chapman CA (2005) Towards an ecological solution to the folivore paradox: patch depletion as an indicator of within-group scramble competition in red colobus. Behav Ecol Sociobiol 59:185–190
Snaith TV, Chapman CA (2007) Primate group size and socioecological models: do folivores really play by different rules? Evol Anthropol 16:94–106
Snaith TV, Chapman CA (2008) Red colobus monkeys display alternative behavioural responses to the costs of scramble competition. Behav Ecol 19:1289–1296

Steenbeek R, van Schaik CP (2001) Competition and group size in Thomas's langurs (*Presbytis thomasi*): the folivore paradox revisited. Behav Ecol Sociobiol 49:100–110
Sterck EHM, Watts DP, van Schaik CP (1997) The evolution of female social relationships in nonhuman primates. Behav Ecol Sociobiol 41:291–309
Struhsaker TT (1975) The red colobus monkey. University of Chicago Press, Chicago
Struhsaker TT (1997) Ecology of an African rain forest: logging in Kibale and the conflict between conservation and exploitation. University of Florida Press, Gainesville
Struhsaker TT (2000) Variation in adult sex ratios of red colobus monkey social groups: implications for interspecific comparisons. In: Kappeler PM (ed) Primate males: causes and consequences of variation in group composition. Cambridge University Press, Cambridge, pp 108–119
Struhsaker T (2008a) *Procolobus rufomitratus* ssp. *tephrosceles*. IUCN 2010 IUCN Red List of Threatened Species, Version 20104
Struhsaker TT (2008b) Demographic variability in monkeys: implication for theory and conservation. Int J Primatol 28:19–34
Struhsaker TT (2010) The red colobus monkeys: variation in demography, behavior, and ecology of endangered species. Oxford University Press, Oxford
Struhsaker TT, Leland L (1987) Colobines: infanticide by adult males. In: Smuts BB, Cheney DL, Seyfarth RM, Wrangham RW, Struhsaker TT (eds) Primate societies. University of Chicago Press, Chicago, pp 83–97
Struhsaker TT, Marshall AR, Detwiler K, Siex K, Ehardt C, Lisbjerg DD, Butynski TM (2004) Demographic variation among Udzungwa red colobus in relation to gross ecological and sociological parameters. Int J Primatol 25:615–658
Terborgh J, Janson CH (1986) The socioecology of primate groups. Annu Rev Ecol Syst 17:111–135
van Schaik CP (1983) Why are diurnal primates living in groups? Behaviour 87:120–144
van Schaik CP (1999) The socioecology of fission–fusion sociality in orangutans. Primates 40:69–86
van Schaik CP, van Hooff JARAM (1983) On the ultimate causes of primate social systems. Behaviour 85:91–117
Van Soest PJ (1994) Nutritional ecology of the ruminant. Cornell University Press, Ithaca
Wasser SK, Bevis K, King G, Hanson E (1997) Noninvasive physiological measures of disturbance in the northern spotted owl. Conserv Biol 11:1019–1022
Wasserman MD, Chapman CA (2003) Determinants of colobine monkey abundance: the importance of food energy, protein and fibre content. J Anim Ecol 72:650–659
Waterman PG, Ross JAM, Bennett EL, Davies AG (1988) A comparison of the floristics and leaf chemistry of the tree flora in two Malaysian rain forests and the influence of leaf chemistry on populations of colobine monkeys in the Old World. Biol J Linn Soc 34:1–32
Wrangham RW (1980) An ecological model of female-bonded primate groups. Behaviour 75:262–300
Wrangham RW (2000) Why are male chimpanzees more gregarious than mothers? A scramble competition hypothesis. In: Kappeler PM (ed) Primate males: causes and consequences of variation in group composition. Cambridge University Press, Cambridge, pp 248–258
Wursig B, Wursig M (1979) Behaviour and ecology of the bottlenose dolphin, *Tursiops truncatus*, in the South Atlantic. Fish Bull 77:399–412
Yeager CP, Kirkpatrick CR (1998) Asian colobine social structure: ecological and evolutionary constraints. Primates 39:147–155
Yeager CP, Kool K (2000) The behavioral ecology of Asian colobines. In: Whitehead PF, Jolly CJ (eds) Old World monkeys. Cambridge University Press, Cambridge, pp 496–521

Chapter 2
Dusky Dolphins: Flexibility in Foraging and Social Strategies

Bernd Würsig and Heidi C. Pearson

A dusky dolphin (*Lagenorhynchus obscurus*) performs a noisy leap in Admiralty Bay, New Zealand. (Photo courtesy of Chris Pearson)

B. Würsig (✉)
Department of Marine Biology, Texas A&M University at Galveston,
200 Seawolf Parkway, Galveston , TX 77553, USA
e-mail: wursigb@tamug.edu

H.C. Pearson
Department of Marine Biology, Texas A&M University at Galveston,
200 Seawolf Parkway, Galveston, TX 77553, USA

University of Alaska Southeast, 11120 Glacier Highway, Juneau, AK 99801, USA

J. Yamagiwa and L. Karczmarski (eds.), *Primates and Cetaceans: Field Research and Conservation of Complex Mammalian Societies*, Primatology Monographs,
DOI 10.1007/978-4-431-54523-1_2,

Abstract Dusky dolphins (*Lagenorhynchus obscurus*) exhibit fission–fusion dynamics as individuals join and split from groups of two to several thousand individuals. During the past three decades, our studies of dusky dolphins in three distinct marine systems have revealed how habitat type, predation risk, and prey availability influence foraging and social strategies. In the large Argentine bay of Golfo San José, fission–fusion dynamics are driven by large group formation during the day to coordinate prey-herding behaviors on southern anchovy (*Engraulis anchoita*) and nighttime resting in small groups near shore for predator avoidance. In the small New Zealand bay of Admiralty Bay, fission–fusion dynamics are also driven by daytime coordinated prey-herding strategies on small schooling fishes, but changes in group size are relatively muted and there is little predation risk. In the open, deep-water environment off Kaikoura, New Zealand, dusky dolphins rest and socialize near shore during the day, moving offshore at night to feed on the deep scattering layer (DSL). Changes in group size are also relatively muted off Kaikoura, and large groups serve to reduce predation risk. Comparisons between the three sites reveal a pattern of increasing group size with increasing openness of habitat. Response to predation pressure includes formation of large groups or formation of small, inconspicuous groups near shore. In the bay systems, fission–fusion dynamics are driven by coordinated foraging strategies on patchily distributed schooling fishes. In the open ocean system, DSL prey resources are more reliable, and fission–fusion dynamics are instead driven by strategies to obtain mates and avoid predators. Despite these differences, dusky dolphins exhibit polygynandry in all three systems. The presence of a single social-sexual system in spite of variability in fission–fusion dynamics has also been documented in chimpanzees (*Pan troglodytes*). In both societies, flexibility in social and foraging strategies enables individuals to respond to changing socioecological conditions. However, traits of the marine environment such as few physical refuges, low cost of transport, the need to herd mobile food resources into tight prey balls, and separation of oxygen and prey may be factors contributing toward differences in fission–fusion dynamics between dolphins and primates.

Keywords Chimpanzee • Dusky dolphin • Fission–fusion • Foraging • *Lagenorhynchus obscurus* • New Zealand • Semi-pelagic • Society

2.1 Introduction

Similar to primates such as chimpanzees (*Pan troglodytes*), bonobos (*Pan paniscus*), and spider monkeys (*Ateles paniscus*), most dolphins form fission–fusion societies (Connor et al. 2000; Gowans et al. 2008). These societies are characterized by dynamic changes in group size and composition as a result of individuals joining and splitting from groups. According to the conceptual framework proposed by Aureli et al. (2008), such societies exhibit a high degree of fission–fusion dynamics. This form of sociality may have evolved in response to patchy and ephemeral food sources, where intragroup competition may be reduced if individuals join and split from parties according to resource availability (Würsig 1978; Wells et al. 1987).

Social and predator pressures have likely contributed to the evolution of highly dynamic fission–fusion societies as well (van Schaik and van Hooff 1983), with the exception of rather closed societies in some killer whale (*Orcinus orca*) populations (Baird 2000) or in environments where other suitable habitats are remote [see Lusseau et al. (2003) for Doubtful Sound, New Zealand bottlenose dolphins (*Tursiops* spp.) and Karczmarski et al. (2005) for atoll-living Hawaiian spinner dolphins (*Stenella longirostris*)].

Although nearshore dolphin species such as common (*T. truncatus*) and Indo-Pacific (*T. aduncus*) bottlenose dolphins have been well studied (see Gowans et al. 2008 for discussion), advancing our knowledge of dolphin fission–fusion dynamics, much less is known about societies of pelagic or semi-pelagic dolphins (i.e., occurring in shallow, coastal areas in addition to deep areas beyond the edge of the continental shelf). The dusky dolphin (*Lagenorhynchus obscurus*) is a semi-pelagic species that occurs in the Southern Hemisphere, primarily off the coasts of southwest Africa (Namibia, South Africa), South America (Peru, Chile, Argentina), and New Zealand (Würsig and Würsig 1980; Würsig et al. 1989, 1997; Cassens et al. 2005) (Fig. 2.1).

Dusky dolphins live in fission–fusion societies of fewer than one dozen to several thousand in one group, and thus their society structure is labile enough to rapidly take advantage of most-efficient prey acquisition (Würsig and Würsig 1980; Dahood 2009; Dahood and Benoit-Bird 2010) and detection and avoidance of predators (Srinivasan 2009; Srinivasan and Markowitz 2010). In shallow-water environments where schooling prey are available as "rare and random" (Poisson-distributed) events, dusky dolphins tend to forage and feed in close coordination (Vaughn et al. 2007). In deep-water environments where prey are associated with the deep scattering layer (DSL), which becomes accessible to small dolphin dive capabilities only at night (Benoit-Bird et al. 2004, 2009), dolphins may simply aggregate for protection against danger and not to coordinate foraging activities.

The dynamic nature of dusky dolphin grouping patterns permits a wide and variable social network, which is likely important in mating, foraging, and calf-rearing strategies. For example, groups that are large or have a rapid turnover in composition may permit individuals to "meet" and interact with various members of the community, perhaps to facilitate coordinated foraging and mating strategies. Nursery groups composed of females and calves may tend to be smaller and more stable, providing opportunities for infant socialization and protection from male harassment (Würsig and Würsig 1980; Markowitz 2004; Weir et al. 2010).

After more than 30 years of observing dusky dolphins in Argentina and New Zealand, we are in a position to describe dusky dolphin sociality in a qualitative and quantitative manner that enables us to draw some comparisons and contrasts with primate societies. In this chapter, we focus on the potential contribution of three factors toward dusky dolphin foraging strategies, social strategies, and fission–fusion dynamics: habitat type, predation risk, and prey type and distribution. We focus on dusky dolphins at three study sites: Golfo San José, Argentina; Kaikoura, New Zealand; and Admiralty Bay, New Zealand (Table 2.1; Fig. 2.2). We discuss how differences in foraging and social strategies drive differences in dusky dolphin fission–fusion dynamics at each site. We then discuss how these

Fig. 2.1 Dusky dolphins (*Lagenorhynchus obscurus*) may form (**a**) large groups, as off Kaikoura, New Zealand or (**b**) small groups, as off Admiralty Bay, New Zealand. (Photographs courtesy of Chris Pearson)

factors may be common versus unique between dolphins and primates, and how they may contribute to similarities and differences in dolphin and primate fission–fusion dynamics.

Table 2.1 Comparison of habitat, food type and distribution, predation risk, group size, and fission–fusion dynamics for dusky dolphins (*Lagenorhynchus obscurus*) in Golfo San José, Argentina and in Kaikoura and Admiralty Bay, New Zealand; and for chimpanzees (*Pan troglodytes*) at Gombe, Tanzania; the Taï Forest, Ivory Coast; and Fongoli, Senegal

	Dusky dolphins[a]			Chimpanzees[b]		
	Golfo San José	Kaikoura	Admiralty Bay	Gombe	Taï	Fongoli
Habitat	Large bay	Coastal waters and open, deep-water canyon	Small, shallow enclosed bay	Semideciduous with riverine forest, woodland, and grassland	Evergreen lowland rainforest	Open woodland and grassland
Primary food type and distribution	Patchily distributed anchovy	More predictable nonschooling fish and squid associated with the deep scattering layer	Patchily distributed pilchard	Patchily distributed fruit	Patchily distributed fruit	Patchily distributed fruit; relatively narrow diet
Relative[c] predation risk	Moderate	High	Low	Low	Moderate	Low
Typical group size	10–12[d] 150–300[e]	10–12[f] 250–1,000[g]	7	6	10	15
Relative fission–fusion fluidity[f]	High	Low	Medium	Medium	High	Low

[a]References: Golfo San José: Würsig and Würsig (1980); Kaikoura: Benoit-Bird et al. (2004), Markowitz (2004), Srinivasan and Markowitz (2010), Würsig et al. (2007); Admiralty Bay: Benoit-Bird et al. (2004), Pearson (2009), Vaughn et al. (2010)

[b]References: Gombe: Goodall (1986), Taï: Anderson et al. (2006), and Boesch and Boesch-Acherman (2000), Fongoli: Pruetz (2006), Pruetz and Bertolani (2009), Pruetz et al. (2002)

[c]Relative to other populations of the same species

[d]Small group "fission"

[e]Large group "fusion" feeding/social

[f]Small nursery and other satellite

[g]Large regular daytime

[h]The degree of group fission and fusion in terms of changes in group size and composition; high fluidity indicates extreme and oftentimes frequent changes in group size and composition

Fig. 2.2 (**a**) Map of New Zealand showing the two main study areas for dusky dolphin (*Lagenorynchus obscurus*) research in Admiralty Bay and Kaikoura, and other localities where dusky dolphins have been sighted in New Zealand. (Modified from Würsig et al. 2007). (**b**) Map of Peninsula Valdés, Argentina, showing two areas where dusky dolphins occur: Golfo San José in the north and Golfo Nuevo in the south. (Map courtesy of Griselda Garaffo)

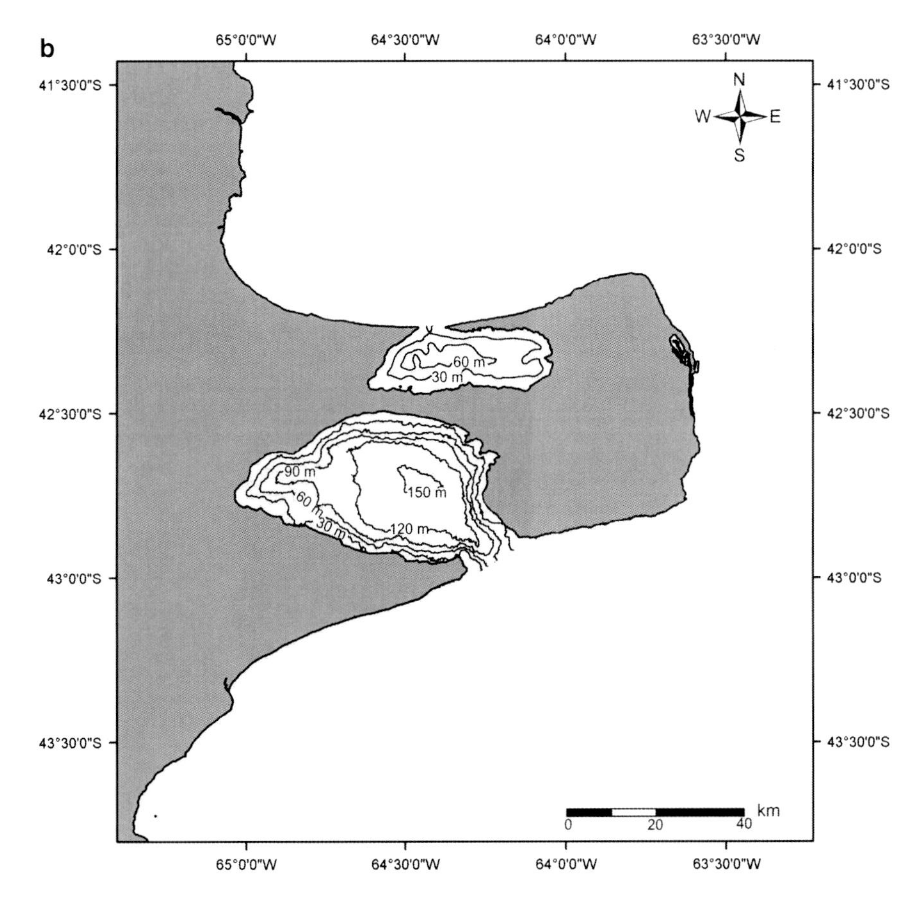

Fig 2.2 (continued)

2.2 Golfo San José

Golfo San José is in the expansive continental shelf region of central Patagonia and measures approximately 750 km² in area with a maximum extent of 50 km east–west and 20 km north–south. Dusky dolphins typically occur less than 5 km from shore at depths less than 200 m (Würsig et al. 1989). Dusky dolphin fission–fusion dynamics are primarily driven by diurnal strategies to search for and capture schooling southern anchovy (*Engraulis anchoita*; Würsig and Würsig 1980; Dans et al. 2010); this type of feeding occurs largely but not exclusively during spring and summer. Over the 24-h cycle, dusky dolphins display a rather predictable cycle of movement and fission–fusion dynamics.

At night, dusky dolphins form subgroups of approximately 8–12 animals and occur less than 2 km from shore in waters less than 40 m deep; subgroups are spatially separated by approximately 300–1,000 m. Radio-tracking data show that dusky dolphins perform shallow dives with little movement during this time, which is indicative of resting behavior (Würsig 1982). Resting in shallow nearshore waters is likely an antipredator strategy against sharks and killer whales. Shallow waters increase the detectability of sharks attacking from below (see Norris and Dohl 1980 and Norris et al. 1994 for similar behavior in spinner dolphins) whereas waters close to shore and the turbulent surf zone enable dusky dolphins to "hide" from killer whales (Würsig and Würsig 1980; Constantine et al. 1998).

In early morning, as light levels cause facultative schoolers such as anchovy to school more tightly, dusky dolphins become more active and begin to travel in search of prey (Würsig 1982). Although dusky dolphin groups may be separated by as much as 5 km (Würsig 1986), groups are usually within visual and acoustic range of one another (Würsig and Würsig 1980). When an anchovy school is detected, dusky dolphins within a group begin to coordinate their feeding behaviors. By using the surface of the water as a wall through which prey cannot escape, and leaping in air to create momentum to reach depth ("head-first re-entry dive") and to create loud slapping noises ("noisy" leap), dusky dolphins herd the anchovy into a tight prey ball (Vaughn et al. 2008; Würsig 1986; Würsig and Würsig 1980), which may be so tight as to become partially anoxic, and cause the fish to become lethargic (B. Würsig, personal observations).

It is difficult for just one group to maintain a prey ball, and feeding stops if other groups do not join (Würsig and Würsig 1980). However, other groups from as much as 8 km away may be attracted to the prey ball by leaps from feeding dusky dolphins and birds circling overhead. In nearshore Patagonia, prey availability for terns, gulls, shearwaters, cormorants, albatrosses, and Magellanic penguins (*Spheniscus magellanicus*) is increased as a result of dolphin prey-herding efforts.

Large feeding groups may contain as many as 300 individuals. Such large group size is beneficial as it increases the duration of the feeding bout, which may last for several hours. After a feeding bout has ended, dusky dolphins are at a high social activity level, which includes acrobatic leaping behavior (e.g., in-air somersaults and rapid twists), high-intensity vocalizations (consisting of frequency- and amplitude-modulated burst pulses; Au et al. 2010), and sexual activities (Würsig and Würsig 1980; Markowitz 2004; Markowitz et al. 2010). Combinations of hetero- and homosexual (or bisexual; Roughgarden 2004) activities occur as well, with males inserting penises into the genital slits and anuses of females and other males, and females inserting rostral and dorsal fins into the genital slits of males and other females. We surmise that social/sexual activities are an important part of greeting ceremonies and social facilitation or "bonding" (Norris and Dohl 1980; Norris et al. 1994).

After the activity level during feeding and socializing has diminished, dusky dolphins split into smaller subgroups once again (Würsig et al. 1989). It is unlikely that membership of any subgroup will be the same as during the night before, except for some longer-term bonds (e.g., mother–calf pairs). We surmise that dusky dolphins regard each other as members of a larger network of associations in one area, and that for the most part it is not important with whom they travel at any one time. We have some indication that there are stabilities of multiyear associations not yet fully described (Würsig and Bastida 1986), and that these may occur along lines of longer-term nongenetic (friendship) and perhaps related (matriarchal) associations (Shelton et al. 2010).

In summary, dusky dolphins in Golfo San José travel in small groups in search of anchovy during the day. When a prey ball is detected, groups converge to coordinate prey-herding behaviors; feeding bouts are followed by sociosexual activity. At night, individuals split into small groups to rest near shore (Würsig et al. 1989).

2.3 Admiralty Bay

Admiralty Bay, at the northern tip of New Zealand's South Island, is an area of approximately 120 km^2 inhabited by dusky dolphins from late fall to early spring. During this time, the abundance of small schooling fishes such as pilchard (*Sardinops neopilchardus*) attracts dusky dolphins to Admiralty Bay. During the winter, there is little predation risk for dusky dolphins because killer whales and sharks are present in Admiralty Bay primarily during the summer (Pearson 2009).

As in Golfo San José, dusky dolphins in Admiralty Bay coordinate foraging activities to herd prey balls. Dusky dolphins typically herd prey by swimming under and around prey balls, then "flashing" their white ventral sides toward the prey just before capture (Vaughn et al. 2008). Mean feeding group size in Admiralty Bay is 8.3 ± 5.0 ($n = 268$), mean feeding bout length is 4.0 ± 6.2 min ($n = 221$), and mean depth of prey during feeding bouts is 3.6 ± 2.7 m ($n = 52$; Vaughn et al. 2007, 2008). As dusky dolphins herd prey toward the surface, prey accessibility is increased for seabirds such as shearwaters (*Puffinus* spp.), Australasian gannets (*Morus serrator*), gulls (*Larus* spp.), spotted shags (*Phalacrocorax punctatus*), and terns (*Sterna* spp.) (Vaughn et al. 2008).

To understand the relationship between coordinated foraging strategies and fission–fusion dynamics of dusky dolphins, three major factors may be examined: (1) the relationship between group size and behavior, (2) the relationship between rate of group size change (i.e., fission and fusion) and behavior, and (3) group composition and social fluidity (strength of bond formation). Data pertaining to the first two factors were collected in Admiralty Bay during 2005–2006 using boat-based behavioral observations. A total of 168 focal group follows were conducted over 168 observation hours. During a focal follow, group size, the proportion of the group engaged in each of four behavioral states (forage, rest, socialize, travel), and location were recorded every 2 min. Group size range was 1–50 individuals, and mean group size was 7.0 ± 6.0 individuals. On average, group size changed during 20 % of the 2-min observation intervals (Pearson 2009).

Generalized estimating equations were used to analyze the influence of behavioral state on group size and rates of group fusion and fission. The proportion of individuals foraging and socializing in a group was significantly related to group size whereas the proportion of individuals resting and traveling in a group had no effect on group size. Specifically, the proportion of individuals foraging was positively related to group size while the proportion of individuals socializing was negatively related to group size. Additionally, the rate of group fusion was positively related to the proportion of individuals foraging in a group while the rate of group fission was positively related to the proportion of individuals resting, socializing, and traveling in a group (Pearson 2009).

Data regarding dusky dolphin group composition were collected in Admiralty Bay from 2001 to 2006. Photo-identification (Würsig and Jefferson 1990) was used to determine group composition, and the program SOCPROG was used to obtain simple ratio association indices (AIs) and to examine behaviorally specific association patterns (Whitehead 2008). During 2001–2006, 228 individuals were sighted five times or more and included in the final sample. The mean AI was 0.04 ± 0.07 SD, indicating that, on average, any 2 individuals spent 4 % of their time together. The mean maximum AI was 0.45 ± 0.16, indicating that, on average, an individual spent 45 % of its time with its closest associate. Association indices were weakest during traveling (mean AI = 0.09 ± 0.04, maximum AI = 0.68 ± 0.24) and highest during foraging (mean AI = 0.13 ± 0.06, maximum AI = 0.83 ± 0.25) and socializing (mean AI = 0.17 ± 0.08, maximum AI = 0.87 ± 0.26) (Pearson 2008).

In summary, large group formation during foraging drives fission–fusion dynamics in Admiralty Bay and Golfo San José. However, fission–fusion dynamics in Admiralty Bay are more "muted" than in Golfo San José because group size does not oscillate as dramatically between large feeding and sociosexual groups and small traveling and resting groups. Dusky dolphins in Golfo San José engage in high-intensity post-feeding socialization whereas dusky dolphins in Admiralty Bay either continue to search for food or go into resting/low-level behavioral modes at the conclusion of a feeding bout. Furthermore, social activity occurs in small groups in Admiralty Bay and large groups in Golfo San José. An explanation for the differences in socializing may be that most data on foraging dusky dolphins in Argentina were gathered in spring, summer, and early autumn, when most mating occurs, whereas in Admiralty Bay, all data were gathered in winter, when mating is not prevalent. Burst pulse vocalizations during foraging were also more numerous per group number per time in Argentina than in Admiralty Bay, indicative of the higher degree of social activity in Argentina (Vaughn-Hirshorn et al. 2012).

Dusky dolphins in Admiralty Bay exhibit a high degree of social fluidity, as indicated by the presence of many weak and few strong AIs. Yet, there is stability as individuals form behaviorally specific preferred associations. In particular, bond formation during foraging and socializing may facilitate coordinated foraging strategies because individuals who are more familiar with each other may be more efficient hunting partners (Pearson 2008). We presently have no similar data for Golfo San José.

2.4 Kaikoura

At least some (if not all) of the dusky dolphins that occur in Admiralty Bay during the winter travel approximately 275 km south to the waters off Kaikoura during the summer (Markowitz 2004; Markowitz et al. 2004; Shelton 2006). Off Kaikoura, dusky dolphins occur along the open coastline of New Zealand, moving between shallow nearshore waters and deep (~2,000 m) waters of the Kaikoura Canyon. In some areas, Kaikoura Canyon extends within 500 m of shore and brings deep productive waters relatively close to shore.

Similar to dusky dolphins in Golfo San José, dusky dolphins in Kaikoura exhibit a rather predictable cycle of movement over a 24-h cycle. However, fission–fusion dynamics are quite different from Golfo San José and Admiralty Bay. At night, dusky dolphins move offshore in large groups to feed. Dusky dolphins move near shore to rest and socialize during the day, oftentimes remaining in large groups. Although large shark predation is no longer a threat off Kaikoura (Srinivasan 2009), occasional and sporadic occurrence of killer whales (Constantine et al. 1998) likely shapes this pattern of offshore feeding and nearshore resting and socializing (Srinivasan and Markowitz 2010).

At night, dusky dolphins move offshore in loose groups of several hundred individuals to feed on lantern fish (family Myctophidae) and squid (*Nototodarus* sp. and *Todaroides* sp.) within the DSL (Cipriano 1992; Benoit-Bird et al. 2004; Dahood 2009). The DSL rises to 29 m or less of the surface during the night and sinks to 200–300 m (and possibly deeper) during the day (Benoit-Bird et al. 2004; Markowitz et al. 2004; Würsig et al. 1989). Dusky dolphins begin to feed on the DSL in the early evening when it is within 130 m of the surface or less (Benoit-Bird et al. 2004).

We surmise that dusky dolphins feeding on the DSL stay together as a loose group of intercommunicating individuals to alert each other to predation threats (Srinivasan and Markowitz 2010). However, within the large foraging group, individuals form subgroups that range in size from one to five individuals. Subgroup size varies according to prey distribution within the DSL and may increase as prey becomes patchier, when it is advantageous for individuals to coordinate foraging behaviors. Subgroup size may also increase with decreasing depth of the DSL, possibly because less time is needed for traveling and more time is available for coordinated foraging at shallow depths (Benoit-Bird et al. 2004). In contrast to coordinated herding of prey balls at Golfo San José and Admiralty Bay, coordinated foraging in Kaikoura most likely serves as information sharing to increase prey-finding abilities (Benoit-Bird et al. 2004; Würsig et al. 1989).

During the day, the large group moves to nearshore waters less than 800 m deep. Instead of feeding (which occurs during 1 % of daytime hours), low-level bouts of social activity and much resting appear to be the norm, although occasions of higher-level social-sexual activities also take place (Dahood 2009; Markowitz et al. 2010). Large groups of 200 to 1,000 tightly spaced individuals are composed of subgroups of approximately 10 individuals each (Würsig et al. 1989, 2007). Within the large groups, individuals exhibit social fluidity with relatively strong social bonding. Markowitz (2004) used the half-weight coefficient index and found mean AI (±SE) to be 0.03 ± 0.0008, indicating a high degree of social "mixing." Mean maximum AI was 0.57 ± 0.0074, indicating that some individuals formed relatively strong social bonds (Markowitz 2004).

Some individuals form smaller "satellite" groups near shore and apart from the large group. Mating groups are typically composed of seven single (i.e., without calves) adults engaged in social-sexual activity. Most sexual behavior involves several males mating with several females, thus likely exhibiting a polygynandrous ("multi-mate") system. There is probably an element of female choice in mating, as sophisticated male–female interactions may allow females to choose their most adroitly maneuvering partners of the moment (see Markowitz et al. 2010 for further discussion).

During the summer, mating activity peaks and dusky dolphin males have engorged testes (Cipriano 1992; Van Waerebeek and Read 1994); this indicates the presence of sperm competition in this species. The peak in sexual activity and testes size coincides with a year-long gestation period and summer calving period (Cipriano 1992; Markowitz et al. 2010). However, some social-sexual activity occurs year round and we surmise that social-sexual activity is an important part of dusky dolphin social life, perhaps helping to establish and maintain bonds.

Satellite groups may also be composed of mothers and calves. Nursery group size ranges from 4 to 100 individuals (i.e., 2–50 mother–calf pairs), with a median group size of 14 individuals (Weir 2007). Most groups occur in waters less than 20 m deep and close to shore, likely for protection against killer whales (Deutsch 2008; Weir et al. 2008, 2010). Nursery groups also provide protection from the high activity levels of the large group and from male harassment. Smaller groups are less detectable (through the encounter effect), and nursery groups are smaller at the beginning of the calving season when calves are the smallest and most vulnerable (Deutsch 2008). In addition to providing protection from predators and conspecific harassment, nursery groups also grant mothers and calves increased time to rest and opportunities for infant socialization and also may enable exploitation of alternative prey sources (Weir et al. 2010).

In summary, dusky dolphin fission–fusion dynamics off Kaikoura are driven by offshore nocturnal feeding on the DSL and diurnal nearshore resting and socializing. Fission–fusion dynamics are present as individuals move between subgroups within the large group while still retaining relatively long-term social bonds. The large group forms a protective "envelope" that most likely serves to reduce predation risk in the open, deep-water environment.

2.5 Discussion

We have information on dusky dolphin occurrence and behavior in three distinct systems: (1) a large Argentine bay (Golfo San José), characterized by daytime prey ball feeding in large groups and nighttime resting close to shore in small groups; (2) a small New Zealand bay (Admiralty Bay), with daytime prey ball feeding and relatively muted changes in group size; and (3) the open ocean (Kaikoura), where a canyon comes close enough to shore for daytime rest and social activities near shore, but nighttime feeding occurs on the DSL in open ocean waters. Off Kaikoura, there are also relatively muted changes in group size, and a large group with small satellite groups of nurseries and social-sexual units is the norm.

In all three systems, dusky dolphins are highly social and usually occur in groups of at least six animals. However, differences in habitat type, predation pressure, and prey type and distribution have led to differences in fission–fusion dynamics in each system. There is a pattern of increasing group size with increasing openness of habitat. The smallest groups are formed within the smaller confines of Admiralty Bay whereas the largest groups are formed in the open-water environment off Kaikoura.

Our results suggest that behavioral responses to predation pressure are a function of habitat type and relative predation risk. Low predation pressure in Admiralty Bay appears to have "released" dusky dolphins from enacting defensive mechanisms such as forming large groups for resting and resting near shore (Pearson 2008). In Golfo San José, where predation risk is moderate, individuals form small groups close to shore at night where they can "hide" from predators. In Kaikoura, where predation risk is high, the largest groups are formed. After feeding offshore during the night, most dusky dolphins off Kaikoura rest in large groups during the day in the relative shallows closer to shore; this offshore to inshore diel shift is likely a strategy to avoid deep-water predation by killer whales.

In the bay systems, schooling fishes are present and daytime bait-balling drives fission–fusion dynamics. In Golfo San José, dusky dolphins split into multiple small groups to find food and aggregate for efficient bait-ball foraging. In Admiralty Bay, dusky dolphins remain in relatively small groups during foraging and feeding. Although group size does increase during foraging in Admiralty Bay, the scale of group fission and fusion is less than in Golfo San José.

It is likely that Admiralty Bay cannot support enough dolphin prey (or dolphins) for large-scale changes in aggregated dolphin numbers to occur. We assume that this relates to lowered efficiency of feeding in small bays and groups. Efficient near-surface bait-ball herding by dusky dolphins in Admiralty Bay may also be deterred by the presence of plunge-diving gannets. Up to 12 gannets may synchronize plunge-diving on a prey ball, an action that subsequently drives the fish deeper in the water column (Vaughn et al. 2010; Machovsky-Capuska et al. 2011). In Golfo San José, none of the birds that take advantage of fish at the surface are plunge divers (Würsig and Würsig 1980), and thus plunge-diving birds do not drive down prey in that system.

In the open ocean system off Kaikoura, dusky dolphins have a reliable food source (the DSL) along the canyon edge and can go to this food source night after night without fail. As it is not necessary to split into small groups to find food, dusky dolphins may remain in the large group envelope for efficient social behavior and—perhaps more importantly—for efficient detection and avoidance of predators.

Although dusky dolphins exhibit differences in foraging strategies, we postulate that the basic social-sexual pattern of polygynandry is the same for dusky dolphins in all three systems: this is apparent because social-sexual activities occur in all three habitats, and there is much multi-mate mating. Even in the seasonal (winter) habitat of Admiralty Bay where more than 50 % of individuals are male (Harlin 2004; Shelton 2006; Shelton et al. 2010), sexual activity still occurs but to a lesser degree than in the other two habitats.

Dusky dolphins exhibit a single sociosexual system that is combined with variability in behavior relative to ecological capabilities and constraints. This adaptability is not surprising, as we know that terrestrial animals are capable of quite similar adaptations to variable habitats (Wrangham 1987). Chimpanzees are polygynandrous in all areas in which they have been studied, but fission–fusion dynamics vary between habitats. In the dense vegetative cover of the Taï rainforest where both food and predators are abundant, mean group size is ten individuals, mean party duration is 24 min, and strong intersexual bonds are formed. In the more open woodland forests of Gombe where food is more widely dispersed and predators are scarce, mean group size is six individuals, mean party duration is 69 min, and there is little intersexual bonding (summarized in Boesch and Boesch-Acherman 2000).

Dusky dolphins, as all delphinid cetaceans, are social creatures par excellence. The basic "unit" of dusky dolphin society may be about six or so animals that know each other well. However, these six individuals need not remain in the same group to retain their affiliation. Individual dusky dolphins are part of a larger social network and associate with a variety of conspecifics as they join and split from groups throughout the course of a day. This social mixing permits individuals to meet and interact with a wide variety of individuals in the community while still retaining strong social bonds. In New Zealand, this may facilitate coordinated foraging strategies in Admiralty Bay during winter, and coordinated mating strategies during summer in Kaikoura. This is the essence of small dolphin fission–fusion dynamics, as described in the 1970s for bottlenose dolphins (Würsig and Würsig 1977; Würsig 1978). This grouping is quite different from the dolphin societies of pilot whales (*Globicephala* spp.), for example, where matriarchy of long-term affiliations of females and their female offspring appears to reign (Kasuya and Marsh 1984; Amos et al. 1993).

2.6 Comparisons with Primates

Dusky dolphin fission–fusion dynamics are similar to those observed in primates such as chimpanzees, bonobos, and spider monkeys (summarized in Pearson 2008). The act of joining and leaving groups according to the shifting balance of costs and benefits associated with foraging, avoiding predation, and finding mates is a behavioral adaptation to rapidly changing socioecological conditions. However, the very different physical environments in which primates and dolphins have evolved have implications for the type and scale of fission–fusion dynamics exhibited by each taxa. Traits of the marine environment such as few physical refuges, a low cost of transport, the need to herd mobile food resources into tight prey balls, and separation of oxygen and prey may be factors contributing towards differences in fission–fusion dynamics between dolphins and primates (Pearson and Shelton 2010).

Primate fission–fusion dynamics also vary widely, and chimpanzee grouping patterns in particular have been studied in a wide variety of habitats, as we have presented for dusky dolphins. A broad comparison of dusky dolphin and chimpanzee fission–fusion dynamics according to habitat type, food availability, and predation risk is presented in Table 2.1.

Primates can engage in foraging, vigilance, and other social interactions in seamless fashion, while breathing in almost unnoticed (and usually truly unnoticed) fashion. For dolphins to operate as members of a social group, they must stop what they are doing at depth, whether it is social foraging or some other activity, and (presumably) consciously leave the underwater activity and rise to the surface to obtain a life-sustaining breath of air. We know that they do so easily, with some dolphins surfacing while others are helping to herd and contain prey (Vaughn et al. 2007, 2008), but we imagine that there are special capability needs connected to this separation of duties between potentially cooperative foraging and the need to leave the immediate society to breathe. The ramifications are many and may be what has led to extreme matriarchy in sperm whales (*Physeter macrocephalus*) and pilot whales (Whitehead 2003).

As a result of their fusiform body shape and low cost of transport (Williams et al. 1992), dolphins require less energy than primates to move though their environment. Compared to primates, dolphin group composition may change more rapidly and individuals may be part of a wider social network because they may encounter dispersed conspecifics at a lower energetic cost. The low cost of transport also facilitates coordinated foraging on patchily distributed prey, as dolphins may spread out in search of prey and then quickly and "cheaply" aggregate when prey is detected (summarized in Pearson 2011).

In the marine environment, light propagates poorly and sound propagates well. Acoustic communication is therefore likely to be a more important component of cetacean societies than of the visually oriented societies of primates. Although primate group members certainly keep in auditory contact when outside of visual range of one another, the scale of cetacean acoustic communication is vast and something that is still unknown for most species. Cetacean social interactions may occur over wider spatial scales than in the terrestrial environment, and the acoustic group size of cetaceans, particularly of large baleen whales, is an exciting avenue of research.

2.7 Conclusion

Primates and dolphins are faced with similar social and ecological pressures and exhibit similarities in fission–fusion dynamics, despite being evolutionarily separated by 95 million years (Bromham et al. 1999) and living in very different environments. As displayed by dusky dolphins in three distinct habitats, flexibility is the rule, enabling individuals to respond to differences in habitat type, predation pressure, and prey type and distribution. The amount of fission–fusion is related to variable needs to find and secure food and to the intensity of predation risk. In the midst of these changes, many individuals form and maintain bonds that are likely important in foraging and mating strategies.

References

Amos B, Schlotterer C, Tautz D (1993) Social structure of pilot whales revealed by analytical DNA profiling. Science 260:670–672

Anderson DP, Nordheim EV, Boesch C (2006) Environmental factors influencing the seasonality of estrus in chimpanzees. Primates 47:43–50

Au WWL, Lammers MO, Yin S (2010) Acoustics of dusky dolphins. In: Würsig B, Würsig M (eds) Dusky dolphins: master acrobats off different shores. Academic/Elsevier, Amsterdam, pp 75–97

Aureli F, Schaffner CM, Boesch C, Bearder SK, Call J, Chapman CA, Connor RC, Di Fiore A, Dunbar RIM, Henzi SP, Holekamp K, Korstjens AH, Layton R, Lee P, Lehmann J, Manson JH, Ramos-Fernandez G, Strier KB, van Schaik CP (2008) Fission–fusion dynamics. Curr Anthropol 49:627–654

Baird RW (2000) The killer whale: foraging specializations and group hunting. In: Mann J, Connor RC, Tyack PL, Whitehead H (eds) Cetacean societies: field studies of whales and dolphins. University of Chicago Press, Chicago, pp 127–153

Benoit-Bird KJ, Würsig B, Mcfadden CJ (2004) Dusky dolphin (*Lagenorhynchus obscurus*) foraging in two different habitats: active acoustic detection of dolphins and their prey. Mar Mamm Sci 20:215–231

Benoit-Bird KJ, Dahood AD, Würsig B (2009) Using active acoustics to compare lunar effects on predator–prey behavior in two marine mammal species. Mar Ecol Prog Ser 395:119–135

Boesch C, Boesch-Acherman H (2000) The chimpanzees of the Taï Forest. Behavioural ecology and evolution. Oxford University Press, Oxford

Bromham L, Phillips MJ, Penny D (1999) Growing up with dinosaurs: molecular dates and the mammalian radiation. Trends Ecol Evol 13:113–118

Cassens I, van Waerebeek K, Best PB, Tzika A, van Helden AL, Crespo EA, Milinkovitch MC (2005) Evidence for male dispersal along the coasts but no migration in pelagic waters in dusky dolphins (*Lagenorhynchus obscurus*). Mol Ecol 14:107–121

Cipriano FW (1992) Behavior and occurrence patterns, feeding ecology and life history of dusky dolphins (*Lagenorhynchus obscurus*) off Kaikoura. Ph.D. Thesis, University of Arizona, Tucson

Connor RC, Mann J, Tyack PL, Whitehead H (2000) The social lives of whales and dolphins. In: Mann J, Connor RC, Tyack PL, Whitehead H (eds) Cetacean societies: field studies of dolphins and whales. University of Chicago Press, Chicago, pp 1–6

Constantine R, Visser I, Buurman D, Buurman R, McFadden B (1998) Killer whale (*Orcinus orca*) predation on dusky dolphins (*Lagenorhynchus obscurus*) in Kaikoura, New Zealand. Mar Mamm Sci 14:324–330

Dahood AD (2009) Dusky dolphin (*Lagenorhynchus obscurus*) occurrence and movement patterns near Kaikoura, New Zealand. M.S. Thesis, Texas A&M University, College Station

Dahood AD, Benoit-Bird KJ (2010) Dusky dolphins foraging at night. In: Würsig B, Würsig M (eds) Dusky dolphins: master acrobats off different shores. Academic/Elsevier, Amsterdam, pp 99–114

Dans SL, Crespo EA, Koen-Alonso M, Markowitz TM, Beron Vera B, Dahood AD (2010) Dusky dolphin trophic ecology: their role in the food web. In: Würsig B, Würsig M (eds) Dusky dolphins: master acrobats off different shores. Academic/Elsevier, Amsterdam, pp 49–74

Deutsch S (2008) Development and social learning of young dusky dolphins. M.Sc. Thesis, Texas A&M University, College Station

Goodall J (1986) The chimpanzees of Gombe: patterns of behavior. Belknap Press of Harvard University Press, Cambridge

Gowans S, Würsig B, Karczmarski L (2008) The social structure and strategies of delphinids: predictions based on an ecological framework. Adv Mar Biol 53:195–294

Harlin AD (2004) Molecular systematic and phylogeography of *Lagenorhynchus obscurus* derived from nuclear and mitochondrial loci. Ph.D. Dissertation. Texas A&M University, College Station

Karczmarski L, Würsig B, Gailey G, Larson KW, Vanderlip C (2005) Spinner dolphins in a remote Hawaiian atoll: social grouping and population structure. Behav Ecol 16:675–685

Kasuya T, Marsh H (1984) Life history and reproductive biology of the short-finned pilot whale, *Globicephala macrorhynchus*. Reports of the International Whaling Commission, Special Edition

Lusseau D, Schneider K, Boisseau OJ, Haase P, Slooten E, Dawson SM (2003) The bottlenose dolphin community of Doubtful Sound features a large proportion of long lasting associations. Can geographic isolation explain this unique trait? Behav Ecol Sociobiol 54:396–405

Machovsky-Capuska GE, Vaughn RL, Würsig B, Katzir G, Raubenheimer D (2011) Dive strategies and foraging effort in the Australasian gannet, *Morus serrator*, revealed by underwater videography. Mar Ecol Prog Ser 442:255–261

Markowitz TM (2004) Social organization of the New Zealand dusky dolphin. Ph.D. Dissertation, Texas A&M University, College Station

Markowitz TM, Harlin AD, Würsig B, McFadden CJ (2004) Dusky dolphin foraging habitat: overlap with aquaculture in New Zealand. Aquat Conserv Mar Freshw Ecosyst 14:133–149

Markowitz TM, Markowitz WJ, Morton LM (2010) Mating habits of New Zealand dusky dolphins. In: Würsig B, Würsig M (eds) Dusky dolphins: master acrobats off different shores. Academic/Elsevier, Amsterdam, pp 151–176

Norris KS, Dohl TP (1980) Behavior of the Hawaiian spinner dolphin, *Stenella longirostris*. Fish Bull US 77:821–849

Norris KS, Würsig B, Wells RS, Würsig M (1994) The Hawaiian spinner dolphin. University of California Press, Berkeley

Pearson HC (2008) Fission-fusion sociality in dusky dolphins (*Lagenorhynchus obscurus*), with comparisons to other dolphins and great apes. Ph.D. Dissertation, Texas A&M University, College Station

Pearson HC (2009) Influences on dusky dolphin (*Lagenorhynchus obscurus*) fission–fusion dynamics in Admiralty Bay, New Zealand. Behav Ecol Sociobiol 63:1437–1446

Pearson HC (2011) Sociability of female bottlenose dolphins (*Tursiops* spp.) and chimpanzees (*Pan troglodytes*): understanding evolutionary pathways toward social convergence. Evol Anthropol 20:85–95

Pearson HC, Shelton DE (2010) A large-brained social animal. In: Würsig B, Würsig M (eds) Dusky dolphins: master acrobats off different shores. Academic/Elsevier, Amsterdam, pp 333–353

Pruetz JD (2006) Feeding ecology of savanna chimpanzees (*Pan troglodytes verus*) at Fongoli, Senegal. In: Hohmann B, Robbins MM, Boesch C (eds) Feeding ecology of great apes and other primates. Cambridge University Press, Cambridge, pp 161–182

Pruetz JD, Bertolani P (2009) Chimpanzee (*Pan troglodytes verus*) behavioral responses to stresses associated with living in a savanna mosaic environment: implications for hominin adaptations to open habitats. Paleoanthropology 2009:252–262

Pruetz JD, Marchant LF, Arno J, McGrew WC (2002) Survey of savanna chimpanzees (*Pan troglodytes verus*) in southeastern Sénégal. Am J Primatol 58:35–43

Roughgarden J (2004) Evolution's rainbow: diversity, gender, and sexuality in nature and people. University of California Press, Berkeley

Shelton DE (2006) Dusky dolphins in New Zealand: group structure by sex and relatedness. M.Sc. Thesis, Texas A&M University, College Station

Shelton DE, Harlin-Cognato AD, Honeycutt RL, Markowitz TM (2010) Dusky dolphin sexual segregation and genetic relatedness in New Zealand. In: Würsig B, Würsig M (eds) Dusky dolphins: master acrobats off different shores. Academic/Elsevier, Amsterdam, pp 195–209

Srinivasan M (2009) Predator influences on the behavioral ecology of dusky dolphins. Ph.D. Dissertation, Texas A&M University, College Station

Srinivasan M, Markowitz TM (2010) Predator threats and dusky dolphin survival strategies. In: Würsig B, Würsig M (eds) Dusky dolphins: master acrobats off different shores. Academic/Elsevier, Amsterdam, pp 133–150

van Schaik CP, van Hooff JARAM (1983) On the ultimate causes of primate social systems. Behaviour 85:91–117

Van Waerebeek K, Read AJ (1994) Reproduction of dusky dolphins, *Lagenorhynchus obscurus*, from coastal Peru. J Mammal 75:1054–1062

Vaughn RL, Shelton DE, Timm LL, Watson LA, Würsig B (2007) Dusky dolphin (*Lagenorhynchus obscurus*) feeding tactics and multi-species associations. N Z J Mar Freshw Res 41:391–400

Vaughn RL, Würsig B, Shelton DE, Timm LL, Watson LA (2008) Dusky dolphins influence prey accessibility for seabirds in Admiralty Bay, New Zealand. J Mammal 89:1051–1058

Vaughn RL, Würsig B, Packard J (2010) Dolphin prey herding: prey ball mobility relative to dolphin group and prey ball sizes, multi-species associates, and feeding duration. Mar Mamm Sci 26:213–225

Vaughn-Hirshorn RL, Hodge KB, Würsig B, Sappenfield RH, Lammers MO, Dudzinski KM (2012) Characterizing dusky dolphin sounds from Argentina and New Zealand. J Acoust Soc Am 132:498–506

Weir JS (2007) Dusky dolphin nursery groups off Kaikoura, New Zealand. M.Sc. Thesis, Texas A&M University, College Station

Weir JS, Duprey NMT, Würsig B (2008) Dusky dolphin (*Lagenorhynchus obscurus*) subgroup distribution: are shallow waters a refuge for nursery groups? Can J Zool 86:1225–1234

Weir JS, Deutsch S, Pearson HC (2010) Dusky dolphin calf rearing. In: Würsig B, Würsig M (eds) Dusky dolphins: master acrobats off different shores. Academic/Elsevier, Amsterdam, pp 177–193

Wells RS, Scott MD, Irvine AB (1987) The social structure of free-ranging bottlenose dolphins. In: Genoways HH (ed) Current mammalogy. Plenum, New York, pp 247–305

Whitehead H (2003) Sperm whales: social evolution in the ocean. University of Chicago Press, Chicago

Whitehead H (2008) Analyzing animal societies: Quantitative methods for vertebrate social analysis. University of Chicago Press, Chicago

Williams TM, Friedl WA, Fong ML, Yamada RM, Sedivy P, Haun JE (1992) Travel at low energetic cost by swimming and wave-riding bottlenose dolphins. Nature (Lond) 355:821–823

Wrangham RW (1987) Evolution of social structure. In: Smuts BB, Cheney DL, Seyfarth RM, Wrangham RW, Struhsaker TT (eds) Primate societies, 1st edn. University of Chicago Press, Chicago, pp 282–296

Würsig B (1978) Occurrence and group organization of Atlantic bottlenose porpoises (*Tursiops truncatus*) in an Argentine bay. Biol Bull 154:348–359

Würsig B (1982) Radio tracking dusky porpoises in the South Atlantic. In: FAO Fisheries Mammals in the Seas, Series No. 5, vol IV. United Nations Food and Agriculture Organization, Rome

Würsig B (1986) Delphinid foraging strategies. In: Schusterman RJ, Thomas JA, Wood FG (eds) Dolphin cognition and behavior: a comparative approach. Erlbaum, London, pp 347–359

Würsig B, Bastida R (1986) Long-range movement and individual associations of two dusky dolphins (*Lagenorhynchus obscurus*) off Argentina. J Mammal 67:773–774

Würsig B, Jefferson TA (1990) Methods of photo-identification for small cetaceans. Rep Int Whaling Comm Special Issue 12:17–78

Würsig B, Würsig M (1977) The photographic determination of group size, composition, and stability of coastal porpoises (*Tursiops truncatus*). Science 198:755–756

Würsig B, Würsig M (1980) Behavior and ecology of the dusky dolphin, *Lagenorhynchus obscurus*, in the South Atlantic. Fish Bull 77:871–890

Würsig B, Würsig M, Cipriano F (1989) Dolphins in different worlds. Oceanus 32:71–75

Würsig B, Cipriano F, Slooten E, Constantine R, Barr K, Yin S (1997) Dusky dolphins (*Lagenorhynchus obscurus*) off New Zealand: status of present knowledge. Rep Int Whaling Comm 47:715–722

Würsig B, Duprey N, Weir J (2007) Dusky dolphins (*Lagenorhynchus obscurus*) in New Zealand waters. Present knowledge and research goals. DOC Res Dev Ser 270:1–28

Chapter 3
Socioecological Flexibility of Gorillas and Chimpanzees

Juichi Yamagiwa and Augustin Kanyunyi Basabose

J. Yamagiwa (✉)
Laboratory of Human Evolution Studies, Department of Zoology, Graduate School of Science, Kyoto University, Sakyo, Kyoto 606-8502, Japan
e-mail: yamagiwa@jinrui.zool.kyoto-u.ac.jp

A.K. Basabose
Centre de Recherche en Sciences Naturelles, Lwiro, D.S. Bukavu, Democratic Republic of Congo

J. Yamagiwa and L. Karczmarski (eds.), *Primates and Cetaceans: Field Research and Conservation of Complex Mammalian Societies*, Primatology Monographs, DOI 10.1007/978-4-431-54523-1_3, © Springer Japan 2014

Abstract The African great apes live in such diverse habitats as lowland tropical forests, montane forests, and dry savannas, and they show great flexibility in their ecological and social features. To identify the factors causing variations in these features, we examined their ranging, grouping, and life history traits in relationship to their dietary preferences. Analysis of their dietary compositions shows that both gorillas and chimpanzees have strong preferences for ripe fruits but with different folivorous and faunivorous diets as their fallback strategies. Dietary variation mainly caused by fluctuation in fruit availability does not constrain group cohesion, but it affects the daily path length of gorillas and influences the party size and fluidity of chimpanzees. High gregariousness among female bonobos and western chimpanzees may be explained by their low feeding competition over supplementary food resources derived from the absence of the sympatric gorillas in their habitats. Although high association and linear hierarchy among female western chimpanzees are attributed to abundant and monopolizable fruit resources, the high female–male association observed in western chimpanzees and bonobos is not explained solely by ecological factors. The higher reliance on fruits leads to a slower life history for gorillas and a faster life history for chimpanzees. Male mating strategies have great influence on the sociality of the female African great apes. Infanticide by males may stimulate female gorillas to join multimale groups and promote rapid reproduction. By contrast, sexual coercion including infanticide by males may prevent female chimpanzees from forming a prolonged association with males, thus promoting slow reproduction. These observations imply that associations between ecological, behavioral, and social features have evolved in different ways with their different life history strategies between the genera *Gorilla* and *Pan*.

Keywords Chimpanzee • Fallback foods • Flexibility • Gorilla • Socioecology • Sympatry

3.1 Introduction

In contrast to cetaceans, which consume various types of animal food irrespective of their body size, foods constrain the body size of primates. Insectivorous primates are relatively smaller than folivorous species, and frugivorous species are intermediate between them (Kay 1984). Primates have also evolved various features of gastrointestinal anatomy and the digestive system to cope with dietary constraints. Consumption of structural carbohydrates and detoxification of secondary compounds have promoted specialization in gut morphology (Milton 1986). Folivorous primates have evolved a sacculated fermenting chamber in the stomach in which microbial fermentation follows digestion and absorption (Chivers and Hladik 1980) or an enlarged cecum or colon in which bacterial fermentation is activated (Stevens and Hume 1995). Frugivorous and faunivorous primates lack these specializations and have evolved locomotive abilities to harvest crops or to prey on insects efficiently (Temerin and Cant 1983; Fleagle 1984; Chivers and Langer 1994; Cannon and Leighton 1994; Isbell et al. 1998)

Despite their large body weight and biomass, the great apes (orangutans, gorillas, and chimpanzees) have a strong preference for fruits and tend to feed regularly on insects. However, they are less able to digest unripe fruit and mature leaves than Old World monkeys (Wrangham et al. 1998; Lambert 1998, 2002; Remis 2000). Consequently, the great apes have broadened their diets to include a highly diverse and flexible range of nonfruit foods (Van Schaik et al. 2004; Yamagiwa 2004). Their behavioral and social flexibilities have possibly evolved to cope with these dietary constraints. Fallback foods (FBFs), characterized by relatively poor nutritional quality and high abundance, take an important role during periods of fruit scarcity (Lambert 2007; Marshall and Wrangham 2007). The nature and availability of FBFs, such as hardness, abundance, and quality, have shaped different behavioral and social features of the great apes (Marshall et al. 2009; Yamagiwa and Basabose 2009; Harrison and Marshall 2011).

The African great apes, classified into two genera with two species in each, have common behavioral tendencies in that females transfer between groups. Female western gorillas (*Gorilla gorilla*), eastern gorillas (*Gorilla beringei*), chimpanzees (*Pan troglodytes*), and bonobos (*Pan paniscus*) tend to emigrate from their natal groups before maturity, and they generally produce their offspring among unrelated females (Nishida and Kawanaka 1972; Harcourt et al. 1976; Kano 1992; Robbins et al. 2004, 2009). These tendencies may contribute to their social flexibilities, because females need to reform their social relationships with unrelated conspecifics. The formation of groups by the African great apes is in marked contrast with those of female-bonded societies (Wrangham 1980), in which females remain within their natal groups and form alliances with related conspecifics by inheriting their dominance ranks from their mothers. Socioecological theory predicts that food availability and predation pressure have stronger influences on female reproduction and association, which in turn influence male movements and association (Wrangham 1987; van Schaik 1989). In contrast to females living in female-bonded societies, female African apes need to seek suitable mates among unrelated females and males after transfer.

The socioecological features of chimpanzees and gorillas have been interpreted as typifying the frugivorous/folivorous dichotomy. Gorillas are regarded as terrestrial folivores (Schaller 1963; Jones and Sabater Pi 1971; Casimir 1975; Watts 1984). Their folivorous diet, continuous availability of resources, and uniform quality of foliage contribute to a low level of feeding competition among individual gorillas and thus enable them to form cohesive groups with egalitarian social relationships: accordingly, observations have revealed non-territoriality between groups and weak site fidelity (Stewart and Harcourt 1987; Harcourt 1992; Watts 1991, 1996, 1998a, b). By contrast, chimpanzees and bonobos have frugivorous diets and engage in arboreal feeding and nesting across various habitats (Wrangham 1977; Baldwin et al. 1982; Nishida and Uehara 1983; Ghiglieri 1984; Doran 1996; Fruth and Hohmann 1996). Heavily frugivorous diets and small high-quality patches of fruits may prevent chimpanzees from forming cohesive groups and promote fission–fusion features in grouping (Nishida 1968; Goodall 1986; Chapman et al. 1995; Wrangham et al. 1996). Bonobos also show fission–fusion features, but maintain larger cohesive groups than chimpanzees, probably because larger fruit patches and terrestrial herbaceous vegetation (THV)) are constantly available in their habitats, thus mitigating conflicts caused by feeding competition (White and Wrangham 1988; Kano 1992). Chimpanzees show territoriality against members of different communities (unit-groups), and male chimpanzees occasionally patrol the peripheral parts of their range, where lethal intercommunity aggressions can occur (Goodall et al. 1979; Nishida et al. 1985; Watts and Mitani 2001; Wilson and Wrangham 2003; Boesch et al. 2007). By contrast, the home range of bonobos extensively overlaps with those of neighboring communities, and intercommunity encounters usually occur peacefully (Kano 1992; Idani 1990; Furuichi 2011).

However, most of the data on gorillas, in particular on the social features of gorillas, have come from long-term studies on a single population of mountain gorillas in the Virunga Volcanoes at the higher altitudes. Such data may not reflect the conditions of the majority of gorilla populations, which inhabit lowland tropical forests. Recent studies on eastern and western lowland gorillas (*Gorilla beringei graueri* and *G. gorilla gorilla*) show their frugivorous features with arboreal feeding and nesting (Tutin and Fernandez 1984; Williamson et al. 1990; Yamagiwa et al. 1994; Kuroda et al. 1996; Remis 1997a; Remis et al. 2001; Doran et al. 2002; Doran-Sheehy et al. 2009; Rogers et al. 2004). Nevertheless, they tend to form cohesive groups as do mountain gorillas (Harcourt et al. 1981; Tutin 1996; Doran and McNeilage 1998; Robbins et al. 2004). Group size and home range size of gorillas are not different between habitats (Doran 2001; Parnell 2002; Yamagiwa et al. 2003; Yamagiwa and Basabose 2006b; Robbins et al. 2004). By contrast, chimpanzees frequently change patterns of association (party size and composition) according to fruit availability and social factors (the presence of estrous females and conflict among males), and their annual range size varies according to the type of habitat (Chapman et al. 1995; Matsumoto-Oda et al. 1998; Yamagiwa 1999; Anderson et al. 2002; Mitani et al. 2002). Gorillas and chimpanzees have high flexibility in their ecological and social features, but they respond in different ways to environmental changes.

Variations in group composition and patterns of association have also been observed within the genera *Gorilla* and *Pan*. Although the median group size of gorillas is similar across habitats, larger group size and a larger proportion of multimale groups in a population are observed for mountain gorillas than for western lowland gorillas (Doran 2001; Yamagiwa et al. 2003; Robbins et al. 2004). Male chimpanzees associate more frequently among themselves than with females, whereas male bonobos associate with females more frequently than with other males (Nishida 1979; Wrangham 1979; Kano 1992; Furuichi and Ihobe 1994; Mitani et al. 2002). Both ecological and social factors may cause these variations. For gorillas, the frugivorous diet may set the upper limit of group size, and the occurrence of infanticide may facilitate formation of multimale groups of gorillas (Harcourt et al. 1981; Watts 1996; Doran and McNeilage 1998; Yamagiwa and Kahekwa 2001; Yamagiwa et al. 2003, 2009; Harcourt and Stewart 2007; Robbins et al. 2007; Stoinski et al. 2009a, b). Stronger intercommunity aggression and hunting of mammals may promote frequent association and coalition among male chimpanzees (Chapman and Wrangham 1993; Newton-Fisher 1999b; Watts and Mitani 2001, 2002; Mitani et al. 2002), whereas the prolonged sexual attractiveness of female bonobos and the indistinct dominance of males over females may prevent males from forming coalitions among themselves and may stimulate stable associations with cycling females (Kano 1992; Parish 1994; Furuichi 1997, 2011).

Ecological and social factors interact in different ways to cause these variations. This chapter aims, therefore, to identify the specific factors forming intra- and interspecific variations in the socioecological features of the African great apes and to elucidate their flexibility to environmental and social changes from an evolutionary perspective. Socioecological features are compared between western gorillas and eastern gorillas as well as between chimpanzees and bonobos, and between subspecies for both *Gorilla* and *Pan*. Because niche separation between sympatric primate species becomes more pronounced during periods of food shortage (Ungar 1996; Tan 1999; Powzyk and Mowry 2003), socioecological features of sympatric gorillas and chimpanzees are discussed in relationship to their foraging strategies.

3.2 Variations in Socioecological Features Within *Gorilla*

The dietary composition of gorillas closely reflects the type of habitat rather than phylogenetic distance. The higher montane forests in equatorial Africa are characterized by a lower diversity of trees and fruits (Hamilton 1975; Sun et al. 1996). Western and eastern gorillas (*G. gorilla gorilla* and *G. beringei graueri*) inhabiting lowland tropical forests consume more types of fruit than eastern gorillas (*G. b. graueri* and *G. b. beringei*) inhabiting montane forests, who in turn consume more kinds of vegetative food (Table 3.1). In particular, the proportion of fruit in the diet is clearly different between gorillas in lowland tropical forests and those in montane forests. Even within a subspecies (*G. b. g*), more than 90 % of plant foods do not overlap and are not available in both lowland (Itebero) and highland (Kahuzi) habitats (Yamagiwa et al. 1994). Moreover, gorillas seasonally change their diet. Eastern gorillas in Kahuzi show a fruigivorous diet in some months but rely completely on vegetative foods in other months (Casimir 1975; Yamagiwa et al. 2005, 2009). These observations suggest a large flexibility in the dietary choices of gorillas according to variations in food availability seasonally and locally.

Frugivorous diet is positively correlated with the daily path length (DPL) of gorillas in both lowland and montane habitats (Table 3.1; Remis 1997b; Goldsmith 1999; Yamagiwa et al. 2003; Doran-Sheehy et al. 2004; Ganas and Robbins 2005). Group size also influences DPLs, and a larger group tends to travel longer distances in the montane forest of Bwindi (Ganas and Robbins 2005). Frugivorous diets may increase scramble competition within groups, and gorillas may respond to this by increasing the number of visited fruit crops. However, such an extension may not result in an expansion of the home range. Based on the general ecological features of primates, a frugivorous species would need a larger home range than a folivorous species of the same group weight (Clutton-Brock and Harvey 1977). In contrast to this assumption, smaller home ranges are found for frugivorous western gorillas than for folivorous eastern gorillas (Table 3.1), which may be caused by the differences in site fidelity between them. Folivorous mountain gorillas use less than 2 km^2 for a monthly range but shift ranges gradually to cover a wider area over the course

Table 3.1 Variations in socioecological features in genus *Gorilla*

	G. g. g.						*G. b. g.*		*G.b.b.*	
	Lopé	Bai Hokou	Mondika	Ndoki	Lossi	Moukalaba	Itebero	Kahuzi	Bwindi	Virungas
Habitat type	Tropical	Tropical	Tropical	Tropical	Tropical	Tropical	Tropical	Montane	Montane	Montane
Number (#) plant foods (number of species)	182 (134)	230 (129)	127 (100)	182 (152)			194 (121)	236 (116)	205 (113)[a]	75 (38)
Percent (%) fruit in plant food species	71	60	70	63			40	20	32[a]	5
Percent (%) fecal samples including fruit remains	98	>99	100	–		99	89	53	47	–
Mean number (#) fruit species per fecal sample	2.7	3.4	3.5			4.1		0.78	1	
Insectivory	Ant, termites	Termites	Termite	Ant, termite		Ant, termite	Ant, termite	Ant	Rare	Rare
Mean daily path length (m)	1,105 m (220–2,790)	2,600 m (300–5,300)	2,014 m (400–4,860)		1,853 m (300–5,500)		1,531 m (142–3,439)	716 m (242–2,055)	547–1,034	472–1,034 (112–2,868)
Home range (study period)	22 km² (10 years)	23 km² (2.2 years)	16 km² (1.3 years)		11 km² (3.2 years)	12 km² (1 year)		42 km² (8 years)	40 km² (3 years)	21–25 km² (5–7 years)
Annual home range	7–14 km²	8–13 km²	15 km²			12 km²		13–18 km²	23–38 km²	9–12 km²
Home range overlap	Extensive	Extensive	Extensive	Extensive	Partly	Extensive	Extensive	Extensive	Extensive	Extensive
Group cohesiveness	High	Subgroup	Subgroup	Subgroup	High but fusion	High	High	High	High	High

Mean/median group size	10			7	14		7	10	10	11
Maximum group size	16			13			17	31	23	65
Proportion of multimale groups	0	<10 %	0	0	0	<10 %	0	8 %	46 %	44 %
Maximum number (#) SB within a group	1	2	2	1	1	2	1	2	3	7
Female transfer	Yes	Yes	Yes	Yes	Yes	Yes	–	Yes	Yes	Yes
Male emigration	Yes	Yes	Yes	Yes	Yes	Yes	–	Yes	Yes	Yes
Male immigration	–	–	–	–	–	–	–	–	–	Rare
Solitary male	Yes	Yes	Yes	Yes	Yes	Yes	Yes	Yes	Yes	Yes
Solitary female	–	Yes	–	–	–	Yes	–	–	–	–
All-male group	–	–	Yes	–	–	–	–	–	Yes	Yes
All-female group	–	–	–	–	–	–	–	Yes	–	–
Group fission	–	Yes	–	–	–	–	–	Yes	Yes	Yes
Infanticide	–	–	–	Possible	–	–	–	Yes	–	Yes

G. g. g., Gorilla gorilla gorilla; G. b. g., Gorilla beringei graueri; G. b. b., Gorilla beringei beringei; SB: Silverback male (mature male)

[a]Data presented from one of three study groups

Sources: Lopé, Williamson et al. (1990), Tutin and Fernandez (1992, 1993), Tutin (1996), Bai Hokou, Remis (1997a, b), Cipoletta (2004), Mondika, Doran et al. (2002), Doran-Sheehy et al. (2004), Ndoki, Nishihara (1994, 1995), Parnell (2002), Stokes et al. (2003), Lossi, Bermejo (2004), Moukalaba, Yamagiwa et al. (2009), Itebero, Yamagiwa et al. (1994), Yamagiwa and Mwanza (1994), Yamagiwa et al. (2003), Kahuzi, Murnyak (1981), Yamagiwa et al. (1993), Yamagiwa et al. (2003, 2005), Yamagiwa and Basabose (2006a, b), Bwindi, McNeilage et al. (2001), Robbins and McNeilage (2003), Ganas et al. (2004), Ganas and Robbins (2005), Virunga, Schaller (1963), Weber and Vedder (1983), Yamagiwa (1986), Watts (1984, 1998a), Gray et al. (2003), Robbins et al. (2004, 2007, 2009), Stoinski et al. (2009a, b), Fawcett et al. (2012)

of a year (Watts 2000). This pattern of ranging may help to avoid overuse of previously trampled areas and allow regeneration of herbaceous vegetation (Vedder 1984; Watts 1998b). Eastern gorillas in the montane forest of Kahuzi tend to shift range seasonally and annually, and no correlation was found between their monthly range size and fruit consumption (Casimir and Butenandt 1973; Goodall 1977; Yamagiwa and Basabose 2006b). By contrast, western gorillas tend to revisit the same area within a short period for feeding or nesting at Mondika and Moukalaba (Doran-Sheehy et al. 2004; Iwata and Ando 2007). The monthly ranges of western gorillas cover a wide area, and the accumulated monthly ranges reach the size of the annual range within a few months at Mondika (Doran-Sheehy et al. 2004). The frequency of nest and nest site reuse is positively correlated with fruit consumption of western gorillas at Moukalaba (Iwata and Ando 2007). Western lowland gorillas tend to harvest fruit crops efficiently as fugivorous primates, in contrast to the folivorous eastern gorillas in the montane forests.

Nevertheless, frugivorous western gorillas do not show territoriality against neighboring groups or the fission–fusion features observed for chimpanzees. Similar to folivorous mountain gorillas in the Virungas (Schaller 1963; Watts 1998a), both eastern and western gorillas in other habitats tend to overlap ranges with neighboring groups irrespective of dietary characteristics (Tutin 1996; Yamagiwa et al. 1996; Bermejo 2004; Remis 1997b; Cipoletta 2004; Ganas and Robbins 2005). Intergroup encounters of mountain gorillas occur aggressively and occasionally involve fierce fights between silverbacks of different groups and infanticides by outside group males (Fossey 1974, 1983; Harcourt 1978; Watts 2003; Harcourt and Stewart 2007; Yamagiwa et al. 2009). By contrast, intergroup encounters of western gorillas may occur more calmly and have no effect on extending their DPL s at Mondika (Doran-Sheehy et al. 2004). At Lossi, an intergroup encounter resulted in the fusion of two groups with peaceful intermingling and co-nesting (Bermejo 2004). Frugivorous diets may not increase feeding competition between groups through territorial defense of home range. Although frugivorous western gorillas tend to spread widely and sometimes to separate in different subgroups while foraging, most subgroupings likely occur in multimale groups, and females usually associate with silverback males (Goldsmith 1999; Doran-Sheehy et al. 2004). Feeding competition possibly induced by frugivorous diets may have no effects on association between females and males.

However, group composition is clearly different between mountain gorillas (*G. b. b.*) and the other subspecies (*G. b. g.* and *G. g. g.*), and this difference may be influenced not by ecological factors but by social factors. In the Virunga mountain gorilla population, many episodes of infanticide have been reported so far (Fossey 1984; Watts 1989). Infanticide is regarded as a male mating tactic of killing suckling infants to stimulate their mothers to resume estrus (Hrdy 1979; van Schaik 2000). Female gorillas are vulnerable to infanticide in the absence of a protector male, and infanticide has great influence on a female's decision of movement (Watts 1989, 1996; Harcourt and Stewart 2007; Yamagiwa et al. 2009). Female mountain gorillas tend to transfer alone to avoid competition with other females over protection from the silverback male, and they generally transfer into multimale groups where they can get more protection (Watts 2000; Robbins and Robbins 2005). These female decisions may in turn influence the males' choice of movement around maturity.

They may change their reproductive strategies, from emigrating out of their natal group and attracting females from other groups during a solitary life, and thus establishing their own reproductive group, to remaining in the natal groups to share the reproductive opportunity with their putative fathers and brothers (Robbins 1999; Robbins and Robbins 2005; Harcourt and Stewart 2007; Stoinski et al. 2009a, b). However, kin-relatedness is not always necessary for coexistence among mature males. Unrelated males sometimes form all-male groups, and some of them continue to stay together in a group after accepting reproductive females (Yamagiwa 1987; Robbins 1995). Demographic history and an analysis of paternity of the habituated groups in the Virungas indicated that association among males was not limited to father–son pairs or half-siblings and that the multimale group structure of mountain gorillas may permit multimale mating (Robbins 2001; Robbins and Robbins 2005; Bradley et al. 2005; Stoinski et al. 2009a).

Infanticide has rarely been reported for other subspecies of gorillas. Female eastern and western lowland gorillas tend to transfer with other females to small groups or solitary males, and maturing silverbacks take females to establish new groups through group fission (Yamagiwa and Kahekwa 2001; Stokes et al. 2003). Most of the groups contain only one silverback, and infanticide may not have appeared as a male mating tactic in the population of eastern and western lowland gorillas. Recent observations of infanticide in the population of eastern lowland gorillas at Kahuzi suggest that rapid changes in gorilla social units and their relations following drastic environmental changes caused by recent human disturbances may increase the probability of infanticide, which possibly increases birthrate (Yamagiwa et al. 2009, 2011).

3.3 Variations in Socioecological Features Within *Pan*

Seasonal fluctuations and local variations in fruit abundance and availability have less influence on the dietary choices of chimpanzees than on those of gorillas. The high similarity in the proportions (51–60 %) among long-term study sites (Bossou, Mahale, Gombe, and Wamba) suggests that the frugivorous feature of the genus *Pan* is consistent across habitats (Table 3.2). Chimpanzees everywhere also allocate more than half of their feeding time to fruits (Hladik 1977; Wrangham et al. 1991; Newton-Fisher 1999a; Morgan and Sanz 2006). Therefore, the difference in fruit availability may affect their searching efforts on time and space. In the tropical forests, fruiting often peaks in the rainy season or during the wettest period (Hilty 1980; Sabatier 1985); a longer dry period may imply a longer period of fruit scarcity. Thus, both western chimpanzees (*Pan troglodytes verus*) and eastern chimpanzees (*Pan troglodytes schweinfurthii*) inhabiting savanna have extraordinary large home ranges (Assirik, 278–330 km^2, Baldwin et al. 1982; Ugalla, 250–560 km^2, Kano 1972). In the montane forest of Kahuzi, a group of chimpanzees has a relatively small home range (16 km^2 for 60 months). Plant phenology characterized by a long fruiting period and high density of THV at higher altitudes may support their nutritional requirements within a small range (Basabose 2005; Yamagiwa et al. 2008).

Table 3.2 Variations in sociecological features in genus *Pan*

	P. t. v.			*P. t. t.*		*P. t. s.*					*P. p.*	
	Assirik	Bossou	Tai	Lopé	Ndoki	Kibale	Budongo	Mahale	Gombe	Kahuzi	Wamba	Lomako
Habitat type	Savanna	L tropical	L tropical	L tropical	L tropical	M tropical	M tropical	M tropical	Woodland	Montane	L tropical	L tropical
Number (#) plant foods (# species)	60 (43)	205 (156)	? (223)	174 (132)	114 (108)	? (112)	146 (103)	328 (198)	205 (147)	156 (110)	133 (114)	113 (81)
Percent (%) fruit in plant food species	79	64	?	84	95		72	51	60	60	59	77
THV feeding	?	+	+	+	+	++	+	+	+	++	?	+
Insectivory	++	++	++	++	++	+	++	++	++	+	+	+
Hunting of mammal prey	++	+	++	+	+	++	++	++	++	+	+	+
Tool use	++	++	++	?	++	++	++	++	++	+	+	+
Home range	278–330 km^2	5–6 km^2	27 km^2	?	?	23–38 km^2	7 km^2	11–34 km^2	26 km^2	16 km^2	58 km^2	
Annual range			14–26 km^2				7 km^2		6–14 km^2	6–8 km^2	12–32 km^2	
Home range overlap	Partly	None	Partly			Partly	Partly	Partly	Partly	Partly	Extensive	
Intercommunity relation-ships	?	?	Antagonistic	?	?	Antagonistic	Antagonistic	Antagonistic	Antagonistic	?	Peaceful	?
Intercommunity killing	–	–	–	–	–	++	–	+	++	–	–	–
Community size	28	20	76	?	?	37	46	29	57	22	58	36
Mean party size	5.3	4	8.3	?	?	5.1	6.3	6.1	5.6	4.4	16.9	5.8
Group fluidity	+	+	+	?	++	++	++	++	++	++	+	+

Association partner	MF>MM>FF	MF>FF	MF>FF	?	?	MM>MF>FF	MM>MF>FF	MM>MF>FF	MM>MF>FF	MM>MF>FF	FF>MF>MM	FF>MF>MM
Grooming partner	?	MF>FF	MF>FF	?	?	MM>MF>FF	MM>MF>FF	MM>MF>FF	MM>MF>FF	?	FF>MF>MM	FF>MF>MM
Female transfer	?	Yes	Yes			Yes	Yes	Yes	Yes	Yes	Yes	Yes
Sociosexual interaction	?	–	+	?	?	–	–	–	–	–	++	++
Male emigration	?	+	–	?	?	–	–	–	–	–	–	–
Male immigration	?	–	–	?	?	–	–	–	–	–	–	–
Infanticide	–	–	(+)	–	–	+	+	+	+	–	–	–

P. t. v., Pan troglodytes verus; P. t. t., Pan troglodytes troglodytes; P. t. s., Pan troglodytes scheinfurthii; P. p., Pan paniscus

Habitat: L tropical, lowland tropical forest; M tropical, medium-altitude tropical forest; THV, terrestrial herbaceous vegetation

Sources: Assirik, Baldwin et al. (1982), Tutin et al. (1983), McGrew et al. (1988), Bossou, Sugiyama and Koman (1979, 1987), Sakura (1994), Muroyama and Sugiyama (1994), Yamakoshi (2004), Tai, Boesch and Boesch-Achermann (2000), Herbinger et al. (2001), Lope, Tutin and Fernandez (1992, 1993), Ndoki, Kuroda et al. (1996), Kibale, Wrangham et al. (1991), Clark and Wrangham (1994), Chapman et al. (1994), Pepper et al. (1999), Budongo, Suzuki (1977), Newton-Fisher et al. (2000), Newton-Fisher (2003), Gombe, Goodall (1977, 1986), Teleki (1973), Wrangham (1977, 1986), Mahale, Nishida (1968, 1979), Nishida and Uehara (1983), Nishida and Hiraiwa-Hasegawa (1987), Nishida and Hosaka (1996), Kahuzi, Basabose (2002, 2004, 2005), Wamba, Kuroda (1979), Kano and Mulavwa (1984), Idani (1990), Hashimoto et al. (1998), Lomako, Badrian and Malenky (1984), White (1996, 1988), Hohmann et al. (1999)

In contrast to gorillas, the spatiotemporal distribution of fruits affects association among group members of chimpanzees. Fruit patch size is positively correlated with the mean party size of chimpanzees and bonobos (Kibale, Chapman et al. 1995; Wrangham et al. 1996; Mitani et al. 2002; Tai, Anderson et al. 2002; Lomako, White 1996). However, the effects of fruit availability are not straightforward. In Budongo, party sizes showed either negative or no relationship with habitat-wide measures of food abundance, although the size of foraging parties fluctuated with the size of food patches (Newton-Fisher et al. 2000). In Kahuzi, party size increased when fruits were clumped and available in large amounts for a prolonged period, while it decreased when fruits were highly available for only a limited period (Basabose 2004). In Mahale, nomadic party size defined as loose aggregations within acoustic range, rather than visible party size, was positively correlated with seasonal fruit availability (Itoh and Nishida 2007). Community size also affects party size. Boesch (1996) compared community and party size among six sites of chimpanzees and two sites of bonobos and found a higher cohesion among members in the smaller community. Lehmann and Boesch (2004) also examined variations in party size in relationship to community size at Tai, and they found that small communities are more cohesive and have a less flexible fission–fusion system. Although the previous studies explained the high gregariousness among bonobos by the abundance and high quality of THV and large fruit patches available throughout the year (Badrian and Malenky 1984; Wrangham et al. 1991; White and Wrangham 1988), bonobos are consistently more gregarious than chimpanzees irrespective of food availability (Kano 1992; Hohmann and Fruth 2003).

The presence of estrous females also increases the party size of chimpanzees and bonobos by attracting adult males (Tai, Anderson et al. 2002; Kibale, Mitani et al. 2002; Mahale, Matsumoto-Oda et al. 1998; Kalinzu, Hashimoto et al. 2003; Wamba, Kano 1980, 1982; Lomako, Hohmann and Fruth 2003). Both female chimpanzees and bonobos show conspicuous swelling of the sexual skin during estrus. Although female bonobos have a lower copulation rate than chimpanzees during estrus, they have longer swelling phase (27 %) than female chimpanzees (4–6 %) during interbirth intervals, and thereby have more sexual interactions with males (Furuichi and Hashimoto 2002; Furuichi 2009). The extended attractivity and receptivity of female bonobos may greatly contribute to the formation of large and stable mixed groups.

Male eastern chimpanzees (*P. t. s.*) tend to associate and groom with other males more frequently than with females, and they form coalitional relationships with each other (Goodall 1986; Nishida et al. 1985; Watts and Mitani 2001; Williams et al. 2002). Adult males are usually dominant over adult females, and the dominance status of males is maintained by aggressive interactions such as explosive displays (Wrangham 1979; Nishida 1979, 1983; Wrangham et al. 1992; Mitani et al. 2000; Muller and Mitani 2005). Males sometimes attacked the members of neighboring communities and killed some of them in most of the long-term study sites of eastern chimpanzees (Gombe, Goodall et al. 1979; Wilson et al. 2004; Mahale, Nishida et al. 1985; Kanyawara, Wilson and Wrangham 2003; Ngogo, Watts et al. 2006). Infanticide by males also occurred in most of the study sites of eastern chimpanzees (Budongo, Kibale, Gombe, and Mahale, Arcadi and Wrangham 1999). Recent DNA analyses suggest that most of the offspring are sired by group males and that coalition among group males is successful for guarding their reproductive priority (Constable et al. 2001; Boesch et al. 2006; Inoue et al. 2008).

By contrast, no episode of infanticide has been reported for western chimpanzees (*P. t. v.*), except for one suspected case at Tai (Boesch and Boesch-Achermann 2000). Although intercommunity relationships are antagonistic and high site fidelity is found in the range use of western chimpanzees at Tai, no killing was observed in intercommunity encounters (Herbinger et al. 2001; Lehmann and Boesch 2004). Because of the difficulty of habituation, very little information has been available for central chimpanzees (*P. t. t.*), except for a case of lethal attack between communities in Loango, Gabon (Boesch et al. 2007). Male bonobos do not form alliances as do male chimpanzees, and they sometimes express submissiveness to females (Kano 1980; Kuroda 1980; Parish 1994, 1996; Furuichi 1997; Vervaecke et al. 1999). Dominance rank among males reflects the dominance relationships among their mothers, and mothers occasionally support their adult sons in agonistic interactions (Kuroda 1979; Kano 1992, 1996).

Female chimpanzees with suckling infants rarely participate in parties (Wrangham 1979), whereas female bonobos usually associate with others irrespective of their reproductive state (Furuichi 1987). Female western chimpanzees tend to associate and groom with each other frequently, at both Bossou and Tai, compared to eastern chimpanzees (Sugiyama 1988; Boesch 1991; Lehmann and Boesch 2008). Doran et al. (2002) attributed the possible explanation of such high female association with males as their counterstrategy to infanticide. The shorter dry season in the habitats of bonobos and western chimpanzees provides abundant and stable fruit food resources, eases feeding competition, and enables them to form mixed parties with a reduction in infant mortality through infanticide. At Tai, female chimpanzees associate among themelves five times frequently than eastern chimpanzees at Budongo, Kibale, Gombe, and Mahale, and they form a distinct linear dominance hierarchy (Wittig and Boesch 2003). They also build long-lasting friendships including food sharing and support, which have rarely been observed for eastern chimpanzees (Boesch and Boesch-Achermann 2000). Such high sociality of female chimpanzees at Tai is explained by their high competition over foods and high predation risk (Wittig and Boesch 2003; Lehmann and Boesch 2008).

3.4 Differences in Socioecological Features Between Sympatric Gorillas and Chimpanzees

Central and eastern chimpanzees are living sympatrically with gorillas in wide areas. However, long-term studies on chimpanzees have been conducted in the sites without sympatric gorillas (Gombe, Mahale, Kibale, and Budongo). Long-term studies on gorillas have also been conducted in the high montane forest of the Virungas, outside the extent of chimpanzee distribution. Until recently, little information has been available on how gorillas and chimpanzees coexist sympatrically, particularly from a comparative perspective in relationship to allopatric populations.

In initial studies, distinct differences in diet and range use between gorillas and chimpanzees were estimated in Kayonza (Bwindi) and Rio Muni (Schaller 1963; Jones and Sabater Pi 1971). Chimpanzees tended to range in primary forest, to stay on dry ridges, and to feed on fruits in trees. By contrast, gorillas tended to range in secondary forest, to stay in wet valleys, and to feed on terrestrial herbs. Dietary differences were considered to promote niche divergence between gorillas and chimpanzees and to shape distinct differentiation in social features between them.

However, recent studies on sympatric gorillas and chimpanzees have reported an extensive overlap in diet and ranging between them (Table 3.3). Fruits constitute the major items overlapping between their diets in both tropical and montane forests (Tutin and Fernandez 1993; Yamagiwa and Basabose 2006a). Significant differences were found in the preference for some fruit species. Fig fruits constitute an important part of the diet of chimpanzees during the entire year in Lopé, Goualougo, and Kahuzi, but the sympatric gorillas consume fewer kinds of fig fruits less often than do chimpanzees (Tutin and Fernandez 1993; Morgan and Sanz 2006; Yamagiwa and Basabose 2006a; Stanford and Nkurunungi 2003). Fecal analyses showed that the number of fruit species and the proportion of fruit remains per fecal sample were positively correlated with fruit abundance for gorillas, whereas not or only weakly correlated for chimpanzees at Goualougo, Kahuzi, and Bwindi (Stanford and Nkurunungi 2003; Yamagiwa and Basabose 2006a; Morgan and Sanz 2006). These observations may reflect the chimpanzee's high selectivity of fruit species during the period of fruit abundance and persistent searching for fruits during the period of fruit scarcity.

In contrast to fruits, chimpanzees tend to eat fewer kinds of leaf than do the sympatric gorillas, and most kinds of leaf eaten by chimpanzees are also eaten by gorillas at any site, suggesting the higher selectivity of leaves by chimpanzees, who may avoid concentration of anti-feedants in leaves. Recent studies suggest that the variations in nutritional quality and chemical defenses of leaves may require highly selective feeding on leaves in unpredictable small patches as with fruits (Harris 2006; Snaith and Chapman 2007). A wide range of leaves in the gorilla diet suggests their specialized digestive physiology (Rogers et al. 1990, 2004; Remis 2003).

A marked divergence in diets was found in animal foods. At Lopé, both gorillas and chimpanzees feed regularly on the weaver ant (*Oecophylla longinoda*), although other species of ants did not overlap between them (Tutin and Fernandez 1992). At Ndoki and Goualougo, gorillas eat termites (*Cubitermes* sp.) by breaking their nests by hand, while chimpanzees feed on another kind of termite (*Macrotermes* sp.) by using a complex set of tools (Suzuki et al. 1995; Kuroda et al. 1996; Morgan and Sanz 2006). The choice of these social insects, feeding techniques, and feeding frequency by both gorillas and chimpanzees differ between habitats, between populations, and even between groups (Yamagiwa et al. 1991; Tutin and Fernandez 1992; Kuroda et al. 1996; Deblauwe et al. 2003; Ganas and Robbins 2004). In general, chimpanzees have a more diverse faunivorous diet than gorillas, and the

Table 3.3 Variations in socioecological features in the sympatric populations of *Gorilla* and *Pan*

	Lopé	Ndoki	Goualougo	Kahuzi	Bwindi
Gorilla	*G. g. g.*	*G. g. g.*	*G. g. g.*	*G. b. g.*	*G. b. b.*
Chimpanzee	*P. t. t.*	*P. t. t.*	*P. t. t.*	*P. t. s.*	*P. t. t.*
Habitat type	L tropical	L tropical	L tropical	Montane	Montane
Number (#) plant foods Gorilla	213	182	107	231	149
Chimpanzee	174	114	158	137	74
Percent (%) overlap species Gorilla	57	37	84	38	42
Chimpanzee	73	60	58	63	85
Number (#) pulp/aril/seed Gorilla	117	133	?	48	47
Chimpanzee	123	103	?	60	46
Percent (%) overlap species Gorilla	79	46	?	73	77
Chimpanzee	82	59	?	58	78
Number (#) leaves Gorilla	56	29	?	81	59
Chimpanzee	30	3	?	43	16
Percent (%) overlap species Gorilla	43	3	?	40	25
Chimpanzee	70	33	?	74	94
Number (#) pith/stem/bark/root Gorilla	28	18	?	92	18
Chimpanzee	11	6	?	31	5
Percent (%) overlap species Gorilla	18	33	?	21	28
Chimpanzee	45	33	?	61	100
Number (#) others Gorilla	12	2	?	10	25
Chimpanzee	7	2	?	3	7
Percent (%) overlap species Gorilla	25	0	?	20	28
Chimpanzee	43	0	?	67	100
Animal foods Gorilla	Ant, termite	Ant, termite, earthworm	Termite	Ant, earthworm	Ant
Chimpanzee	Ant, termite, bee	Termite, mammal, birds	Termite	Ant, bee, mammal	Ant, bee, mammal

Habitat: L tropical, lowland tropical forest; M tropical, medium-altitude tropical forest

Sources: Assirik, Baldwin et al. (1982), Tutin et al. (1983), McGrew et al. (1988), Bossou, Sugiyama and Koman (1979, 1987), Sakura (1994), Muroyama and Sugiyama (1994), Yamakoshi (2004), Tai, Boesch and Boesch-Achermann (2000), Herbinger et al. (2001), Lope, Tutin and Fernandez (1992, 1993), Ndoki, Nishihara and Kuroda (1991), Nishihara (1995), Suzuki et al. (1995), Kuroda et al. (1996), Goualougo, Morgan and Sanz (2006), Kahuzi, Basabose (2002, 2004, 2005), Yamagiwa et al. (2005), Yamagiwa and Basabose (2006a, b), Bwindi, Stanford and Nkurunungi (2003), Ganas and Robbins (2004), Ganas et al. (2004)

hunting of mammals and use of tools for extracting animal foods, which are usually observed for chimpanzees across habitats, have never been observed for the sympatric gorillas.

Bonobos rarely feed on insects or prey on mammals, and no feeding-related tool use has been reported in their wild populations (Kano 1992; McGrew 1992; Ingmanson 1996; Grubera et al. 2010). Instead, bonobos feed regularly on the pith of *Zingiberacea* and *Marantacea*, as observed for gorillas (Malenky and Stiles 1991). Wrangham (1986) postulated that use of THV as FBFs mitigates feeding competition during a period of fruit scarcity and leads to the high cohesiveness of bonobos. The presence of sympatric gorillas may prevent chimpanzees from extensive use of THV, whereas the absence of gorillas may have enabled bonobos to use THV at any time in their habitats. Yamakoshi (2004) examined this "THV hypothesis" by comparing THV consumption among bonobos, gorillas, and chimpanzees in allopatric habitats with those in sympatric habitats. Although bonobos consume more THV than chimpanzees, they eat THV regardless of season, and no clear correlation was found between gregariousness of female bonobos, fruit availability, and their THV feeding rate. Sympatric gorillas and chimpanzees may not compete exclusively for feeding on THV. Although the foregoing evidence seems to refute the "THV hypothesis," Yamakoshi found significant differences in social features between subspecies of chimpanzees in relationship to the presence/absence of the sympatric gorillas, as observed between bonobos and chimpanzees (Yamakoshi 2004) The high cohesiveness among female western chimpanzees is attributed to the abundance of FBFs in their habitats, such as palm, pith, and nuts. These foods are available throughout the year, and chimpanzees use tools for processing them (Yamakoshi 1998). Both western chimpanzees and bonobos are genetically distant from central and eastern chimpanzees (0.5 and 1.5 million years after separation, respectively: Morin et al. 1994; Gagneux et al. 1999; Gondera et al. 2010). The Dahomey Gap and the Congo River may have constituted the barriers between them. Gorillas have not expanded their distribution into habitats beyond the barriers. It is possible that the absence of gorillas may have enabled bonobos and western chimpanzees to obtain free access to plenty of supplemental food resources and to develop different social features, away from eastern and central chimpanzees, through maintaining female gregariousness.

Recent studies on the sympatric gorillas and chimpanzees also indicate that both frequently range in the same type of habitat (Tutin and Fernandez 1984, 1985; Kuroda et al. 1996; Yamagiwa et al. 1996; Furuichi et al. 1997; Stanford and Nkurunungi 2003). An extensive overlap in ranging areas between gorillas and chimpanzees may provide them with frequent encounters with each other. A large overlap in fruit diets also predicts their frequent encounters at the fruiting crops. However, very few encounters between them have actually been observed in either tropical forests or montane forests. Most encounters occurred in fig trees with large fruit crops in both habitats. Cofeeding of gorillas and chimpanzees was observed in the tropical forests at Ndoki and Goualougo (Kuroda et al. 1996; Morgan and Sanz 2006), whereas mutual avoidance and aggressive interactions were observed in the montane forests at Kahuzi and Bwindi (Yamagiwa et al. 1996; Stanford and

Nkurunungi 2003). These findings suggest that interspecies feeding competition (possibly over fruits) is stronger in the montane forest than in the tropical forests.

Both preferred foods and FBFs influence ranging of gorillas and chimpanzees and consequently reduce the opportunity of their encounters in the sympatric areas. FBFs tend to shape processing adaptations, whereas preferred foods tend to shape harvesting adaptations (Marshall and Wrangham 2007). Relatively abundant and low-quality FBFs drive specialized adaptations toward processing, whereas relatively rare and high-quality FBFs drive behavioral adaptations such as fission–fusion grouping and tool use (Lambert 2007). In the montane forest of Kahuzi, gorillas take the former fallback strategy, feeding on THV and leaves and barks of woody plants during the whole year, and chimpanzees take the latter strategy, feeding usually on fruits with searching for animal foods as the high-quality FBFs (Yamagiwa and Basabose 2009). These differences in fallback strategies may lead to even use of a wide range by gorillas and to frequent reuse of particular areas within a small range by chimpanzees (Yamagiwa et al. 2012). Fruit scarcity and rich THV in the montane forest may promote a folivorous or herbivorous strategy of gorillas and enable chimpanzees to maintain a small range. On the other hand, fruit abundance and poor THV in a lowland tropical forest may promote different ranging strategies. Gorillas tend to revisit fruit crop, forming stable groups, while chimpanzees tend to disperse, harvesting fruit crops individually or forming small parties. High processing and digestive abilities enable gorillas to maintain a similar range to that in montane forest. High locomotive ability and fission–fusion grouping enable chimpanzees to change their range size according to fruit availability. Their tool using ability also promotes insectivory by chimpanzees, such as termite fishing or ant dipping, to enable them to expand their range into arid areas. Such differences in fallback strategies between gorillas and chimpanzees permit similar range size across the habitats of gorillas (Yamagiwa 1999) and large variations in range size among the habitats of chimpanzees, from 6.8 km^2 at Budongo (Newton-Fisher 2003) to 250–560 km^2 at Ugalla (Kano 1972).

3.5 Determinant Factors of the Social Organizations of Gorillas and Chimpanzees

Growing evidence from field studies suggests that ecological and social factors may interact in different ways between *Gorilla* and *Pan*. The similarity of their preferred foods may promote range overlap, and scarcity of their preferred foods may increase intraspecies competition over food. In the mid- to late Miocene, climatic shifts led to a large-scale forest reduction and the preferred fruits of apes became increasingly rare (Andrews et al. 1997; Potts 2004). The extended period of fruit scarcity may have forced both gorillas and chimpanzees to find alternative food resources, such as FBFs. Their original FBFs might have been leaves and pith of THV because the fewer secondary compounds lead to easy digestion. Gorillas use barks and leaves of woody plants as their staple FBFs, which require more processing and promote

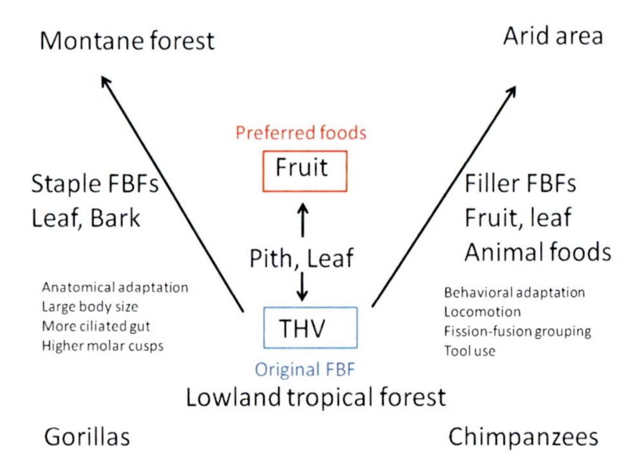

Fig. 3.1 Fallback strategies of *Gorilla* and *Pan*. *FBF*, fallback foods

anatomic adaptation, such as large body size, a larger and more ciliated gut, and higher molar-shearing blades and cusps for efficient digestion of higher-fiber foods (Collet et al. 1984; Remis et al. 2001). By contrast, chimpanzees use less nutritious but abundant fruits, such as fig fruits, as filler FBFs (Harrison and Marshall 2011). In contrast to gorillas, which always swallow all parts of plants, chimpanzees usually chew fibrous parts and spit them out as a wadge without passing them through the gut. Therefore, *Pan* has not developed the special digestive system. *Pan* also use seeds and animal foods as filler FBFs, which are often mechanically protected and difficult to find, requiring behavioral innovation and tool use (Potts 2004). Among *Pan*, bonobos are living in the tropical forest only with the absence of gorillas. They use rich THV and form cohesive groups whose home ranges overlap extensively with those of their neighbors, as observed for gorillas (White and Wrangham 1988; Kano 1992). In the presence of sympatric gorillas, chimpanzees rarely feed on THV, and instead use animal foods as FBFs (Yamagiwa and Basabose 2009). Their fallback strategy requires flexible grouping and ranging according to fruit availability, and stronger territoriality, which consequently enabled them to expand their range into arid areas where gorillas could not survive (Fig. 3.1).

Recent findings on socioecological features of the African great apes also suggest that female transfer is their common status, irrespective of habitat type or dietary composition (Yamagiwa 1999, 2004). Ecological factors may not shape social structure but may influence association patterns within each genus. The degree of frugivory may not increase contest competition among gorillas but increases scramble competition, which results in extension of daily path length without changes in group cohesion (Goldsmith 1999; Doran 2001; Yamagiwa et al. 2003). Seasonal fluctuation and local differences in fruit abundance may increase both scramble and contest competition among chimpanzees and bonobos, which influences the size and fluidity of their temporal parties (Chapman et al. 1995; White 1996; Anderson et al. 2002; Mitani et al. 2002). High association and affiliation among female western chimpanzees and among female bonobos are possibly formed as a strategy against infanticide or predation or both in the presence of abundant and stable fruit resources and the absence of sympatric gorillas (Doran et al. 2002; Yamakoshi 2004; Lehmann and Boesch 2008). However, male dominance over females is consistent across all populations of chimpanzees and differs from the indistinct dominance between male and female bonobos irrespective of ecological factors. Sociosexual interactions including GG (Genito-genital) rubbing frequently observed among bonobos are rarely observed among both eastern and western chimpanzees. Such social features could not be explained solely by ecological factors.

The mating strategies of the male African great apes have a strong influence on the movements and associations of females. The occurrence of infanticide affects a female's movement decisions and the choice of which group she joins in the population of Virunga mountain gorillas (Watts 1989, 1996; Robbins 2001; Robbins and Robbins 2005; Harcourt and Stewart 2007; Robbins et al. 2009; Yamagiwa et al. 2009). The females' choices also influence the males' movements, and the female preference for multimale groups prevents male emigration from their natal groups after maturity and promotes association among males. Interbirth intervals in the Virunga population, where infanticide occurs, are shorter than those of the eastern gorilla population at Kahuzi, where infanticide was not observed (Yamagiwa et al. 2003). Males remaining in their natal groups tend to start active reproduction earlier than males emigrating from their natal groups at Virungas (Robbins and Robbins 2005; Harcourt and Stewart 2007). Infanticide may promote rapid reproduction through formation of multimale groups. Ecological factors may also constrain group size and the group composition of gorillas. A large group size of more than 30 individuals and multimale group composition are mostly limited to populations of eastern gorillas inhabiting montane forests (Yamagiwa et al. 2003). A frugivorous diet and limited distribution of fruits may prevent western gorillas from forming large cohesive groups, including multiple males. Recent studies at several sites (Mbeli Bai, Maya Nord, Lossi, and Bai Hokou) suggest that frugivorous western lowland gorillas show slower physical maturation and longer interbirth intervals than folivorous mountain gorillas (Robbins et al. 2004; Breuer et al. 2009). DNA analysis suggests that related males stay in separate groups or as solitaries but at close proximity in the population of western gorillas at Mondika (Bradley et al. 2004). These observations suggest that male gorillas have two types of social organization: association within groups, such as mountain gorillas, and dispersed networks, as in western gorillas (Fig. 3.2). Each type can shift to the other in response to a combination of ecological and social factors.

Sexual coercion of male chimpanzees also has a large influence on the associations of females. In the populations of eastern chimpanzees, males tend to form a strong coalition to guard their reproductive priority. Intercommunity killing, infanticides, and other violent interactions involving adult males did not occur in conflicts over foods but in social contexts (Newton-Fisher 1999a; Watts and Mitani 2001; Wilson et al. 2004). The solitary nature of female eastern chimpanzees seems to avoid such violent interactions caused by males over estrous females (Fig. 3.3). By contrast, male western chimpanzees and male bonobos do not have violent interactions within or between communities. The extended estrus of females and the

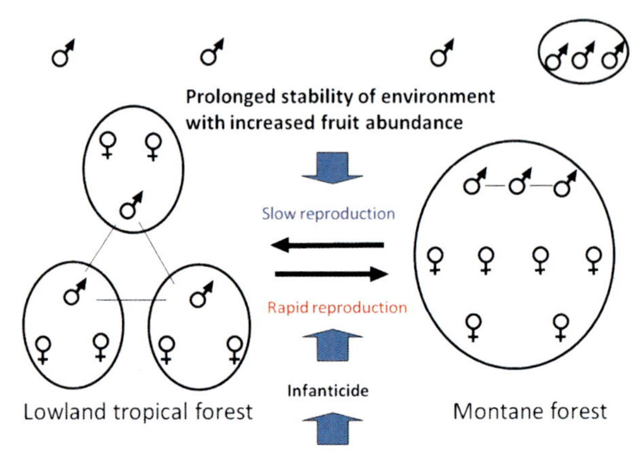

Fig. 3.2 Variation in social structure of *Gorilla*

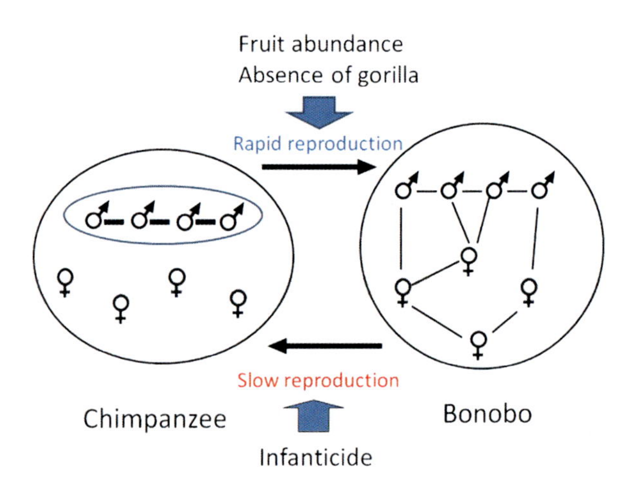

Fig 3.3 Variation in social structure of *Pan*

social use of sexual behavior may explain the peaceful associations among male and female bonobos in mixed parties (Kano 1992; Furuichi and Hashimoto 2002). However, the high sociality of female western chimpanzees cannot be fully interpreted as the result of an extended estrus, because they have reproductive features similar to those of eastern chimpanzees. Recent DNA analysis suggests that male western chimpanzees also sired most of the offspring within their communities at Tai (Vigilant et al. 2001). It remains unknown how they attained high reproductive success through peaceful interactions within and between communities. Ecological constraints may not constitute the primary factors but possibly promote different types of sociality between western and eastern chimpanzees through differences in the influence of sympatric gorillas.

The analysis of this report suggests a more complex integration between social and ecological factors than that previously estimated for coping with the phylogenetic inertia of the African great apes. Their common feature of natal female dispersal may constitute the basic factors causing such social complexity. Variations in their social features generally follow the phylogenetic distance, as pointed out by Doran et al. (2002). However, it should be noted that behavioral and social features of the African great apes are too greatly diverse compared to their morphological or physiological traits, and that male mating strategy has strong influence on female sociality (Harcourt and Stewart 2007). This study suggests that infanticide by males may stimulate female gorillas to join multimale groups and promote rapid reproduction. By contrast, infanticide tends to occur in the populations of eastern chimpanzees in which females less frequently associate with males. Female western chimpanzees showing high association with males tend to resume estrus within 1 year or have a short interbirth interval, as observed for bonobos (Boesch and Boesch-Achermann 2000; Sugiyama 2004). In contrast to gorillas, infanticide by male chimpanzees may prevent the females from forming a prolonged association with males and promote slow reproduction. These findings imply that associations between ecological, behavioral, and social features have evolved in different ways in the life history strategies between the genera *Gorilla* and *Pan*.

3.6 Conclusion

1. The African great apes (gorillas and chimpanzees) have not developed a specialized digestive system and have the common dietary feature of preferring ripe fruits.
2. Distinct dietary differences are found in fallback foods. Gorillas use the bark and leaves of woody plants as their staple FBFs, which require more processing and promote anatomic adaptation, whereas chimpanzees use less nutritious but abundant fruits, seeds, and animal foods as their filler FBFs and have developed behavioral innovation and tool use.
3. Such different abilities might have enabled gorillas and chimpanzees to expand their range into different habitats in the mid- to late Miocene, when climatic shifts led to a large-scale forest reduction and the preferred fruits of apes became increasingly rare. Gorillas expanded into higher montane forests where plenty of THV are available, whereas chimpanzees expanded into arid areas where fig fruits and animal foods are available.
4. Sympatric conditions might have promoted niche separation between gorillas and chimpanzees. Bonobos and western chimpanzees who live with abundant available fruits and in the absence of gorillas tend to form cohesive groups and have a shorter life history compared to eastern and central chimpanzees with fission–fusion social features.
5. Male sexual coercion such as infanticide and intercommunity killing has had a different influence on the life history strategy of female apes. This coercion leads to a fast life history and to multimale groups for gorillas, although it leads to a slow life history and fission–fusion social features for chimpanzees.

Acknowledgments This study was financed by in part by the Grants-in-Aid for Scientific Research by the Ministry of Education, Culture, Sports, Science, and Technology, Japan (Nos. 162550080, 19107007, and 24255010 to J. Yamagiwa), the Global Environmental Research Fund by Japanese Ministry of Environment (F-061 to T. Nishida, Japan Monkey Centre), and the Kyoto University Global COE Program "Formation of a Strategic Base for Biodiversity and Evolutionary Research," and was conducted in cooperation with CRSN and ICCN. We thank Dr. S. Bashwira, Dr. B. Baluku, Mr. M.O. Mankoto, Mr. B. Kasereka, Mr. L. Mushenzi, Ms. S. Mbake, Mr. B.I. Iyomi, Mr. C. Schuler, and Mr. R. Nishuli for their administrative help. We appreciate to Ms. Chieko Ando for her kind permission for the usage of a picture. We are also greatly indebted to Mr. M. Bitsibu, Mr. S. Kamungu, and all the guides, guards, and field assistants in the Kahuzi-Biega National Park for their technical help and hospitality throughout the fieldwork.

References

Anderson DP, Nordheim EV, Boesc C, Moermond TC (2002) Factors influencing fission–fusion grouping in chimpanzees in the Tai National Park, Cote d'Ivoire. In: Boesch C, Hohmann G, Marchant L (eds) Behavioral diversity in chimpanzees and bonobos. Cambridge University Press, Cambridge, pp 90–101

Andrews P, Begun DR, Zylstra M (1997) Interrelationships between functional morphology and paleoenvironments in Miocene hominoids. In: Began DR, Ward CV, Rose MD (eds) Function, phylogeny and fossils: miocene hominoid evolution and adaptation. Plenum Press, New York, pp 29–58

Arcadi AC, Wrangham RW (1999) Infanticide in chimpanzee: review of cases and a new within-group observation from the Kanyawara study group in Kibale National Park. Primates 40:337–351

Badrian N, Malenky R (1984) Feeding ecology of *Pan paniscus* in the Lomako Forest, Zaire. In: Susman RL (ed) The pygmy chimpanzee: evolutionary biology and behavior. Plenum, New York, pp 275–299

Baldwin PJ, McGrew WC, Tutin CEG (1982) Wide-ranging chimpanzees at Mt. Assirik, Senegal. Int J Primatol 3:367–385

Basabose AK (2004) Fruit availability and chimpanzee party size at Kahuzi montane forest, Democratic Republic of Congo. Primates 45:211–219

Basabose AK (2005) Ranging patterns of chimpanzees in a montane forest of Kahuzi, Democratic Republic of Congo. Int J Primatol 26:31–52

Bermejo M (2004) Home-range use and intergroup encounters in western gorillas (*Gorilla g. gorilla*) at Lossi Forest, North Congo. Am J Primatol 64:223–232

Boesch C (1991) The effects of leopard predation on grouping patterns in forest chimpanzees. Behaviour 117:220–242

Boesch C (1996) Social grouping in Tai chimpanzees. In: McGrew WC, Marchant LF, Nishita T (eds) Great ape societies. Cambridge University Press, Cambridge, pp 101–113

Boesch C, Boesch-Achermann H (2000) The chimpanzees of the Taï Forest: behavioural ecology and evolution. Oxford University Press, Oxford

Boesch C, Kohou G, Néné H, Vigilant L (2006) Male competition and paternity in wild chimpanzees of the Taï Forest. Am J Phys Anthropol 130:103–115

Boesch C, Head J, Tagg N, Arandjelovie M, Vigilant L, Robbins MM (2007) Fatal chimpanzee attack in Loango National Park, Gabon. Int J Primatol 28:1025–1034

Bradley BJ, Doran-Sheehy DM, Lukas D, Boesch C, Vigilant L (2004) Dispersed male networks in western gorillas. Curr Biol 14:510–513

Bradley BJ, Robbins MM, Williamson EA, Steklis HD, Steklis NG, Eckhardt N, Boesch C, Vigilant L (2005) Mountain gorilla tug-of-war: silverbacks have limited control over reproduction in multimale groups. Proc Natl Acad Sci USA 102:9418–9423

Breuer T, Hockemba MBM, Olejniczak C, Parnell R, Stokes E (2009) Physical maturation, life-history classes, and age estimates of free-ranging western gorillas: insights from Mbeli Bai, Republic of Congo. Am J Primatol 71:106–119
Cannon CH, Leighton M (1994) Comparative locomotor ecology of gibbons and macaques: selection of canopy elements for crossing gaps. Am J Phys Anthropol 93:505–524
Casimir MJ (1975) Feeding ecology and nutrition of an eastern gorilla group in the Mt. Kahuzi region (République du Zaire). Folia Primatol (Basel) 24:1–36
Casimir MJ, Butenandt E (1973) Migration and core area shifting in relation to some ecological factors in a mountain gorilla group (*Gorilla gorilla beringei*) in the Mt. Kahuzi region (Republique du Zaire). Z Tierpsychol 33:514–522
Chapman CA, Wrangham RW (1993) Range use of the forest chimpanzees of Kibale: implications for the understanding of chimpanzee social organization. Am J Primatol 31:263–273
Chapman CA, White FJ, Wrangham RW (1994) Party size in chimpanzees and bonobos: a reevaluation of theory based on two similarly forested sites. In: Wrangham RW, McGrew WC, de Waal FBM (eds) Chimpanzee cultures. Harvard University Press, Cambridge, pp 41–58
Chapman C, Wrangham RW, Chapman L (1995) Ecological constraints on group size: an analysis of spider monkey and chimpanzee subgroups. Behav Ecol Sociobiol 36:59–70
Chivers DJ, Hladik CM (1980) Morphology of the gastrointestinal tract in primates: comparisons with other mammals in relation to diet. J Morphol 116:337–386
Chivers DJ, Langer P (1994) The digestive system in mammals: food, form and function. Cambridge University Press, Cambridge
Cipoletta C (2004) Effects of group dynamics and diet on the ranging patterns of a western gorilla group (*Gorilla gorilla gorilla*) at Bai Hokou, Central African Republic. Am J Primatol 64: 193–205
Clark AP, Wrangham RW (1994) Chimpanzee arrival pant-hoots: do they signify food or status? Int J Primatol 15:185–205
Clutton-Brock TH, Harvey PH (1977) Species differences in feeding and ranging behaviour in primates. In: Clutton-Brock TH (ed) Primate ecology. Academic, London, pp 557–584
Collet J, Bourreau E, Cooper RW, Tutin CEG (1984) Experimental demonstration of cellulose digestion by *Troglodytella gorilla*, an intestinal ciliate of lowland gorillas. Int J Primatol 5:328
Constable JL, Ashley MV, Goodall J, Pusey AE (2001) Noninvasive paternity assignment in Gombe chimpanzees. Mol Ecol 10:1279–1300
Deblauwe I, Dupain J, Nguenang GM, Werdenich D, Van Elsacker L (2003) Insectivory by *Gorilla gorilla gorilla* in southeast Cameroon. Int J Primatol 24:493–502
Doran DM (1996) Comparative positional behavior of the African apes. In: McGrew WC, Marchant LF, Nishida T (eds) Great ape societies. Cambridge University Press, Cambridge, pp 213–224
Doran DM (2001) Subspecies variation in gorilla behavior: the influence of ecological and social factors. In: Robbins MM, Sicotte P, Stewart KJ (eds) Mountain gorillas: three decades of research at Karisoke. Cambridge University Press, Cambridge, pp 123–149
Doran DM, McNeilage A (1998) Gorilla ecology and behavior. Evol Anthropol 6:120–131
Doran DM, McNeilage A, Greer D, Bocian C, Mehlman P, Shah N (2002) Western lowland gorilla diet and resource availability: new evidence, cross-site comparisons, and reflections on indirect sampling methods. Am J Primatol 58:91–116
Doran-Sheehy DM, Greer D, Mongo P, Schwindt D (2004) Impact of ecological and social factors on ranging in western gorillas. Am J Primatol 64:207–222
Doran-Sheehy D, Mongo P, Lodwick J, Conklin-Brittan NL (2009) Male and female western gorilla diet: preferred foods, use of fallback resources, and implications for ape versus Old World monkey foraging strategies. Am J Phys Anthropol 140:727–738
Fawcett K, Bush G, Seimon A, Phillips GP, Tuyisingize D, Uwingeli P (2012) Long term changes in the Virunga Volcanoes. In: Plumptre AJ (ed) The ecological impact of long-term changes in Africa's Rift Valley. Nova Science, New York, pp 141–166
Fleagle JG (1984) Primate locomotion and diet. In: Chivers DJ, Wood BA, Bilsborough A (eds) Food acquisition and processing in primates. Plenum Press, New York, pp 105–117

Fossey D (1974) Observations on the home range of one group of mountain gorillas (*Gorilla gorilla beringei*). Anim Behav 22:568–581
Fossey D (1983) Gorillas in the mist. Houghton Mifflin, Boston
Fossey D (1984) Infanticide in mountain gorillas (*Gorilla gorilla beringei*) with comparative notes on chimpanzees. In: Hausfater G, Hrdy S (eds) Infanticide: comparative and evolutionary perspectives. Aldine, Hawthorne, pp 217–236
Fruth B, Hohmann G (1996) Nest building behavior in the great apes: the great leap forward? In: McGrew WC, Marchant LF, Nishida T (eds) Great ape societies. Cambridge University Press, Cambridge, pp 225–240
Furuichi T (1987) Sexual swelling, receptivity and grouping of wild pygmy chimpanzee females at Wamba, Zaire. Primates 28:309–318
Furuichi T (1997) Agonistic interactions and matrifocal dominance rank of wild bonobos (*Pan paniscus*) at Wamba. Int J Primatol 18:855–875
Furuichi T (2009) Factors underlying party size differences between chimpanzees and bonobos: a review and hypotheses for future study. Primates 50:197–209
Furuichi T (2011) Female contributions to the peaceful nature of bonobo society. Evol Anthropol 20:131–142
Furuichi T, Hashimoto C (2002) Why female bonobos have a lower copulation rate during estrus than chimpanzees. In: Boesch C, Hohmann G, Marchant LF (eds) Behavioural diversity in chimpanzees and bonobos. Cambridge University Press, Cambridge, pp 156–167
Furuichi T, Ihobe H (1994) Variation in male relationships in bonobos and chimpanzees. Behaviour 130:211–228
Furuichi T, Inagaki H, Angoue-Ovono S (1997) Population density of chimpanzees and gorillas in the Petit Loango Reserve, Gabon: employing a new method to distinguish between nests of the two species. Int J Primatol 18:1029–1046
Gagneux P, Willis C, Gerloff U, Tautz D, Morin PA, Boesch C, Fruth B, Hohmann G, Ryder OA, Woodruff DS (1999) Mitochondrial sequences show diverse evolutionary histories of African hominoids. Proc Natl Acad Sci USA 96:5077–5082
Ganas J, Robbins MM (2004) Intrapopulation differences in ant eating in the mountain gorillas of Bwindi Impenetrable National Park, Uganda. Primates 45:275–278
Ganas J, Robbins MM (2005) Ranging behavior of the mountain gorilla (*Gorilla beringei beringei*) in Bwindi Impenetrable National Park, Uganda: a test of the ecological constraints model. Behav Ecol Sociobiol 58:277–288
Ganas J, Robbins MM, Nkurunungi JB, Kaplin BA, McNeilage A (2004) Dietary variability of mountain gorillas in Bwindi Impenetrable National Park, Uganda. Int J Primatol 25:1043–1072
Ghiglieri MP (1984) The chimpanzees of Kibale Forest: a field study of ecology and social structure. Columbia University Press, New York
Goldsmith ML (1999) Ecological constraints on the foraging effort of western gorillas (*Gorilla gorilla gorilla*) at Bai Hokou, Central African Republic. Int J Primatol 20:1–23
Gondera MK, Locatellia S, Ghobriala L, Mitchella MW, Kujawskib JT, Lankesterc FJ, Stewarta C-B, Tishkoffd SA (2010) Evidence from Cameroon reveals differences in the genetic structure and histories of chimpanzee populations. Proc Natl Acad Sci USA 108:4761–4771
Goodall AG (1977) Feeding and ranging behavior of a mountain gorilla group (*Gorilla gorilla beringei*) in the Tshibinda-Kahuzi region (Zaire). In: Clutton-Brock TH (ed) Primate ecology. Academic, New York, pp 450–479
Goodall J (1986) The chimpanzees of Gombe. Belknap, Cambridge
Goodall J, Bandora A, Bergmann E, Busse C, Matama H, Mpongo E, Pierce A, Riss D (1979) Intercommunity interactions in the chimpanzee population of the Gombe National Park. In: Hamburg DA, McCown ER (eds) The great apes. Benjamin/Cummings, Menlo Park, pp 13–53
Gray M, McNeilage A, Fawcett K, Robbins M, Ssebide B, Mbula D, Uwingeli P (2003) Virunga Volcanoes range and census. Joint Organizers Report; UWA, ORTPN, ICCN
Grubera T, Claya Z, Zuberbühlera K (2010) A comparison of bonobo and chimpanzee tool use: evidence for a female bias in the *Pan* lineage. Anim Behav 80:1023–1033

Hamilton AC (1975) A quantitative analysis of altitudinal zonation in Uganda forest. Vegetation 30:99–106

Harcourt AH (1978) Strategies of emigration and transfer by primates, with particular reference to gorillas. Z Tierpsychol 48:401–420

Harcourt AH (1992) Coalitions and alliances: are primates more complex than non-primates? In: Harcourt AH, de Waal EBM (eds) Coalitions and alliances in human and other animals. Oxford University Press, Oxford, pp 445–472

Harcourt AH, Stewart KJ (2007) Gorilla society: conflict, compromise, and cooperation between sexes. The University of Chicago Press, Chicago

Harcourt AH, Stewart KJ, Fossey D (1976) Male emigration and female transfer in wild mountain gorillas. Nature (Lond) 263:226–227

Harcourt AH, Fossey D, Pi S (1981) Demography of *Gorilla gorilla*. J Zool Soc Lond 195:215–233

Harris TR (2006) Between-group contest competition for food in a highly folivorous population of black and white colobus (*Colobus guereza*). Behav Ecol Sociobiol 61:317–329

Harrison ME, Marshall AJ (2011) Strategies for the use of fallback foods in apes. Int J Primatol 32:531–565

Hashimoto C, Suzuki S, Takenoshita Y, Yamagiwa J, Basabose AK, Furuichi T (2003) How fruit abundance affects the chimpanzee party size: a comparison between four study sites. Primates 44:77–81

Hashimoto C, Tashiro Y, Kimura D, Enomoto T, Ingmanson EJ, Idani G, Furuichi T (1998) Habitat use and ranging of wild bonobos (*Pan paniscus*) at Wamba. Int J Primatol 19:1045–1060

Herbinger L, Boesch C, Rothe H (2001) Territory characteristics among three neighboring chimpanzee communities in the Tai National Park, Ivory Coast. Int J Primatol 32:143–167

Hilty SL (1980) Flowering and fruiting periodicity in a premontane rain forest in Pacific Colombia. Biotropica 12:292–306

Hladik CM (1977) Chimpanzees of Gabon and chimpanzees of Gombe: some comparative data on diet. In: Clutton-Brock TH (ed) Primate ecology. Academic, London, pp 481–501

Hohmann G, Fruth B (2003) Intra- and inter-sexual aggression by bonobos in the context of mating. Behaviour 136:1219–1235

Hohmann G, Gerloff U, Tautz D, Fruth B (1999) Social bonds and genetic ties: kinship, association and affiliation in a community of bonobos (*Pan paniscus*). Behaviour 136:1219–1235

Hrdy SB (1979) Infanticide among animals: a review, classification, and examination of the implications for the reproductive strategies of females. Ethol Sociobiol 1:13–40

Idani G (1990) Relations between unit-groups of bonobos at Wamba, Zaire: encounters and temporary fusions. Afr Stud Monogr 11:153–156

Ingmanson EJ (1996) Tool-using behavior in wild *Pan paniscus*: social and ecological considerations. In: Russon AE, Bard KA, Parker ST (eds) Reaching into thought: the minds of great apes. Cambridge University Press, Cambridge, pp 190–210

Inoue E, Inoue-Murayama M, Vigilant L, Takenaka O, Nishida T (2008) Relatedness in wild chimpanzees: influence of paternity, male philopatry, and demographic factors. Am J Primatol 137:256–262

Isbell LA, Pruetz JD, Lewis M, Young TP (1998) Locomotor activity differences between sympatric patas monkeys (*Erythrocebus patas*) and vervet monkeys (*Cercopithecus aethiops*): implications for the evolution of long hindlimb length in *Homo*. Am J Phys Anthropol 105:199–207

Itoh N, Nishida T (2007) Chimpanzee grouping patterns and food availability in Mahale Mountains National Park, Tanzania. Primates 48:87–96

Iwata Y, Ando C (2007) Bed and bed-site reuse by western lowland gorillas (*Gorilla g. gorilla*) in Moukalaba-Doudou National Park, Gabon. Primates 48:77–80

Jones C, Sabater Pi J (1971) Comparative ecology of *Gorilla gorilla* (Savage and Wyman) and *Pan troglodytes* (Blumenbach) in Rio Muni, West Africa. Bibl Primatol 13:1–96

Kano T (1972) Distribution and adaptation of the chimpanzees on the eastern shore of Lake Tanaganyika. Kyoto Univ Afr Stud 7:37–129

Kano T (1980) Social behavior of wild pygmy chimpanzees (*Pan paniscus*) of Wamba: a preliminary report. J Hum Evol 9:243–260
Kano T (1982) The social group of pygmy chimpanzees (*Pan paniscus*) of Wamba. Primates 23:171–188
Kano T (1992) The last ape: pygmy chimpanzee behavior and ecology. Stanford University Press, Palo Alto
Kano T (1996) Male rank order and copulation rate in a unit-group of bonobos at Wamba, Zaire. In: McGrew WC, Marchant LF, Nishida T (eds) Great ape societies. Cambridge University Press, Cambridge, pp 135–145
Kano T, Mulavwa M (1984) Feeding ecology of the pygmy chimpanzees (*Pan paniscus*) of Wamba. In: Susman RL (ed) The pygmy chimpanzee: evolutionary biology and behavior. Plenum, New York, pp 233–274
Kay RF (1984) On the use of anatomical features to infer foraging behavior in extinct primates. In: Rodman PS, Cant JGH (eds) Contributions to an organismal biology of prosimians, monkey and apes. Smithsonian Institution Press, Washington, DC, pp 173–191
Kuroda S (1979) Grouping of the pygmy chimpanzees. Primates 20:161–183
Kuroda S (1980) Social behavior of the pygmy chimpanzees. Primates 21:181–197
Kuroda S, Nishihara T, Suzuki S, Oko RA (1996) Sympatric chimpanzees and gorillas in the Ndoki forest, Congo. In: McGrew WC, Marchant LF, Nishida T (eds) Great ape societies. Cambridge University Press, Cambridge, pp 71–81
Lambert JE (1998) Primate digestion: interactions among anatomy, physiology, and feeding ecology. Evol Anthropol 7:8–20
Lambert JE (2002) Digestive retention times in forest guenons with reference to chimpanzees. Int J Primatol 26:1169–1185
Lambert JE (2007) Seasonality, fallback strategies, and natural selection: a chimpanzee and cercopithecoid model for interpreting the evolution of hominin diet. In: Unger PS (ed) Evolution of the human diet: the known, the unknown and the unknowable. Oxford University Press, Oxford, pp 324–343
Lehmann J, Boesch C (2004) To fission or to fusion: effects of community size on wild chimpanzee (*Pan troglodytes verus*) social organization. Behav Ecol Sociobiol 56:207–216
Lehmann J, Boesch C (2008) Sexual differences in chimpanzee sociality. Int J Primatol 29:65–81
Malenky RW, Stiles EW (1991) Distribution of terrestrial herbaceous vegetation and its consumption by *Pan paniscus* in the Lomako forest, Zaire. Am J Primatol 23:153–169
Marshall AJ, Wrangham R (2007) Evolutionary consequences of fallback foods. Int J Primatol 28:1219–1235
Marshall AJ, Boyko CM, Feilin KL, Boyko RH, Leighton M (2009) Defining fallback foods and assessing their importance in primate ecology and evolution. Am J Phys Anthropol 140: 603–614
Matsumoto-Oda A, Hosaka K, Huffman MA, Kawanaka K (1998) Factors affecting party size in chimpanzees of the Mahale mountains. Int J Primatol 19:999–1011
McGrew WC (1992) Chimpanzee material culture: implications for human evolution. Cambridge University Press, Cambridge
McGrew WC, Baldwin PJ, Tutin CEG (1988) Diet of wild chimpanzees (*Pan troglodytes verus*) at Mt. Assirik, Senegal: 1. Composition. Am J Primatol 16:213–226
McNeilage A, Plumptre AJ, Brock-Doyle A, Vedder A (2001) Bwindi Impenetrable National Park, Uganda: Gorilla census 1997. Oryx 35:39–47
Milton K (1986) Digestive physiology in primates. News Physiol Sci 1:76–79
Mitani JC, Merriwether DA, Zhang C (2000) Male affiliation, cooperation, and kinship in wild chimpanzees. Anim Behav 59:885–893
Mitani JC, Watts DP, Pepper JW, Merriwether DA (2002) Demographic and social constraints on male chimpanzee behaviour. Anim Behav 64:727–737
Morgan D, Sanz C (2006) Chimpanzee feeding ecology and comparisons with sympatric gorillas in the Goualougo triangle, Republic of Congo. In: Hohmann G, Robbins MM, Boesch C (eds)

Feeding ecology in apes and other primates: ecological, physical and behavioral aspects. Cambridge University Press, Cambridge, pp 97–122

Morin PA, Moore JJ, Chakraborty R, Jin L, Goodall J, Woodruff DS (1994) Kin selection, social structure, gene flow, and evolution of chimpanzees. Science 265:1193–1201

Murnyak DF (1981) Censusing the gorillas in Kahuzi-Biega National Park. Biol Conserv 21:163–176

Muroyama Y, Sugiyama Y (1994) Grooming relationships in two species of chimpanzees. In: Wrangham RW, McGrew WC, de Waal FBM (eds) Chimpanzee cultures. Harvard University Press, Cambridge, pp 169–180

Muller MN, Mitani JC (2005) Conflict and cooperation in wild chimpanzees. Adv Study Behav 35:275–331

Newton-Fisher NE (1999a) The diet of chimpanzees in the Budongo forest reserve, Uganda. Afr J Ecol 37:344–354

Newton-Fisher NE (1999b) Association by male chimpanzees: a social tactic? Behaviour 136:705–730

Newton-Fisher NE (2003) The home range of the Sonso community of chimpanzees from the Budongo Forest, Uganda. Afr J Ecol 41:150–156

Newton-Fisher NE, Reynolds V, Plumptre AJ (2000) Food supply and chimpanzee (*Pan troglodytes schweinfurthii*) party size in the Budongo forest reserve, Uganda. Int J Primatol 21:613–628

Nishida T (1968) The social group of wild chimpanzees in the Mahali mountains. Primates 9:167–224

Nishida T (1979) The social structure of chimpanzees of the Mahale mountains. In: Hamburg DA, McCown ER (eds) The great apes. Benjamin/Cummings, Menlo Park, pp 73–121

Nishida T (1983) Alpha status and agonistic alliance in wild chimpanzees (*Pan troglodytes schweinfurthii*). Primates 24:318–336

Nishida T, Kawanaka K (1972) Inter-unit-group relationships among wild chimpanzees of the Mahali mountains. Kyoto Univ Afr Stud 7:131–169

Nishida T, Uehara S (1983) Natural diet of chimpanzees (*Pan troglodytes schweinfurthii*): long-term record from the Mahale mountains, Tanzania. Afr Stud Monogr 3:109–130

Nishida T, Hiraiwa-Hasegawa M, Hasegawa T, Takahata Y (1985) Group extinction and female transfer in wild chimpanzees in the Mahale mountains. Z Tierpsychol 67:284–301

Nishida T, Hiraiwa-Hasegawa M (1987) Chimpanzees and bonobos: cooperative relationships among males. In: Smuts BB, Cheney DL, Seyfarth RM, Wrangham RW, Struhsaker TT (eds) Primate societies. University of Chicago Press, Chicago, pp 165–177

Nishida T, Hosaka K (1996) Coalition strategies among adult male chimpanzees of the Mahale mountains, Tanzania. In: McGrew WC, Marchant LF, Nishida T (eds) Great ape societies. Cambridge University Press, Cambridge, pp 114–134

Nishihara T (1994) Population density and group organization of gorillas (*Gorilla gorilla gorilla*) in the Nouabalé-Ndoki National Park, Congo. Africa Kenkyu 44:29–45 (in Japanese with English summary)

Nishihara T (1995) Feeding ecology of western lowland gorillas in the Nouabale-Ndoki National Park, Congo. Primates 36:151–168

Nishihara T, Kuroda S (1991) Soil-scratching behavior by western lowland gorillas. Folia Primatol 57:48–51

Parish AR (1994) Sex and food control in the "uncommon chimpanzee:" how bonobo females overcome a phylogenetic legacy of male dominance. Ethol Sociobiol 15:157–179

Parish AR (1996) Female relationships in bonobos (*Pan paniscus*). Hum Nat 7:61–96

Parnell RJ (2002) Group size and structure in western lowland gorillas (*Gorilla gorilla gorilla*) at Mbeli Bai, Republic of Congo. Am J Primatol 56:193–206

Pepper JW, Mitani JC, Watts DP (1999) General gregariousness and specific social preferences among wild chimpanzees. Int J Primatol 20:613–632

Potts R (2004) Paleoenvironments and the evolution of adaptability in great apes. In: Russon A, Begun DR (eds) The evolution of thought: evolutionary origins of great ape intelligence. Cambridge University Press, Cambridge, pp 237–259

Powzyk JA, Mowry CB (2003) Dietary and feeding differences between sympatric *Propithecus diadema diadema* and *Indri indri*. Int J Primatol 24:1143–1162

Remis MJ (1997a) Western lowland gorillas (*Gorilla gorilla gorilla*) as seasonal frugivores: use of variable resources. Am J Primatol 43:87–109

Remis MJ (1997b) Ranging and grouping patterns of a western lowland gorilla group at Bai Hokou, Central African Republic. Am J Primatol 43:111–133

Remis MJ (2000) Initial studies on the contribution of body size and gastrointestinal passage rates to dietary flexibility among gorillas. Am J Phys Anthropol 112:171–180

Remis MJ (2003) Are gorillas vacuum cleaners of the forest floor? The roles of body size, habitat, and food preferences on dietary flexibility and nutrition. In: Taylor AB, Goldsmith ML (eds) Gorilla biology: multidisciplinary perspective. Cambridge University Press, Cambridge, pp 385–404

Remis MJ, Dierenfeld ES, Mowry CB, Carroll RW (2001) Nutritional aspects of western lowland gorilla (*Gorilla gorilla gorilla*) diet during seasons of fruit scarcity at Bai Hokou, Central African Republic. Int J Primatol 22:807–836

Robbin MM (2001) Variation in the social system of mountain gorillas: the male perspective. In: Robbins MM, Sicotte P, Stewart KJ (eds) Mountain gorillas: three decades of research at Karisoke. Cambridge University Press, Cambridge, pp 29–58

Robbins MM (1995) A demographic analysis of male life history and social structure of mountain gorillas. Behaviour 132:21–47

Robbins MM (1999) Male mating patterns in wild multimale mountain gorilla groups. Anim Behav 57:1013–1020

Robbins MM, McNeilage A (2003) Home range and frugivory patterns of mountain gorillas in Bwindi Impenetrable National Park, Uganda. Int J Primatol 24:467–491

Robbins AM, Robbins MM (2005) Fitness consequences of dispersal decisions for male mountain gorillas (*Gorilla gorilla beringei*). Behav Ecol Sociobiol 58:295–309

Robbins MM, Bermejo M, Cipolletta C, Magliocca F, Parnell RJ, Stokes E (2004) Social structure and life-history patterns in western gorillas (*Gorilla gorilla gorilla*). Am J Primatol 64: 145–159

Robbins MM, Robbins AM, Gerald-Steklis N, Steklis HD (2007) Socioecological influences on the reproductive success of female mountain gorillas (*Gorilla beringei beringei*). Behav Ecol Sociobiol 61:919–931

Robbins AM, Stoinski TS, Fawcett KA, Robbins MM (2009) Socio-ecological influences on the dispersal of female mountain gorillas: evidence of a second folivore paradox. Behav Ecol Sociobiol 63:477–489

Rogers ME, Maisels F, Williamson EA, Fernandez M, Tutin CEG (1990) Gorilla diet in the Lopé reserve, Gabon: a nutritional analysis. Oecologia (Berl) 84:326–339

Rogers ME, Abernethy K, Bermejo M, Cipolletta C, Doran D, McFarland K, Nishihara T, Remis M, Tutin CEG (2004) Western gorilla diet: a synthesis from six sites. Am J Primatol 64:173–192

Sabatier D (1985) Saisonalite et determinisme du pie de fructification en foret guyanaise. Terre Vie 40:289–320

Sakura O (1994) Factors affecting party size and composition of chimpanzees (*Pan troglodytes verus*) at Bossou, Guinea. Int J Primatol 15:167–183

Schaller GB (1963) The mountain gorilla: ecology and behavior. University of Chicago Press, Chicago

Snaith TV, Chapman CA (2007) Primate group size and socioecological models: do folivores really play by different rules? Evol Anthropol 16:94–106

Stanford CB, Nkurunungi JB (2003) Behavioral ecology of sympatric chimpanzees and gorillas in Bwindi Impenetrable National Park, Uganda: diet. Int J Primatol 24:901–918

Stevens CE, Hume ID (1995) Comparative physiology of the vertebrate digestive system, 2nd edn. Cambridge University Press, Cambridge

Stewart KJ, Harcourt AH (1987) Gorillas: variation in female relationships. In: Smuts BB, Cheney DL, Seyfarth RM, Wrangham RW, Struhsaker TT (eds) Primate societies. University of Chicago Press, Chicago, pp 155–164

Stoinski TS, Rosenbaum S, Ngaboyamahina T, Vecellio V, Ndagijimana F, Fawcett K (2009a) Patterns of male reproductive behavior in multi-male groups of moluntain gorillas: examining theories of reproductive skew. Behaviour 146:1193–1215

Stoinski TS, Rosenbaum S, Ngaboyamahina T, Vecellio V, Ndagijimana F, Fawcett KA (2009b) Proximate factors influencing dispersal decisions in male mountain gorillas, *Gorilla beringei beringei*. Anim Behav 77:1155–1164

Stokes EJ, Parnell RJ, Olejniczak C (2003) Fmale dispersal and reproductive success in wild western lowland gorillas (*Gorilla gorilla gorilla*). Behav Ecol Sociobiol 54:329–339

Sugiyama Y (1988) Grooming interactions among adult chimpanzees at Bossou, Guinea, with special reference to social structure. Int J Primatol 9:393–407

Sugiyama Y (2004) Demographic parameters and life history of chimpanzees at Bossou, Guinea. Am J Primatol 124:154–166

Sugiyama Y, Koman J (1979) Social structure and dynamics of wild chimpanzeews at Bossou, Guinea. Primates 20:323–339

Sugiyama Y, Koman J (1987) A preliminary list of chimpanzees' alimentation at Bossou, Guinea. Primates 28:133–147

Sun C, Kaplan BA, Kristensen KA, Munyaligoga V, Mvukiyumwami J, Kajango KK, Moermond TC (1996) Tree phenology in a tropical montane forest in Rwanda. Biotropica 26:668–681

Suzuki A (1977) Primate ecology. In: Itani J (ed) Lectures in anthropology. Yuzankaku Shuppan, Tokyo, pp 147–194

Suzuki S, Kuroda S, Nishihara T (1995) Tool-set for termite fishing by chimpanzees in the Ndoki forest, Congo. Behaviour 132:219–235

Tan CL (1999) Group composition, home range size, and diet of three sympatric bamboo lemur species (genus *Hapalemur*) in Ranomafana National Park, Madagascar. Int J Primatol 20:547–566

Teleki G (1973) The omnivorous chimpanzee. Sci Am 228:33–42

Temerin LA, Cant JGH (1983) The evolutionary divergence of Old World monkeys and apes. Am Nat 122:335–351

Tutin CEG (1996) Ranging and social structure of lowland gorillas in the Lopé Reserve, Gabon. In: McGrew WC, Marchant LF, Nishida T (eds) Great ape societies. Cambridge University Press, Cambridge, pp 58–70

Tutin CEG, Fernandez M (1984) Nationwide census of gorilla (*Gorilla g. gorilla*) and chimpanzee (*Pan t. troglodytes*) populations in Gabon. Am J Primatol 6:313–336

Tutin CEG, Fernandez M (1985) Food consumed by sympatric populations of *Gorilla g. gorilla* and *Pan t. troglodytes* in Gabon: some preliminary data. Int J Primatol 6:27–43

Tutin CEG, Fernandez M (1992) Insect-eating by sympatric lowland gorillas (*Gorilla g. gorilla*) and chimpanzees (*Pan t. troglodytes*) in the Lopé Reserve, Gabon. Am J Primatol 28:29–40

Tutin CEG, Fernandez M (1993) Composition of the diet of chimpanzees and comparisons with that of sympatric lowland gorillas in the Lopé Reserve, Gabon. Am J Primatol 30:195–211

Tutin CEG, McGrew WC, Baldwin PJ (1983) Social organization of savanna-dwelling chimpanzees, *Pan troglodytes verus*, at Mt. assirik, Senegal. Primates 24:154–173

Ungar P (1996) Feeding height and niche separation in sympatric monkeys and apes. Folia Primatol 67:163–168

van Schaik CP (1989) The ecology of social relationships among female primates. In: Standen V, Foley RA (eds) Comparative socioecology: the behavioural ecology of humans and other mammals. Blackwell, Oxford, pp 195–218 van Schaik CP (2000) Infanticide by male primates: the sexual selection hypothesis revisited. In: Janson CH, van Schaik CP (eds) Infanticide by males and its implications. Cambridge University Press, Cambridge, pp 27–60

van Schaik CP, Preuschoft S, Watts DP (2004) Great ape social system. In: Russon AE, Begun DR (eds) The evolution of thought: evolutionary origins of great ape intelligence. Cambridge University Press, Cambridge, pp 190–209

Vedder AL (1984) Movement patterns of a group of free-ranging mountain gorillas (*Gorilla gorilla beringei*) and their relation to food availability. Am J Primatol 7:73–88

Vervaecke HD, Vries H, van Elsacker L (1999) An experimental evaluation of the consistency of competitive ability and agonistic dominance in different social contexts in captive bonobos. Behaviour 136:423–442

Vigilant L, Hofreiter M, Siedel H, Boesch C (2001) Paternity and relatedness in wild chimpanzee communities. Proc Natl Acad Sci USA 98:12890–12895

Watts DP (1984) Composition and variability of mountain gorilla diets in the Central Virungas. Am J Primatol 7:323–356

Watts DP (1989) Infanticide in mountain gorillas: new cases and a reconsideration of the evidence. Ethology 81:1–18

Watts DP (1991) Strategies of habitat use by mountain gorillas. Folia Primatol (Basel) 56:1–16

Watts DP (1996) Comparative socio-ecology of gorillas. In: McGrew WC, Marchant LF, Nishida T (eds) Great ape societies. Cambridge University Press, Cambridge, pp 16–28

Watts DP (1998a) Long-term habitat use by mountain gorillas (*Gorilla gorilla beringei*). 1. Consistency, variation, and home range size and stability. Int J Primatol 19:651–680

Watts DP (1998b) Long-term habitat use by mountain gorillas (*Gorilla gorilla beringei*). 2. Reuse of foraging areas in relation to resource abundance, quality, and depletion. Int J Primatol 19:681–702

Watts DR (2000) Mountain gorilla habitat use strategies and group movements. In: Boinski S, Garber PA (eds) On the move: how and why animals travel in groups. The University of Chicago Press, Chicago, pp 351–374

Watts DP (2003) Gorilla social relationships: a comparative overview. In: Taylor AB, Goldsmith ML (eds) Gorilla biology: a multidisciplinary perspective. Cambridge University Press, Cambridge, pp 302–327

Watts DP, Mitani JC (2001) Boundary patrols and intergroup encounters in wild chimpanzees. Bahaviour 138:299–327

Watts DP, Mitani JC (2002) Hunting behavior of chimpanzees at Ngogo, Kibale National Park, Uganda. Int J Primatol 23:1–28

Watts DP, Muller M, Amsler S, Mbabazi G, Mitani JC (2006) Lethal intergroup aggression by chimpanzees in the Kibale National Park, Uganda. Am J Primatol 68:161–180

Weber AW, Vedder A (1983) Population dynamics of the Virunga gorillas 1959–1978. Biol Conserv 26:341–366

White FJ (1988) Party composition and dynamics in *Pan paniscus*. Int J Primatol 9:179–193

White FJ (1996) *Pan paniscus* 1973 to 1996: twenty-three years of field research. Evol Anthropol 5:11–17

White FJ, Wrangham RW (1988) Feeding competition and patch size in the chimpanzee species *Pan paniscus* and *Pan troglodytes*. Behaviour 105:148–163

Williams JM, Liu H-Y, Pusey AE (2002) Costs and benefits of grouping for female chimpanzees at Gombe. In: Boesch C, Hohmann G, Marchant LF (eds) Behavioural diversity in chimpanzees and bonobos. Cambridge University Press, Cambridge, pp 192–203

Williamson RW, Tutin CEG, Rogers ME, Fernandez M (1990) Composition of the diet of lowland gorillas at Lopé in Gabon. Am J Primatol 21:265–277

Wilson ML, Wrangham RW (2003) Intergroup relations in chimpanzees. Annu Rev Anthropol 32:363–392

Wilson ML, Wallauer WR, Pusey AE (2004) Intergroup violence in chimpanzees: new cases from Gombe National Park, Tanzania. Int J Primatol 25:523–550

Wittig RM, Boesch C (2003) Food competition and linear dominance hierarchy among female chimpanzees in the Tai National Park. Int J Primatol 24:847–867

Wrangham RW (1977) Feeding behaviour of chimpanzees in Gombe National Park, Tanzania. In: Clutton-Brock TH (ed) Primate ecology: studies of feeding and ranging behaviour in lemurs, monkeys, and apes. Academic, London, pp 503–538

Wrangham RW (1979) Sex differences in chimpanzee dispersion. In: Hamburg DA, McCown ER (eds) The great apes. Benjamin/Cummings, Menlo Park, California, pp 480–489

Wrangham RW (1980) An ecological model of female-bonded primate groups. Behaviour 75:262–300

Wrangham RW (1986) Ecology and social relationships in two species of chimpanzees. In: Rubenstein DI, Wrangham RW (eds) Ecological aspects of social evolution: birds and mammals. Princeton University Press, Princeton, pp 352–378

Wrangham RW (1987) Evolution of social structure. In: Smuts BB, Wrangham RW, Cheney DL, Struhsaker TT, Seyfarth RM (eds) Primate societies. University of Chicago Press, Chicago, pp 282–296

Wrangham RW, Conklin NL, Chapman CA, Hunt KD (1991) The significance of fibrous foods for Kibale chimpanzees. Philos Trans R Soc Lond 334:171–178

Wrangham RW, Clark AP, Isabirye-Basuta G (1992) Female social relationships and social organization of Kibale Forest chimpanzees. In: Nishida T, McGrew WC, Marler P, Pickford M, de Waal FBM (eds) Topics in primatology, vol 1, Human origins. The University of Tokyo Press, Tokyo, pp 81–98

Wrangham RW, Chapman CA, Clark-Arcadi AP, Isabirye-Basuta G (1996) Social ecology of Kanyawara chimpanzees: implications for understanding the costs of great ape groups. Cambridge University Press, Cambridge, pp 45–57

Wrangham RW, Conklin-Brittain NL, Hunt KD (1998) Dietary response of chimpanzees and cercopithecines to seasonal variation in fruit abundance. I. Antifeedants. Int J Primatol 19: 949–970

Yamagiwa J (1986) Activity rhythm and ranging of a solitary male mountain gorilla (*Gorilla gorilla beringei*). Primates 27:273–282

Yamagiwa J (1987) Intra- and inter-group interactions of an all-male group of Virunga mountain gorillas (*Gorilla gorilla beringei*). Primates 28:1–30

Yamagiwa J (1999) Socioecological factors influencing population structure of gorillas and chimpanzees. Primates 40:87–104

Yamagiwa J (2004) Diet and foraging of the great apes: ecological constraints on their social organizations and implications for their divergence. In: Russon AE, Begun DR (eds) The evolution of thought: evolutionary origins of great ape intelligence. Cambridge University Press, Cambridge, pp 210–233

Yamagiwa J, Basabose AK (2006a) Diet and seasonal changes in sympatric gorillas and chimpanzees at Kahuzi-Biega National Park. Primates 47:74–90

Yamagiwa J, Basabose AK (2006b) Effects of fruit scarcity on faraging strategies of sympatric gorillas and chimpanzees. In: Hohmann G, Robbins MM, Boesch C (eds) Feeding ecology in apes and other primates. Cambridge University Press, Cambridge, pp 73–96

Yamagiwa J, Basabose AK (2009) Fallback foods and dietary partitioning among *Pan* and *Gorilla*. Am J Phys Anthropol 140:739–750

Yamagiwa J, Kahekwa J (2001) Dispersal patterns, group structure and reproductive parameters of eastern lowland gorillas at Kahuzi in the absence of infanticide. In: Robbins MM, Sicotte P, Stewart KJ (eds) Mountain gorillas: three decades of research at Karisoke. Cambridge University Press, London, pp 89–122

Yamagiwa J, Mwanza N (1994) Day-journey length and daily diet of solitary male gorillas in lowland and highland habitats. Int J Primatol 15:207–224

Yamagiwa J, Mwanza N, Spangenberg A, Maruhashi T, Yumoto T, Fischer A, Steinhauer-Burkart B (1993) A census of the eastern lowland gorillas *Gorilla gorilla graueri* in Kahuzi-Biega National Park with reference to mountain gorilla *G.g. beringei* in the Virunga Region, Zaire. Biol Conserv 64:83–89

Yamagiwa J, Mwana N, Yumoto T, Maruhashi T (1991) Ant eating by eastern lowland gorillas. Primates 32:247–253

Yamagiwa J, Mwanza N, Yumoto Y, Maruhashi T (1994) Seasonal change in the composition of the diet of eastern lowland gorillas. Primates 35:1–14

Yamagiwa J, Maruhashi T, Yumoto T, Mwanza N (1996) Dietary and ranging overlap in sympatric gorillas and chimpanzees in Kahuzi-Biega National Park, Zaire. In: McGrew WC, Marchant LF, Nishida T (eds) Great ape societies. Cambridge University Press, Cambridge, pp 82–98

Yamagiwa J, Kahekwa J, Basabose AK (2003) Intra-specific variation in social organization of gorillas: implications for their social evolution. Primates 44:359–369

Yamagiwa J, Basabose AK, Kaleme K, Yumoto T (2005) Diet of Graue's gorillas in the montane forest of Kahuzi, Democratic Republic of Congo. Int J Primatol 26:1345–1373

Yamagiwa J, Basabose AK, Kaleme KP, Yumoto T (2008) Phenology of fruits consumed by aympatric population of gorillas and chimpanzees in Kahuzi-Biega National Park, Democratic Republic of Congo. Afr Stud Monogr (Suppl) 39:3–22

Yamagiwa J, Kahekwa J, Basabose AK (2009) Infanticide and social flexibility in the genus *Gorilla*. Primates 50:293–303

Yamagiwa J, Basabose AK, Kahekwa J, Bikaba D, Ando C, Matsubara M, Iwasaki N, Sprague DS (2011) Long-term research on Grauer's gorillas in Kahuzi-Biega National Park, DRC: life history, foraging strategies, and ecological differentiation from sympatric chimpanzees. In: Kappeler PM, Watts DP (eds) Long-term field studies of primates. Springer, New York, pp 385–412

Yamagiwa J, Basabose AK, Kahekwa J, Bikaba D, Matsubara M, Ando C, Iwasaki N, Sprague DS (2012) Long-term changes in habitats and ecology of African apes in Kahuzi-Biega National Park, Democratic Republic of Congo. In: Plumptre AJ (ed) The ecological impact of long-term changes in Africa's Rift Valley. Nova Science, New York, pp 175–193

Yamakoshi G (1998) Dietary responses to fruit scarcity of wild chimpanzees at Bossou, Guinea: possible implications for ecological importance of tool use. Am J Phys Anthropol 106:283–295

Yamakoshi G (2004) Food seasonality and socioecology in pan: are West African chimpanzees another bonobos? Afr Stud Monogr 25:45–60

Chapter 4
You Are What You Eat: Foraging Specializations and Their Influence on the Social Organization and Behavior of Killer Whales

John K.B. Ford and Graeme M. Ellis

J.K.B. Ford (✉) • G.M. Ellis
Fisheries and Oceans Canada, Pacific Biological Station, Nanaimo, BC, Canada V9T 6N7
e-mail: John.K.Ford@dfo-mpo.gc.ca

J. Yamagiwa and L. Karczmarski (eds.), *Primates and Cetaceans: Field Research and Conservation of Complex Mammalian Societies*, Primatology Monographs,
DOI 10.1007/978-4-431-54523-1_4,

Abstract The feeding ecology of predators can have a profound effect on their life history and behaviour. The killer whale—the apex marine predator—has a cosmopolitan distribution throughout the world's oceans. Globally, it is a generalist predator with a diverse diet, but regionally, different socially and genetically isolated killer whale populations can have highly specialized foraging strategies involving only a few types of prey. In the eastern North Pacific, the three sympatric killer whale lineages have distinct dietary specializations: one feeds primarily on marine mammals, another on salmon, and the third appears to specialize on sharks. These ecological specializations are associated with distinct patterns of seasonal distribution, group size, social organization, foraging behavior, and acoustic activity. Divergent foraging strategies may have played a major role in the social isolation and genetic divergence of killer whale populations.

Keywords Apex predator • Feeding ecology • *Orcinus orca*

4.1 Introduction

Ecological specialization is an important factor promoting the evolution of biological diversity and speciation (Futuyma and Moreno 1988; Robinson et al. 1996; Dieckmann and Doebeli 1999; Schluter 2001; Via 2001). Optimal foraging theory predicts that selection will generally favor dietary specialization, as specialists have a competitive advantage over generalists in foraging efficiency (as in the adage, "the jack-of-all-trades is the master of none") (Stephens and Krebs 1986; Futuyma and Moreno 1988; Robinson et al. 1996). Such selection may drive the divergent evolution of a wide variety of adaptive traits involving morphology, physiology, and behavior of populations or subpopulations with different foraging strategies or in contrasting environments. Divergent selection between sympatric populations may lead to assortative mating, reproductive isolation, and, ultimately, speciation (Dieckmann and Doebeli 1999; Schluter 2001; Via 2001; McKinnon et al. 2004).

Killer whales, the largest of the dolphins (family Delphinidae), provide an exceptional opportunity to gain insight into the processes and outcomes of ecological specialization and divergence in a highly social and versatile mammalian predator. This species (only a single species, *Orcinus orca*, is currently recognized) is one of the most widely distributed mammals on the planet. It has a cosmopolitan distribution in all the world's oceans, from the pack ice edges in both the Northern and Southern Hemispheres through the equatorial tropics (Ford 2002). Although rare in many regions, it is relatively common in cool, productive, high-latitude waters, particularly in nearshore areas. Despite their wide distribution, killer whales are not abundant, with a minimum estimated global population of 50,000, but probably not greatly more (Forney and Wade 2006). Killer whales occupy the top trophic position in the oceans and have no predators. As a species, killer whales could be considered generalist predators, with an extremely diverse array of more than 140 species of vertebrates and invertebrates—from small schooling fish to the largest of the cetaceans—recorded as prey (Ford 2009). However, field studies in several global regions have revealed that local populations can have remarkably specialized diets and may forage selectively for only a very small subset of the prey species that the predator is capable of consuming. In this chapter, we provide a description of three distinct killer whale lineages that co-occur in coastal waters of the northeastern Pacific, focusing in particular on the influence that ecological specialization appears to have had on their divergent lifestyles, including habitat use patterns, social structure, behavior, and use of underwater sound. We also provide a brief overview of how these lineages came to be identified and known in these waters and of recent work in other regions that suggests that ecological specialization is characteristic of this apex social predator.

4.2 Discovery of Killer Whale Lineages in the Eastern North Pacific

Before the 1970s, scientific understanding of the killer whale was poor and was based almost entirely on anecdotal or opportunistic observations rather than on dedicated scientific studies (Martinez and Klinghammer 1970). However, a live-capture fishery for killer whales that developed during the late 1960s in nearshore waters of southern British Columbia, Canada, and northern Washington State, USA, highlighted the need for basic abundance and life history data for management. As a result, in 1972 our late colleague, Michael Bigg, initiated field studies of killer whales in this area based primarily on the identification of individuals from photographs of natural markings on the whales' dorsal fin and grey "saddle patch" at the base of the fin. This technique was considered quite novel and unproven at the time, but Bigg quickly showed that it was an effective means of collecting reliable population abundance and life history data on these difficult-to-study animals (Bigg et al. 1976). We joined this field effort at different points in the 1970s and, working together with Bigg and our colleague Ken Balcomb in Washington State, broadened the study's scope to include social organization, foraging ecology, behavior, and vocalizations (Bigg et al. 1987).

By the late 1970s, it was apparent that two different types of killer whales coexisted in the region. One type, named "residents," lived in stable groups of 10 to 25 and were found reliably in predictable "core areas" throughout at least summer and fall. A second type was found in the same waters but only rarely and sporadically. These whales were observed alone or in small groups of 2 to 6, tended to swim close along shorelines, often erratically, and were never seen to mix with the larger "resident" groups. As it was thought that these whales were merely passing through the home ranges of the residents, they were named "transients" (Bigg 1982). Resident and transient killer whales were occasionally observed within a few hundred meters of each other but showed no obvious reaction to the presence of the other whales and did not intermingle. However, resident groups frequently mixed with other residents and transients with other transients. Although residents and transients were clearly socially isolated, it was not certain what these two types represented. Initially it was thought that transients were individuals that had dispersed from resident groups, possibly in other regions, and were adopting a "low profile" behavior while transiting core areas of residents. However, subtle differences in dorsal fin shape and pigmentation suggested an underlying genetic distinction between them. As the number of observations of feeding grew in the early 1980s, evidence mounted that residents and transients were distinct ecotypes with fundamentally different diets—residents prey on fish and transients on marine mammals (Bigg et al. 1985, 1987). That these two types of whales specialize on such different kinds of prey helped explain the growing number of differences we observed in the movement patterns, social structure, vocalizations, and behavior of residents and transients.

To our surprise, in the early 1990s we discovered a third type of killer whale, named "offshores," in British Columbian waters (Ford et al. 1992; Ford et al. 2000). These whales have slightly different fin shapes than residents and transients and appear to be somewhat smaller in body size. Offshore killer whales generally prefer the outer continental shelf, and it was only when we expanded our study area to include these waters that we found these whales. Residents and transients also use these outer waters, and offshores have recently made more frequent appearances in nearshore areas (Dahlheim et al. 2008). Despite their mostly sympatric distribution, all three killer whale types maintain social isolation from each other (Ford et al. 2000). From the few available observations of predation by offshore killer whales and their patterns of behavior and vocal activity, it appears that they are primarily or entirely fish feeders with a probable specialization on sharks (Ford et al. 2000, 2011; Jones 2006; Dahlheim et al. 2008).

In addition to our own long-term studies in British Columbia and Washington State, numerous other researchers have undertaken fieldwork on various aspects of the life history, ecology, and behavior of killer whales, both in our study area and in adjacent coastal waters. Over the years, these efforts have together provided a much improved understanding of the divergent ecological specializations of residents and transients and the role these have played in defining the lifestyles of these lineages.

4.3 Population Delineation of Lineages

Resident, transient, and offshore killer whale lineages are sympatric in coastal waters of the eastern North Pacific from California to the Aleutian Islands in Alaska. Molecular studies have confirmed what earlier observations suggested—that the three lineages are genetically distinct and gene flow between them is minimal or absent (Stevens et al. 1989; Hoelzel et al. 1998; Barrett-Lennard 2000; Morin et al. 2010). At least two of these lineages—residents and transients—are represented by multiple discrete populations of typically a few hundred individuals. Four populations have been described for residents (Matkin et al. 1999; Ford et al. 2000; Matkin et al. 2007a). Each population ranges over roughly 1,300- to

1,800-km sections of coastline that overlap substantially. Despite overlapping distribution, each population generally occupies rather discrete areas, especially during summer and fall. Groups of resident whales from adjacent populations have been observed in close proximity on a few occasions, but no intermingling has taken place. However, groups belonging to the same population frequently join and travel together, occasionally forming large multigroup aggregations that may persist for several days. DNA fingerprinting indicates that mating takes place between groups within each resident population, and intermating between populations is extremely rare (Barrett-Lennard 2000).

Transient killer whales are subdivided into at least five regional populations, each typically composed of 100–300 individuals (Bigg et al. 1987; Black et al. 1997; Ford and Ellis 1999; Matkin et al. 1999, 2007a). Groups of transients within each population regularly intermingle and, in contrast to residents, they will also associate with members of adjacent transient populations during the infrequent occasions when they roam into the range of another population (Ford and Ellis 1999; Ford et al. 2007). Offshore killer whales appear to consist of a single population of at least 250 animals that ranges widely over the continental shelf, from southern California to the eastern Aleutian Islands, Alaska (Ford et al. 2000; Matkin et al. 2007a; Dahlheim et al. 2008). The extent of potential movements beyond the continental shelf for any of these whales is unknown because of the lack of field effort in offshore waters.

4.4 Dietary Specialization

Gaining insight into the feeding habits of free-ranging cetaceans is difficult because predation usually takes place underwater and out of sight. We have studied the diets of resident, transient, and offshore killer whales using three different methods: (1) direct observation of predation when it takes place at the surface, (2) collection of prey fragments left in the water column following a kill, and (3) recovery of prey remains from the stomachs of beach-cast carcasses. Others have also used chemical analyses of skin and blubber biopsy samples collected from killer whales to infer diet from stable isotope ratios, fatty acids, and levels of various types of contaminants (e.g., Krahn et al. 2007).

Surface observations and identification of prey fragments from kills indicate that the diet of resident killer whales in British Columbia (Fig. 4.1) consists primarily of teleost fishes, in particular the Pacific salmonids (*Oncorhynchus* spp.) (Fig. 4.2: Ford et al. 1998; Saulitis et al. 2000; Ford and Ellis 2006). Non-salmonid fishes such as lingcod (*Ophiodon elongatus*), Dover sole (*Microstomus pacificus*), and Pacific halibut (*Hippoglossus stenolepis*) have also been identified from predation events, but these represent less than 3 % of observed kills. A surprising result of our prey fragment sampling has been the pronounced preference that residents have for Chinook salmon (*Oncorhynchus tshawytscha*). In total, more than 70 % of identified salmonid kills have been Chinook, despite this species being one of the least

Fig. 4.1 A male resident killer whale surfaces following capture of a Chinook salmon, the primary prey species of this ecotype. (Photograph by M. Malleson)

common of the five salmonid species available in the whales' habitat (Ford and Ellis 2006). Chinook predominated in our samples even when other salmonids, such as sockeye (*O. nerka*) and pink (*O. gorbuscha*) salmon, were far more abundant in foraging areas during summer spawning migrations, outnumbering Chinook by as many as 500 fish to 1 (Ford et al. 1998; Ford and Ellis 2006). Chum salmon (*O. keta*) are significant prey during a short period in the fall, but Chinook still appear to be taken preferentially. Prey remains recovered from beach-cast carcasses of residents are generally consistent with our observations of predation. Chinook salmon has been identified in most stomach contents to date, and various non-salmonids and squid have also been represented occasionally (Ford et al. 1998).

It is most probable that the whales' preference for different salmonids—and other prey species for that matter—is proportional to their relative profitability. Chinook are by far the largest of the Pacific salmon, commonly reaching sizes of more than 20 kg, and they tend to have the highest lipid content of the salmonids, enhancing their net energy density. Chum salmon are the second largest salmonid and can reach 10 kg or more. The much smaller sockeye and pink salmon seem to be of little interest to the whales, despite their brief but often great abundance during summer.

In striking contrast to resident killer whales, transient killer whales (Fig. 4.3) have only been observed to hunt and consume endothermic prey, primarily marine mammals and occasionally seabirds. In British Columbia, Washington State, and Southeast Alaska, the most frequent prey species by far (about 50 % of kills) is the harbour seal (*Phoca vitulina*), a small (average, 60–80 kg) pinniped that is common throughout nearshore waters of the region (Fig. 4.2) (Ford et al. 1998; Matkin et al.

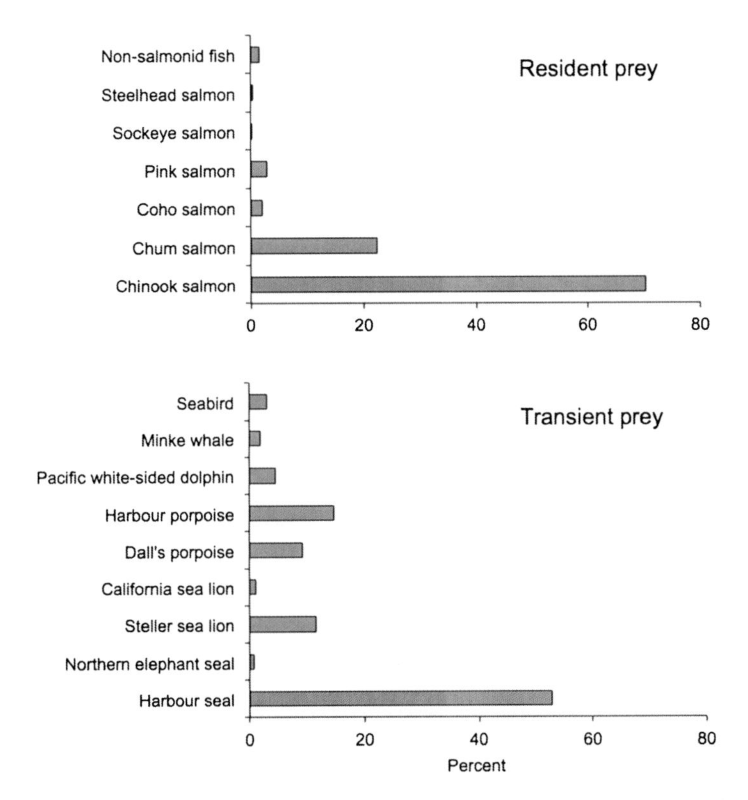

Fig. 4.2 Frequency distribution of prey species observed to be consumed by resident (*top*, $n=439$ kills) and transient (*bottom*, $n=251$ kills) killer whales in coastal waters of British Columbia, Washington State, and southeastern Alaska. [Data from Ford et al. (1998), Ford and Ellis (2006), and Ford and Ellis (unpublished data)]

2007a, b). Harbour porpoise (*Phocoena phocoena*) and Dall's porpoise (*Phocoenoides dalli*) together make up about one-quarter of observed kills, with the remainder composed of Steller sea lions (*Eumetopias jubatus*), California sea lions (*Zalophus californianus*), Pacific white-sided dolphins (*Lagenorhynchus obliquidens*), minke whales (*Balaenoptera acutorostrata*), northern elephant seals (*Mirounga angustirostris*), and various seabird species (Ford et al. 1998, 2005). Swimming deer (*Odocoileus hemionus*) and moose (*Alces alces*) have on rare occasions been reported to be killed by killer whales in the region, almost certainly transients (Pike and MacAskie 1969; Matkin et al. 1999). Seabirds do not seem to be an important prey item of transient killer whales. Only a minority of seabirds that are harassed and killed by transients are ultimately consumed: most are abandoned. Interaction with seabirds usually involves juvenile whales and may represent play behavior that ultimately functions to develop prey handling skills (Ford et al. 1998; Saulitis et al. 2000). Transients have not been observed to take any fish species, nor have any fish remains been identified in stomach contents of beach-cast carcasses of transients (Ford et al. 1998; Saulitis et al. 2000; Heise et al. 2003).

Fig. 4.3 A female transient killer whale hunting for the preferred prey of this ecotype, harbour seals. (Photograph by J. Towers)

There is little evidence that transient individuals or matrilines specialize on particular types or species of marine mammals, despite the very different tactics needed to capture and kill them (harbour seals versus Dall's porpoise, for example; Ford et al. 1998). Our long-term monitoring of transient predation has shown that the variety of prey species taken by particular individuals or groups is strongly correlated with the cumulative number of predation events documented for those animals (Ford et al. 1998). Predation of minke whales by transients in our study area is uncommon, but a particular matriline (the T18 group) has been involved in more cases than one would expect by chance (Ford et al. 2005; J.K.B.F. and G.M.E., unpublished data). This matriline also hunts more typical prey, such as harbour seals and porpoises. There are no records of transients in our study area having successfully killed large whales such as adult gray (*Eschrichtius robustus*), humpback (*Megaptera novaeangliae*), fin (*Balaenoptera physalus*), or blue (*Balaenoptera musculus*) whales. Indeed, foraging transients rarely show any reaction to these potential prey species despite their frequent presence in their vicinity (Jefferson et al. 1991). This indifference is likely related to the difficulty in catching the fast-swimming fin and blue whales and the risk of injury posed by defensive responses from gray and, especially, humpback whales (Ford and Reeves 2008). Gray whale calves and juveniles, however, are frequently targeted by foraging transients in central California (Ternullo and Black 2002) and around the eastern Aleutian Islands, Alaska (Barrett-Lennard et al. 2005; Matkin et al. 2007).

Offshore killer whales are the least known of the three lineages in the region. They have been observed consuming a probable Pacific halibut (Jones 2006) and possibly blue sharks (*Prionace glauca*) and Chinook salmon (Dahlheim et al. 2008).

Stomach contents of a killer whale identified as an offshore by mtDNA analysis included two carcharinid sharks and two opah (*Lampris regius*, a large pelagic teleost fish; Morin et al. 2006). Recently, we observed offshore killer whales feeding on multiple Pacific sleeper sharks (*Somniosus pacificus*) (Ford et al. 2011). A diet consisting largely of sharks, with their abrasive skin, might explain the extreme tooth wear that appears to be common in offshore killer whales (Ford et al. 2011). Stable isotope ratios and fatty acid profiles determined from skin and blubber biopsy samples also suggest that the diet of offshore killer whales is distinct from that of either resident or transient lineages (Herman et al. 2005; Krahn et al. 2007).

4.5 Social Organization

Similar to most delphinids, killer whales are highly social, group-living animals. However, the social structure of resident, transient, and offshore killer whales differs considerably, and these differences appear to be related to and are likely determined by their respective ecological specializations. Resident killer whales live in matrilines that are exceptionally stable in composition. A typical matriline is composed of an older female, her sons and daughters, and the offspring of her daughters. Because longevity of females can reach 80 years and females have their first viable calf at about 14 years (Olesiuk et al. 2005), a matriline may contain as many as four generations of maternally related individuals. More than 30 years of demographic data have demonstrated that dispersal from the matriline is virtually absent in resident killer whales—both males and females remain in their natal group for life (Bigg et al. 1990; Ford et al. 2000; Ellis et al. 2007). In no case has an individual whale been observed to leave its matriline and join another on a long-term basis, other than in a few rare cases involving orphans.

Members of resident matrilines travel together and they seldom separate by more than a few kilometers or for more than a few hours. Contact is maintained among matriline members by the exchange of discrete, stereotyped underwater calls that are unique to the group (Ford 1989, 1991; Miller et al. 2004). Matrilines frequently travel in the company of certain other matrilines that are closely related, based on high degrees of call similarity, and likely shared a common maternal ancestor in the recent past. Matrilines that spend the majority of their time together are designated as pods (Bigg et al. 1990). Pods are less stable than matrilines, and member matrilines may spend days or weeks apart. However, matrilines still spend more time with others from their pod than with those from other pods. In British Columbia, resident pods are on average composed of three matrilines (range = 1–11; Ford et al. 2000), with a mean total size of 18 whales (range = 2–49; Ford et al. 2000). Residents often form large temporary aggregations involving multiple matrilines and pods, especially at times when prey densities are high.

A level of social structure above the resident pod is the clan, which is defined by patterns of call similarity. Clans are composed of pods that share a portion of their repertoire of stereotyped calls. Different clans have no calls in common. Pods belonging to a clan are likely descendants of an ancestral pod, and their acoustic similarities reflect this common heritage. Call repertoires are traditions passed on across generations by vocal learning, and calls actively or passively change in structure or use over time. Calls are retained within the lineage because of the lack of dispersal from matrilines. Clans are sympatric, and the two to nine pods that make up each clan frequently travel together as well as with pods from different clans (Ford 1991; Yurk et al. 2002).

Transient killer whale society lacks the closed, strictly matrilineal structure seen in residents. Transients usually travel in groups of two to six individuals, much smaller than the typical size of resident matrilines and pods. In contrast to residents, offspring often disperse from the natal matriline for extended periods or permanently (Bigg et al. 1987; Ford and Ellis 1999; Baird and Whitehead 2000). Female offspring usually leave their natal group around the time of sexual maturity and travel with other transient groups. These young females usually give birth to their first calf shortly after dispersing. Once dispersed, these females may rejoin their natal matriline occasionally, but generally only for brief periods after they have calves of their own. Male dispersal does take place, but the pattern is less clear because of uncertainty in the status of many individuals in the population. The range of transients appears to extend beyond our study area, possibly into offshore waters, and gaps of many years can occur between sightings of individuals (Ford and Ellis 1999; Ford et al. 2007). There are numerous cases of mothers and a single adult son staying together for decades, but few where a mother and more than one adult son have persisted. Male siblings may disperse from these groups at puberty, but if so they must leave our study area as none has been resighted after disappearing from the natal group, either as a member of another group or as a lone individual. All lone adult males found in the study area appear to have lost their mothers through mortality. These individuals often travel alone or associate with a variety of different transient matrilines, but rarely with other lone males. The associations of transient matrilines are very dynamic, and they do not form consistent groupings equivalent to resident pods. Also, in contrast to residents, transient populations do not seem to be acoustically subdivided into clans. Instead, all transients in a population share a distinctive set of calls, although some additional calls or variants of shared calls may be specific to a subregion or portion of the population (Ford 1984; Deecke et al. 2005).

The typically small size of transient groups is likely a result of the foraging strategy of this lineage. Transients generally hunt other marine mammals with stealth: they swim quietly to prevent detection by their acoustically sensitive prey, and attack using the element of surprise (Ford 1984; Barrett-Lennard et al. 1996). This strategy no doubt constrains group size, as larger groups such as those of residents would increase the probability of the predators being detected by their prey. Small groups may also be most energetically efficient for transients when hunting smaller marine mammals such as harbour seals (Baird and Dill 1996).

As with most details of their life history and behavior, the social organization of offshore killer whales is poorly understood. Their group sizes tend to be relatively large, certainly much larger than those of transients, and possibly larger on average than residents. Groups of 2 to 100 or more individuals have been documented in encounters with offshores off the coast of British Columbia, with about half involving 20 or more individuals. These larger groups probably represent temporary gatherings of smaller social units, possibly related to prey density as in residents. We have documented persistent bonds lasting more than a decade between females and adult males, which likely represent mothers and their adult sons. However, we have not observed long-term associations between reproductive females, as seen in the multi-generation matrilines of residents. This finding suggests a dynamic society with dispersal from the natal matriline as in transients, but frequent formation of larger aggregations as in residents.

4.6 Seasonality and Habitat Use

All three lineages of killer whales are found in coastal waters of the northeastern Pacific throughout the year, but there are significant differences in their seasonality and patterns of habitat use. The seasonal movements of resident killer whales are closely tied to those of their primary prey. Several studies have demonstrated correlations between resident whale occurrence in nearshore waters and the aggregate abundance of multiple salmon species migrating through nearshore waters to coastal spawning rivers in British Columbia and Washington State (Heimlich-Boran 1986; Guinet 1990; Nichol and Shackleton 1996). However, these analyses were undertaken before it was known that these whales forage selectively for Chinook salmon and shun the smaller but much more abundant pink and sockeye salmon (Ford et al. 1998; Ford and Ellis 2006). Correlations of whale occurrence with these abundant salmonids are thus incidental, and the whales are instead attracted by migrating Chinook salmon, which pass through these migratory corridors in lower numbers but concurrently with the smaller species. Movements of resident killer whales in this area during October and November are clearly associated with fall migrating chum salmon, which the whales do consume (Nichol and Shackleton 1996; Ford and Ellis 2006). Interestingly, a different population of resident killer whales in south-central Alaska moves into Prince William Sound during midsummer, where they forage extensively for coho salmon (*Oncorhynchus kisutch*) (Saulitis et al. 2000). Neither Chinook nor chum salmon are common in this area at this time of year, although these same whales feed on Chinook and chum salmon in other areas and times of year (C. Matkin, personal communication).

During winter and spring, resident whales mostly vacate their summer habitat in nearshore waters and appear to range widely along the outer exposed coast. It is likely that the whales maintain their focus on Chinook salmon prey during this time of year. Most other salmonid species are pelagic and unavailable to the whales during this time of year, but nonmigratory or early spawning runs of Chinook are found in these outer coast waters (Ford and Ellis 2006). Residents may also increase their consumption of non-salmonid species such as Pacific halibut during winter and spring.

Compared to residents, transient killer whales have a relatively uniform pattern of occurrence in nearshore waters throughout the year, likely because their primary prey species—harbour seal, harbour porpoise, Dall's porpoise, and Steller sea lion—are nonmigratory and available in all months of the year. However, there is an interesting seasonal peak in local occurrence along the west coast of North America that appears to coincide with the pupping season of harbour seals. In Glacier Bay, the northern limit of the range of the so-called "West Coast" transient population (~58°30′N latitude), transient whale occurrence peaks in June and July (Matkin et al. 2007). Near the southern extent of their range, around Vancouver Island (~48° to 51°N latitude), there is an obvious peak in occurrence during August and September (Baird and Dill 1995). Both these periods coincide with local peaks of pupping and weaning of harbour seals, which exhibits a latitudinal cline in timing along the West Coast (Temte et al. 1991). Pups are likely easy and abundant prey for transients, and the whales appear to move in accordance with their seasonal availability.

Offshore killer whales appear to exhibit a diffuse seasonal shift in distribution along the West Coast of North America. The majority of sightings in the southern portion of their known range, off central and south California, have been recorded during fall and winter (September to March; Dahlheim et al. 2008). Sightings in Alaska, the northern portion of the range of offshore killer whales, have taken place only during April to September, but there is minimal observer effort during winter in this area (Dahlheim et al. 2008; C. Matkin, personal communication). Off British Columbia, roughly the latitudinal midpoint of their range, sightings of offshores have been recorded in all months. Without a better understanding of the primary prey species of offshore killer whales, it is not possible to interpret the significance of this apparent seasonal distribution shift.

Differences are also apparent in finer-scale patterns of habitat use by the three killer whale lineages. Residents congregate during summer and fall in core feeding areas in locations where geography and tidal currents act to concentrate migratory salmon (Heimlich-Boran 1988; Nichol and Shackleton 1996; Saulitis et al. 2000; Ford 2006). During the peak of salmon abundance, the majority of matrilines in a resident population may gather in these core areas, and individual matrilines or pods may spend weeks in a relatively restricted area that the whales could transit in a day or two. There are distinctions among the movement patterns of different resident pods within a population's overall range. Although most resident groups may be observed in most parts of the range, particular pods and matrilines have preferred areas that they frequent more often than other groups (Osborne 1999; Ford 1991, 2006; Ford et al. 2000; Hauser et al. 2007), likely because of the benefit of foraging in familiar areas where individuals have experience in locating local concentrations of prey.

While in their core summer feeding areas, resident killer whales spend 50–65 % of their time foraging (Heimlich-Boran 1988; Ford 1989; Morton 1990). Between foraging bouts, the whales group together and socialize or rest, which together represent about 30–40 % of their time. In at least two resident populations, the whales may also spend considerable time rubbing their bodies on certain shelving, pebble beaches that have been used traditionally for many years (Ford 1989; Matkin et al. 1999).

In contrast to resident whales, transient killer whales typically do not remain for long in any particular location. They are almost constantly on the move, swimming from one prey hotspot to the next. Because of their apparent reliance on stealth for capturing marine mammals, it is no doubt more productive for transients to hunt elsewhere once potential prey is alert to their presence. By covering 75–150 km of coastline per day, transients tend to undergo more frequent extensive travel throughout their range than do residents. Nonetheless, as with residents, at least some transient groups have preferred areas within the overall population range, where local knowledge of the location of pinniped haulouts or predictable concentrations of small cetaceans may serve to improve hunting efficiency (Ford and Ellis 1999). Compared to residents, transients dedicate considerably more time to foraging and traveling (>75 % of their activity budget: Morton 1990; Baird and Dill 1995). Socializing and resting activities, which comprise about one-third of the activity budget for residents, are seldom exhibited by transients (<10 % of activities; Morton 1990; Baird and Dill 1995; Barrett-Lennard et al. 1996; Deecke et al. 2005). Beach rubbing has not been reported for transients.

Details of habitat use by offshore killer whales are not yet clear because of the comparatively infrequent encounters with this population. Long-distance movements appear to be undertaken frequently by offshore whales. Several identified individuals have been observed at the extremities of the population's known range, which extends more than 4,000 km from the Aleutian Islands to Southern California (Dahlheim et al. 2008). Any potential habitats that may be used preferentially by a subset of the offshore population, and what prey species may drive their movements, have yet to be described.

4.7 Foraging Behavior

The distinct diets of killer whale lineages are associated with corresponding contrasts in their foraging behavior. When foraging, members of a resident killer whale matriline or pod spread out, often over areas of several square kilometers, with individuals or small subgroups diving and surfacing independently while swimming generally in the same direction. They maintain contact and likely coordinate movements through the frequent exchange of loud underwater calls, which are effective to ranges of 10–25 km (Ford 1989; Miller et al. 2004; Miller 2006). When foraging in coastal inlets, channels, and straits, individuals and small maternal groups usually forage along the shoreline, while other whales, particularly mature males, forage alone farther from shore and in deeper water. Foraging resident whales dive for 2–3 min (Ford 1989; Morton 1990) to depths typically less than 30 m, but occasionally to more than 150 m (Baird et al. 2005a). These depths are similar to those used by their primary prey species, Chinook salmon (Candy and Quinn 1999).

Foraging resident killer whales find prey using echolocation, which may be effective for detecting Chinook salmon at ranges of 100 m or more (Au et al. 2004). By foraging in loosely dispersed groups, the detection rate of scattered salmon is likely enhanced. However, residents whales do not appear to cooperatively herd or capture prey. Rather, prey capture is undertaken primarily by individuals with occasional cooperation from offspring, siblings, or other close matrilineal kin. The majority of salmonid prey items captured by adult females and subadults are brought to the surface, where they are broken up for sharing within the matriline or for provisioning young offspring (Ford and Ellis 2006). Adult males usually capture and consume salmonid prey alone.

In contrast to residents, transient killer whales forage in near silence in an apparent attempt to minimize detection by their acoustically sensitive marine mammal prey (Ford 1984; Morton 1990; Barrett-Lennard et al. 1996; Deecke et al. 2005). Transients rarely exchange underwater calls while hunting for prey (Deecke et al. 2005), and echolocation click production is also greatly suppressed (Barrett-Lennard et al. 1996). Both pinniped and cetacean prey have excellent hearing abilities at the frequencies used by killer whales for calling and echolocation and could detect and potentially evade approaching transients if they were to vocalize (Barrett-Lennard et al. 1996; Deecke et al. 2005). As vocalizing would likely incur high costs in terms of reduced rates of prey capture, transients appear to depend on passive listening to detect and approach prey from a distance, likely cueing on the animals' vocalizations or swimming noises (Barrett-Lennard and Heise 2006). There is little cost associated with the production of underwater sounds for resident whales because salmonids and most other fish have relatively low hearing sensitivity to such frequencies and are unlikely to detect approaching whales at a distance (Barrett-Lennard et al. 1996; Deecke et al. 2005).

Transient killer whales employ two fairly distinct modes of foraging: nearshore and open water. When foraging nearshore, the whales swim in relatively tight groups and follow the contour of the shoreline, round headlands, and enter bays without hesitation (Morton 1990; Barrett-Lennard et al. 1996). They often circle small islets and reefs, particularly those that serve as pinniped haulouts. Resident whales, in contrast, forage along more direct routes, usually swimming from headland to headland. Dive durations of foraging transient whales are typically twice the duration of the 2- to 3-min dives of residents, and may exceed 10 min (Morton 1990). Nearshore foraging is generally associated with capture of pinniped prey, particularly harbour seals (Baird and Dill 1995; Barrett-Lennard et al. 1996; Saulitis et al. 2000). When foraging in open water, transient groups spread out over a larger area, with individuals swimming several hundred meters apart, often roughly abreast. Most prey captured during open water foraging are porpoises or dolphins, but seals or sea lions may also be taken (Barrett-Lennard et al. 1996; Saulitis et al. 2000).

Transients share the majority of their prey (Baird and Dill 1995), likely to an even greater extent than do residents because of the larger body masses of most marine mammal prey items. Transient group members frequently use cooperative hunting tactics to catch and subdue their prey (Baird and Dill 1995; Ford et al. 1998). Predation on Steller sea lions, for example, can be extended events that may entail risk of injury to the attacking whales. These prey can be large (up to 1,000 kg in males) with sizeable canine teeth that can inflict significant wounds during defensive or retaliatory actions. Groups of transient killer whales attack single sea lions in open water by circling the animal so as to prevent it from reaching shore, while individuals take turns rushing toward the prey and ramming it or striking it with their tail flukes. This action may continue for 1–2 h until the animal is sufficiently debilitated so that it can be safely grasped, drowned, and shared among group members. Transients may also hunt fast-swimming Dall's porpoise using a cooperative "tag team" tactic where individuals take turns chasing the prey animal to exhaustion. Transients have been also been observed to herd groups of 50+ Pacific white-sided dolphins into confined or shallow bays where individuals can be readily captured. Transients hunt these difficult-to-capture species in significantly larger groups than when foraging for the smaller harbour seals (Ford et al. 1998). These groups often represent temporary associations of smaller, stable social units.

4.8 Acoustic Communication

As do most delphinids, killer whales have a well-developed acoustic communication system. However, as noted earlier, the types and extent of vocalization show major differences among lineages. Resident killer whales frequently exchange strident calls from stable repertoires of a dozen or more call types. These learned call types or their variants are specific to clans, pods, and matrilines, and thus encode the matrilineal genealogy of individuals (Ford 1991). This specificity likely enhances the effectiveness of these calls as intragroup contact signals, especially when whales are dispersed and traveling in association with other matrilines or pods. These group-specific dialects may also play a role as a behavioral mechanism to prevent inbreeding. As there is no dispersal from the natal matriline, resident killer whales would be at considerable risk of inbreeding without a reliable means of distinguishing between kin and non-kin mating partners. Group-specific call repertoires appear to serve such a function (Ford 1991), and genetic studies have shown that resident whales mate with individuals that are outside the pod or clan and are acoustically dissimilar (Barrett-Lennard 2000).

Although transient killer whales spend much of their time foraging for marine mammals in silence, they become highly vocal while attacking and consuming their prey (Ford 1984; Deecke et al. 2005). Calling at such times likely carries little cost as stealth is no longer needed, and it may help coordinate cooperative attack tactics within the group or serve other social functions after the kill is made. Similar to resident killer whales, transients have repertoires of distinctive stereotyped call types. Unlike residents, however, these repertoires generally do not differ among groups. As there is dispersal from the natal matriline in this ecotype, group-specific calls would not be expected. Also, dispersal reduces the risk of inbreeding, so the requirement for an acoustic outbreeding mechanism may be reduced in transients.

The fish-eating offshore killer whales are as vocal as resident killer whales. Preliminary analyses indicate that offshores produce stereotyped calls that are distinct from any of those of residents or transients, but it is not yet known whether any calls are specific to particular groups. As our understanding of the social dynamics of this poorly known lineage improves, patterns of call use should become clearer.

4.9 Specializations in Other Regions

Field studies in other global regions have provided additional evidence that ecological specializations are typical of most killer whale populations. Although these populations are not as well known as resident and transients in the eastern North Pacific, it is apparent that at least in some cases their specializations have had similar influences on patterns of social structure, behavior, and vocal activity. Off the northern coast of Norway, a population of killer whales moves seasonally in relation to their primary prey, the Atlantic herring (*Clupea harengus*) (Similä 1997). In coastal fjords where herring congregate in high densities during fall and winter, the whales employ a cooperative foraging tactic known as "carousel feeding" to capture these small schooling fishes: this involves a group of whales encircling and herding a school of herring into a tight ball close to the surface. Once the school is concentrated, individuals dive under the school and strike it with their tail flukes. Fish stunned directly by the physical blow from the flukes or the associated loud cavitation sound are then eaten individually (Similä and Ugarte 1993; Simon et al. 2005). These herring-eating killer whales appear to live in matrilineally organized pods similar in size to those of fish-feeding resident killer whales, but it is not known whether they share the same extreme stability (Similä 1997). They are highly vocal and have pod-specific call repertoires as observed in resident killer whales (Strager 1995), which would suggest a stable pod structure.

In the Strait of Gibraltar, a small population of killer whales appears to specialize on predation of bluefin tuna (*Thunnus thynnus*) as the fishes enter and exit the Mediterranean Sea during their breeding migration (Reeves and Notarbartolo di Sciara 2006). To catch these swift tuna, the whales employ an endurance-exhaustion technique involving protracted chases at swimming speeds of 12–14 km/h for periods of 30–40 min (Guinet et al. 2007). Killer whales can sustain sufficient swimming speeds necessary to catch small to medium (0.8–1.5 m) tuna using this technique but appear unable to match the swimming ability of larger fish.

On the coast of Patagonia, Argentina, a small population of killer whales uses a novel, but risky, hunting technique that involves intentional stranding in the shallows to capture young southern sea lions (*Otaria flavescens*) and southern elephant seals (*Mirounga leonina*) at the water's edge (Lopez and Lopez 1985). Whales hunt cooperatively and share their prey with others in the group (Hoelzel 1991). A similar beaching tactic is used by killer whales in the sub-Antarctic Crozet Islands when hunting southern elephant seal pups (Guinet 1992). As do mammal-hunting transients in the northeastern Pacific, whales in both these Southern Hemisphere locations have small group sizes, hunt mostly in silence, and appear to locate prey by passive listening (Guinet 1992; J.K.B.F., unpublished data).

Three distinct forms of killer whales—known as types A, B, and C—have been described in circumpolar waters of the Antarctic (Pitman and Ensor 2003). These sympatric forms differ in pigmentation patterns, genetic structure (mtDNA sequences), patterns of habitat use, and diet (Pitman and Ensor 2003; Krahn et al. 2008; LeDuc et al. 2008). Type A killer whales are found mostly in ice-free waters where they apparently feed mainly on cetaceans, particularly Antarctic minke whales (*Balaenoptera bonaerensis*). Type B whales forage primarily in loose pack ice and appear to specialize on seals. These whales exhibit a novel hunting tactic in which group members coordinate their swimming movements to create a large wave that washes seals off ice floes (Visser et al. 2008). Type C whales inhabit dense pack ice and appear to be fish feeders, having been observed preying on Antarctic toothfish (*Dissostichus mawsoni*) (Pitman and Ensor 2003; Krahn et al. 2008). They are substantially smaller than other Antarctic killer whales, with adults approximately 1–3 m shorter in length than type A individuals (Pitman et al. 2007). Type C whales tend to have larger group sizes than mammal-hunting types A and B, which is consistent with the pattern of group sizes versus prey type in the northeastern Pacific. Unfortunately, too little is known about these Antarctic ecotypes to determine whether ecological specialization has influenced their social structure, behavior, and acoustics in ways similar to those of lineages in other regions.

4.10 Conclusions

The killer whale is a highly versatile social predator that has evolved to successfully occupy a variety of specialized ecological niches in the world's oceans. In so doing, this species has assumed a variety of distinct lifestyles that have been shaped by these ecological specializations. In the eastern North Pacific, the three killer whale lineages have distinct patterns of seasonal distribution, group size, social organization, foraging behavior, and acoustic activity, which can be related to their preferred type of prey and the strategies the animals use to acquire it. Some similar patterns are apparent among killer whales in other regions, although a lack of field data prevents a more complete assessment of the parallels between ecotype and life history or behavior for these populations.

Although different killer whale lineages may be genetically distinct, there is no evidence that dietary preferences result from any genetic predisposition. Globally, there is no congruence between killer whale ecotype and genotype (Hoelzel et al. 2002; LeDuc et al. 2008). Instead, ecological specializations appear to represent behavioral traditions that likely evolved independently in different regions. It is plausible that ecological divergence could arise in sympatry with, for example, the innovation of a novel foraging tactic in a particular matriline that allowed predation on a new type of prey. If this matriline and its descendants became further specialized on this prey type, rates of association with other groups that do not adopt this new diet may diminish over time, leading to social segregation and reproductive isolation. Such a process could lead ultimately to speciation. Resident and transient killer whale lineages in the northeastern Pacific have been suggested to represent incipient species (Baird et al. 1992) and Antarctic type A, B, and C killer whales to represent distinct species (LeDuc et al. 2008).

There are still many questions concerning ecological specialization in killer whales that remain to be answered. For example, to what extent might specializations constrain a lineage's ability to switch to alternative prey species in a changing environment? The preferences for fish and marine mammal prey exhibited by resident and transient killer whales, respectively, are extremely strong, and there is no evidence that one ecotype ever switches to the prey type of the other or has the behavioral flexibility to do so. Marine mammals in coastal waters of the northeastern Pacific can discriminate between lineages and will flee from transients but show indifference to residents (Ford and Ellis 1999; Deecke et al. 2002), suggesting that if residents ever hunt marine mammals, it must occur extremely rarely. The suite of specialized behaviors that make resident killer whales adept at locating and catching Chinook salmon likely would be ineffective for hunting marine mammals. Transients would similarly be ill equipped to adopt a fish-feeding lifestyle.

The extent of dietary flexibility of killer whales has implications for their potential role in driving marine ecosystem dynamics. It has been proposed that a shift to sea otter predation by mammal-hunting killer whales in the Aleutian Islands resulted in a precipitous decline in sea otter abundance that started in the mid-1980s (Estes et al. 1998). This shift is thought to be a response to reduced availability of the whale's presumed primary prey in the region, harbour seals and Steller sea lions. In an extension of this hypothesis, Springer et al. (2003) postulated that the decline of sea otters was the last in a series of population collapses of prey species of mammal-hunting killer whales in the northern Gulf of Alaska that was triggered by the decimation of the great whales by industrial whaling in the nineteenth and twentieth centuries. This hypothesis has been challenged on various grounds (Trites et al. 2007; Wade et al. 2007), particularly because there is no evidence that the great whales (especially adults) have ever played an important role in the diet of killer whales (Mizroch and Rice 2006; Ford and Reeves 2008). Although it may be possible that predation by killer whales could result in depletion of targeted prey species, dietary specializations could have significant constraints on the directions that subsequent prey shifts may take.

To date, most ecologically specialized killer whale populations, including sympatric fish-eating and mammal-eating ecotypes, have been described in highly productive cold temperate or polar waters, likely the result of the diversity of abundant prey types available in these high latitudes, which has provided the opportunity for niche partitioning. It may well be that killer whales in less productive tropical or subtropical waters are generalist predators that include a greater variety of prey in their diets (Baird et al. 2005b). For example, a high incidence of killer whale teeth scars on humpback whales using breeding grounds off the west coast of Mexico suggests that predation in this area, especially on calves, may be extensive (Steiger et al. 2008). This prey resource is seasonal, however, as humpbacks only occupy these breeding grounds for 3 to 5 months in winter. Because there is no evidence that killer whales follow migrating humpback whale mothers and calves to their high-latitude feeding grounds, it is likely that the predators shift to alternative prey species for the remainder of the year.

Globally, killer whales form a mosaic of distinct populations, some overlapping and others geographically discrete, that are ecologically specialized to greater or lesser degrees. Each population is likely to have foraging tactics, activity patterns, social organization, and acoustic behavior that have been shaped by its dietary specialty. Highly specialized populations can be expected to have lifestyles that are closely adapted to their foraging strategy, whereas more generalist populations may be relatively less constrained by any particular prey type. In certain regions, such as the northeastern Pacific, some parts of this mosaic are becoming fairly clear. In other regions, such as the Antarctic, a fascinating picture is emerging but significant knowledge gaps remain to be filled. In regions where killer whales are little studied, such as in sparsely inhabited tropical waters, there is much yet to be discovered. Only when all the components of this global mosaic of killer whale populations have been described will we have a complete appreciation of the range of ecological specializations and lifestyles of this multifaceted and resourceful predator.

Acknowledgments We thank R. Baird, C. Chapman, V. Iriarte, C. Matkin, and J. Watson for helpful comments on earlier drafts of this chapter, and M. Malleson and J. Towers for kindly allowing us to use their photographs.

References

Au WWL, Ford JKB, Horne JK, Newman Allman KA (2004) Echolocation signals of free-ranging killer whales (*Orcinus orca*) and modeling of foraging for Chinook salmon (*Oncorhynchus tshawytscha*). Acoust Soc Am 115(2):901–909

Baird RW, Abrams PA, Dill LM (1992) Possible indirect interactions between transient and resident killer whales: implications for the evolution of foraging specializations in the genus Orcinus. Oecologia, 89:125–132

Baird RW, Dill LM (1995) Occurrence and behavior of transient killer whales: seasonal and pod-specific variability, foraging, and prey handling. Can J Zool 73:1300–1311

Baird RW, Dill LM (1996) Ecological and social determinants of group size in transient killer whales. Behav Ecol 7:408–416

Baird RW, Hanson MB, Dill LM (2005a) Factors influencing the diving behaviour of fish-eating killer whales: sex differences and diel and interannual variation in diving rates. Can J Zool 83:257–267

Baird RW, McSweeney DJ, Bane C, Barlow J, Salden DR, Antoine LK, LeDuc RG, Webster DL (2005b) Killer whales in Hawaiian waters: information on population identity and feeding habits. Pac Sci 60:523–530

Baird RW, Whitehead H (2000) Social organisation of mammal eating killer whales: group stability and dispersal patterns. Can J Zool 78:2096–2105

Barrett-Lennard LG (2000) Population structure and mating systems of northeastern pacific killer whales. Ph.D. Thesis, University of British Columbia, Vancouver

Barrett-Lennard LG, Ford JKB, Heise KA (1996) The mixed blessing of echolocation: differences in sonar use by fish-eating and mammal-eating killer whales. Anim Behav 51:553–565

Barrett-Lennard LG, Heise K (2006) The natural history and ecology of killer whales. In: Estes JA, DeMaster DP, Doak DF, Williams TM, Brownell RL (eds) Whales, whaling and ocean ecosystems. University of California Press, Berkeley, pp 163–173

Barrett-Lennard LG, Matkin CO, Ellifrit DK, Durban J, Mazzuca L (2005) Black and white versus gray: estimating kill rates, consumption rates, and population-level impacts of transient killer whales feeding on gray whales. Abstract, 16th Biennial Conference on the Biology of Marine Mammals, San Diego, 12–16 December 2005, p 26

Bigg MA (1982) An assessment of killer whale (*Orcinus orca*) stocks off Vancouver Island, British Columbia. Rep Int Whal Comm 32:655–666

Bigg MA, Ellis GM, Ford JKB, Balcomb KC (1987) Killer whales: a study of their identification, genealogy and natural history in British Columbia and Washington State. Phantom Press, Nanaimo

Bigg MA, Ford JKB, Ellis GM (1985) Two sympatric forms of killer whales off British Columbia and Washington. In: Sixth Biennial Conference on the Biology of Marine Mammals, Vancouver, Canada, November 1985

Bigg MA, MacAskie IB, Ellis G (1976) Abundance and movements of killer whales off eastern and southern Vancouver Island with comments on management. Unpublished report, Arctic Biological Station, Department of Fisheries and Environment, Ste Anne-de-Bellevue

Bigg MA, Olesiuk PF, Ellis GM, Ford JKB, Balcomb KC III (1990) Social organization and genealogy of resident killer whales (*Orcinus orca*) in the coastal waters of British Columbia and Washington State. Rep Int Whal Comm Spec Issue 12:383–405

Black NA, Schulman-Janiger A, Ternullo RL, Guerrero-Ruiz M (1997) Killer whales of California and western Mexico: a catalog of photo-identified individuals. NOAA Technical Memorandum NOAA-TM-NMFS-SWFSC-247

Candy JR, Quinn TP (1999) Behaviour of adult Chinook salmon (*Oncorhynchus tshawytscha*) in British Columbia coastal waters determined from ultrasonic telemetry. Can J Zool 77: 1161–1169

Dahlheim ME, Schulman-Janiger A, Black N, Ternullo R, Ellifrit D, Balcomb KC (2008) Eastern temperate North Pacific offshore killer whales (*Orcinus orca*): occurrence, movements, and insights into feeding ecology. Mar Mamm Sci 24:719–729

Deecke VB, Ford JKB, Slater P (2005) The vocal behaviour of mammal-eating killer whales: communicating with costly calls. Anim Behav 69:395–405

Deecke VB, Slater P, Ford JKB (2002) Selective habituation shapes acoustic predator recognition in harbour seals. Nature (Lond) 420:171–173

Dieckmann U, Doebeli M (1999) On the origin of species by sympatric speciation. Nature (Lond) 400:354–357

Ellis GM, Ford JKB, Towers JR (2007) Northern resident killer whales in British Columbia: photo-identification catalogue 2007. Pacific Biological Station, Fisheries and Oceans Canada, Nanaimo. http://www.pac.dfo-mpo.gc.ca/sci/sa/cetacean/default_e.htm

Estes JA, Tinker MT, Williams TM, Doak DF (1998) Killer whale predation on sea otters linking oceanic and nearshore ecosystems. Science 282:473–476

Ford JKB (1984) Call traditions and dialects of killer whales (*Orcinus orca*) in British Columbia. Ph.D. Dissertation, University of British Columbia, Vancouver

Ford JKB (1989) Acoustic behaviour of resident killer whales (*Orcinus orca*) off Vancouver Island, British Columbia. Can J Zool 67:727–745

Ford JKB (1991) Vocal traditions among resident killer whales (*Orcinus orca*) in coastal waters of British Columbia. Can J Zool 69:1454–1483

Ford JKB (2002) Killer whale *Orcinus orca*. In: Perrin WF, Wursig B, Thewissen JGM (eds) The encyclopedia of marine mammals. Academic, San Diego, pp 669–676

Ford JKB (2006) An assessment of critical habitats of resident killer whales in waters off the Pacific coast of Canada. Fisheries and Oceans Canada, Canadian Science Advisory Secretariat, Ottawa. http://www.dfo-mpo.gc.ca/csas/

Ford JKB (2009) Killer whales *Orcinus orca*. In: Perrin WF, Würsig B, Thewissen JGM (eds) The encyclopedia of marine mammals, 2nd edn. Academic, San Diego, pp 650–657

Ford JKB, Ellis FM, Nichol LM (1992) Killer whales of the Queen Charlotte Islands: a preliminary study of the abundance, distribution, and population identity of *Orcinus orca* in the waters of Haida Gwaii. Unpublished report, South Moresby/Haida Gwaii Haanas National Parks Reserve, Canadian Parks Service

Ford JKB, Ellis GM (1999) Transients: mammal-hunting killer whales of British Columbia, Washington, and Southeastern Alaska. UBC Press, Vancouver

Ford JKB, Ellis GM (2006) Selective foraging by fish-eating killer whales *Orcinus orca* in British Columbia. Mar Ecol Prog Ser 316:185–199

Ford JKB, Ellis GM, Balcomb KC (2000) Killer whales: the natural history and genealogy of *Orcinus orca* in the waters of British Columbia and Washington. UBC and University of Washington Press, Vancouver

Ford JKB, Ellis GM, Barrett-Lennard LG, Morton AB, Palm RS, Balcomb KC III (1998) Dietary specialization in two sympatric populations of killer whales (*Orcinus orca*) in coastal British Columbia and adjacent waters. Can J Zool 76:1456–1471

Ford JKB, Ellis GM, Durban JW (2007) An assessment of the potential for recovery of West Coast transient killer whales using coastal waters of British Columbia. Fisheries and Oceans Canada, Canadian Science Advisory Secretariat, Ottawa. http://www.dfo-mpo.gc.ca/csas/

Ford JKB, Ellis GM, Matkin DR, Balcomb KC, Briggs D, Morton AB (2005) Killer whale attacks on minke whales: prey capture and antipredator tactics. Mar Mamm Sci 21:603–618

Ford JKB, Ellis GM, Matkin CO, Wetklo MH, Barrett-Lennard LG, Withler RE (2011) Shark predation and tooth wear in a population of northeastern Pacific killer whales. Aquat Biol 11:213–224

Ford JKB, Reeves RR (2008) Fight or flight: antipredator strategies of baleen whales. Mammal Review, 38:50–86

Forney KA, Wade PR (2006) Worldwide distribution and abundance of killer whales. In: Estes JA, Demaster DP, Doak DF, Williams TM, Brownell RL Jr (eds) Whales, whaling, and ocean ecosystems. University of California Press, Berkeley, pp 145–173

Futuyma DJ, Moreno G (1988) The evolution of ecological specialization. Annu Rev Ecol Syst 19:207–233

Guinet C (1990) Sympatrie des deux categories d'orques dans le detroit de Johnstone, Columbie Britannique. Rev Ecol (Terre Vie) 45:25–34

Guinet C (1992) Hunting behaviour of killer whales (*Orcinus orca*) around Crozet Islands. Can J Zool 70:1656–1667

Guinet C, Domenici P, de Stephanis R, Barrett-Lennard LG, Ford JKB, Verborgh P (2007) Killer whale predation on bluefin tuna: exploring the hypothesis of the endurance-exhaustion technique. Mar Ecol Prog Ser 347:111–119

Hauser DDW, Logsdon MG, Holmes EE, VanBlaricom GR, Osborne RW (2007) Summer distribution patterns of southern resident killer whales *Orcinus orca*: core areas and spatial segregation of social groups. Mar Ecol Prog Ser 351:301–310

Heimlich-Boran J (1986) Fishery correlations with the occurrence of killer whales in greater Puget Sound. In: Kirkevold BC, Lockard JS (eds) Behavioral biology of killer whales. Liss, New York, pp 113–131

Heimlich-Boran JR (1988) Behavioural ecology of killer whales (*Orcinus orca*) in the Pacific Northwest. Can J Zool 66:565–578

Heise K, Barrett-Lennard LG, Saulitis E, Matkin C, Bain D (2003) Examining the evidence for killer whale predation on Steller sea lions in British Columbia and Alaska. Aquat Mamm 29:325–334

Herman DP, Burrows DG, Wade PR, Durban JW, Matkin CO, LeDuc R, Barrett-Lennard LG, Krahn MM (2005) Feeding ecology of eastern North Pacific killer whales *Orcinus orca* from fatty acid, stable isotope, and organochlorine analyses of blubber biopsies. Mar Ecol Prog Ser 302:275–291

Hoelzel AR (1991) Killer whale predation on marine mammals at Punta Norte, Argentina: food sharing, provisioning and foraging strategy. Behav Ecol Sociobiol 29:197–204

Hoelzel AR, Dahlheim ME, Stern SJ (1998) Low genetic variation among killer whales (*Orcinus orca*) in the eastern North Pacific, and differentiation between foraging specialists. J Hered 89:121–128

Hoelzel AR, Natoli A, Dahlheim ME, Olavarria C, Baird RW, Black NA (2002) Low worldwide genetic diversity in the killer whale: implications for demographic history. Proc R Soc Lond Ser B 269:1467–1473

Jefferson TA, Stacey PF, Baird RW (1991) A review of killer whale interactions with other marine mammals: predation to co-existence. Mamm Rev 21:151–180

Jones IM (2006) A northeast Pacific offshore killer whale (*Orcinus orca*) feeding on a Pacific halibut (*Hippoglossus stenolepis*). Mar Mamm Sci 22:198–200

Krahn MM, Herman DP, Matkin CO, Durban JW, Barrett-Lennard L, Burrows DG, Dahlheim ME, Black N, Leduc RG, Wade PR (2007) Use of chemical tracers in assessing the diet and foraging regions of eastern North Pacific killer whales. Mar Environ Res 63:91–114

Krahn MM, Pitman RL, Burrows DG, Herman DP, Pearce RW (2008) Use of chemical tracers to assess diet and persistent organic pollutants in Antarctic type C killer whales. Mar Mamm Sci 24:643–663

LeDuc RG, Robertson KM, Pitman RL (2008) Mitochondrial sequence divergence among Antarctic killer whale ecotypes is consistent with multiple species. Biol Lett 4:426–429

Lopez JC, Lopez D (1985) Killer whales (*Orcinus orca*) of Patagonia, and their behavior of intentional stranding while hunting nearshore. J Mammal 66:181–183

Martinez DR, Klinghammer E (1970) The behavior of the whales, *Orcinus orca*: a review of the literature. Z Tierpsychol 27:828–839

Matkin C, Ellis G, Saulitis E, Barrett-Lennard L, Matkin D (1999) Killer whales of Southern Alaska. North Gulf Oceanic Society, Homer

Matkin CO, Barrett-Lennard LG, Yurk H, Ellifrit D, Trites AW (2007a) Ecotypic variation and predatory behavior among killer whales (*Orcinus orca*) off the eastern Aleutian Islands. Alaska Fish Bull 105:74–87

Matkin DR, Straley JM, Gabriele CM (2007b) Killer whale feeding ecology and non-predatory interactions with other marine mammals in the Glacier Bay region of Alaska. In: Piatt JF, Gende SM (eds) Proceedings of the Fourth Glacier Bay Science Symposium, pp 155–158. http://pubs.usgs.gov/sir/2007/5047/

McKinnon JS, Mori S, Blackman BK, David U, Kingsley DM, Jamieson L, Chou J, Schluter D (2004) Evidence for ecology's role in speciation. Nature (Lond) 429:294–298

Miller PJO, Shapiro AD, Tyack PL, Solow AR (2004) Call-type matching in vocal exchanges of free-ranging resident killer whales, *Orcinus orca*. Anim Behav 67:1099–1107

Miller PJO (2006) Diversity in sound pressure levels and estimated active space of resident killer whale vocalizations. J Comp Physiol A 192:449–459

Mizroch SA, Rice DW (2006) Have North Pacific killer whales switched prey species in response to depletion of the great whale populations? Mar Ecol Prog Ser 310:235–246

Morin PA, Leduc RG, Robertson KM, Hedrick NM, Perrin WF, Etnier M, Wade P, Taylor BL (2006) Genetic analysis of killer whale (*Orcinus orca*) historical bone and tooth samples to identify western U.S. ecotypes. Mar Mamm Sci 22:897–909

Morin PA, Archer FI, Foote AD, Vilstrup J, Allen EE, Wade P, Durban J, Parsons KL, Pitman RL, Li L, Bouffard P, Abel Nielsen SC, Rasmussen M, Willerslev E, Gilbert MTP, Harkins T (2010) Complete mitochondrial genome phylogeographic analysis of killer whales (*Orcinus orca*) indicates multiple species. Genome Res 20:908–916

Morton AB (1990) A quantitative comparison of the behavior of resident and transient forms of the killer whale off the central British Columbia coast. Rep Int Whal Comm Spec Issue 12:245–248

Nichol LM, Shackleton DM (1996) Seasonal movements and foraging behaviour of northern resident killer whales (*Orcinus orca*) in relation to the inshore distribution of salmon (*Oncorhynchus* spp.) in British Columbia. Can J Zool 74:983–991

Olesiuk PF, Ellis GM, Ford JKB (2005) Life history and population dynamics of northern resident killer whales (*Orcinus orca*) in British Columbia. Canadian Science Advisory Secretariat, Fisheries and Oceans, Canada. http://www.dfo-mpo.gc.ca/csas/
Osborne RW (1999) A historical ecology of Salish Sea "resident" killer whales (*Orcinus orca*), with implications for management. Ph.D. Thesis, University of Victoria, British Columbia
Pike GC, MacAskie IB (1969) Marine mammals of British Columbia. Fish Res Board Can Bull 71:1–54
Pitman RL, Ensor P (2003) Three forms of killer whales (*Orcinus orca*) in Antarctic waters. J Cetacean Res Manag 5:131–139
Pitman RL, Perryman WL, LeRoi D, Eilers E (2007) A dwarf form of killer whale in Antarctica. J Mammal 88:43–48
Reeves RR, Notarbartolo Di Sciara G (2006) The status and distribution of cetaceans in the Black Sea and Mediterranean Sea. IUCN Centre for Mediterranean Cooperation, Malaga, Spain. http://cmsdata.iucn.org/downloads/status_distr_cet_blac_med.pdf
Robinson BW, Wilson DS, Shea GO (1996) Trade-offs of ecological specialization: an intraspecific comparison of pumpkinseed sunfish phenotypes. Ecology 77:170–178
Saulitis E, Matkin C, Barrett-Lennard L, Heise K, Ellis G (2000) Foraging strategies of sympatric killer whale (*Orcinus orca*) populations in Prince William Sound. Alaska Mar Mamm Sci 16:94–109
Schluter D (2001) Ecology and the origin of species. Trends Ecol Evol 16(7):372–380
Similä T (1997) Behavioral ecology of killer whales in Northern Norway. Norwegian College of Fisheries Science. Ph.D. Thesis, University of Tromso, Tromso
Similä T, Ugarte F (1993) Surface and underwater observations of cooperatively feeding killer whales in northern Norway. Can J Zool 71:1494–1499
Simon M, Ugarte F, Wahlberg M, Miller LA (2005) Icelandic killer whales *Orcinus orca* use a pulsed call suitable for manipulating the schooling behaviour of herring *Clupea harengus*. Bioacoustics 16:57–74
Springer AM, Estes JA, van Vliet GB, Williams TM, Doak DF, Danner EM, Forney KA, Phister B (2003) Sequential megafaunal collapse in the North Pacific Ocean: an ongoing legacy of industrial whaling? Proc Natl Acad Sci USA 100:12223–12228
Steiger GH, Calambokidis J, Straley JM, Herman LM, Cerchio S, Salden DR, Urbán J, Jacobsen RJK, von Ziegesar O, Balcomb KC, Gabriele CM, Dahlheim ME, Uchida S, Ford JKB, Ladron de Guevara P, Yamaguchi M, Barlow J (2008) Geographic variation in killer whale attacks on humpback whales in the North Pacific: implications for predation pressure. Endang Species Res 4:247–256
Stephens DW, Krebs JR (1986) Foraging theory. Princeton University Press, Princeton
Stevens TA, Duffiled DA, Asper ED, Hewlett KG, Bolz A, Gage LJ, Bossart GD (1989) Preliminary findings of restriction fragment differences in mitochondrial DNA among killer whales. Can J Zool 67:2592–2595
Strager H (1995) Pod-specific call repertoires and compound calls of killer whales, *Orcinus orca* Linnaeus, 1758, in the waters of northern Norway. Can J Zool 73:1037–1047
Temte JL, Bigg MA, Wiig O (1991) Clines revisited: the timing of pupping in the harbour seal (*Phoca vitulina*). J Zool (Lond) 224:617–632
Ternullo R, Black N (2002) Predation behavior of transient killer whales in Monterey Bay, California. In: Fourth International Orca Symposium and Workshop, CEBC-CNRS, France, pp 156–159
Trites AW, Deecke VB, Gregr EJ, Ford JKB, Olesiuk PF (2007) Killer whales, whaling, and sequential megafaunal collapse in the North Pacific: a comparative analysis of the dynamics of marine mammals in Alaska and British Columbia following commercial whaling. Mar Mamm Sci 23:751–765
Via S (2001) Sympatric speciation in animals: the ugly duckling grows up. Trends Ecol Evol 16:381–390

Visser IN, Smith TG, Bullock ID, Green GD, Carlsson OGL, Imberti S (2008) Antarctic peninsula killer whales (*Orcinus orca*) hunt seals and a penguin on floating ice. Mar Mamm Sci 24:225–234

Wade PR, Barrett-Lennard LG, Black NA, Burkanov VN, Burdin AM, Calambokidis J, Cerchio S, Dahlheim ME, Ford JKB, Friday NA, Fritz LW, Jacobsen JK, Loughlin TR, Matkin CO, Matkin DR, McCluskey SM, Mehta AV, Mizroch SA, Muto MM, Rice DW, Robe P, Clapham P (2007) Killer whales and marine mammal trends in the North Pacific: a re-examination of evidence for sequential megafauna collapse and the prey-switching hypothesis. Mar Mamm Sci 23:766–802

Yurk H, Barrett-Lennard LG, Ford JKB, Matkins C (2002) Cultural transmission within maternal lineages: vocal clans in resident killer whales in southern Alaska. Anim Behav 63:1103–1119

Chapter 5
Japanese Macaques: Habitat-Driven Divergence in Social Dynamics

Goro Hanya

G. Hanya (✉)
Primate Research Institute, Kyoto University, 41-2 Kanrin, Inuyama 484-8506, Japan
e-mail: hanya@pri.kyoto-u.ac.jp

J. Yamagiwa and L. Karczmarski (eds.), *Primates and Cetaceans: Field Research and Conservation of Complex Mammalian Societies*, Primatology Monographs,
DOI 10.1007/978-4-431-54523-1_5, © Springer Japan 2014

Abstract Japanese macaques (*Macaca fuscata*), among the most intensively studied nonhuman primates in the world, live in a wide range of habitats in the Japanese archipelago. They offer us interesting examples on how habitat affects the social and population dynamics of long-lived animals. Studies of provisioned groups up to the 1970s revealed the basic social structure of Japanese macaques, characterized by a female-philopatric matrilineal society. Subsequently, two long-term study sites were established to study the nonprovisioned wild population in warm-temperate evergreen and cool-temperate deciduous forests in lowland Yakushima and Kinkazan, respectively. In both sites, a population increase was observed during the first decade of the long-term study, which was accompanied by group fission. An abrupt population decline resulting from external and environmental changes was then observed in both sites. The biggest difference between lowland Yakushima and Kinkazan is the inequality among groups and the stability of groups, which results from differences in the intensity of intergroup competition. In lowland Yakushima, macaques are under intense intergroup competition, and small groups suffer from low birthrate; finally, they may become extinct. In Kinkazan, intergroup competition is not intense, and there are no group size-dependent population fluctuations. This difference is believed to be a result of the more clumped distribution of high-quality foods in Yakushima compared to Kinkazan. In Yakushima, another long-term study site has been established recently in the high-altitude coniferous forest. In the future, Yakushima may offer us a rare opportunity to study the long-term social and population dynamics and within-population interchange of groups in a heterogeneous habitat.

Keywords Birthrate • Fruit production • Group extinction • Habitat • Japanese macaques • Mass mortality • Population dynamics • Socioecology

5.1 Introduction

Japanese macaques (*Macaca fuscata*) live over a wide range of habitats in Honshu, Shikoku, and Kyushu Islands and some small islands in the Japanese archipelago habitats (Fig. 5.1). In the lowland forest of Yakushima, which is the southern limit of the distribution of the species, macaques live in an evergreen forest mixed with subtropical species. The temperature rarely drops below 10 °C, even in winter. In contrast, in the snowy Shiga Heights in Nagano Prefecture, which is probably the coldest habitat for wild primates, the temperature often drops below −20 °C and the snowfall reaches several meters in depth. Shimokita Peninsula is the northern limit of distribution of not only Japanese macaques but also all the nonhuman primates. Japanese macaques also live in high mountains: they use alpine grasslands around the summit of Mt. Yarigatake (3,050 m) (Izumiyama et al. 2003). Some of the populations are highly dependent on crops (Izumiyama et al. 2003). Considering that primates are originally tropical animals, the extensiveness of the habitat occupied by Japanese macaques is surprising.

Fig. 5.1 Study sites of Japanese macaques

Japanese macaques are undoubtedly the species that has been studied for the longest period among primates. The study of wild Japanese macaques started on 3 December 1948. On that day, Kinji Imanishi and his two undergraduate students at Kyoto University, Shunzo Kawamura and Jun'ichiro Itani, conducted an expeditionary survey in Koshima, a small islet in Miyazaki Prefecture, Kyushu. Since then, field studies of Japanese macaques have been conducted in various sites in Japan. Japanese primatologists were particularly interested in the evolution of social structure, and they have accumulated data based on individual identification and long-term observations. The long-term data set of various study sites, combined with the extensive habitat diversity of this species, offers us a rare opportunity to examine the effect of habitat on long-term social dynamics. Studies on intraspecies variation are important to assess how flexibly animals can match their social behavior to the current environment (Nakagawa et al. 2010). Socioecological models mainly focus on interspecies variations (Sterck et al. 1997), but it remains unclear how much those models are applicable to explain intraspecies variations.

Here, I compare the social dynamics of Japanese macaques in various sites in Japan, including both provisioned and nonprovisioned populations. In particular, I examine the two long-term study sites in detail: Yakushima and Kinkazan. These two sites are among the habitats for Japanese macaques where any form of artificial habitat disturbance is minimal, such as provisioning, deforestation/aforestation, hunting, and crop raiding. In both sites, multiple groups of Japanese macaques have been individually identified and observed for more than 30 years. The two habitats are contrasting: warm-temperate evergreen forest in Yakushima and cool-temperate deciduous forest in Kinkazan, which are the two main types of Japanese macaque habitat. First, I summarize the social organizations and social dynamics of provisioned Japanese macaques. Second, I describe the social dynamics observed in nonprovisioned Yakushima and Kinkazan in detail and examine the similarities and differences. Third, I explore the habitat characteristics that may affect the differences in the social dynamics between the two sites. Finally, I introduce an ongoing project to compare the spatial variations in the social and population dynamics of Japanese macaques in Yakushima that are living at different altitudinal zones.

5.2 Social Dynamics of Provisioned Japanese Macaques

Most Japanese macaque studies during the early period of their research were conducted among provisioned groups. Japanese macaques were hunted as game until 1947, so they were afraid of humans when researchers began observations in late 1940s. Provisioning was the only way to habituate the macaques to human observers. By the 1970s, Japanese macaques were provisioned in more than 30 sites, including the long-term study sites in Koshima, Takasakiyama, Arashiyama, and Katsuyama. Most of the provisioning was performed by cities, prefectures, or travel companies to attract tourists.

The researchers found a similar social organization in the various study sites of provisioned Japanese macaques. After the 1980s, these similarities were found to be largely applicable to nonprovisioned groups. Japanese macaques form matrilineal social groups (Kawamura 1958; Furuichi 1985). Females stay in their natal group for all their life (Yamagiwa and Hill 1998). There are linear and stable dominance hierarchies within both sexes, and females inherit their social rank from their mothers (Koyama 1967; Hill and Okayasu 1995). Females usually confine their daily social interactions, such as grooming, to their maternal kin (Yamada 1963; Takahashi and Furuichi 1998). When the group fissions, females persistently associate with their kin (Furuya 1969; Koyama 1970; Oi 1988). In contrast, males disperse their natal groups during puberty, and adult males tend to stay in a group for only a few years (Fukuda 1982; Sprague et al. 1998). Males can avoid inbreeding by repeated emigration and immigration.

As a consequence of their enhanced food availability, provisioned Japanese macaques increased in population size. The population size in Takasakiyama was around 160 before provisioning began in 1953 and increased 1.093 times every year during the period from 1953 to 1975 (Sugiyama et al. 1995). The population reached more than 2,000 individuals, forming three groups, in 1979; the largest group included more than 1,200 animals. In many other provisioned populations, such as Arashiyama and Shiga Heights, group size increased to more than 200, which was larger than the maximum size reported for nonprovisioned groups (160 in Takasakiyama before provisioning started) (Takasaki and Masui 1984). Some of the sons of high-ranking females did not emigrate from their natal groups after they reached maturity (Kutsukake and Hasegawa 2005). The extreme concentration of high-quality foods enabled the macaques to maintain extraordinarily large group sizes and allowed some males to remain with their natal groups.

Because of the difficulty in managing extraordinarily large groups, the amount of provisioned foods was decreased by the managers around 1970. Abrupt changes in food availability significantly affected the group dynamics of the provisioned groups. In Mt. Ryozen, central Japan, one group of Japanese macaques was provisioned from 1966 to 1973. After provisioning ended, the birthrate decreased, infant mortality increased (Sugiyama and Ohsawa 1982b), and group desertion of females was frequently observed (Sugiyama and Ohsawa 1982a). If female desertion is defined as leaving the original group for more than 1 month without either serious illness or injury, there were 14 cases and all the 22 female deserters were 5 or more years old. The proportion of female deserters was 9.48 %/year. Several females left their group for 1 year and were sporadically observed to range alone. Some of them returned to their group, but some of them did not. In May 1978, 11 orphan subadult and juvenile females, who were deprived of their mothers by large-scale capture in November 1977, deserted the group and formed a new home range 4 km away from their original home range. The new group was composed of both high- and low-ranking individuals, and some of them left their maternal siblings in the original group. Thus, it was different from the usual group fission, in which monkeys separate with their kin. Sugiyama and Ohsawa discuss that female desertion of the group occurs when food supply becomes insufficient compared to group size.

Studies of provisioned groups revealed the basic social structure of Japanese macaques, characterized by a female-philopatric matrilineal society. The studies also clarified that macaques respond to abrupt external changes, such as large-scale capture and sudden decrease of food availability, in a different way from their ordinary pattern, such as philopatry and strong bonds with maternal kin. However, all these changes were artificial, so it remained unknown whether these changes occur in natural conditions.

5.3 Social Dynamics in the Lowland Forest of Yakushima

Yakushima is an island in southwestern Japan (30°N, 131°E) that occupies an area of 503 km^2 (Fig. 5.2). The highest peak is Mt. Miyanouradake (1,936 m a.s.l.), which is the second highest mountain in western Japan. The mean annual temperatures are 20 °C and 12 °C, and the mean temperatures of the coldest month (February) are 11 °C and 3.4 °C in the forests at altitudes of 100 m and 1,050 m, respectively (Tagawa 1980; Hanya 2004a). Kimura and Yoda (1984) classified the vegetation of Yakushima into five zones. (1) In the subtropical warm-temperate transitional zone (0–100 m a.s.l.), subtropical plants such as strangler figs (*Ficus superba* and *Ficus microcarpa*) are mixed with warm-temperate evergreen broad-leaved trees. (2) In the warm-temperate evergreen broad-leaved forest zone (100–800 m a.s.l.), warm-temperate evergreen broad-leaved trees (e.g., *Castanopsis cuspidata, Quercus salicina, Distylium racemosum*) are dominant. (3) In the warm-temperate/cool-temperate transitional forest zone (800–1,200 m a.s.l.), warm-temperate evergreen broad-leaved trees such as *Quercus acuta*, *Q. salicina*, and *D. racemosum* are mixed with

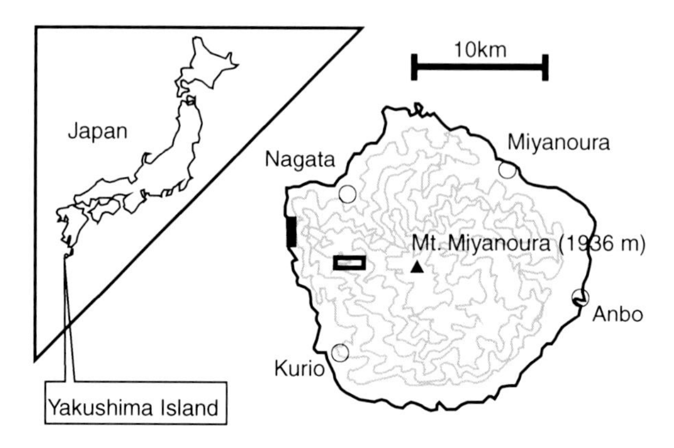

Fig. 5.2 Yakushima Island. *Closed square* indicates the long-term study site in the western lowland forest. *Open square* indicates the newly established study site in the coniferous forest. *Open circles* are major villages. Contours were drawn every 300 m in elevation

conifers such as *Cryptomeria japonica*, *Abies firma*, and *Tsuga sieboldii*. (4) In the cool-temperate zone (1,200–1,700 m a.s.l.), conifers such as *C. japonica*, *A. firma*, and *T. sieboldii* are dominant. (5) In the summit dwarf scrub (1,700 m), tall trees cannot grow, and a bamboo, *Pseudosasa owatarii*, covers the summit area. Japanese macaques inhabit all these zones, but their population density is highest in the lowland forest (<400 m a.s.l.), and did not differ among other zones. Difference in annual fruit production is the main factor affecting the altitudinal variations in macaque density (Hanya et al. 2004b).

After the pioneering survey by Kawamura and Itani in 1952, long-term study of Japanese macaques in Yakushima started in 1974, in the western lowland forest of Yakushima. Although large-scale forest development was ongoing all over Japan at that time, wide areas of natural vegetation still remained in Yakushima. At the same time, thanks to the hunting tradition using of traps, not guns, to capture macaques, Yakushima macaques were not afraid of humans even before intensive observation began. In 1974–1976, young Japanese primatologists studying at various field sites gathered in Yakushima and conducted a census of Japanese macaques three times. They found that the population density in this area was highest for this species (33/km^2) (Maruhashi 1982), and macaques were relatively habituated to humans. Among them, Tamaki Maruhashi, a graduate student at Kyoto University, and his colleagues habituated and identified all members of the Ko group in 1974, which was the first successful case for nonprovisioned Japanese macaques. Since then, intensive observation of multiple groups of Japanese macaques has continued.

Japanese macaque groups in the western lowland forest of Yakushima changed in a very dynamic way (Fig. 5.3). The Ko group, which was first habituated by Maruhashi, contained 47 animals in 1976. The Ko group fissioned twice within 3 years of the beginning of the research period. One of the daughter groups fissioned again in 1987, forming four groups. Six group fissions were observed among the Ko lineage and their neighboring groups during the period 1974–1987 (Sugiura et al. 2002). Increase

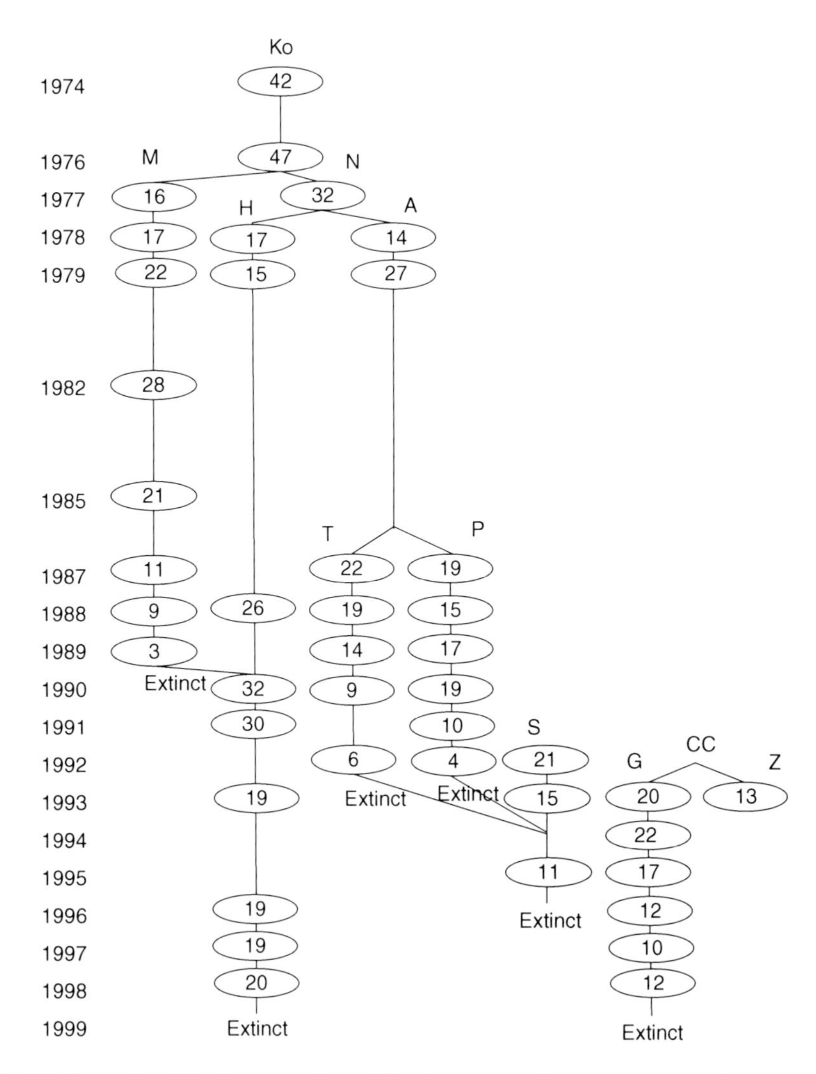

Fig. 5.3 Dynamics of Ko lineage and its neighboring groups of Japanese macaques in the western lowland forest of Yakushima. *Encircled numbers* are numbers of individuals. (Modified from Hanya 2002)

of group numbers probably resulted from the increase in population density. The population density at the onset of the study was 33 macaques/km^2 (Maruhashi 1982); however, it increased to 62–100/km^2 in 1993 (Yoshihiro et al. 1999). Yakushima macaques were hunted until the end of the 1960s for biomedical experiments, maintaining the population below the carrying capacity of their habitat. The population was probably recovering during the early period of the long-term research.

In contrast, after 1988, group fission rarely occurred, and some groups became smaller and finally became extinct. The first group extinction occurred in the M group,

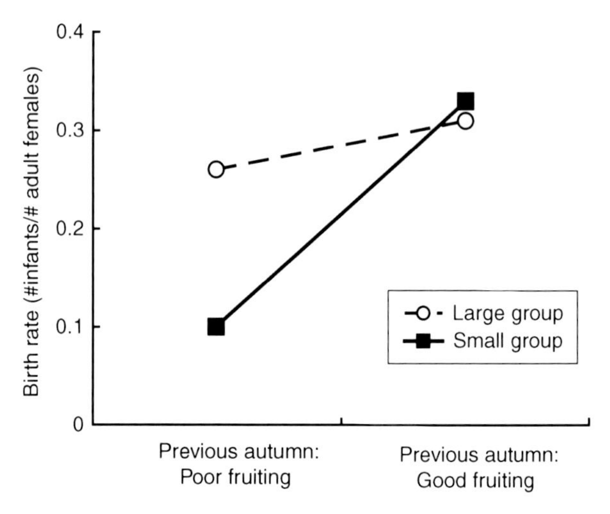

Fig. 5.4 Effect of group size and fruit production on birthrate of Japanese macaques in the western lowland forest of Yakushima. (Modified from Suzuki et al. 1998)

one of the daughter groups of the Ko group. Since the fission from the Ko group in 1977, the M group increased to 28 individuals in 1982. Then, the group size decreased gradually, and its home range also became smaller. In 1989, one adult male, one adult female, and her adolescent daughter were the only members of the M group. During the mating season of this year, the male emigrated, and the two remaining females joined the neighboring H group as the lowest-ranking animals (Takahata et al. 1994). Japanese macaque females were believed to stay in the natal group for all their life except when the group fissions; this was the first case of group fusion observed in Japanese macaques. In the early 1990s, other daughter groups of the Ko lineage, T and P, also decreased in number, and the few remaining females fused with the neighboring S group. In 1995, the S group also decreased in size and then disappeared (Sugiura et al. 2002).

Group extinction was influenced by intergroup competition. In Yakushima, intergroup relationships were antagonistic (Saito et al. 1998), and a dominant–subordinate relationship was apparent when the group size differed considerably (Sugiura et al. 2000). When the two different-sized groups encounter each other, the smaller group usually flees only when they notice the larger group by vocalization. Before their extinction, the home range of the M group was so small that there was no home range that they could use exclusively. They wandered as if they were escaping from the larger neighboring H group (Takahata et al. 1994). The T and P groups were also driven away by the neighboring CC group, which included more than 30 individuals and had migrated from an east mountainous area (Sugiura et al. 2002). As a result of intergroup competition, the birthrate of small-sized groups was smaller than in large-sized groups (Takahata et al. 1998). This difference was intensified when fruit production was poor (Fig. 5.4) (Suzuki et al. 1998). When the group becomes small, few infants are born, and the group becomes even smaller, and finally it vanishes.

In early 1999, a mass mortality of Japanese macaques occurred in the long-term study site in Yakushima. At that time, all individuals in five groups were identified. During the absence of observers from January to April of that year, 56 % of the animals disappeared, including all the members of the H (the last remaining Ko lineage) and G groups (Hanya et al. 2004a). Mass mortality among Japanese macaques had been reported in northern Japan as a result of heavy snow and an extremely cold winter (Izawa 1988). However, the lowland forest of Yakushima is the warmest habitat, harboring rich food resources and the highest population density for this species (Hanya et al. 2004b).

Although the direct cause of this mass mortality is not known, exceptionally poor fruit production in the preceding autumn certainly had an effect. During the 14 years from 1988 to 2001, fruit production in the autumn of 1998 was the poorest, only one-tenth of that in 1993, when the fruit production was the greatest. In fact, fruits produced in autumn usually remain until January and February. However, in this year, all fruits were consumed in December, and the macaques then ate mature leaves (Hanya et al. 2004a). Mature leaves are nutritionally lower in quality than fruits, and it is difficult for the macaques to satisfy their energy requirements even if they eat up to their gut capacity (Mori 1979). The fresh carcasses collected at the early stage of mass mortality contained little deposited fat (Hanya et al. 2004a), suggesting that the nutritional condition was bad in this year. Although it is very likely that there were other direct causes, such as disease, the poor fruit crop and the resulting poor nutritional conditions were the important background of this mass mortality.

Another important aspect of this mass mortality was local concentration (Fig. 5.5). The two extinct groups were neighbors to each other, and there was a pattern that mortality decreased with increasing distance from the two extinct groups (Hanya et al. 2004a). Although the cause of this pattern remains unknown, prevalence of epidemic disease may explain this. Whatever the cause, the imbalance of mortality among groups has changed intergroup relationships. For example, the K group, which was not affected by the mass mortality, expanded their home range northward, into the center of the mass mortality area, and fissioned a few years later. The home ranges of the extinct groups were occupied by other groups within only 2 years.

As a result of these social and population changes, all the Ko lineage groups have disappeared. Now the home range is occupied by other groups, which migrated from the east (vertical migration from higher altitude) or south (horizontal migration from the lowland). The long-term study of Yakushima revealed that the Japanese macaque groups are under intense competitive relationships. Japanese macaque groups fluctuate from both external (e.g., mass mortality) and internal (e.g., intergroup competition) causes and are not stable over the long term.

5.4 Social Dynamics in Kinkazan

Kinkazan is an island that lies 700 m offshore Oshika Peninsula, Miyagi Prefecture, northern Honshu. Its area is 10 km^2 with the highest peak of 445 m a.s.l. The mean annual temperature is 11 °C. The island is rarely covered with snow, although it

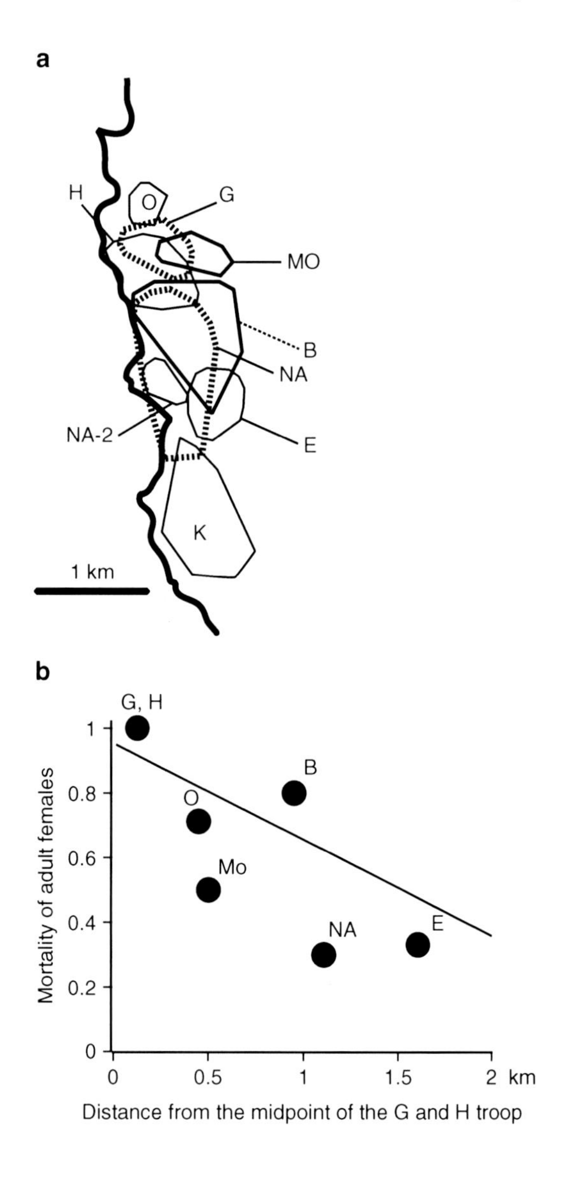

Fig. 5.5 (**a**) Distribution of Japanese macaque groups in the western lowland forest of Yakushima in 1998, before mass mortality occurred. (**b**) Relationships between the adult female mortality of each group during the mass mortality in 1999 and its distance from the two extinct groups (G and H). (Modified from Hanya et al. 2004a)

occasionally snows on cold winter days. The island is covered with a mixed forest of deciduous and coniferous trees, such as *Fagus crenata*, *Abies firma*, and *Pinus thunbergii*. However, saplings of woody plants have rarely developed into mature trees recently because of high feeding pressure by sika deer. Grasslands of *Zoysia japonica*, *Miscanthus sinensis*, and other grasses widely cover some parts of the island (Agetsuma and Nakagawa 1998).

In Kinkazan, some short-term surveys of Japanese macaques were conducted in the 1960s and 1970s. In 1982, Kosei Izawa started a long-term study of Japanese macaques on this island when he was assigned to a professor at Miyagi University of Education, a nearby university from Kinkazan. With the aid of his colleagues and his undergraduate students, he has conducted a census of Japanese macaques three times a year. Because of the small area and clear visibility, it was possible to count all the population in this island. Detailed behavioral observations based on individual identification were also conducted by various researchers.

Population changes in Kinkazan have been summarized by Izawa (2005). In Kinkazan, there was only one group in the 1960s. If the 1960s and 1970s are included, when only intermittent data are available, five fissions occurred in 42 years. Among them, two group fissions occurred after the long-term study started in 1982. No group extinction was observed. Mass mortality occurred in the winter of 1984 winter as a result of heavy snow and exceptional cold. During this year, the population decreased from 270 to 180. The population gradually recovered after that, reaching 294 in 1994. Subsequently, the population began decreasing gradually, to 217 individuals in 2003. Even when the population size decreased, there was no tendency for only a particular group to decrease in size. Intergroup relationships are not antagonistic. The frequency of intergroup encounters was one third of that in Yakushima, in accord with the differences in group density (Sugiura et al. 2000). When two groups encountered, no apparent social interactions were observed (Saito et al. 1998). There was no tendency for the birthrate to be smaller for small groups than large groups (Takahata et al. 1998).

5.5 Similarities and Differences in the Social Dynamics in Yakushima and Kinkazan

There are both similarities and contrasts in the social and population dynamics of Japanese macaques between the two study sites. In both sites, population increase was accompanied by group fissions. In Yakushima, group fission occurred frequently during the period when population density doubled (1970s–1980s). In Kinkazan, population size was less than 70 in 1962 and increased to almost 300 in 1994. The number of groups also increased, from one to six, during that period. It is suggested that there is a limit to the maximum group size, which is probably around 50 in Yakushima and 80 in Kinkazan. These numbers are much smaller than the group size of most of the provisioned groups. Increased within-group competition and the difficulty in maintaining group spread are likely to be key factors limiting maximum group size; however, there are no quantitative data to suggest why the maximum group size differs among habitats. Another similarity is that a sudden population decrease from external and environmental changes can occur over a long time. The effect can be as great as killing one third of the entire population, as in Kinkazan, or local but so large as to make multiple groups extinct within a few months, as in Yakushima. Both the long-term studies tell us that the effects of these rarely occurring events are not negligible over the long term.

The biggest difference between Yakushima and Kinkazan is the inequality among groups and the stability of the groups, which result from the difference in the intensity of intergroup competition. In Yakushima, macaques are under intense intergroup competition, and small groups suffer from low birthrate and finally may disappear. In Kinkazan, intergroup competition is not intense, and there are no group size-dependent population fluctuations. Maruhashi et al. (1998) compared the home range structure between the two habitats. In Yakushima, (1) food tree density was higher, (2) interfeeding bout site distance was shorter, (3) daily travel distance was shorter, (4) home range size was smaller, and (5) the macaque groups shared a greater proportion of their home range with neighboring groups compared to Kinkazan. Consequently, food distribution is more clumped, and thus the quality of the home range is more worth defending in Yakushima than in Kinkazan. Intergroup competition is enhanced in Yakushima compared to Kinkazan by the high frequency of intergroup encounters that result from higher group density and greater overlap in home range between neighboring groups.

5.6 Linking Environmental, Population, and Social Changes: Commencement of Another Long-Term Research Project

The two long-term study sites successfully revealed the variability of social dynamics of Japanese macaques under natural conditions. However, we still cannot understand the social dynamics as an ecological process of population dynamics because quantitative data on habitat changes are lacking. In addition, continuous data on population density are not available in Yakushima.

In both Yakushima and Kinkazan, the increase of population size up to the early 1980s was probably related to past hunting pressure. In Yakushima, for example, it is said that 950 macaques were exported from Yakushima during the period of 1950–1969 (Azuma 1984). However, no record remains where in Yakushima and how many were captured each year, so it is difficult to estimate how much impact the hunting had on macaque populations. In Yakushima, it is also said that secondary vegetation along a road that was opened in 1967, a few years before the start of the long-term research, changed in succession (Maruhashi 1984). It is possible that food availability, and thus carrying capacity, changed with vegetational succession, but there are no quantitative data to examine. In Kinkazan, the population has been gradually decreasing since the late 1990s (Izawa 2005). It is believed that high grazing pressure by sika deer is degrading the island vegetation, and recent strong typhoons, which damaged many old large trees, have accelerated the deterioration (Izawa 2005). However, there are no quantitative data on the vegetation changes for the past few decades.

In Yakushima, in spite of the diversity of habitat along the elevational gradient, studies of Japanese macaques were largely conducted only in the western lowland forest until the 1980s. In 1989, Shinichi Yoshihiro organized a census team (Yakushima Macaque Research Group; Yakuzaru-Chosa-Tai) to study distribution of macaque groups in various areas in Yakushima, including higher mountainous zones and around the coastal villages, where macaques raid crops. Every summer, more than 40 volunteers, who are mostly inexperienced undergraduate students, joined the census. From 1990 to 1993, they studied the lowland and clarified that population size in that area (127 km^2, 1–2 km from the coast) was 2,000–3,850 macaques (Yoshihiro et al. 1998). From 1994 to 1997, they studied the vertical distribution of Japanese macaques in the western area, which is the only area where natural vegetation is preserved from the coast to the summit (Yoshihiro et al. 1999; Hanya et al. 2004b).

After the completion of the island-wide distribution survey, the census team set up a new long-term study site in the coniferous forest in the western area. They established a census area of 7.5 km^2 and have studied group density by a modified point census, a method that they devised (Hanya et al. 2003b). They also studied the composition of several identified groups in their study site. In 2000, four groups were identified, and one of them fissioned in 2005. Based on the results of the census team, I identified all the individuals in one group among them (HR group) and conducted detailed behavioral observation for 1 year from April 2000. I also set a permanent plot in both primary and logged forests in 1999 and 2002, respectively, to study vegetation, fruit production, and its supra-annual changes. Thanks to the lessons of the other long-term study sites, we realize that, to study social dynamics, we have to systematically monitor changes in the habitat and population simultaneously with group composition and distribution.

Although the long-term study in the coniferous forest is still in its infancy, we found that the macaques are so different from their lowland counterparts in various interesting points. They are much more folivorous: 38 % of their annual feeding time was spent for mature leaves, which was much longer than for fruits (13 %) or seeds (4 %) (Hanya 2004a). Fruit production in the coniferous forest was only one third that of the lowland forest (Hanya et al. 2003a), and most of their main food trees were small-sized, high-density trees (Hanya 2004b, 2009), suggesting that competition is unlikely to occur in the coniferous forest. In fact, intergroup encounters were infrequent in the coniferous forest and not antagonistic when occurring (Hanya et al. 2008). There was no size-dependent difference in birthrate, such as in the lowland forest (Hanya et al. 2008). With respect to the intergroup relationships, the coniferous forest of Yakushima was more similar to Kinkazan, than to coastal forest, although those two forests are only 7 km apart, and there is no genetic differentiation (Hayaishi and Kawamoto 2006). Interestingly, however, female social relationships within a group were quite similar between the coniferous and lowland groups (Hanya et al. 2008). This observation suggests that social behaviors of female Japanese macaques are robust and do not change in response to the current environment.

Another important aspect of Yakushima is that macaque individuals, or even macaque groups, can move between the two study sites. Migration of macaque groups from upward (from east) is observed in the western lowland forest (Sugiura et al. 2002). On the other hand, in the coniferous forest, distribution of the four or five identified groups has been stable for the past 10 years, except the one case of group fission. It is estimated that seven to eight macaque groups are distributed between the two study sites (Yoshihiro et al. 1999), so it may be possible that social dynamics in one of the areas affects the other, at least indirectly. Every summer, similar monitoring of group density, composition, and distribution of multiple identified groups and fruit production is conducted in both these study sites. In the future, Yakushima may offer us a rare opportunity to study the long-term social and population dynamics and within-population interchange of groups in a heterogeneous habitat.

Long-term study of Japanese macaques has revealed a complex and diverse picture of social dynamics. It is now evident that long-term ecological monitoring of the habitat is indispensable to clarify the interrelationships between ecology and society for this species, and we have just started meeting the challenge.

References

Agetsuma N, Nakagawa N (1998) Effects of habitat differences on feeding behaviors of Japanese monkeys: comparison between Yakushima and Kinkazan. Primates 39:275–289

Azuma S (1984) Monkey, forest, and human: a history of biological nature (in Japanese). Monkey 283-285:94–102

Fukuda F (1982) Male movement of transfer between groups of Japanese macaques. Jpn J Ecol 32:491–498

Furuichi T (1985) Inter-male associations in a wild Japanese macaque troop on Yakushima Island, Japan. Primates 26:219–237

Furuya Y (1969) On the fission of troops of Japanese monkeys. II. General view of troop fission of Japanese monkeys. Primates 10:47–69

Hanya G (2002) Island in the southern limit of distribution (in Japanese). In: Oi T, Masui K (eds) Natural history of Japanese macaques: their ecological varieties and conservation. Tokai University Press, Tokyo, pp 229–250

Hanya G (2004a) Diet of a Japanese macaque troop in the coniferous forest of Yakushima. Int J Primatol 25:55–71

Hanya G (2004b) Seasonal variations in the activity budget of Japanese macaques in the coniferous forest of Yakushima: effects of food and temperature. Am J Primatol 63:165–177

Hanya G (2009) Effects of food type and number of feeding sites in a tree on aggression during feeding in wild *Macaca fuscata*. Int J Primatol 30:569–581

Hanya G, Noma N, Agetsuma N (2003a) Altitudinal and seasonal variations in the diet of Japanese macaques in Yakushima. Primates 44:51–59

Hanya G, Yoshihiro S, Zamma K, Kubo R, Takahata Y (2003b) New method to census primate groups: estimating group density of Japanese macaques by point census. Am J Primatol 60:43–56

Hanya G, Matsubara M, Sugiura H, Hayakawa S, Goto S, Tanaka T, Soltis J, Noma N (2004a) Mass mortality of Japanese macaques in a western coastal forest of Yakushima. Ecol Res 19:179–188

Hanya G, Yoshihiro S, Zamma K, Matsubara H, Ohtake M, Kubo R, Noma N, Agetsuma N, Takahata Y (2004b) Environmental determinants of the altitudinal variations in relative group densities of Japanese macaques on Yakushima. Ecol Res 19:485–493

Hanya G, Matsubara M, Hayaishi S, Zamma K, Yoshihiro S, Kanaoka MM, Sugaya S, Kiyono M, Nagai M, Tsuriya Y, Hayakawa S, Suzuki M, Yokota T, Kondo D, Takahata Y (2008) Food conditions, competitive regime, and female social relationships in Japanese macaques: within-population variation on Yakushima. Primates 49:116–125

Hayaishi S, Kawamoto Y (2006) Low genetic diversity and biased distribution of mitochondrial DNA haplotypes in the Japanese macaque (*Macaca fuscata yakui*) on Yakushima Island. Primates 47:158–164

Hill DA, Okayasu N (1995) Absence of youngest ascendancy in the dominance relations of sisters in wild Japanese macaques (*Macaca fuscata yakui*). Behaviour 132:367–379

Izawa K (1988) The ecological study of wild Japanese monkeys living in Kinkazan Island, Miyagi Prefecture: on the population change and the group division (in Japanese). Bull Miyagi Univ Edu 23:1–9

Izawa K (2005) Population changes of Japanese macaques in Kinkazan in 1982–2003 (in Japanese). Miyagiken no Nihonzaru (Jpn Macaques Miyagi Prefect) 19:1–10

Izumiyama S, Mochizuki T, Shiraishi T (2003) Troop size, home range area and seasonal range use of the Japanese macaque in the Northern Japan Alps. Ecol Res 18:465–474

Kawamura S (1958) The matriarchal social order in the Minoo-B Group. Primates 1:149–156

Kimura K, Yoda K (1984) Structure and regeneration process of evergreen conifers and broad-leaved trees in the Yaku-shima Wilderness Area, Yaku-shima. In: Nature Conservation Bureau EA, Japan (ed) Conservation Reports of the Yaku-shima Wilderness Area, Kyushu, Japan. Environment Agency, Tokyo, pp 399–436

Koyama N (1967) On dominance rank and kinship of a wild Japanese monkey troop in Arashiyama. Primates 8:189–216

Koyama N (1970) Changes in dominance rank and division of a wild Japanese monkey troop in Arashiyama. Primates 11:335–390

Kutsukake N, Hasegawa T (2005) Dominance turnover between an alpha and a beta male and dynamics of social relationships in Japanese macaques. Int J Primatol 26:775–800

Maruhashi T (1982) An ecological study of troop fissions of Japanese monkeys (*Macaca fuscata yakui*) on Yakushima Island, Japan. Primates 23:317–337

Maruhashi T (1984) Observe groups through individuals: forest of Yakushima macaques (in Japanese). Monkey 283-285:18–25

Maruhashi T, Saito C, Agetsuma N (1998) Home range structure and inter-group competition for land of Japanese macaques in evergreen and deciduous forests. Primates 39:291–301

Mori A (1979) An experiment on the relation between the feeding speed and the caloric intake through leaf eating in Japanese monkeys. Primates 20:185–195

Nakagawa N, Nakamichi M, Sugiura H (2010) The Japanese macaques. Springer, Tokyo

Oi T (1988) Sociological study on the troop fission of wild Japanese monkeys (*Macaca fuscata yakui*) on Yakushima Island. Primates 29:1–19

Saito C, Sato S, Suzuki S, Sugiura H, Agetsuma N, Takahata Y, Sasaki C, Takahashi H, Tanaka T, Yamagiwa J (1998) Aggressive intergroup encounters in two populations of Japanese macaques (*Macaca fuscata*). Primates 39:303–312

Sprague DS, Suzuki S, Takahashi H, Sato S (1998) Male life history in natural populations of Japanese macaques: migration, dominance rank, and troop participation of males in two habitats. Primates 39:351–363

Sterck EHM, Watts DP, van Schaik CP (1997) The evolution of female social relationships in nonhuman primates. Behav Ecol Sociobiol 41:291–309

Sugiura H, Saito C, Sato S, Agetsuma N, Takahashi H, Tanaka T, Furuichi T, Takahata Y (2000) Variation in intergroup encounters in two populations of Japanese macaques. Int J Primatol 21:519–535

Sugiura H, Agetsuma N, Suzuki S (2002) Troop extinction and female fusion in wild Japanese macaques in Yakushima. Int J Primatol 23:69–84

Sugiyama Y, Ohsawa H (1982a) Population dynamics of Japanese macaques at Ryozenyama: III. Female desertion of the troop. Primates 23:31–44

Sugiyama Y, Ohsawa H (1982b) Population dynamics of Japanese monkeys with special reference to the effect of artificial feeding. Folia Primatol (Basel) 39:238–263

Sugiyama Y, Iwamoto T, Ono Y (1995) Population control of artificially provisioned Japanese monkeys. Primate Res 11:197–207

Suzuki S, Noma N, Izawa K (1998) Inter-annual variation of reproductive parameters and fruit availability in two populations of Japanese macaques. Primates 39:313–324

Tagawa H (1980) Vegetation on the western slope of Mt. Kuniwaridake, Yakushima Island. Sci Rep Kagoshima Univ 29:121–137

Takahashi H, Furuichi T (1998) Comparative study of grooming relationships among wild Japanese macaques in Kinkazan A troop and Yakushima M troop. Primates 39:365–374

Takahata Y, Suzuki S, Okayasu N, Hill D (1994) Troop extinction and fusion in wild Japanese macaques of Yakushima Island, Japan. Am J Primatol 33:317–322

Takahata Y, Suzuki S, Okayasu N, Sugiura H, Takahashi H, Yamagiwa J, Izawa K, Agetsuma N, Hill D, Saito C, Sato S, Tanaka T, Sprague D (1998) Does troop size of wild Japanese macaques influence birth rate and infant mortality in the absence of predators? Primates 39:245–251

Takasaki H, Masui K (1984) Troop composition data of wild Japanese macaques reviewed by multivariate methods. Primates 25:308–318

Yamada M (1963) A study of blood-relationship in the natural society of the Japanese macaque. Primates 4:43–65

Yamagiwa J, Hill D (1998) Intraspecific variation in the social organization of Japanese macaques: past and present scope of field studies in natural habitats. Primates 39:257–273

Yoshihiro S, Furuichi T, Manda M, Ohkubo N, Kinoshita M, Agetsuma N, Azuma S, Matsubara H, Sugiura H, Hill D, Kido E, Kubo R, Matsushima K, Nakajima K, Maruhashi T, Oi T, Sprague D, Tanaka T, Tsukahara T, Takahata Y (1998) The distribution of wild Yakushima macaque (*Macaca fuscata yakui*) troops around the coast of Yakushima Island, Japan. Primate Res 14:179–187

Yoshihiro S, Ohtake M, Matsubara H, Zamma K, Hanya G, Tanimura Y, Kubota H, Kubo R, Arakane T, Hirata T, Furukawa M, Sato A, Takahata Y (1999) Vertical distribution of wild Yakushima macaques (*Macaca fuscata yakui*) in the western area of Yakushima Island, Japan: preliminary report. Primates 40:409–415

Chapter 6
Shark Bay Bottlenose Dolphins: A Case Study for Defining and Measuring Sociality

Margaret A. Stanton and Janet Mann

M.A. Stanton (✉)
Department of Biology, Georgetown University, Washington, DC, USA

Department of Anthropology, The George Washington University, Washington, DC, USA
e-mail: mastanton@gwu.edu

J. Mann
Department of Biology, Georgetown University, Washington, DC, USA

Department of Psychology, Georgetown University, Washington, DC, USA

J. Yamagiwa and L. Karczmarski (eds.), *Primates and Cetaceans: Field Research and Conservation of Complex Mammalian Societies*, Primatology Monographs, DOI 10.1007/978-4-431-54523-1_6, © Springer Japan 2014

Abstract Bottlenose dolphins are attractive candidates for the application of social network analysis (SNA), in part because of their complex fission–fusion social organization characterized by dynamic, temporally variable groups. In Shark Bay, Western Australia, researchers have studied the resident bottlenose dolphins since 1982. Using data on two calves from the Shark Bay dataset, here we present a case study to provide an example of the variety of social measures available to researchers, including both traditional measures as well as network metrics. In particular, this example case study advocates the use of multiple measures of sociality with careful consideration of what dimensions were captured before making inferences.

Keywords Association • Bottlenose dolphins • Fission–fusion • Interaction • Primates • Social metrics • Social network analysis

6.1 Introduction

Similar to primates such as humans (*Homo sapiens*), chimpanzees (*Pan troglodytes*), bonobos (*Pan paniscus*), and spider monkeys (*Ateles* spp.), bottlenose dolphins (*Tursiops* sp.) and other delphinids exhibit an intrinsically complex fission–fusion social organization characterized by the dynamic nature of compositionally and temporally variable groups (Goodall 1986; Symington 1988; Connor et al. 2000; Brager 1999; Coscarella et al. 2011). Not surprisingly, measuring sociality in these complex societies is no easy task and often requires a multifaceted approach with careful consideration of what inferences may be drawn from each available social metric. In this chapter, we use our long-term study of bottlenose dolphin mothers and calves to demonstrate the range of measures that can be used to capture some aspect of dolphin social life, particularly those achieved by employing social network analysis. This innovative technique has rapidly increased in popularity because of its ability to quantify multi-actor interactions, thereby providing more complete descriptions of complex societies. We provide examples of both association-based social networks and interaction-based social networks that are more analogous to the grooming networks of chimpanzees.

6.1.1 Bottlenose Dolphins of Shark Bay

An important distinction between the foregoing primate fission–fusion systems and that of bottlenose dolphins is the openness of bottlenose dolphin communities (Smolker et al. 1992). Although the subgroups of chimpanzee, spider monkey, and most other fission–fusion species are composed of members from a larger closed social unit, bottlenose dolphin communities exist on an open–closed continuum. At some sites, bottlenose dolphin communities are closed or semiclosed (e.g., Wells et al. 1987; Lusseau et al. 2003), but in Shark Bay, Australia, the community is unbounded with an overlapping mosaic of hundreds to thousands of individuals (Mann et al. 2012). A consequence of openness is that the potential relationships are not constrained by social unit size. Additionally, although the fissions and fusions of terrestrial social groups are limited by the cost of locomotion, this constraint is considerably less restrictive in the aquatic environment of the bottlenose dolphin (Williams et al. 1992), facilitating more frequent interaction with larger groups of individuals on an irregular basis. As a consequence, variation in patterns of association within a population of bottlenose dolphins is exceptionally large (Smolker et al. 1992; Gibson and Mann 2008a). Average group size among bottlenose dolphins in Shark Bay is 4.8 individuals; however, the size and composition of these groups is likely dependent on social context (Smolker et al. 1992). Male bottlenose dolphins in Shark Bay form hierarchical alliances cooperating to obtain and sequester females for mating. "First-order alliances" consist of pairs or trios of individual males, whereas teams of these first-order alliances, referred to as "second-order alliances," cooperate to steal female consorts from other alliances or prevent thefts (Connor et al. 1992). Males in first and second order alliances are more highly related than expected by chance, suggesting inclusive fitness benefits to alliance formation (Krützen et al. 2003). An alternative strategy, termed a "super-alliance," is a second-order alliance consisting of labile first-order alliances whose members frequently switch partners (Connor et al. 2001). Interestingly, members of super-alliances appear no more related to each other than expected by chance (Krützen et al. 2003). Recent research suggests a third level of alliance formation, and the nested nature of male bottlenose dolphin alliances is arguably more complex than cooperation behavior in any nonhuman mammal (Connor et al. 2011).

In contrast to males, female bottlenose dolphins of Shark Bay do not form alliances and vary widely in degree of sociality, forming loose social networks with the number of known lifetime associates ranging from 1 to 160 (Smolker et al. 1992; Gibson and Mann 2008a, b). In a recent comparison of male and female social network metrics, we found that males and females do not differ in the their total number of associates (degree), but as expected given male alliance formation, males have stronger associations and are more cliquish (Mann et al. 2012). That said, females do appear to have preferred associates, but typically spend less than 30 % of their time with these top associates (Smolker et al. 1992). Interestingly, female dolphins depend on nondefensible ephemeral food patches (e.g., schools of fish) and are thus tolerant, yet selfish, about access to food (Mann et al. 2007); therefore, defense of resources does not explain patterns of female sociality. Predation on calves is also unlikely to be the main cause of these groups as shark predation does not appear to be a primary predictor of calf mortality (Mann and Watson-Capps 2005), although group sizes are larger in the newborn period (Mann et al. 2000). In Shark Bay females give birth to a single calf after a 12-month gestation period. Calves are weaned at an average age of 4 years, but females do not have their first calf until age 11–12 years (Mann et al. 2000). In contrast to primates who spend their extended developmental period buffered by their natal social group

(Leigh and Blomquist 2007), bottlenose dolphins do not spend the juvenile period in stable groups and must negotiate a complex social environment in the absence of direct maternal care (Mann et al. 2000; Tsai and Mann 2013). A recent examination of the possible function of female bottlenose dolphin social groups in Shark Bay found some support for the protection of young calves (first year of life) from predators because mothers with young calves tended to form larger groups. However, the formation of mother–calf groups was better explained overall by the hypothesis that grouping enables calves, particularly males, to develop social skills before the lack of social savvy incurs a reproductive cost (Gibson and Mann 2008b). This hypothesis was borne out by a subsequent study showing that early (pre-weaning) social networks predict juvenile (post-weaning to age 10 years) male mortality (Stanton and Mann 2012).

Interestingly, bottlenose dolphin calves also vary in degree of sociality ($N_{\text{associates}} = 1–77$) and have the ability to separate from their mothers and form unique associates. Because bottlenose dolphins show bisexual philopatry, calf social relationships often persist into adulthood (Tsai and Mann 2013), but despite the attention given to the adult bottlenose dolphin fission–fusion society, there are few in-depth investigations into bottlenose dolphin calf social development. Using the number of associates and the proportion of time spent in groups when together and separated from each other as measures of sociality, Gibson and Mann (2008a) assessed predictors of individual variation in the social patterns of Shark Bay mothers and calves. Not surprisingly, the results of this study indicate that the number of associates, time spent in groups, and time spent separated from their mothers changes as calves approach weaning. The researchers also found differences based on calf sex and maternal sociality. With age, males increased their time in groups during separations whereas this measure decreased in females. In addition, the number of calf associates was strongly related to their mother's number of associates, especially for females (Gibson and Mann 2008a). We recently employed social network analysis to further investigate calf social networks during temporary mother–calf separations and found that calves had larger, less dense ego networks than their mothers. Additionally, male calves formed stronger bonds with other male calves during separations (Stanton et al. 2011). These results suggest that during separations calves are independently developing the social skills and bonds necessary for future success, particularly males who rely on alliance formation for mating opportunities as adults. Juvenile males, however, appear to harass male calves and may be detrimental to male calf future fitness (Stanton and Mann 2012). The function and consequences of individual variation in calf sociality, which are just beginning to be explored, are critical for understanding both prolonged development and social complexity in bottlenose dolphins. The next step is to examine these patterns in greater depth. To highlight individual social variation as well as some of the numerous methods with which social patterns can be quantified, we present a series of social measures calculated for two Shark Bay bottlenose dolphin calves.

6.2 Method

Researchers have studied the bottlenose dolphin females, calves, and their associates ($N > 1{,}500$) of Shark Bay, Australia, since 1988. This rcscarch is facilitated by a large number of identifiable individuals and an extensive 30-year dataset. Existing Shark Bay data include both "snapshot" survey data and more intensive focal follow data. Boat-based focal follows of specific mother–calf pairs provide detailed behavioral information including group composition, activity, location, and specific social interactions using standard quantitative sampling techniques including point, scan, and continuous sampling (Altmann 1974). Party composition is scanned for every minute during a focal follow, and association is conservatively determined using a 10-m chain rule where one dolphin is considered to be in a group with another dolphin if they are separated by 10 m or less. Individuals are identified by dorsal fin using photo-identification techniques (Smolker et al. 1992). Focal follows of individuals involve intensive sampling, but provide greater detail and precision in terms of individual social variation, particularly when examining mother–calf pairs, by allowing for more reliable identification of young calves and better assessment of calf behavior during temporary long-distance separations from their mothers (Gibson and Mann 2009). Two calves, one male (MIG) and one female (LEN), were observed for ~33 h and ~40 h, respectively, during their first 4 years of life. These calves were chosen because both were observed for 4 years and both possess similarly sized networks, which facilitates comparison. Using MIG's and LEN's focal follow party composition data, we first calculated a variety of traditional, non-network measures of individual sociality as described in Table 6.1.

To employ social network analysis on this dataset, we used SocProg 2.3 (Whitehead 2009) and UCINET6 (Borgatti et al. 2002) software to construct the ego networks of LEN and MIG from focal follow party composition data (Fig. 6.1). An ego network is a type of social network consisting of a focal individual or "ego" and only those individuals directly connected to the focal. All networks were drawn in NetDraw using the spring-embedding algorithm (Borgatti 2002). Two individuals were connected to each other by an edge if they were observed in the same group, and the strength of their relationship was calculated by taking the average proportion of observations (APO) when two individuals were observed together. The average is necessary to account for biases based on sampling effort. For example, if two dolphins, SMO and COO, were observed together for a total of 120 min and SMO was observed for 180 min total, although COO was observed for 480 min total, then SMO spent 0.75 of his time with COO, whereas COO spent 0.25 of his time with SMO. To create a symmetrical sociomatrix so as not to imply a false sense of directionality in the relationship, these two proportions would be averaged for an APO = 0.5. It is important to note that this measure does not directly translate into the percent of time two individuals were seen together. An APO of 0.5 does not indicate that two animals were observed together 50 % of the time. However, higher APOs are considered indicative of stronger relationships.

Table 6.1 Non-network social measure definitions

Measure of sociality	Description
Average group size	Average size of groups in which the calf was observed defined by 10-m chain rule; includes mother and calf
Time alone (%)	Percent of observation time in which the calf was not in a group with any other individual
Time in groups (%)	Percent of observation time during which the calf was observed in a group containing an individual other than the calf's mother
Time socializing (%)	Percent of observation time in which the calf was actively socializing
Time in group (%) with	Percent of observation time in which the calf was observed in a group consisting of
Mother only	Mother only
All females	One or more females excluding the mother
All males	One or more males
Mixed sex	Both males and females excluding the mother
Time (%) associated with	Percent of observation time in which the calf was observed in a group consisting of at least one of the following age-sex classes:
Adult female	Adult female excluding mother
Adult male	Adult male
Juvenile female	Juvenile female
Juvenile male	Juvenile male
Calf, female	Calf, female
Calf, male	Calf, male
Average fission–fusion rate	The average number of times per hour the calf's group composition changes, including the mother

Association, however, is not the only social measure from which social networks may be constructed; indeed, measuring association is generally considered a proxy for interaction data because interactions are often difficult to observe in the field. Grooming in primates and petting (an affiliative behavior where one dolphin actively moves the pectoral fin on a body part of another dolphin; Fig. 6.2) in dolphins provide excellent interaction data from which to build social networks. We constructed social networks based on petting interaction events observed during all focal follows of Shark Bay mother–calf pairs during the first 4 years of the lives of LEN and MIG (Fig. 6.3). Because of the difficulty of obtaining these data, we did not wish to assign too much meaning to the number of observed interactions; therefore, these petting networks are binary, meaning a line between two individuals indicates the presence of a relationship but contains no information about strength. These interaction networks provide an additional dimension to the investigation of social patterns provided by association networks that assume that associated individuals interact with each other. It is important to note at this juncture that the networks presented here are static and were constructed by combining 4 years of data to create a single network. Although multiple years provide more data with which to determine associations, it is likely that each calf's social network differs from year to year, with

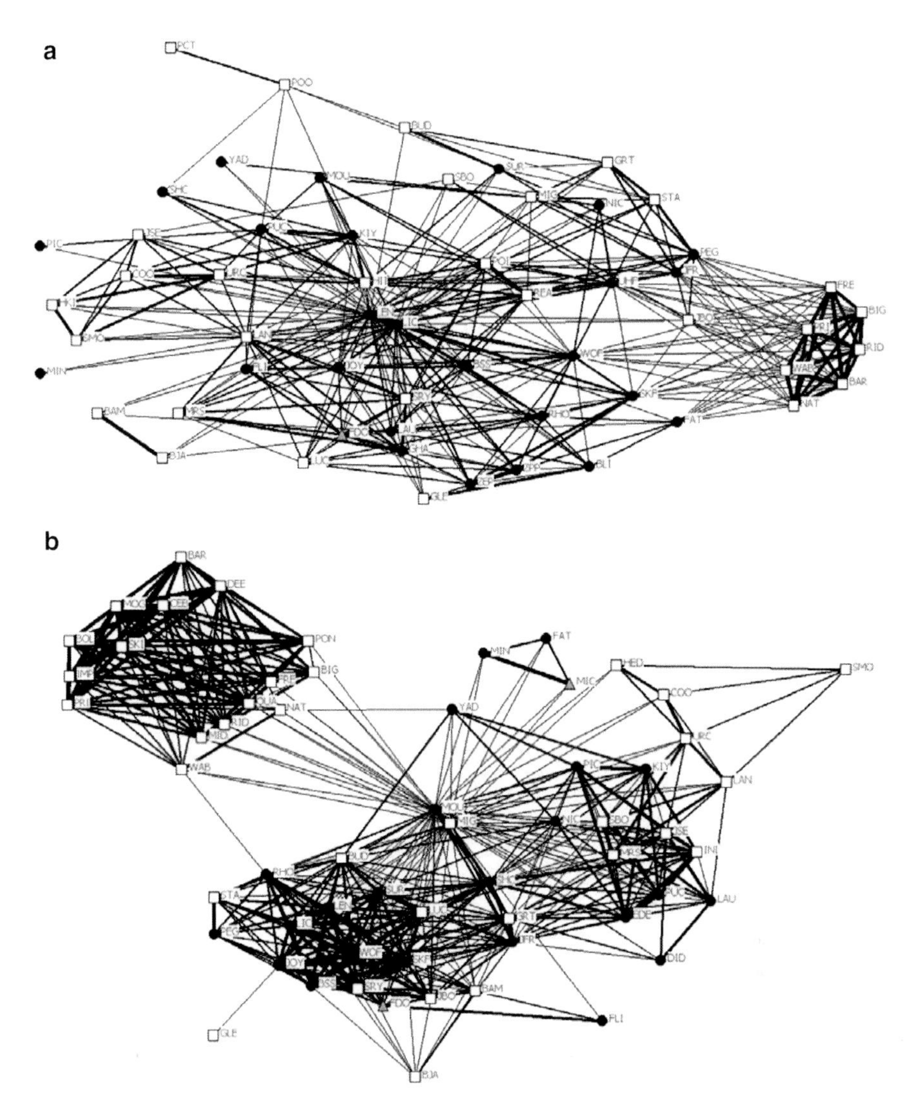

Fig. 6.1 Weighted ego networks of the calves LEN (**a**) and MIG (**b**). *Thicker edges* indicate stronger relationships. Only those edges with an average proportion of observations (APO) >0.50 are shown for clarity; however, all associations were included in the analysis. The mother is the closest node to the focal calf found near the center of the graphs. Mothers are LIC and MOU, respectively. *Circles*, females; *squares*, males; *triangles*, unknown

relationships forming and fading over time. Dynamic social network analysis, however, presents novel methodological obstacles that are beyond the scope of this case study. More detailed descriptions of the metrics calculated from both the association and petting networks are available in Chap. 10 of this volume.

Fig. 6.2 Two juvenile dolphins petting

6.3 Results and Discussion

The results of various traditional non-network measures of individual sociality are presented in Table 6.2, and the social network analysis results are presented in Table 6.3.

Although we cannot draw inferences from the analysis of two calves, our intention here is to emphasize varying aspects of sociality and the measures with which to address them. For example, MIG spends a larger portion of his time alone than LEN, which may lead to the conclusion that MIG is less social than LEN. However, while in a group MIG spends more than twice as much time socializing with other dolphins. Additionally, as expected by her greater amount of time in groups, LEN spends more time with every age-sex class than MIG, with the exception of male calves. It is interesting that MIG, a male calf, spends considerably more time with other male calves than does LEN, a female calf. Finally, although fission–fusion social systems receive a great deal of attention in the literature, the rate of change in group composition is rarely reported. In this case, LEN's fission–fusion rate is greater than that of MIG, which is also not surprising given the difference in time spent alone versus time spent in groups.

As for the association-based ego networks, LEN and MIG had a similar number of associates, at 57 and 62, respectively. Visual inspection of these graphs suggest that LEN and her mother are in the center of a large subgroup, while MIG and his mother are more peripherally connected to a couple of subgroups (Fig. 6.1).

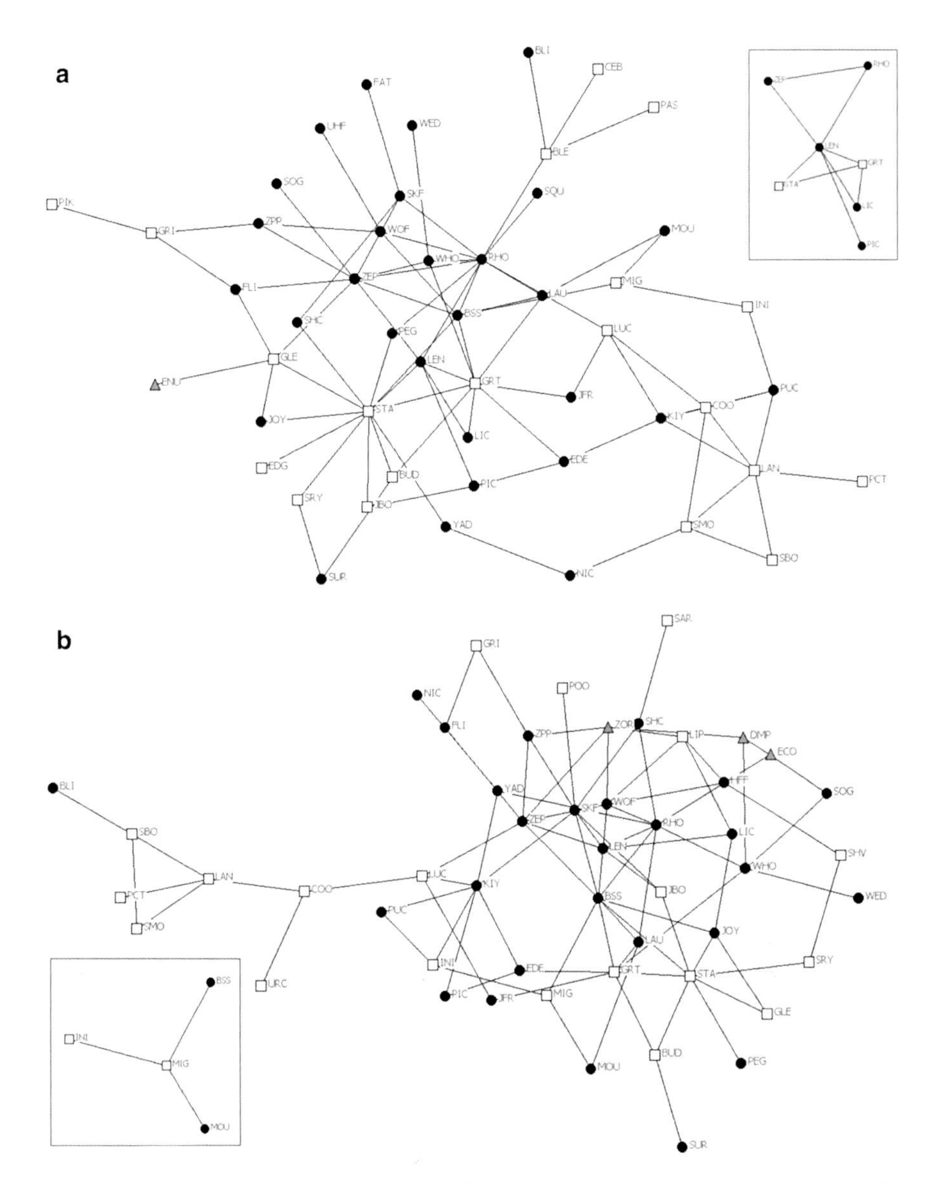

Fig. 6.3 Main component of social networks built from petting interaction data for (**a**) years LEN was a calf (2002–2005) and (**b**) years MIG was a calf (2004–2007). LEN's and MIG's petting ego networks appear in the *insets*. *Circles*, females; *squares*, males; *triangles*, unknown

Thus, MIG's ego network also appears to contain more clusters, some of which are adult males likely consorting with MIG's mother near the end of his infancy. However, although visual inspection of networks is a useful investigation technique, observed patterns should be verified using appropriate network metrics. For example, most network metrics at both the individual and the whole ego network

Table 6.2 Non-network social measures results based on focal follow data for LEN and MIG

	LEN ♀	MIG ♂
Mean group size	6.9	5.3
Time alone (%)	4.6	16.1
Time in groups (%)	70.4	54.1
Time socializing (%)	2.5	5.8[a]
Time (%) in group with		
Mother only	25.0	30.0
All females	21.2	4.7
All males	0.0	3.5[a]
Mixed sex	47.8	42.9
Time (%) associated with		
Adult female	63.8	49.0
Adult male	31.4	18.5
Juvenile female	52.5	23.8
Juvenile male	23.4	9.8
Calf female	40.0	16.7
Calf male	27.1	42.3[a]
Mean fission–fusion rate (number/h)	7.5	5.9

[a]LEN had higher levels of association overall, but MIG associated more often with young males and spent a greater percentage of observation time socializing

Table 6.3 Social network metrics calculated from the association ego networks of LEN and MIG at both individual and ego network levels

	LEN		MIG	
	Individual	Ego network average	Individual	Ego network average
Strength	31.45 (0.06)	12.70 (0.14)	33.15 (0.06)	12.48 (0.17)
Eigenvector centrality	0.26 (0.01)	0.12 (0.01)	0.24 (0.01)	0.11 (0.01)
Weighted clustering coefficient	0.23 (0.01)	0.44 (0.01)	0.20 (0.01)	0.57 (0.01)

Individual metrics refer to those of LEN and MIG whereas ego network metrics are the average of all individuals in the ego network. Metrics were calculated in SOCPROG 2.3 using all available associations. Square brackets contain bootstrap standard errors using 1,000 replicates. Strength indicates how connected an individual is to others by summing the weights of his/her associations. Eigenvector centrality is an additional measure of connectedness, but also considers the associations of an individual's neighbors (e.g., an individual may have high eigenvector centrality by being strongly linked to many individuals or by being linked to fewer well-connected individuals). Weighted clustering coefficients show how 'cliquish' or tight the sub-networks are (all individuals within a clique are also tightly associated). More detailed descriptions of these metrics are available in Chap. 10 of this volume or in Whitehead (2008)

level are similar between the two calves with the exception of the network-wide clustering coefficient, which is higher in MIG's ego network.

The most obvious differences between the association-based ego networks (Fig. 6.1) and the petting networks (Fig. 6.3) are size and density, defined as the number of actual edges divided by the number of possible edges in the network.

LEN's and MIG's ego networks have unweighted densities of 0.55 and 0.36, respectively, but the entire petting networks containing LEN as a calf and MIG as a calf have much lower densities, of 0.03 and 0.04, respectively. Although LEN and MIG were associated with 57 and 62 other dolphins, respectively, petting was only observed between LEN and 6 others, and between MIG and 3 others, which may suggest stronger social relationships between these individuals; however, considerably more data are necessary to draw any conclusions.

The aim of this case study is to illustrate some of the diverse social measures available to researchers and the desirability of using multiple measures to discover those features most important to a given society or research query. We particularly advocate capitalizing on recent advances in social network analysis that allow for the quantification of multi-actor interactions. A thorough investigation including multiple dimensions of sociality coupled with careful consideration of the inferences drawn from each measure is necessary to provide the detail required for a more complete understanding of animal societies.

References

Altmann J (1974) Observational study of behavior: sampling methods. Behaviour 49:227–266

Borgatti SP (2002) NetDraw. http://www.analytictech.com/netdraw/netdraw.htm

Borgatti SP, Everett MG, Freeman LC (2002) UCINET for Windows: software for social network analysis. Analytic Technologies, Harvard, MA.

Brager S (1999) Association patterns in three populations of Hector's dolphin, *Cephalorhynchus hectori*. Can J Zool 77:13–18

Connor RC, Smolker RA, Richards AF (1992) Two levels of alliance formation among male bottlenose dolphins (*Tursiops* sp.). Proc Natl Acad Sci USA 89:987–990

Connor RC, Wells R, Mann J, Read A (2000) The bottlenose dolphin: social relationships in a fission–fusion society. In: Mann J, Connor RC, Tyack P, Whitehead H (eds) Cetacean societies: field studies of dolphins and whales. The University of Chicago Press, Chicago, pp 91–126

Connor RC, Heithaus MR, Barré LM (2001) Complex social structure, alliance stability and mating access in a bottlenose dolphin "super-alliance". Proc R Soc B Biol Sci 268:263–267

Connor RC, Watson-Capps JJ, Sherwin WB, Krützen M (2011) A new level of complexity in the male alliance networks of Indian Ocean bottlenose dolphins (*Tursiops* sp.). Biol Lett 7: 623–626

Coscarella MA, Gowans S, Pedraza SN, Crespo EA (2011) Influence of body size and ranging patterns on delphinid sociality: associations among Commerson's dolphins. J Mammal 92: 544–551

Gibson QA, Mann J (2008a) Early social development in wild bottlenose dolphins: sex differences, individual variation and maternal influence. Anim Behav 76:375–387

Gibson QA, Mann J (2008b) The size, composition and function of wild bottlenose dolphin (*Tursiops* sp.) mother–calf groups in Shark Bay, Australia. Anim Behav 76:389–405

Gibson QA, Mann J (2009) Do sampling method and sample size affect basic measures of dolphin sociality? Mar Mamm Sci 25:187–198

Goodall J (1986) The chimpanzees of Gombe: patterns of behavior. Harvard University Press, Cambridge

Krützen M, Sherwin WB, Connor RC, Barré LM, Van de Casteele T, et al (2003) Contrasting relatedness patterns in bottlenose dolphins (*Tursiops* sp.) with different alliance strategies. Proc R Soc B Biol Sci 270:497–502

Leigh SR, Blomquist GE (2007) Life history. In: Campbell C, Fuentes A, MacKinnon KC, Panger M, Bearder S (eds) Primates in perspective. Oxford University Press, Oxford, pp 396–407

Lusseau D, Schneider K, Boisseau OJ, Haase P, Slooten E, Dawson SM (2003) The bottlenose dolphin community of Doubtful Sound features a large proportion of long-lasting associations. Behav Ecol Sociobiol 54:396–405

Mann J, Watson-Capps JJ (2005) Surviving at sea: ecological and behavioural predictors of calf mortality in Indian Ocean bottlenose dolphins (*Tursiops* sp.). Anim Behav 69:899–909

Mann J, Connor RC, Barre LM, Heithaus MR (2000) Female reproductive success in bottlenose dolphins (*Tursiops* sp.): life history, habitat, provisioning, and group-size effects. Behav Ecol 11:210–219

Mann J, Sargeant BL, Minor M (2007) Calf inspections of fish catches in bottlenose dolphins (*Tursiops* sp.): opportunities for oblique social learning? Mar Mamm Sci 23:197–202

Mann J, Stanton MA, Patterson EM, Bienenstock EJ, Singh LO (2012) Social networks reveal cultural behaviour in tool using dolphins. Nat Commun 3:980

Smolker R, Richards A, Connor RC, Pepper J (1992) Sex differences in patterns of association among Indian Ocean bottlenose dolphins. Behaviour 123:38–69

Stanton MA, Mann J (2012) Early social networks predict survival in wild bottlenose dolphins. PLoS One 7(10):e47508

Stanton MA, Gibson QA, Mann J (2011) When mum's away: a study of mother and calf ego networks during separations in wild bottlenose dolphins (*Tursiops* sp.). Anim Behav 82:405–412

Symington MM (1988) Demography, ranging patterns, and activity budgets of black spider monkeys (*Ateles paniscus chamek*) in the Manu National Park, Peru. Am J Primatol 15:45–67

Tsai Y, Mann J (2013) Dispersal, philopatry and the role of fission–fusion dynamics in bottlenose dolphins. Mar Mamm Sci 29:261–279

Wells RS, Scott MD, Irvine AB (1987) The social structure of free-ranging bottlenose dolphins. Curr Mamm 1:247–305

Whitehead H (2008) Analyzing animal societies: Quantitative methods for vertebrate social analysis. University of Chicago Press, Chicago

Whitehead H (2009) SOCPROG programs: analysing animal social structures. Behav Ecol Sociobiol 63:765–778

Williams T, Friedl W, Fong M, Yamada R, Sedivy P, Haun J (1992) Travel at low energetic cost by swimming and wave-riding bottlenose dolphins. Nature (Lond) 355:821–823

Part II
Life History and Social Evolution

Chapter 7
Female Coexistence and Competition in Ringtailed Lemurs: A Review of a Long-Term Study at Berenty, Madagascar

Yukio Takahata, Naoki Koyama, Shin'ichiro Ichino, Naomi Miyamoto, Takayo Soma, and Masayuki Nakamichi

Y. Takahata (✉)
School of Policy Studies, Kwansei Gakuin University, Gakuen 2-1, Sanda 669-1337, Japan
e-mail: z96014@kwansei.ac.jp

J. Yamagiwa and L. Karczmarski (eds.), *Primates and Cetaceans: Field Research and Conservation of Complex Mammalian Societies*, Primatology Monographs, DOI 10.1007/978-4-431-54523-1_7, © Springer Japan 2014

Abstract Ringtailed lemurs (*Lemur catta*) form female-bonded/matrilineal social groups. In this review, we summarize our long-term field study carried out at Berenty, Madagascar to discuss the balance between female coexistence and competition in this prosimian primate. In our study population, females cooperatively competed against the females of neighboring groups; however, they also displayed persistent aggression toward females within their own group, which occasionally resulted in group fission. The correlation between female fecundity and group size generally agreed with the intergroup feeding competition (IGFC) model. No significant differences in reproductive success were seen among rank categories of females in the medium- and small-sized groups. In contrast, low-ranked females of large-sized groups exhibited lower reproductive success, which may have been the result of within-group competition. Males appeared to be "parasites" on the female groups.

Keywords Berenty • Female coexistence • Female competition • Ringtailed lemur

7.1 Introduction

Life histories and reproductive features of female primates have been studied by numerous primatologists to (1) evaluate sexual selection hypotheses (e.g., Small 1989), (2) assess socioecological hypotheses (Sterck et al. 1997), and (3) analyze the balance between within-group competition and between-group competition (Wrangham 1980; van Schaik 1983). However, relatively few long-term data are available for wild prosimians.

The ringtailed lemur (*Lemur catta*) is a diurnal group-living prosimian that inhabits the dryland of southern Madagascar (Sussman et al. 2006). Current studies include ongoing observations of two wild ringtailed lemur populations at the Berenty and Beza Mahafaly Reserves, Madagascar (Jolly et al. 2006a). Ringtailed lemur populations generally form discrete female-bonded/matrilineal social groups, although some cases of female transfers between groups have occurred (Sauther et al. 1999). Group size usually ranges from 3 to more than 20 lemurs (Mittermeier et al. 1994). A linear dominance rank order exists among adult members within a group; however, occasionally the rank can become convoluted and change abruptly (Sauther et al. 1999; Ichino 2004). Adult females are socially dominant over adult males (Jolly

N. Koyama • N. Miyamoto • T. Soma
Center for African Area Studies, Kyoto University, Yoshida Shimoadachicho 46, Sakyo-Ku, Kyoto 606-8501, Japan

S. Ichino
Primate Research Institute, Kyoto University, Kanrin, Inuyama 484-8506, Japan

M. Nakamichi
Graduate School of Human Sciences, Osaka University, Yamadagaoka 1-2, Suita 565-0871, Japan

1984), and they actively defend their home range against neighboring groups (Mertl-Millhollen et al. 2006). Females sometimes exhibit severe aggression toward other females in their own group (Vick and Pereira 1989), which may result in group fission (Ichino 2006; Ichino and Koyama 2006).

In a review discussing the evolution of female social relationships in nonhuman primates, Sterck et al. (1997) considered the ringtailed lemur a "dispersal-egalitarian" species, although field data indicate that they could be regarded as a "resident-nepotistic" species, such as the *Macaca* spp. (Koyama et al. 2002). Nevertheless, several distinct differences exist between ringtailed lemurs and cercopithecoid species (Sauther et al. 1999). For example, matrilineal "inheritance" of dominance status is rare (Pereira 1995). Severe within-group female competition ("targeting aggression," as defined by Vick and Pereira 1989) frequently results in "female eviction:" that is, victimized females are driven from natal groups. In addition, ringtailed lemurs do not maintain complicated social relationships in multimale and multifemale groups, as do cercopithecine primates (Sauther et al. 1999). Thus, the ringtailed lemur may represent a translational stage in the evolution of the sophisticated female-bonded/matrilineal society maintained by cercopithecoid species. Ringtailed lemurs therefore appear to be a suitable species with which to analyze hypotheses concerning the balance between female coexistence and female competition.

We have observed a wild population of ringtailed lemurs at Berenty Reserve, Madagascar since 1989, with regard to social communication (Oda 1996; Oda and Masataka 1996), social relationships and behavior (Nakamichi and Koyama 1997, 2000; Nakamichi et al. 1997), female reproductive parameters (Koyama et al. 2001), population density (Koyama et al. 2002), long-term dominance relationships (Koyama et al. 2005), body mass and tick infection (Koyama et al. 2008), group fission, female eviction, range takeover (Ichino and Koyama 2006), and feeding ecology (Soma 2006). In this review, we summarize these observations and discuss how ringtailed lemurs maintain a balance between coexistence and competition.

7.2 Background

7.2.1 History of the Berenty Reserve

The population of ringtailed lemurs inhabiting Berenty Reserve in southern Madagascar has been studied by Alison Jolly and her colleagues since the 1960s (Jolly 1967; Jolly et al. 2006b). Berenty Reserve is in the "Berenty Estate" located next to the Mandrare River (24°98′S–46°28′E). The estate was founded by the de Heaulme family in 1936 as a sisal (*Agave rigida*) plantation (Jolly 2004). It is characterized by a semiarid climate, with temperatures ranging from 40 °C at midday to 10 °C at night (Jolly et al. 2006b). Most rainfall occurs from November to February, and mean annual rainfall during 1989–1998 was 580.6 mm (Koyama et al. 2001).

Creation of the sisal plantation has resulted in extensive disturbance of the natural vegetation in this area (Jolly et al. 2006b). Berenty Reserve represents a fragment of a protected forest (approximately 200 ha), comprising four zones: (1) "Ankoba," which is a replanted forest with nonnative trees; (2) the "Tourist Front," which is part of the western boundary of the reserve that is studded with tourist bungalows; (3) the "Gallery" forest, which is a natural forest dominated by *Tamarindus indica*, with a canopy covering 50 % or more of the sky; and (4) the "Scrub" forest, which is a drier, natural forest with 50 % or more open sky (Jolly et al. 2002). This reserve also serves as habitat for potential predators, such as raptors (e.g., *Polyboroides radiatus*), domestic dogs, and cats. Wild populations of Verreaux's sifaka (*Propithecus verreauxi*), white-footed sportive lemur (*Lepilemur leucopus*), and gray mouse lemur (*Microcebus murinus*) also inhabit this reserve. In addition, an artificially introduced hybrid population of red-fronted brown lemur (*Eulemurluluvusrufus*) and collared brown lemur (*E. collaris*) shares the reserve (Pinkus et al. 2006).

A 100-ha area, excluding Ankoba, is called "Malaza" and has been the focus of four decades of research on ringtailed lemurs (Jolly et al. 2006b). This area supported 17 groups of ringtailed lemurs in 1974, comprising 153 individuals, whereas 25 groups comprising 292 individuals were observed in 2005 (Jolly et al. 2006c). This population increase may be partially attributable to artificial influences, for example, an off-season food supply derived from artificially introduced nonnative trees, water basins, and banana feeding by guides and tourists (Jolly et al. 2006c). Banana feeding has been discouraged since 1999, and the artificial water supply was eliminated to control the lemur population in 2007.

Intergroup relationships seem to have undergone a transformation with the increase in population and group densities (Jolly et al. 2006c). Group ranges overlapped somewhat in the 1960s, but each group maintained and defended their own exclusive territory (Mertl-Millhollen et al. 1979). Home range boundaries remained stable over time in the 1970s, but intergroup aggression increased. Jolly (1972) reported that different groups replaced each other in every favored spot on the basis of a "time plan" rather than by spatially exclusive possession. Groups attempted to defend their ranges, but the defense did not ensure exclusivity of the use of ranges (Mertl-Millhollen et al. 1979). From 1999 to 2000. Pride et al. (2006) found that each group experienced intergroup conflicts 2.7 times/day in the "Tourist Front," 1.7 times/day in the "Gallery" forest, and 0.4 times/day in the "Scrub" forest. Pride et al. also pointed out that the outcomes of these conflicts depended not on group size/number of adult females, but on location, with groups tending to win conflicts within their core ranges. Thus, ringtailed lemurs exhibit high adaptability to environments: they can be territorial or not, corresponding to the population density (Sauther et al. 1999).

7.2.2 *Our Study Area and Population*

We began long-term observations on a population of ringtailed lemurs within an area of 14.2 ha of "Malaza" in September 1989 (Fig. 7.1). This area corresponds to the "Gallery" forest and the "Tourist Front." The most abundant tree species is

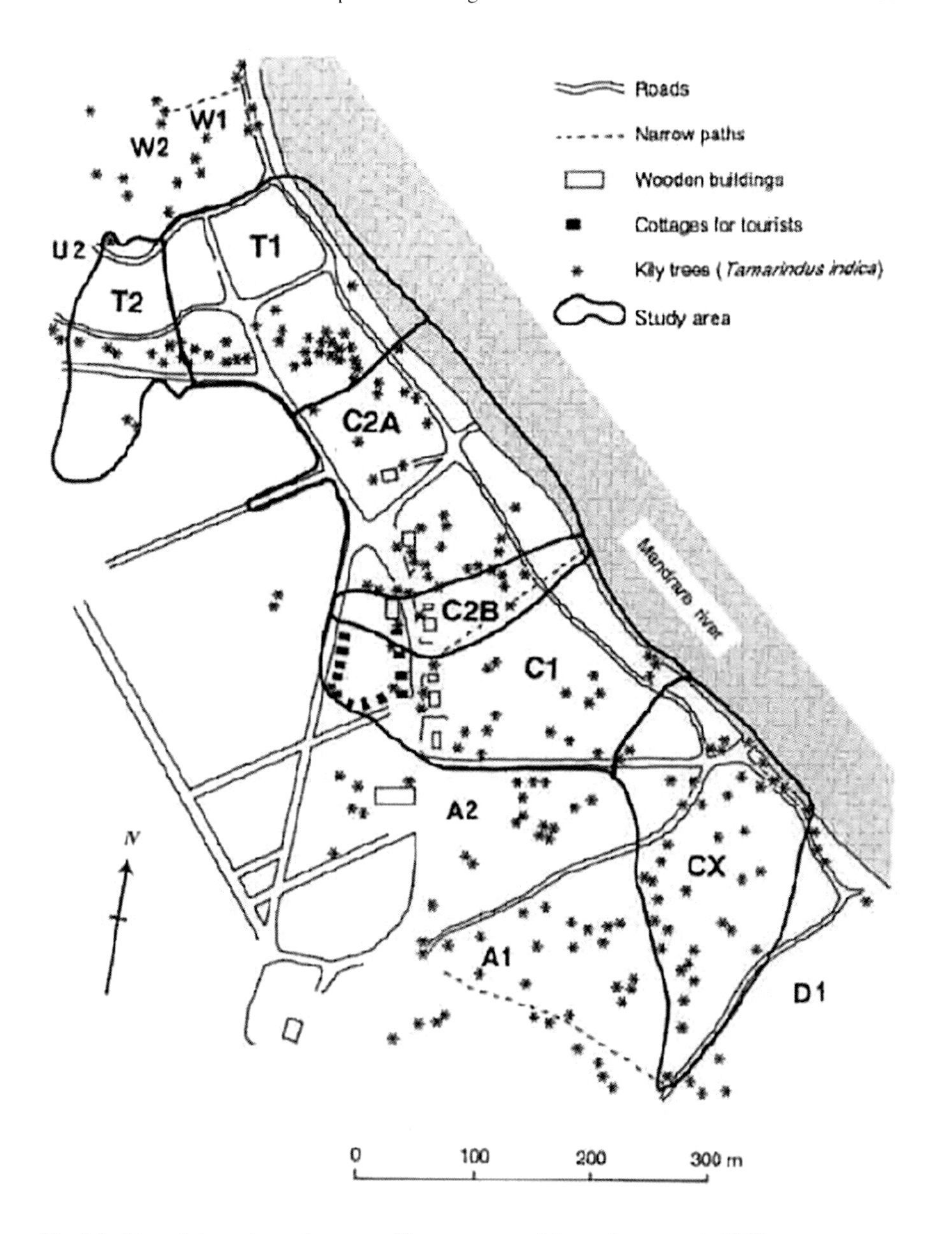

Fig. 7.1 Map of the main study area and home ranges of the study groups in 1999

Tamarindus indica, a keystone species for ringtailed lemurs (Simmen et al. 2006). *T. indica* intake alone accounts for 34.9 % of the ringtailed lemur's feeding time in our study population (Soma 2003).

The *T. indica* population at Berenty Reserve has declined, although the reason is uncertain (Blumenfeld-Jones et al. 2006). The number of *T. indica* trees per ring-tailed lemur has decreased in the study area, from 2.8 in 1989 to 1.8 in 2000 (Koyama et al. 2006). In addition, the artificially introduced hybrid population of

Table 7.1 Observation periods

N. Koyama	September–December 1989, August–December 1990, August–December 1991, August–September 1992, August–December 1993, August–December 1994, September–December 1995, September–December 1996, August–September 1997, July–December 1998, November–December 1999, November 2000
M. Nakamichi	August–November 1994
Y. Takahata	August–October 1997, August–October 1998
N. Miyamoto	August 1997–June 1998, August–October 1999
S. Ichino	August 1998–September 1999, August–November 2000, March 2001–January 2002
T. Soma	August 2000–January 2002

brown lemurs has increasingly emerged as a food competitor (Pinkus et al. 2006). Inevitably, ecological conditions became worse because of both intraspecies and interspecies competition over food resources.

We have identified all the ringtailed lemurs in the study population by facial features, pelage color, and idiosyncratic markings (hair dying). We have also chronicled their troop affiliations over time. Table 7.1 shows our observation periods. Female ringtailed lemurs at Berenty have a short mating season, which occurs around April or May. For example, the females of Troop A were receptive for only 2 consecutive days during the 1982 mating season (Koyama 1988). Thus, most births occur around September. Therefore, we checked group composition and changes in membership (birth, death, immigration, and emigration) from August to November, before and after every year's birth season.

7.3 Life History and Social Relationships in the Study Population

7.3.1 Reproduction and Development

As mentioned earlier, ringtailed lemurs breed seasonally (Jolly 1967; Sauther 1991). According to data recorded in the study population from 1989 to 1999, 82 % of births occur in September, corresponding with the end of the dry season (Koyama et al. 2001), and mean birthrate (percentage of adult females who gave birth in a breeding season) is 75.0 %. Infant mortality rate within 1 year after birth is 37.7 %; possible reasons include malnutrition, infanticide (Ichino 2005), predation, and accidents (e.g., drowning and straying from mothers). The birth sex ratio (female:male) is 1:1.19, which is not significantly skewed in relationship to the population level.

In contrast to most anthropoid primates, ringtailed lemurs exhibit no sexual dimorphism. In 1999, we captured 101 ringtailed lemurs and measured their body mass (Koyama et al 2008). The mean body mass of adult females was 2.27 kg and that of adult males was 2.22 kg. Body mass increases up to 3 years of age in both sexes, at which time growth appears to reach a growth plateau. Males do not display extended growth (bimaturism), nor do they grow faster than females (rate dimorphism).

Approximately 60 % of the infants survive up to 2 years, with no sexual differences in survival rate observed during 1989–1999 (Koyama et al. 2002). Females remain within their natal groups and begin to breed when they are about 2 years old (i.e., female philopatry) (Koyama et al. 2001). Most females give birth at the age of 3 years (45.8 % of all first births) or 4 years (37.5 %), after which the birthrate fluctuates by 12 years of age (approximately 75 %). However, insufficient information is available to estimate the entire reproductive lifespan of female ringtailed lemurs.

In contrast to females, all males leave their natal groups by the time they are 5 years old, close to puberty (Koyama et al. 2002). Thereafter, males change non-natal groups every several years. Male transfer frequently occurs around the mating season, indicating that transfer may be proximately driven by sexual competition and mate choice, as Sussman (1992) reported for the Beza Mahafaly population. Males migrate from one group to another in pairs or groups, that is, "migration partners," as described by Gould (1997). Such males maintained affiliative relationships within a group at both Beza Mahafaly and Berenty (Nakamichi and Koyama 1997).

The longevity of ringtailed lemurs is still unknown. Several females may have lived to become at least 13 years old at Berenty (Koyama et al. 2001, 2002). Such females occasionally distance themselves from groups, ranging alone, probably because of senility. In contrast, few reliable longevity data exist for males because of their dispersal from the study population.

7.3.2 Dominance Rank Order Among Adult Ringtailed Lemurs

Adult members within a group display a linear dominance rank order, although rank can be convoluted (see tables 1 and 2 in Koyama et al. 2005). Such dominance ranks are determined based on (1) approach–retreat interactions while feeding and drinking and (2) submissive vocalizations (spat calls) (Koyama et al. 2005).

Adult females are dominant over adult males in the study population (Koyama et al. 2005), as Jolly (1984) pointed out. This "female dominance" has been a puzzle for primatologists, and numerous hypotheses have been proposed to account for this behavior, which is still an open question (Sauther et al. 1999; Rasamimanana et al. 2006).

Mothers tend to be dominant over their daughters, irrespective of their ages (Nakamichi et al. 1997). Even when old females are ranked in the lowest position, they are still dominant over their daughters, suggesting the existence of long-term psychological bonding between old mothers and mature daughters.

Males reach their highest ranks at 8–9 years of age in the study population (Koyama et al. 2005), and alpha males hold their rank for an average of 2.2 years. Although no significant correlation is found between female rank and body mass, higher-ranked males tend to be heavier than lower-ranked ones (Koyama et al. 2008). Heavier males should have an advantage over others in male-to-male competition for mates (Koyama 1988), so it is puzzling why males have not developed sexual dimorphism despite the polygynous mating system.

7.3.3 Social Relationships Among Adult Ringtailed Lemurs

Adult females are the "core" of the social relationships within a group (Oda 1996; Nakamichi and Koyama 1997). In our observations, close proximity and affiliative interactions (e.g., grooming) occur more frequently between related females within second degrees of consanguinity than between unrelated females. Adult ringtailed lemurs usually bilaterally groom each other (mutual grooming); however, they occasionally exhibit unilateral grooming as a type of "greeting behavior." In such cases, subordinates are likely to groom dominants more frequently than vice versa.

Newborn infants influence the social relationships between mothers and unrelated females during the birth season (Nakamichi and Koyama 2000). Mothers with newborn infants receive a number of affiliative contacts (e.g., grooming and infant-licking) from other group members, including unrelated adult males, during the first and second months of motherhood. Infant-licking behavior may enable adult females and males to create or maintain stable social relationships with one another.

Nakamichi and Koyama (1997) found that adult males form affiliative partnerships with all ranks and age classes, as reported for a wild group of Japanese macaques (Furuichi 1985). Gould (1997) found similar relationships among adult males at Beza Mahafaly, particularly between "migration partners." It is possible that such high-level associations may confer benefits (predator protection, health, and enhanced detection of attacks by group males). Indeed, group males constantly attempt to drive away nongroup males, probably to prevent nongroup males from joining the group and to keep them away from group females (Nakamichi and Koyama 1997).

There appears to be no consistent tendency in social relationships between adult males and females. Nakamichi and Koyama (1997) reported that most adult males of Troop T1 had frequent proximity relationships with one or more adult females during the 1994 birth season. In contrast, only the alpha male had frequent proximity relationships with some adult females within Troop C2, whereas other males did not have such relationships with females (see fig. 2 of Nakamichi and Koyama 1997). At Beza Mahafaly, Sauther and Sussman (1993) reported that only a single nonnatal "central" male is likely to monopolize social interactions with adult females in groups studied intensively, whereas Gould (1996) observed that males of all ranks and of variable tenure have social interactions with both high- and low-ranking females. These results suggest that ringtailed lemurs possess an unexpected social ability to manipulate social relationships among themselves according to the social situation, or it may simply indicate undeveloped intelligence.

7.4 Population and Group Changes in the Study Population: Between-Group Competition

7.4.1 Increases in Population Density and Immigration/Emigration

When we began our observations in September 1989, three groups of ringtailed lemurs consisting of 63 individuals existed within our study area of 14.2 ha (4.4 lemurs/ha) (Koyama et al. 2002). Six groups comprising 82 lemurs (5.8 lemurs/ha) were present by September 1999. The mean annual rate of population increase was 2.7 %, mainly the result of the high fecundity of adult females. During the 10-year time period, 204 infants were born, 125 lemurs died, 58 lemurs immigrated into the study population, and 118 emigrated, bringing a net of 19 new lemurs into the study population.

Immigration and emigration play a large role in the study population size, and these movements fall into several categories (Ichino and Koyama 2006), as follows. (1) The most usual case is "male transfer." Male lemurs change groups after a tenure that varies from 1–7 years (Koyama et al. 2002). The mean length of stay in a group is 3.1 years. Thus, males disperse from natal sites, although it is unknown how far. From 1989 to 1999, 90 males left the study population and 53 males immigrated into it. Thus, the high population in the study area may have biased the entire population structure of the Malaza area as a "male source." (2) "Female transfer" infrequently occurs. Two unidentified adult females immigrated into Troop C2A in 1998 (KN-group in Fig. 7.2), possibly because they had been evicted from other groups beyond the study area. (3) "Group invasion" or "group dismissal/eviction" to and from the study area occasionally occurs. One group invaded the study area (Troop U2 in Fig. 7.2), and three groups left the study area (Troops B and U2, and HSK-group in Fig. 7.2), from 1989 to 1999.

The adult sex ratio (adult females vs. adult males) during the birth season fluctuated from 1:0.615 to 1:1.22 between 1989 and 1999 but without consistent correlation with time or population density (Koyama et al. 2002). Pooled data show that the number of adult males per adult females is 0.968, suggesting that every adult male principally belongs to one of the social groups during the birth season. Meanwhile, the mean group size decreased from 21.0 in 1989 to 13.7 in 1999.

7.4.2 Female Eviction and Group Fission

"Female eviction" is a common social phenomenon in ringtailed lemurs (Vick and Pereira 1989; Ichino and Koyama 2006). It is defined as follows. One or several females become the target of persistent aggression by other females (targeting aggression, defined by Vick and Pereira 1989) and are eventually evicted from their group. Female eviction primarily occurs around birth seasons and in large-sized

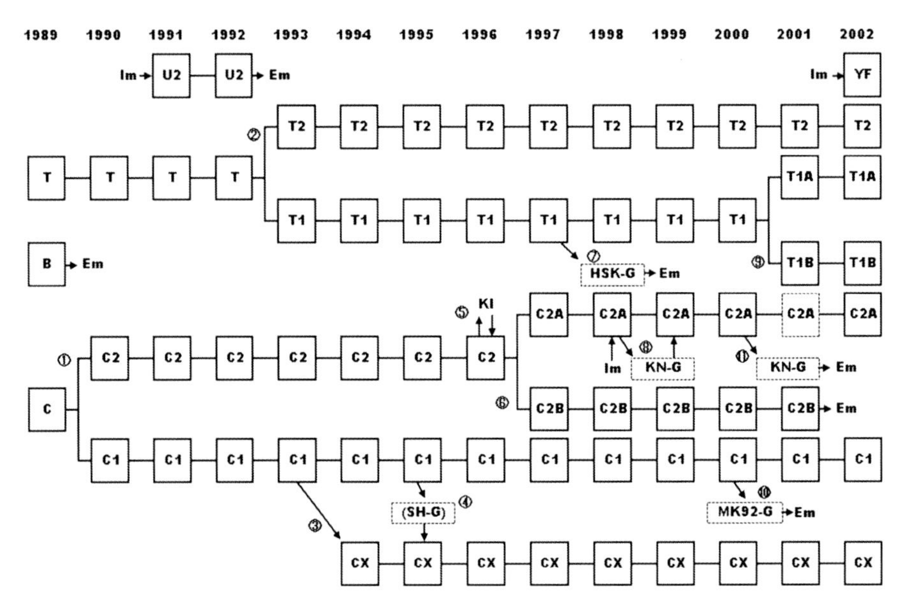

Fig. 7.2 Female eviction and group fission events that occurred from September 1989 to January 2002. Each *square* indicates groups living within the study area. The *dotted squares* indicate nomadic groups. *Im*, shifting of a group to the study area; *Em*, shifting of a group beyond the study area. *Circled numbers* indicate cases of female eviction/group fission

groups comprising 16 or more lemurs with 7 or more adult females (Ichino and Koyama 2006).

Figure 7.2 shows the troop histories of the study population from September 1989 to January 2002. Six cases of female eviction were directly observed during the 12.5 years. In one case, the evicted females were able to rejoin the original groups (KN-G in 1998; Fig. 7.2). In another case, the evicted females established a new home range, and mature males joined them, thereby forming a new social group, that is, "group fission" (Troop T1B in 2000). In three cases, the evicted females could not establish a stable range within the study area and eventually disappeared from the study area, as "nomadic groups" without definite home ranges (HSK-G in 1997, MK92-G in 2000, and KN-G in 2001). Evicted females would be expected to encounter numerous difficulties establishing new ranges and securing mating partners because of high group density. In particular, the number of adult males per adult female tends to be low in these newly formed groups, which suggests that males hesitated to join small groups. In the remaining case, the evicted females sporadically fought with females of another group and dominated them, eventually forming one group. Such a case can be termed "group fusion" (SH-G in 1995). If evicted females join other groups without aggressive fighting, it is called "female transfer".

Female eviction usually occurs among matrilineal kin groups (Ichino and Koyama 2006), and group males rarely play a dominant role, indicating that female

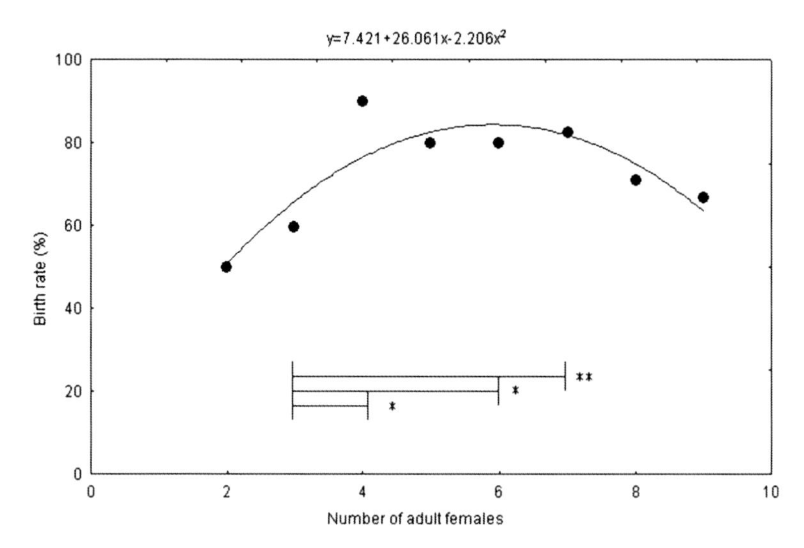

Fig. 7.3 Regression between the number of adult females and birthrate. Birthrate generated an inverted U-shaped curve, which approximated a second-degree curve ($y=7.42+26.06x-2.21x^2$) ($r^2=0.774$, $P<0.03$). The group with three adult females had a lower birthrate than those with four, six, and seven adult females ($\chi^2=5.92$, $df=1$, $P<0.02$; $\chi^2=5.09$, $df=1$, $P<0.03$; and $\chi^2=6.8$, $df=1$, $P<0.01$, respectively)

eviction is the result of female within-group competition through kin selection; these evictions may function to decrease the intensity of within-group competition by reducing the number of group members.

In addition to these cases of female eviction, four group fission events occurred in the study population between 1989 and 2002 (Troops C1 and C2 in 1989, Troops C1 and CX in 1993, Troops T1 and T2 in 1993, Troops C2A and C2B in 1997; Fig. 7.2). As the processes leading to these specific group fission events were not directly observed, it was uncertain whether the females were evicted by other females or if they voluntarily left their groups. As a result of these female evictions or group fissions, the number of groups in the study area increased from three in 1989 to seven in 2002 (Fig. 7.2).

7.4.3 Between-Group Competition

High group density (42.2 groups/km^2 in 1999) should have intensified "scramble competition" over food resources in the study population. In such territorial defenses, adult females, not males, play the most active role in intergroup confrontations, regardless of dominance rank (Nakamichi and Koyama 1997). Groups occasionally take over the secure ranges of other groups, and such cases are termed "range take-over" (Ichino and Koyama 2006). During our observations, several groups lost their ranges and were thought to have become "nomadic groups" (e.g., Troop B in Fig. 7.2).

Intergroup relationships may influence female fecundity (Takahata et al. 2006). Based on the 60 group-years of data recorded from 1989 to 2001 in the study population, the birthrate follows an inverted U-shaped curve against the number of adult females (Fig. 7.3). These data agree with the assumption of the intergroup feeding competition (IGFC) hypothesis proposed by Wrangham (1980), and not with that of the predation-feeding competition (PFC) hypothesis put forth by van Schaik (1983). In contrast, infant mortality rate is not consistently correlated with group size/number of adult females (see figs. 1b and 2b in Takahata et al. 2006), as Jolly et al. (2002) pointed out. Figure 7.3 suggests that females of small-sized groups suffer from the stress of between-group competition. When we captured lemurs and measured their body mass in 1999, the members of a small group (Troop CX) exhibited the smallest body mass values and the heaviest infection by ticks (*Haemaphysalis (Rhipistoma) lemuris*), which may have been related to their environmental and social conditions, as Troop CX inhabits the most humid area of the gallery forest and is subordinate to neighboring groups.

7.5 Female Dominance Rank and Reproductive Success: Within-Group Competition

7.5.1 Dominance Rank Order and Targeting Aggression Among Females

Adult females within a group usually display a linear dominance rank order. Differing from cases reported in cercopithecoid species (e.g., *Macaca fuscata*; Koyama 1967), young females tend to occupy the lowest ranks, irrespective of their mother's rank (Koyama et al. 2005). Furthermore, older sisters are dominant over the younger sisters. Thus, matrilineal "inheritance" of dominance rank is rare. These tendencies are probably the result of the rarity of alliances, that is, support from mothers or other kin. Indeed, of the 1,137 agonistic interactions recorded in Troops C2 and T1 in the 1994 birth season, 1,114 (98.0 %) were decidedly dyadic interactions (Nakamichi and Koyama 1997). Specifically, submissive vocalizations did not function to recruit support from allies against opponents, and no solicitation behavior occurred to form alliances.

Young females acquire higher ranks by outranking older/dominant females and reach the highest ranks by 7–9 years of age (Koyama et al. 2005). Females occasionally display persistent aggression (targeting behavior) toward dominant females. Females direct aggression at handicapped individuals in some cases, for example, those exhibiting senility or those in labor/delivery. In September 1997, a female (MW-911) in Troop CX suddenly attacked a dominant female (SH-92), who had given birth 6 h earlier, and eventually MW-911 outranked SH-92 (Takahata et al. 2001). Then, MW-911 continued to threaten SH-92 for several weeks. A similar attack by other females on a mother and newborn infant was reported in Troop C2 (Okamoto 1998).

As stated earlier, alliances occur infrequently among female ringtailed lemurs; however, Nakamichi et al. (1997) reported a case in which alliances may have affected the rank changes among adult females in Troop T1. In this case, after the alpha female fell to the lowest ranking position, she went on to outrank another female with support from her adult daughter and another unrelated female. This case may correspond to the "germ" of stable dominance rank order maintained by maternal kin relationships found among cercopithecoid primates.

7.5.2 Do High-Ranking Females Attain High Reproductive Success?

Based on the reproductive data recorded in the study population from 1989 to 2001, we analyzed the correlations between female rank and reproductive parameters (Takahata et al. 2008). As group size affects female fecundity (Takahata et al. 2006), we divided the entire data set into three groups of different sizes, based on the number of adult females: large-sized (8–9 adult females), medium-sized (4–7 adult females), and small-sized (2–3 adult females). In general, high-ranked females did not always attain high reproductive success among the size group (Fig. 7.4). In particular, no significant differences were observed in the number of surviving infants per female among female rank categories in medium-sized and small-sized groups. This result is contrary to expectations, because females frequently fight for higher rank.

In contrast, low-ranked females had a smaller number of surviving infants than mid-ranked females in large-sized groups. It is probable that severe within-group competition in large-sized groups lowers the reproductive success of low-ranked females.

7.6 Conclusions

7.6.1 The Balance Between Female Coexistence and Competition

The female-bonded/matrilineal group of ringtailed lemurs should have evolved to collectively secure local resources (food and water) against competitors (other groups or species). However, it is not always advantageous for every female to coexist in a large group in the high group/population density of Berenty Reserve, (Figs. 7.3, 7.4). Although large-sized groups confer resource defense advantages during the food-scarce weaning season, each member does not always gain foraging benefits, probably because of crowding and a lower proportion of individuals eating at any given time during foraging bouts (Pride et al. 2006).

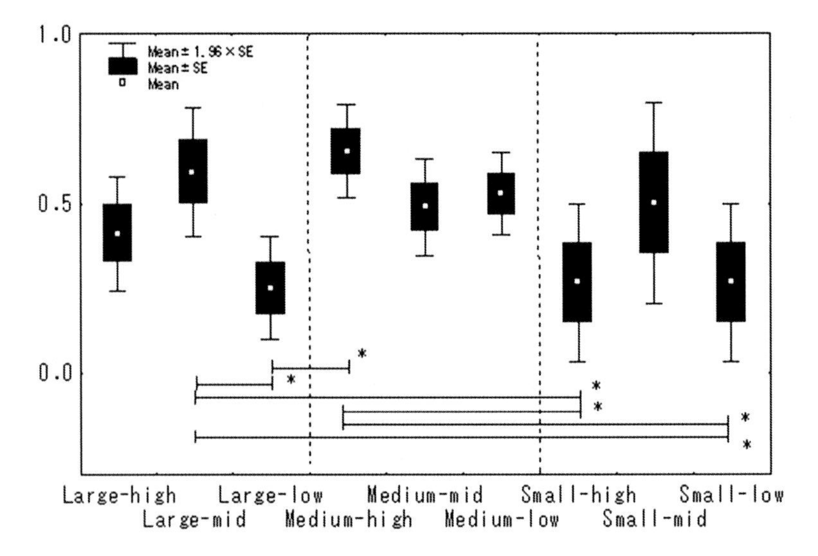

Fig. 7.4 Female rank categories in each group size and mean number of surviving infants 1 year after birth. *Large-high*, *large-mid*, and *large-low* indicate the high-ranked, mid-ranked, and low-ranked females, respectively, of large-sized groups with eight or more females. *Medium-high*, *medium-mid*, and *medium-low* are the high-ranked, mid-ranked, and low-ranked females, respectively, of medium-sized groups with four to seven adult females. *Small-high*, *small-mid*, and *small-low* are the high-ranked, mid-ranked, and low-ranked females, respectively, of small-sized groups with two to three adult females. $*P<0.05$; $**P<0.01$

Pride (2005) found that cortisol concentrations are lowest in medium-sized groups of ringtailed lemurs, and that cortisol levels do not differ between dominant and subordinate females within a group at Berenty, indicating that adult females belonging to large-sized groups suffer from higher levels of stress, which may have resulted in low reproductive success. This finding may also help to explain why female eviction occurs frequently in large groups. When a medium-sized group increases in number, the reproductive success of each female must decrease for reasons of stress (Fig. 7.4). In such a situation, dominant females attempt to reduce group size by evicting subordinates through targeting aggression, or subordinate females may attempt to outrank dominant females to avoid eviction.

In contrast, females in small groups suffer from greater stress because of between-group competition, indicating that it is less advantageous for females to be evicted from groups. Irrespective of frequent group fission, groups rarely admit evicted females from other groups as new members (Fig. 7.2). Evicted females probably encounter numerous difficulties establishing new ranges because of high group density in the study population.

A group of ringtailed lemurs typically exhibits the following cycle (fig. 14.2 in Ichino and Koyama 2006). First, a group gains some advantages in between-group competition and increases in size. However, within-group competition increases when group size exceeds its optimal level. Several females are then compelled to leave by other females (female eviction). If such females can establish new home ranges and gain mating partners (males), a new group may be formed (group fission). If there is no available space, the evicted females become a nomadic group (group dismissal/eviction), or at worst, they die (group extinction). In some cases, evicted females fight with another group, take over the range, and form one group (group fusion), as reported for wild toque macaques (Dittus 1987). Otherwise, they may transfer into another group (female transfer), as reported for wild Japanese macaques (Takahata et al. 1994) and vervet monkeys (Isbell et al. 1991), although such cases appear to be rather rare.

7.6.2 Why Do Females Coexist Within a Group?

A female ringtailed lemur rarely forages by herself, except when she is very old. Numerous costs and risks must be considered when leaving a group (e.g., acquisition of a new home range and mating partners, and avoidance of predation). These considerations may restrain females from voluntarily deserting a group. Even in the case of an eviction, these females usually range with relatives. Thus, group-living appears to be an essential lifestyle for them.

There are several differences between ringtailed lemur society and the female-bonded/matrilineal societies of the cercopithecine species, as indicated by preceding studies (Jolly 1984; Sauther et al. 1999). These differences provide hints for understanding the evolution of primate societies. For example, females of cercopithecine species develop a stable dominance rank system. Daughters are ranked immediately beneath their mothers, as though rank were bestowed through inheritance (Koyama 1967). In contrast, the rank order of female ringtailed lemurs frequently fluctuates, and their kin relationships do not always affect their rank order, probably because of insufficiency in cognitive and behavioral abilities (Pereira 1993). Undoubtedly, the ability of ringtailed lemurs to form alliances with kin-related individuals is limited to the degree of second consanguinity, which is much lower than anthropoid primates (Nakamichi and Koyama 1997). The bond between mother and daughter, as well as sisters, may be the only definite long-lasting psychological bond for them.

However, a few young females attain higher relative ranks (Koyama et al. 2005), and all these females are daughters of alpha females, so this may represent the archetype of the "maternal kin-selected society." Anthropoid primate females have developed such social abilities that enable alliance among members of more distant kin, or even with non-kin, to obtain advantages in between-group competition.

Why do female ringtailed lemurs compete for high rank within a group? It is possible that rank correlates with reproductive success (Fedigan 1983). Undoubtedly, low-ranked females exhibit a lower value of reproductive success than other females in large-sized groups, suggesting that the intensity of within-group competition is intercorrelated with group size and rank, as indicated by Van Noordwijk and van Schaik (1999) for wild long-tailed macaques.

No significant differences were observed in reproductive success among rank categories of females in medium- and small-sized groups in the study population, thereby differing from large-sized groups and some reports on cercopithecoid species (Cheney et al. 2004; however, see Kümmerli and Martin 2005). In addition, female eviction has rarely been observed in medium- and small-sized ringtailed lemur groups, probably because females require a sufficient number of group members to compete with neighboring groups. Within-group competition may result merely in outranking, not eviction, in such groups.

7.6.3 Why Do Male Ringtailed Lemurs Coexist Within a Group?

Male ringtailed lemurs within a group usually exhibit a linear dominant rank order. Males occasionally perform "stink fights" with each other as a type of ritualized aggressive behavior. Why do males coexist within a group? Similar questions have been discussed for other male primates from various points of view (Pereira et al. 2000): theories include communal defense of female groups, counterstrategies against infanticide, and antipredator behavior.

Based on our data, adult males appear to be "parasites" on female groups. Although adult females play the lead during group encounters, males also confront and attack the males of rival groups, but not so aggressively. Furthermore, males rarely take the initiative during female eviction or group fission.

Notably, most males belong to a group during the birth season. It is profitable for males to belong to a group, as the days on which females are receptive to mating are rather few. Because high-ranking males are not always selected by females as mating partners (Koyama 1988), any male may have a chance to mate if he belongs to the group.

Sussman (1992) pointed out that male migration tends to equalize the sex ratios in groups at Beza Mahafaly. However, our data show that the proportion of adult males tends to be low in newly formed small groups, indicating that males appear to hesitate to participate in small-sized groups of evicted females. Such groups are not favorable for males, because females may suffer increased stress caused by between-group competition. Males may select medium-sized or large groups with stable home ranges to secure reproductive success, despite opposition from resident males. Thus, the social groups of ringtailed lemurs may be characterized by "female competition" and "male choice of female groups" in the high group and population density of the study population.

Acknowledgments We express our gratitude to A. Randrianjafy, former Director of the Botanical and Zoological Park of Tsimbazaza, the de Heaulme family, A. Jolly, Y. Kawamoto, and H. Hirai for their kind help and cooperation. We also thank two reviewers for their useful comments on the manuscript, and L. Karczmarski and J. Yamagiwa for giving us the opportunity to write this review. This work was supported by a Grant-in-Aid for Scientific Research to N. Koyama (No. 06610072 and No. 05041088) and to Y. Takahata (No. 12640700 and No. 21405015).

References

Blumenfeld-Jones K, Randriamboavonjy TM, Williams G, Mertl-Millhollen AS, Pinkus S, Rasamimanana H (2006) Tamarind recruitment and long-term stability in the gallery forest at Berenty, Madagascar. In: Jolly A, Sussman RW, Koyama N, Rasamimanana H (eds) Ringtailed lemur biology. Springer, New York, pp 69–85

Cheney DL, Seyfarth RM, Fischer J, Beehner J, Bergman T, Johnson SE, Kitchen DM, Palombit RA, Rendall D, Silk JB (2004) Factors affecting reproduction and mortality among baboons in the Okavango Delta, Botswana. Int J Primatol 25:401–428

Dittus WPJ (1987) Group fusion among wild toque macaques: an extreme case of inter-group resource competition. Behaviour 100:247–291

Fedigan LM (1983) Dominance and reproductive success. Yearb Phys Anthropol 26:91–129

Furuichi T (1985) Inter-male associations in a wild Japanese macaque troop on Yakushima Island, Japan. Primates 26:219–237

Gould L (1996) Male–female affiliative relationships in naturally occurring ringtailed lemurs (*Lemur catta*) at the Beza-Mahafaly Reserve, Madagascar. Am J Primatol 39:63–78

Gould L (1997) Intermale affiliative behavior in ringtailed lemurs (*Lemur catta*) at the Beza-Mahafaly Reserve, Madagascar. Primates 38:15–30

Ichino S (2004) Socio-ecological study of ring-tailed lemurs (*Lemur catta*) at Berenty Reserve, Madagascar. Ph.D. dissertation, Kyoto University, Kyoto

Ichino S (2005) Attacks on a wild infant ring-tailed lemur (*Lemur catta*) by immigrant males at Berenty, Madagascar: interpreting infanticide by males. Am J Primatol 67:267–272

Ichino S (2006) Troop fission in wild ring-tailed lemurs (*Lemur catta*) at Berenty, Madagascar. Am J Primatol 68:97–102

Ichino S, Koyama N (2006) Social changes in a wild population of ringtailed lemurs (*Lemur catta*) at Berenty, Madagascar. In: Jolly A, Sussman RW, Koyama N, Rasamimanana H (eds) Ringtailed lemur biology. Springer, New York, pp 233–243

Isbell LA, Cheney DL, Seyfarth RM (1991) Group fusions and minimum group sizes in vervet monkeys (*Cercopithecus aethiops*). Am J Primatol 25:57–65

Jolly A (1967) Breeding synchrony in wild *Lemur catta*. In: Altman SA (ed) Social communication among primates. University of Chicago Press, Chicago, pp 3–14

Jolly A (1972) Troop continuity and troop spacing in *Propithecus verreauxi* and *Lemur catta* at Berenty (Madagascar). Folia Primatol (Basel) 17:335–362

Jolly A (1984) The puzzle of female feeding priority. In: Small MF (ed) Female primates. Liss, New York, pp 197–215

Jolly A (2004) Lords and lemurs: mad scientists, kings with spears, and the survival of diversity in Madagascar. Houghton Mifflin, Boston

Jolly A, Dobson A, Rasamimanana HM, Walker J, Solberg M, Perel V (2002) Demography of *Lemur catta* at Berenty Reserve, Madagascar: Effects of troop size, habitat and rainfall. Int J Primatol 23:327–353

Jolly A, Sussman RW, Koyama N, Rasamimanana H (2006a) Ringtailed lemur biology: *Lemur catta* in Madagascar. Springer, New York, p 376

Jolly A, Koyama N, Rasamimanana H, Crowley H, Williams G (2006b) Berenty Reserve: a research site in southern Madagascar. In: Jolly A, Sussman RW, Koyama N, Rasamimanana H (eds) Ringtailed lemur biology. Springer, New York, pp 32–42

Jolly A, Rasamimanana H, Braun M, Dubovick T, Mills C, Williams G (2006c) Territory as bet-hedging: *Lemur catta* in a rich forest and an erratic climate. In: Jolly A, Sussman RW, Koyama N, Rasamimanana H (eds) Ringtailed lemur biology. Springer, New York, pp 187–207

Koyama (1967) On dominance rank and kinship of a wild Japanese monkey troop in Arashiyama. Primates 8:189–216

Koyama N (1988) Mating behavior of ring-tailed lemurs (*Lemur catta*) at Berenty, Madagascar. Primates 29:163–175

Koyama N, Nakamichi M, Oda R, Miyamoto N, Ichino S, Takahata Y (2001) A ten-year summary of reproductive parameters for ring-tailed lemurs at Berenty, Madagascar. Primates 42:1–14

Koyama N, Nakamichi M, Ichino S, Takahata Y (2002) Population and social dynamics changes in ring-tailed lemur troops at Berenty, Madagascar between 1989–1999. Primates 43:291–314

Koyama N, Ichino S, Nakamichi M, Takahata Y (2005) Long-term changes in dominance ranks among ring-tailed lemurs at Berenty Reserve, Madagascar. Primates 46:225–234

Koyama N, Soma T, Ichino S, Takahata Y (2006) Home ranges of ringtailed lemur troops and the density of large trees at Berenty Reserve, Madagascar. In: Jolly A, Sussman RW, Koyama N, Rasamimanana H (eds) Ringtailed lemur biology. Springer, New York, pp 86–101

Koyama N, Aimi M, Kawamoto Y, Hirai H, Go Y, Ichino S, Takahata H (2008) Body mass of wild ring-tailed lemurs in Berenty Reserve, Madagascar, with reference to tick infestation: a preliminary analysis. Primates 49:9–15

Kümmerli R, Martin RD (2005) Male and female reproductive success in *Macaca sylvanus* in Gibraltar: no evidence for rank dependence. Int J Primatol 26:129–1249

Mertl-Millhollen AS, Gustafson HI, Budnitz N, Dinis K, Jolly A (1979) Population and territory stability of the *Lemur catta* at Berenty, Madagascar. Folia Primatol (Basel) 31:106–122

Mertl-Millhollen AS, Rambeloarivony H, Miles W, Kaiser VA, Loretta LG, Dorn T, Williams G, Rasamimanana H (2006) The influence of tamarind tree quality and quantity on *Lemur catta* behavior. In: Jolly A, Sussman RW, Koyama N, Rasamimanana H (eds) Ringtailed lemur biology. Springer, New York, pp 102–118

Mittermeier RA, Tattersall I, Konstant WR, Meyers DM, Mast RB (1994) Lemurs of Madagascar. Conservation International, Washington, DC

Nakamichi M, Koyama N (1997) Social relationships among ring-tailed lemurs (*Lemur catta*) in two free-ranging troops at Berenty Reserve, Madagascar. Int J Primatol 18:73–93

Nakamichi M, Koyama N (2000) Intra-troop affiliative relationships of females with newborn infants in wild ring-tailed lemurs (*Lemur catta*). Am J Primatol 50:187–203

Nakamichi M, Rakototiana MLO, Koyama N (1997) Effects of spatial proximity and alliances on dominance relations among female ring-tailed lemurs (*Lemur catta*) at Berenty Reserve, Madagascar. Primates 38:331–340

Oda R (1996) Effects of contextual and social variables on contact call production in free-ranging ringtailed lemurs (*Lemur catta*). Int J Primatol 17:191–205

Oda R, Masataka N (1996) Interspecific responses of ringtailed lemurs to playback of antipredator alarm calls given by Verreaux's sifakas. Ethology 102:441–453

Okamoto M (1998) The birth of wild ring-tailed lemurs at Berenty Reserve, Madagascar. Primate Res 14:25–34 (in Japanese with English summary)

Pereira ME (1993) Agonistic interaction, dominance relation, and ontogenetic trajectories in ring-tailed lemurs. In: Pereira ME, Fairbanks LA (eds) Juvenile primates. Oxford University Press, New York, pp 285–305

Pereira ME (1995) Development and social dominance among group-living primates. Am J Primatol 37:143–175

Pereira ME, Clutton-Brock TH, Kappeler PM (2000) Understanding male primates. In: Kappeler PM (ed) Primate males. Cambridge University Press, Cambridge, pp 271–277

Pinkus S, Smith JNM, Jolly A (2006) Feeding competition between introduced *Eulemur fulvus* and native *Lemur catta* during the birth season at Berenty Reserve, Southern Madagascar. In: Jolly A, Sussman RW, Koyama N, Rasamimanana H (eds) Ringtailed lemur biology. Springer, New York, pp 119–140

Pride RE (2005) Optimal group size and seasonal stress in ring-tailed lemurs (*Lemur catta*). Behav Ecol 16:550–560

Pride RE, Felantsoa D, Randriamboavonjy T, Randriambelona (2006) Resource defense in *Lemur catta*: the importance of group size. In: Jolly A, Sussman RW, Koyama N, Rasamimanana H (eds) Ringtailed lemur biology. Springer, New York, pp 208–232

Rasamimanana H, Andrianome VN, Rambeloarivony H, Pasquet P (2006) Male and female ring-tailed lemur's energetic strategy does not explain female dominance. In: Jolly A, Sussman RW, Koyama N, Rasamimanana H (eds) Ringtailed lemur biology. Springer, New York, pp 271–295

Sauther ML (1991) Reproductive behavior of free-ranging *Lemur catta* at Beza Mahafaly Special Reserve, Madagascar. Am J Phys Anthropol 84:463–477

Sauther ML, Sussman RW (1993) A new interpretation of the social organization and mating system of the ringtailed lemur (*Lemur catta*). In: Kappeler PM, Ganzhorn JU (eds) Lemur social systems and their ecological basis. Plenum Press, New York, pp 111–121

Sauther ML, Sussman RW, Gould L (1999) The socioecology of the ringtailed lemur: thirty-five years of research. Evol Anthropol 8:120–132

Simmen F, Sauther ML, Soma T, Rasamimanana H, Sussman RW, Jolly A, Tarnaud L, Hladik A (2006) Plant species fed on by *Lemur catta* in gallery forests of the southern domain of Madagascar. In: Jolly A, Sussman RW, Koyama N, Rasamimanana H (eds) Ringtailed lemur biology. Springer, New York, pp 55–68

Small MF (1989) Female choice in nonhuman primates. Yearb Phys Anthropol 32:103–127

Soma T (2003) Feeding ecology of ring-tailed lemurs (*Lemur catta*) at Berenty Reserve, Madagascar. Master's thesis, Graduate School of Asian and African Area Studies, Kyoto University, Kyoto

Soma T (2006) Tradition and novelty: *Lemur catta* feeding strategy on introduced tree species at Berenty Reserve. In: Jolly A, Sussman RW, Koyama N, Rasamimanana H (eds) Ringtailed lemur biology. Springer, New York, pp 141–159

Sterck EHM, Watts DF, van Schaik CP (1997) The evolution of female social relationships in nonhuman primates. Behav Ecol Sociobiol 41:291–309

Sussman RW (1992) Male life history and intergroup mobility among ringtailed lemurs (*Lemur catta*). Int J Primatol 13:395–413

Sussman RW, Sweeney S, Green GM, Porton I, Andrianasolondraibe OL, Patsirarson J (2006) A preliminary estimate of *Lemur catta* population density using satellite imagery. In: Jolly A, Sussman RW, Koyama N, Rasamimanana H (eds) Ringtailed lemur biology. Springer, New York, pp 16–31

Takahata Y, Suzuki S, Okayasu N, Hill D (1994) Troop extinction and fusion in wild Japanese monkeys of Yakushima Island, Japan. Am J Primatol 33:317–322

Takahata Y, Koyama N, Miyamoto N, Okamoto M (2001) Daytime deliveries observed for the ring-tailed lemurs of the Berenty Reserve, Madagascar. Primates 42:267–271

Takahata Y, Koyama N, Ichino S, Miyamoto N, Nakamichi M (2006) Influence of group size on reproductive success of female ring-tailed lemurs: distinguishing between IGFC and PFC hypotheses. Primates 47:383–387

Takahata Y, Koyama N, Ichiro S, Miyamoto N, Nakamichi M, Soma T (2008) The relationship between female rank and reproductive parameters of ringtailed lemur: a preliminary analysis. Primates 49:135–138

van Noordwijk MA, van Schaik CR (1999) The effects of dominance rank and group size on female lifetime reproductive success in wild long-tailed macaques, *Macaca fascicularis*. Primates 20:105–130

van Schaik CP (1983) Why are diurnal primates living in groups? Behaviour 87:120–144

Vick LG, Pereira ME (1989) Episodic targeting aggression and the histories of lemur social groups. Behav Ecol Sociobiol 25:3–12

Wrangham RW (1980) An ecological model of female-bonded primate groups. Behaviour 75:262–300

Chapter 8
Social Structure and Life History of Bottlenose Dolphins Near Sarasota Bay, Florida: Insights from Four Decades and Five Generations

Randall S. Wells

Abstract Studies of social ecology can benefit from long-term observations, as these provide researchers with opportunities to distinguish between the relative contributions of life history, demographics, and ecological pressures to the development of social patterns. Long-term study can provide the means of interpreting changes in stable social patterns relative to changes in environmental factors or availability of members of specific age-sex classes. The strength of social patterns

R.S. Wells (✉)
Chicago Zoological Society, c/o Mote Marine Laboratory 1600 Ken Thompson Parkway, Sarasota, FL 34236, USA
e-mail: rwells@mote.org

J. Yamagiwa and L. Karczmarski (eds.), *Primates and Cetaceans: Field Research and Conservation of Complex Mammalian Societies*, Primatology Monographs, DOI 10.1007/978-4-431-54523-1_8, © Springer Japan 2014

can be measured by their persistence from one generation to the next, and as individuals pass life history milestones.

The Sarasota Dolphin Research Program has been engaged in studies of bottlenose dolphins along the central west coast of Florida, including Sarasota Bay, since 1970. The research includes focal animal behavioral observations, photographic identification surveys, biopsy darting for genetic and contaminant samples, and occasional capture–release efforts to examine the animals' behavior, ecology, life history, population biology, health, and concentrations and effects of environmental contaminants. More than 4,800 individuals have been identified in the bays and coastal Gulf of Mexico waters of the region, including the approximately 160 dolphins using Sarasota Bay on a regular basis. A mosaic of adjacent, often slightly overlapping dolphin communities has been identified based on sighting locations and social associations, and genetic findings support these designations. These communities are genetically distinguishable but not isolated.

The communities of dolphins residing in and around Sarasota Bay, the most intensively studied animals, are characterized by a high level of multigenerational site fidelity and low levels of emigration and immigration. The social structure includes three basic components: nursery groups built around females with young of similar age, juvenile groups, and adult males, mostly in strongly bonded, long-term male pairs or sometimes as single individuals. This overall structure has remained relatively stable through five generations; however, core area use, group size, and some social association patterns show variability over time. Paternity testing suggests that male pair-bonding may improve reproductive success. Female reproductive success appears to be related to mother's age, experience, and environmental contaminant residues. Older, more experienced mothers are more successful in rearing young over the typical 3- to 6-year period of association; these females have also previously depurated organochlorine contaminants that otherwise might have influenced reproduction and health.

Keywords Bottlenose dolphins • Life history • Reproductive success • Site philopatry • Social structure

8.1 Introduction

Studies of social ecology can benefit from long-term observations, as these provide researchers with opportunities to distinguish between the relative contributions of life history, demographics, and ecological pressures to the development of social patterns. Long-term study can provide the means of interpreting changes in stable social patterns relative to changes in environmental factors or availability of members of specific age-sex classes. The strength of social patterns can be measured by their persistence from one generation to the next, and as individuals pass life history milestones.

Many delphinid cetaceans live in complex societies (Norris and Dohl 1980a; Wells et al. 1999; Mann et al. 2000). Long lifespans, in some species exceeding 50 years, likely contribute to the observed social complexity. To fully understand the social structure of these animals and the ecological influences on social patterns, it is beneficial to observe individuals throughout the course of their lives. Several field studies of dolphins have begun to approach this goal. Hawaiian spinner dolphin (*Stenella longirostris*) studies initiated by Ken Norris and Tom Dohl in the late 1960s have been continued at intervals over several decades, providing insights into the structure and dynamics of the fluid societies of these small delphinids (Norris and Dohl 1980b; Norris et al. 1994). Studies of killer whales (*Orcinus orca*) initiated by Michael Bigg near Vancouver Island in the early 1970s and continued by others to the present (Bigg 1982; Ford and Fisher 1983; Parsons et al. 2009; Foster et al. 2012) have led to detailed descriptions of the workings of these highly stable societies. The social behavior of dusky dolphins (*Lagenorhynchus obscurus*) has been studied in depth in Argentina and later in New Zealand by Bernd and Melany Würsig and colleagues since the mid-1970s (Würsig and Würsig 1980; Benoit-Bird et al. 2004; Weir et al. 2008).

One of the first of the long-term studies of delphinid social ecology began in Sarasota Bay, Florida, in 1970, with common bottlenose dolphins (*Tursiops truncatus*) (Scott et al. 1990a; Wells 1991, 2003, 2009a). The study reported in this chapter has followed individually identified resident dolphins for more than 42 years, in some cases, and across at least five maternally related generations. Long-term observations combined with information on life history and ecology have begun to provide an understanding of the social structure of this dolphin community and the environmental and demographic factors influencing this structure.

8.2 Background

8.2.1 Study Area

The study area includes the inshore and coastal waters along the central west coast of Florida, from Tampa Bay southward through Charlotte Harbor and Pine Island Sound, and the Gulf of Mexico to about 5–10 km offshore (Fig. 8.1). Most of the research effort has been concentrated in Sarasota Bay and adjacent bays, sounds, and Gulf waters within 1 km of the shore because of initial findings of dolphin residency to these waters (Irvine and Wells 1972) and their proximity to our base of operations at Mote Marine Laboratory in Sarasota. The shallow (<4 m deep), sheltered bay and estuarine waters are separated from the Gulf of Mexico by a series of barrier islands, communicating with the Gulf through narrow, deeper (up to ~10-m-deep) passes. The bays contain areas of shallow seagrass meadows and are fringed by mangroves, along with manmade features such as bridges, piers, and seawalls. Natural or dredged channels 3–4 m deep run through seagrass meadows and sand or mud flats. A gently sloping, shallow sandy bottom extends

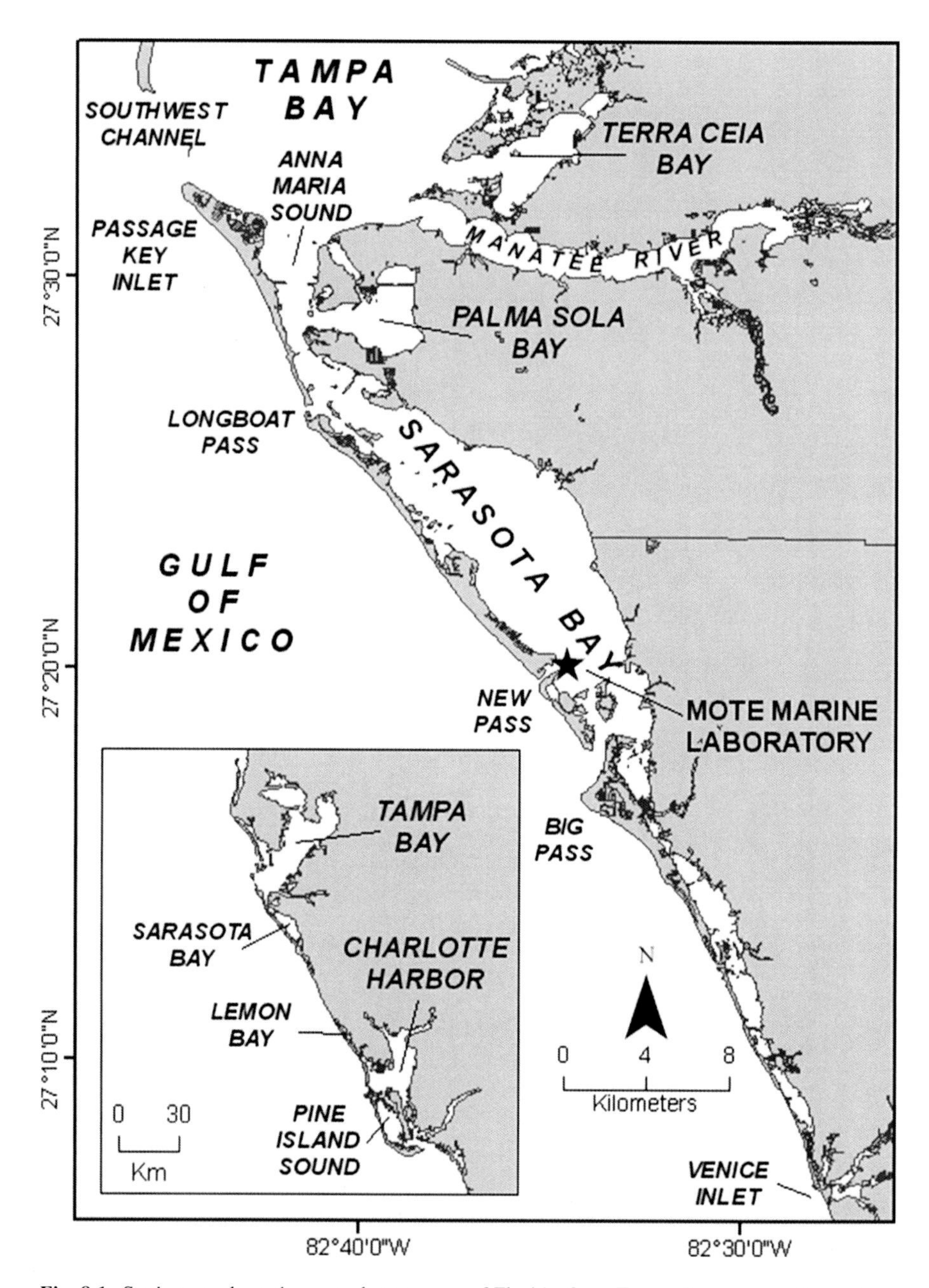

Fig. 8.1 Study area along the central west coast of Florida, from Tampa Bay through Pine Island Sound. The Sarasota dolphin community range extends from southern Tampa Bay to Venice Inlet

offshore from the beaches on the Gulf sides of the barrier islands. Sarasota and Manatee Counties, which encompass the Sarasota Bay study area, are heavily populated, with more than 687,000 people (as of 2008), and more than 45,000 registered vessels (as of 2008). Beach- and boat-based tourism is of major importance to the region.

8.2.2 History of the Sarasota Dolphin Research Program

A pilot tagging study conducted through Mote Marine Laboratory during 1970–1971 to investigate movements and activities of common bottlenose dolphins (*Tursiops truncatus*) along the central west coast of Florida, and to test tag designs, initially identified patterns of residency for dolphins in this region, setting the stage for continuing research (Irvine and Wells 1972). Additional research, including radio-tagging and radio-tracking, along with observational studies and some photographic identification efforts through the University of Florida during 1975–1978, confirmed previous residency findings. During this work and subsequent opportunistic surveys in 1979, re-identifications of 92 % of the dolphins tagged during 1970–1971 suggested that residency might be long term (Irvine et al. 1981). The high frequency of resightings of identifiable individuals led to an initial description of home range and social patterns (Wells et al. 1980).

Seasonal, systematic photographic identification surveys were initiated in 1980, and continued through 1989 through the University of California, Santa Cruz, and Dolphin Biology Research Institute. Since 1989, the program has been coordinated through the Chicago Zoological Society. Seasonal surveys continued until 1993, when year-round, monthly surveys were implemented and continued through the present (Scott et al. 1990a; Wells 1991, 2003). These surveys are conducted from small (<8-m-long) outboard-powered vessels following standard routes through the study area, selected daily depending on previous coverage and conditions. When dolphin groups are encountered, data are collected on location, time, environmental parameters, dolphin activities, numbers of dolphins, calves, and young-of-the-year (YOY). Photographs are taken of dolphin dorsal fins to identify individuals from distinctive markings (Scott et al. 1990b; Würsig and Jefferson 1990); the resulting identification catalog included more than 4,800 dolphins as of the end of 2013. Through 2013, the sighting database included more than 113,000 individual identifications from more than 41,000 dolphin group records collected since 1970, with some individuals having been resighted more than 1,400 times each.

Life history and health data are obtained through occasional capture–release sessions, in which small groups of selected dolphins are encircled with a 500-m-long, 4-m-deep seine net in shallow water (<2 m deep). Each individual is brought aboard a specialized veterinary examination vessel, where sex is determined; it is weighed, measured for a standard suite of lengths and girths (Read et al. 1993; Tolley et al. 1995), measured ultrasonically for blubber thickness (Wells 1993; Noren and Wells 2009), examined by a veterinarian externally and through ultrasonography, sampled, marked if necessary (Wells 2009b), photographed, and released (Wells et al. 2004). Samples are collected for basic health profiles (Wells et al. 2004; Hall et al. 2007; Schwacke et al. 2009), which are used as reference values for comparison with other populations experiencing unusual mortality events, abnormal environmental conditions, or anthropogenic stressors (St. Aubin et al. 2013; Schwacke et al. 2010, 2011). Samples are also collected for microbiology and disease processes (Buck et al. 2006; Burdett Hart et al. 2010, 2011; Nollens et al. 2009; Rowles et al. 2011; Hart et al. 2012), immune system function (Lahvis et al. 1995; Ruiz et al. 2009), serology (Duignan et al. 1996; Venn-Watson et al. 2008), reproductive hormones (Wells et al. 1987), biotoxins (Fire et al. 2008; Twiner et al. 2011), kidney health (Venn-Watson et al. 2010), and environmental contaminant concentrations (Schwacke et al. 2002; Wells et al. 2005; Houde et al. 2005, 2006; Hall et al. 2006; Bryan et al. 2007; Woshner et al. 2008; Yordy et al. 2010a,b,c,d; Kucklick et al. 2011; Miller et al. 2011). When ages are not known from long-term observations of mothers with calves, age is determined from examination of growth layer groups in a sectioned tooth (Hohn et al. 1989) and, experimentally, through telomere analyses (Dunshea et al. 2011). Genetic samples are collected for evaluating relationships, including paternity (Duffield and Wells 1991, 2002; Sellas et al. 2005). Since 1984, more than 700 sets of measurements have been collected from more than 225 individuals, with some measured up to 15 times. Since 1988, more than 700 blood samples have been collected from more than 230 individuals, some sampled as many as 15 times.

Capture–release sessions also provide opportunities to measure hearing abilities (Mann et al. 2010) and obtain acoustic recordings for studies of communication, including signature whistle characteristics and development (Fripp et al. 2005; Sayigh et al. 1990, 1995, 2007; Watwood et al. 2004, 2005; Esch et al. 2009), as well as to perform acoustic playback experiments to examine whistle function (Sayigh et al. 1999; Janik et al. 2006, 2013). Individual dolphins have been recorded during capture–release sessions since 1975, mostly via a suction cup-mounted hydrophone, and during focal animal behavioral follows, resulting in recordings from 225 individuals, some recorded up to 16 times and/or spanning more than 30 years.

In combination, the observational and capture–release datasets have led to the compilation of reproductive histories of more than 100 mothers with 300 of their calves. Some females have been observed with as many as 10 calves during the course of our research. Data include birthdates of calves, calf sex, mother's age at time of birth (including age at first birth in some cases), duration of the mother–calf association, and circumstances leading to separation. These datasets also provide crucial background data in support of focal animal behavioral observations. Since 1992, more than 2,073 focal follows have been conducted on more than 143 different individuals followed up to 61 times each.

Ecological studies involve collaborative efforts. Carcasses recovered by Mote Marine Laboratory's Stranding Investigations Program are examined and necropsied for determination of cause of death and for collection of standardized measurements and biological samples, including stomach contents (Barros and Wells

1998; Wells et al. 2008; Fauquier et al. 2009; DeLynn et al. 2011). Since 1985, more than 65 dolphins with sighting histories in our database have been recovered, and stomach contents have been collected from 33 Sarasota Bay residents. Stomach content data are examined relative to data from quantitative purse seine survey operations conducted during winter and summer field seasons to determine the abundance, distributions, length frequencies, body conditions, and species assemblages of fish using Sarasota Bay (Gannon et al. 2009; Berens McCabe et al. 2010). During more than 1,189 sets of the purse seine since 2004, more than 480,790 fish of 132 species have been caught, examined, measured, and released. Data on injuries from shark bites and stingray barbs are collected during necropsies and health assessments.

8.2.3 Study Population

Strong site fidelity has been demonstrated by dolphins using Sarasota Bay. Dolphins were considered to be residents of Sarasota Bay if they were seen at least ten times and more than half of their sighting records occurred within the core study area, the region bounded by Tampa Bay to the north and Venice Inlet to the south, and the barrier island chain to the west (Fig. 8.1). For 2007, the most recent year for which population analyses have been completed, 155 identifiable dolphins (and their dependent calves) met these criteria. On average, 89 % (±12 % SD) of the sightings of these animals occurred within Sarasota Bay. Of the 67 dolphins present in 2007 known to be at least 15 years old, 96 % had been observed in the area over a span of at least 15 years, with some observed for as many as 37 years.

The 155 identifiable dolphins comprised 96 % of the dolphins seen in Sarasota Bay, indicating that the total number of dolphins using the bay in 2007, including unmarked animals, was about 163. In 2007, about 84 % of the resident dolphins were of known sex (52 % F and 48 % F), and about 90 % were of known age class (42 % subadult and 58 % adult). As of 2013, the oldest resident male recorded to date was 50 years old, and the oldest female was 63 years old, based on long-term observations and growth layer groups in teeth (Hohn et al. 1989). The number of identifiable dolphins using Sarasota Bay on a regular basis has varied over time, ranging between 111 and 166 during the 15-year period of consistent survey effort from 1993 through 2007. Variations in abundance appear to have occurred at least partially in response to changes in commercial fishing regulations, resulting in increased prey abundance, and the periodic occurrence of severe harmful algal blooms (*Karenia brevis* red tides). Annual fecundity rate averaged about 0.14, recruitment rate to age 1 year averaged about 0.05, loss rate including known mortalities plus disappearances averaged about 0.04–0.06, and annual immigration rates were about 0.03 and emigration rates about 0.03–0.05 during 1980–1987 (Wells and Scott 1990) and during 1993–2007 (unpublished data).

8.3 Life History and Social Relationships in the Study Population

8.3.1 Reproduction and Development

The resident female bottlenose dolphins of Sarasota Bay tend to remain associated with the Bay throughout their lives, facilitating monitoring of reproductive success. Most calves are born during late spring or early summer and remain with their mothers for the next 3–6 years. Typically, separation from the mother occurs before the birth of her next calf. Males reach sexual maturity at 10–13 years of age, whereas females mature at 5–12 years of age (Wells and Scott 1999; Wells 2003). Many females give birth to their first calf at 8 to 10 years of age, following a 12-month gestation period. The female reproductive lifespan is prolonged, with a few females as old as 48 years successfully producing and rearing calves. Paternity tests have demonstrated that males in the age range of 13–40 years of age at least sire calves (Duffield and Wells 2002). Physical maturity is reached by females by about 12 years of age, and males by about 20 years of age, leading to significant sexual dimorphism in body length, mass, and other features (Read et al. 1993; Tolley et al. 1995).

8.3.2 Basic Societal Components

The bottlenose dolphins of Sarasota Bay are distributed at any given time in units referred to as groups (= school or sighting; Wells et al. 1987), defined operationally as "cohesive collections of conspecifics in a limited area (typically within several hundred meters), often engaged in similar activities and moving in the same general direction, maintained by social factors as a unit; groups may be stable over long periods of time or may change composition over periods ranging from minutes to weeks" (Wells et al. 1999). For Sarasota Bay bottlenose dolphins, these units are generally more similar to the small, changeable "parties" of chimpanzees (Goodall 1983) than to the permanent "pods" of killer whales (Bigg 1982). This working definition is a useful and replicable classification tool for the biologist in the field, and long-term observations of repeated patterns suggest that the observed groupings have biological meaning as well. However, our definition likely does not accurately reflect the dolphins' full perspective on what constitutes an interacting social unit. In the murky estuarine waters of Sarasota Bay and vicinity, acoustic communication plays an important role in dolphin interactions. Signature whistles are believed to be used as contact calls in these environments (Watwood et al. 2005), and the active space for these whistles has been estimated to range from hundreds of meters to kilometers, depending on the sound attenuation characteristics of the habitat (Quintana-Rizzo et al. 2006). Thus, dolphins beyond the researcher's sight may be interacting with the dolphins under observation. For example, on occasions when a member of a strongly bonded adult male pair is observed alone, the pair is often seen together again a short time later, and it is suspected that the males were in acoustic contact while "separated" (Watwood et al. 2005). Consideration of bottlenose dolphin groups should specify the kinds of interactions of interest.

The Sarasota Bay dolphins live in a fission–fusion society exhibiting the full spectrum of the term "group" as already defined (Wells et al. 1987; Connor et al. 2000; Wells 2003). They swim in small groups composed typically of 5 to 7 dolphins, ranging on rare, brief occasions to as many as 30 individuals (Wells et al. 1980, 1987). Variations in group size likely reflect demographic conditions as well as the ways in which the dolphins balance taking advantage of the benefits of group formation, such as protection from predation, while minimizing the costs, for example, from feeding competition (Wells et al. 1980). With a few exceptions, observed group composition changes frequently, over minutes to hours. Several basic group categories based on age, sex, and reproductive state can be described for Sarasota Bay dolphins, including (1) nursery groups, (2) juvenile groups of young dolphins independent of their mothers, and (3) adult males, typically as strongly bonded pairs.

Nursery groups, consisting of females with their most recent offspring, are the largest groups in the area. The reproductive state of the female and the age of her calf appear to be more important determinants of group composition than other factors such as relatedness (Wells et al. 1987). Group size decreases with calf age (Wells et al. 1987). Females with calves at similar levels of dependency tend to swim together, presumably because they must contend with similar needs for increased feeding to support lactation and because they must adjust their swimming patterns to facilitate frequent bouts of nursing. Female associates tend to be drawn from a pool of other mothers who inhabit significantly overlapping ranges. Wells (1991) referred to these female groups involving recurring associations as "bands," reflecting the long-term social and geographic relationships among the females. Because of the extended reproductive lifespan of the Sarasota Bay dolphins, the pool of potential associates may include multiple generations of related and unrelated females who may swim together if they are in reproductive synchrony. The band structure that was well defined in the 1980s has changed during the past 20 years. Wells (2003) described the initial stages of these changes as involving the reduction of the more northerly Anna Maria band through mortality and lack of recruitment, and fissioning of the more southerly Palma Sola Band following growth from successful recruitment of female offspring during the 1980s and the loss of two of the oldest members in the early 1990s. The Anna Maria Band has ceased to exist, the previous members of the Palma Sola band continue to swim in smaller groups of only a few females, and a band of Tampa Bay females regularly summers in the northern portion of the Sarasota community's range.

The most stable components of the nursery groups are the individual mother–calf pairs, with associations typically lasting 3–6 years. The period of mother–calf association typically extends well beyond the time of nutritional weaning and appears to provide opportunities for calves to learn important skills, such as feeding techniques (Wells 2003). Nutritional weaning is believed to occur during the second year of life, even though lactation and nursing may continue at a reduced level. Some calves orphaned in their second year of life have survived without being adopted by other resident females, even when related adult females are in the area. Older calves,

Fig. 8.2 Presumed post-reproductive females swimming together in 2010. From *left* to *right*: Blacktip Doubledip (57 years), Squiggy (54 years), and Nicklo (60 years). (Photograph by Sarasota Dolphin Research Program; taken under National Marine Fisheries Service Scientific Research Permit No. 522-1785)

especially females, will sometimes maintain close associations with their mothers and new siblings for months or more, presumably learning about calf rearing, and perhaps providing relief to the mother if she can share some rearing responsibilities with her older daughter.

Females without calves are often found together. In some cases, these associations involve females in their fifties and are believed to be post reproductive as they have not given birth for 13–20 years (Fig. 8.2). These presumed post-reproductive females associate with younger mothers and their calves, including kin, from time to time, but the relationships and presumed functions are not nearly so consistent or clear as those reported for killer whales (Foster et al. 2012).

The separation of calves from their mothers can occur abruptly, as is typically the case with male offspring, or it can be gradual, involving an incremental reduction in frequency and duration of associations over a number of months, as is observed more commonly for female calves (Wells et al. 1987; Wells 1991, 2003; McHugh et al. 2011a). Most newly independent calves join others in juvenile groups, while others may remain mostly alone within a very limited range for months before assimilating into groups with other juveniles. Juvenile groups are fluid in composition from day to day, include both sexes, and may include a broad range of ages up to early or mid-teens, in some cases reflecting delayed social maturity after individuals become sexually mature. These appear to be important formative years for the social development of young dolphins, as juveniles often engaged in social interactions that will take on greater importance later in life, such as copulations, affiliative behaviors, and agonistic behaviors. Juveniles interact with a large number of individuals of all age and sex classes, suggesting that this is a period of social exploration (McHugh et al. 2011a). Associations occurring during this phase often are maintained or recur throughout the individual's life.

The duration of involvement in juvenile groups varies by sex. Females tend to leave juvenile groups before males. Female association patterns begin to change in association with the birth of a female's first calf, but stronger associations with experienced mothers typically occur with subsequent calves (Owen 2001). Males mature later than females, and often associate with juveniles until they develop a strong pair bond with another male of similar age (Wells et al. 1987).

Alliances between adult males, in the form of long-term stable pair bonds, are among the strongest features of the Sarasota Bay bottlenose dolphin social structure, with average half-weight association coefficients of 0.753 (Wells 1991, 2003; Owen et al. 2002). Pair-bond formation is the norm for males in Sarasota Bay, with more than 93 % forming an alliance by age 20 (Owen et al. 2002). At any given time, about 57 % of adult or potentially adult males are paired, and 72 % of males 20 years or more of age are paired (Owen et al. 2002). The remaining males appear to be in transition, developing alliances or having lost an alliance partner. The average minimum age for first-time pair-bond formation is 11 years (Owen et al. 2002). Alliances are usually formed by individuals within less than 4 years of age of one another and who have been among the top five associates of each other within the 5 years preceding pair formation (Owen 2003). As has been noted by Goldberg and Wrangham (1997) for chimpanzees, males do not preferentially form alliances with close relatives; alliance partners are no more related to one another than they are to non-alliance males (Owen 2003). Some alliances have been observed over more than two decades, and about half of alliances end because of the loss of a partner (Owen 2003). Alliances may provide enhanced predator protection, which will grant pairs access to habitats where prey and predators may be more abundant (Owen 2003). Adult male pairs rarely associate with other males. They commonly move between groups of adult females, and may spend days to a week or more engaged in mate guarding with reproductively receptive females, engaging in sequential female defense polygyny (Moors 1997; Owen et al. 2002). Males play no role in calf rearing.

8.3.3 Social Matrix

Geography is the key defining feature of the Sarasota Bay bottlenose dolphin society. The inshore region from southern Tampa Bay to Venice Inlet and within several kilometers of the Gulf shore (Fig. 8.1) is the stage upon which the lives and social interactions of the resident dolphins are played out over decades and across generations. Sarasota Bay dolphins are constantly on the move through this region, in small groups that encounter other groups, and often change composition through joinings and separations as a result of these encounters. The resident dolphins interact with, and occasionally interbreed with, dolphins in adjacent ranges, but to a lesser extent than with the dolphins who share a Sarasota Bay home range (Wells 1986; Duffield and Wells 1991, 2002). The nature of these encounters and associations within a long-established range is reminiscent of the communities of chimpanzees (Goodall 1983), leading to adoption of this term as a descriptor of the kind of bottlenose dolphin social unit evident through much of the species inshore range in the southeastern United States (Wells 1986, Wells et al. 1987). As applied to dolphins, the term is defined as a regional society of animals sharing ranges and social associates, but exhibiting genetic exchange with other social units (Wells et al. 1999). A community is distinguished from the similar concept of a "population" by the fact that the latter is typically defined as a closed reproductive unit.

Geographic and physiographic features help to define the community range. More than 89 % of the resident sightings occur inshore of the barrier island chain bounding Sarasota Bay to the west and south, and south of an extensive shallow sandbank that delineates Tampa Bay from Sarasota Bay. Sighting frequencies for Sarasota Bay residents in the Gulf of Mexico decrease with distance from passes leading into Sarasota Bay, in contrast to the pattern of even distribution along the coast as exhibited by dolphins who rarely enter Sarasota Bay (Fazioli et al. 2006). Most Sarasota Bay residents have been recorded from all parts of the community range at some time in their lives, but individuals tend to frequent specific core areas, often characterized by habitat type. For example, some individuals spend most of their time in the vicinity of shallow seagrass meadows, while others emphasize the deeper, more open waters of Sarasota Bay proper in their daily movements. Upon reaching independence, calves often occupy all or part of their mother's core area.

Our understanding of Sarasota dolphin community parameters has evolved over time, with expanded regional survey coverage beginning in the 1980s and with the accumulation of long-term individual sighting records. The size of the community range is currently estimated at about 125 km^2 (Wells 2003; Urian et al. 2009), up from the 85 km^2 reported from the much more limited dataset from the early years of the research program in the 1970s (Wells et al. 1980). Adult males tend to range farther than females (including up to 150 km^2 or more), occasionally leaving the community range for months or more before returning, presumably in search of breeding opportunities (Wells 1991; Urian et al. 2009). At least in part as a result of increased survey coverage (spatial and temporal) and incorporation of improvements in photographic identification techniques, estimates of the numbers of dolphins using Sarasota Bay have increased from about 100 for the early years (Wells et al. 1980; Wells and Scott 1990) to about 160 residents in recent years (Wells 2009a, b).

Bottlenose dolphins are distributed continuously along the central west coast of Florida, including waters adjacent to Sarasota Bay. Consideration of genetics, ranging patterns, social associations, and stable isotope analyses has demonstrated the existence of a mosaic of communities in this region (Wells 1986; Wells et al. 1987; Duffield and Wells 1991, 2002; Sellas et al. 2005; Urian et al. 2009; Barros et al. 2010; Bassos-Hull et al., 2013). These communities are not isolated, behaviorally or genetically. About 15 % of calves born to Sarasota Bay resident mothers were sired by nonresident males (Duffield and Wells 2002). Community ranges sometimes overlap, and mixing occurs where community ranges are in close proximity. About 14 % to 17 % of groups including Sarasota dolphins also include dolphins from other communities (Wells et al. 1987; Fazioli et al. 2006). Agonistic interactions between dolphins from different communities occur, but are not frequently observed.

8.4 Factors Associated with Long-Term Social Stability and Variability

8.4.1 Site Fidelity

Strong long-term site fidelity ensures the possibility of frequent encounters with other community members. Social relationships, once developed, can be maintained through repeated contact. Dolphins born to community members tend to remain in the community, leading to the concurrent existence within the community of matrilineally related individuals spanning as many as five generations. Dolphins originally marked during the initial 1970–1971 tagging project were observed in the Sarasota Bay area for 27 years, on average, before they died or disappeared, and two of the original individuals were still seen in 2013, 42 years after their initial marking.

Natal site philopatry of both sexes appears to be the rule, but it is not absolute. Dispersal outside the community is not common (Wells 2003; Sellas et al. 2005), but can involve either males or females. Individuals may also leave the community range temporarily for periods of months, or in rare cases years, and some of these have been observed in nearby communities. Within the Sarasota community range, core areas may shift for some individuals over time. For example, a summer influx of females from southern Tampa Bay into the northern waters of the Sarasota community range beginning in the 1990s coincided with a southward shift in the core areas of several lifelong Sarasota Bay resident females (Wells 2003).

Overall, the Sarasota community range has exhibited great stability over four decades of observations, across multiple generations of residents. Similarly, dolphins seen primarily in Gulf coastal waters, Tampa Bay, or Charlotte Harbor/Pine Island Sound have been observed in those same waters over several decades (Urian et al. 2009; Wells 2009a, b; Bassos-Hull et al. 2013). Sellas et al. (2005) found strong genetic subdivision between dolphins inhabiting the coastal Gulf of Mexico and inshore waters including Sarasota Bay, in spite of the lack of isolation. The observed genetic distinctions support the idea that inshore communities may have existed for more than a few generations, perhaps as far back as the geological formation of the bays themselves (Sellas et al. 2005). The communities along the central west coast of Florida have continued to exist in spite of catastrophic environmental perturbations, including harmful algal blooms and hurricanes. Red tides from the toxic dinoflagellate *Karenia brevis* occur every few years along the central west coast of Florida, killing large numbers of marine vertebrates, including fish and sometimes dolphins (Fire et al. 2007, 2008). A severe red tide lasting 11 months in 2005 resulted in the loss of more than 70 % of primary dolphin prey fish in the Sarasota Bay area (Barros and Wells 1998; Gannon et al. 2009). The red tide resulted in temporary shifts in group size and habitat use (McHugh et al. 2011b), but the long-term residents remained within the long-established community range. In 2004 Category 4 Hurricane Charley struck Charlotte Harbor, immediately south of Sarasota Bay. In spite of tremendous coastal devastation and extensive pollution, the long-term resident dolphins remained in the area, and overall dolphin abundance appeared unchanged (Bassos-Hull et al. 2013). The concept of a stable, long-term, geographically based bottlenose dolphin community appears to be sufficiently robust to allow differentiation of units through consideration of a variety of parameters, including genetics (Duffield and Wells 1991, 2002; Sellas et al. 2005), ranging and social association patterns (Wells 1986; Wells et al. 1987; Urian et al. 2009), and stable isotopes (Barros et al. 2010). Strong attachment to a long-term community range facilitates development and maintenance of social relationships with other residents.

8.4.2 Life History

Protracted maternal investment and long lifespans also contribute to the development and maintenance of social relationships. The continued association of mothers and calves well beyond nutritional weaning suggests the importance of this relationship for calf learning. Mothers interact with a large number of associates (Wells et al. 1987), exposing their calves, in a protected context, to individuals with whom they may interact for decades to come. Associations in juvenile groups lead to male pair bonds that can last for decades (Wells et al. 1987; Owen et al. 2002). Repeated associations of adult females over the course of rearing as many as ten calves through a reproductive lifespan of four decades can improve the females' probabilities for successful calf rearing (Owen 2001; Wells 2003).

8.5 Factors Associated with Reproductive Success

8.5.1 Female Reproductive Success

Female reproductive success is related to the mother's age and level of experience (Wells 2000, 2003). Fewer than half of mothers less than 10 years of age successfully rear their calves through the first year (Wells 2003). Considering calf parity regardless of age, 58 % of first-time mothers, including some presumably primiparous females in their early teens, are successful through their first year of calf-rearing (Fig. 8.3). This percentage increases over the next two calves before declining somewhat with subsequent calves.

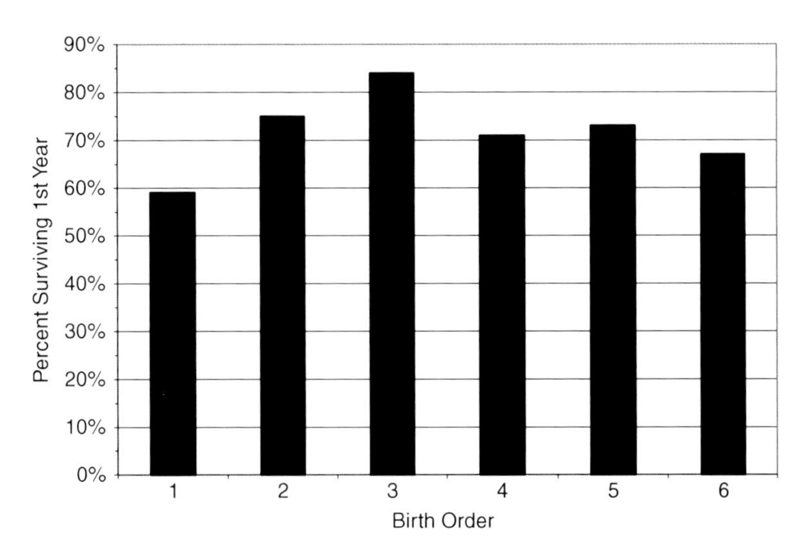

Fig. 8.3 First-year calf survival relative to birth order for 114 calves of known parity, or presumed parity based on mother's age at birth of first observed calf

Age-related female reproductive success likely results from a combination of factors. Primiparous mothers are significantly smaller on average (length and mass) than multiparous mothers, suggesting that some of the younger mothers may not be fully developed as they attempt to rear their first calf. First-time mothers also have higher concentrations of lipohilic organochlorine pollutants such as PCBs and pesticides in their tissues, which they transfer to their calves via their fat-rich milk (Wells et al. 2005; Yordy et al. 2010b). Estimates suggest that a mother may transfer 80 % of her body burden of these organochlorine contaminants to her calf in the first few months of lactation (Cockcroft et al. 1989). Concentrations in Sarasota Bay mothers before lactation, accumulated over the first 6–13 years of life, exceed hypothesized thresholds for health and reproductive impacts, placing the first-born calves at higher risk for survival (Schwacke et al. 2002; Hall et al. 2006). First-born calves in Sarasota Bay exhibit higher concentrations than subsequent calves as a legacy of their mother's original contaminant load (Wells et al. 2005). This process of depuration reduces the available contaminants for subsequent calves, reducing risks. However, as calving intervals and associated intervals between lactation periods lengthen later in a female's life (Wells 2000), tissue concentrations of contaminants increase (Wells et al. 2005), perhaps explaining at least in part the decline in first-year survival of calves after the third birth (Fig. 8.3).

Social factors also influence female reproductive success and may be related to maternal experience. Calves raised in larger and more stable groups demonstrated the highest survival (Wells 2000). These kinds of groups likely provide enhanced protection from threats such as predation, aggressive conspecifics, or boat collisions (Wells and Scott 1997), and would provide opportunities for learning through observation, allomaternal care, and socialization. Owen (2001) found that experienced mothers tended to include other mothers with calves as close associates more frequently than did first-time mothers. Multiparous mothers also demonstrated increased control over their calf's environment as compared to first-time mothers, by maintaining greater synchrony and keeping them closer (Owen 2001).

8.5.2 Male Reproductive Success

Social factors appear to play a stronger role in male reproductive success than do biological factors such as age or size for Sarasota Bay dolphins. Males associate preferentially with breeding females well before the beginning of the breeding season, perhaps to influence female choice later (Owen et al. 2002). Genetic paternity tests have shown that the Sarasota dolphins do not engage in monogamy, although some males may sire more than one calf through a particular female (Duffield and Wells 2002). Sires may be young or old, ranging in age from 13 to 40 years, and they may be larger or smaller individuals as compared to other adult males, including their alliance partners (Duffield and Wells 2002; Wells 2003). However, paired males sire disproportionately more calves than do unpaired males, suggesting an evolutionary basis for the development of these cooperative alliances. Potentially receptive females are the nearest neighbors of male alliances significantly more often, and for longer periods of time, than they are with unpaired males, providing paired males with greater access to mating opportunities (Owen 2003). Aggressive interactions between males and females in reproductive contexts appear to be much less common in Sarasota Bay than at other sites where male alliances have been observed, such as Shark Bay, Western Australia, suggesting either a greater role for female choice, or that control of females may be more subtle, perhaps influenced by the significant sexual dimorphism observed in Sarasota Bay (Wells et al. 1987; Tolley et al. 1995; Moors 1997; Connor et al. 2000; Owen 2003; Wells 2003).

8.6 Conclusions

Bottlenose dolphin social systems are the result of at least 10 to 12 million years (Myr) of delphinid evolution (Barnes 2002). The species has faced a wide range of environmental changes during its evolutionary history, and has adapted to these changes in part through the development of a high degree of behavioral plasticity, including variability in social structure (Wells et al. 1999; Mann et al. 2000; Reynolds et al. 2000). The occurrence of bottlenose dolphins in Sarasota Bay is a relatively recent phenomenon, because it has only been a few thousand years since the barrier islands and shallow bays of Florida's west coast appeared in their current configuration. Disentangling the basic, core features of bottlenose dolphin societies from the range of variability that provides the species with crucial evolutionary resiliency in the face of environmental change is challenging. Long lifespans and social changes associated with life history milestones provide additional complications, but through long-term study they also offer opportunities for developing an understanding of the factors influencing social structure. Research carried out over much of the lifespan of an individual allows observations through changing environmental conditions, leading to identification of persistent patterns. Similarly, repeated observations across multiple generations allow the identification of age-related social patterns. Variations on these general themes provide indications of the potential range of responses to environmental changes.

Strong site philopatry over multiple decades, in combination with long lifespans and the co-occurrence of as many as five generations of related individuals, provide a solid basis for repeated interactions with familiar individuals, leading to long-term social relationships and contributing to a relatively stable social system. Patterns of social associations relative to age, sex, and reproductive status have been repeated consistently across generations and through dramatic environmental changes, allowing a description of the fundamental social structure. A stable society facilitates cultural transmission of knowledge, as has been noted for Sarasota Bay dolphins relative to feeding behaviors, for example (Wells 2003).

Although strong site fidelity establishes conditions supporting the development of a stable, long-term social system, it can also create problems for the animals because it exposes them to localized threats. Coastal bottlenose dolphins are facing increasing threats of human origin from such sources as environmental contaminants (Schwacke et al. 2002; Wells et al. 2005; Woshner et al. 2008; Yordy et al. 2010a), recreational and commercial fishing gear ingestion and entanglement (Wells and Scott 1994; Wells et al. 1998, 2008; Powell and Wells 2011), boat traffic and collisions (Wells and Scott 1997; Nowacek et al. 2001; Buckstaff 2004; Wells et al. 2008), and provisioning by humans (Cunningham-Smith et al. 2006; Powell and Wells 2011). The cumulative effects of anthropogenic and natural threats can place the continued survival of the long-term resident community at risk. For example, in 2006 about 2 % of the Sarasota Bay community died from ingestion of recreational fishing gear following the severe red tide of 2005 that depleted available prey, a level of additional mortality that was unsustainable (and fortunately did not continue). To date, the dolphins of Sarasota Bay have not demonstrated a capacity for shifting their community range in response to dramatic environmental changes such as severe red tides. If these dolphins occupy an ecological "cul-de-sac" where range shifts are precluded, then this raises important concerns for the future, when global climate disruption will likely alter the local environment significantly (Wells 2010). How much capacity will these animals have to respond to environmental changes?

From an applied perspective, a stable, geographically based community can serve as a biologically meaningful unit for wildlife management purposes (Wells 1986; Urian et al. 2009; Bassos-Hull et al. 2013). The ability to relate community exposure to specific local anthropogenic threats facilitates development and implementation of mitigation measures. Mitigation of known anthropogenic threats will become increasingly important as the animals face new threats such as global climate disruption. The behavioral plasticity and long reproductive lifespan of the species may provide a high degree of resiliency, but the capacity of the animals to respond to existing and emerging threats is likely not without bounds. Successful human mitigation of anthropogenic threats will provide the animals with increased capacity to respond to changes in their environment and will help to provide opportunity for the continued long-term stability of the community.

Acknowledgments Many people over the past 43 years have contributed to the information presented in this chapter. Without the initial efforts of Blair Irvine and Michael Scott in the 1970s (and continuing today), there would have been no long-term study to report. Over the years the program has benefited greatly from the dedicated services of our laboratory managers and field coordinators, including Kim Urian, Sue Hofmann, Kim Bassos-Hull, Stephanie Nowacek, and Jason Allen, along with myriad staff, students, colleagues, and volunteers. Crucial information on ages has been provided by Aleta Hohn, and on genetic relationships by Debbie Duffield. Major support for ongoing operations has been provided by the Chicago Zoological Society, the Batchelor Foundation, NOAA's Fisheries Service, Disney, Earthwatch Institute, the U.S. Marine Mammal Commission, Dolphin Quest, and Mote Marine Laboratory. Many thanks to Katherine McHugh for her review of an early draft.

References

Barnes LG (2002) Dephinoids, evolution of the modern families. In: Perrin WF, Würsig B, Thewissen JGM (eds) Encyclopedia of marine mammals. Academic, San Diego, pp 314–316

Barros NB, Wells RS (1998) Prey and feeding patterns of resident bottlenose dolphins (*Tursiops truncatus*) in Sarasota Bay, Florida. J Mammal 79(3):1045–1059

Barros NB, Ostrom P, Stricker C, Wells RS (2010) Stable isotopes differentiate bottlenose dolphins off west central Florida. Mar Mamm Sci 26:324–336

Bassos-Hull K, Perrtree R, Shepard C, Schilling S, Barleycorn A, Allen J, Balmer B, Pine W, Wells R (2013) Long-term site fidelity and seasonal abundance estimates of common bottlenose dolphins (*Tursiops truncatus*) along the southwest coast of Florida and responses to natural perturbations. J Cetacean Res Manag 13:19–30

Benoit-Bird KJ, Würsig B, McFadden CJ (2004) Dusky dolphin (*Lagenorhynchus obscurus*) foraging in two different habitats: active acoustic detection of dolphins and their prey. Mar Mamm Sci 20:215–231

Berens McCabe E, Gannon DP, Barros NB, Wells RS (2010) Prey selection in a resident common bottlenose dolphin (*Tursiops truncatus*) community in Sarasota Bay, Florida. Mar Biol 157(5):931–942

Bigg M (1982) An assessment of killer whale (*Orcinus orca*) stocks off Vancouver Island, British Columbia. Rep Int Whaling Comm 32:655–666

Bryan CE, Christopher SJ, Balmer BC, Wells RS (2007) Establishing baseline levels of trace elements in blood and skin of bottlenose dolphins in Sarasota Bay, Florida: implications for non-invasive monitoring. Sci Total Environ 388:325–342

Buck JD, Wells RS, Rhinehart HL, Hansen LJ (2006) Aerobic microorganisms associated with free-ranging bottlenose dolphins in coastal Gulf of Mexico and Atlantic Ocean waters. J Wildl Dis 42:536–544

Buckstaff KC (2004) Effects of watercraft noise on the acoustic behavior of bottlenose dolphins, *Tursiops truncatus*, in Sarasota Bay, Florida. Mar Mamm Sci 20:709–725

Burdett Hart L, Wells RS, Adams JD, Rotstein DS, Schwacke LH (2010) Modeling lacaziosis lesion progression in common bottlenose dolphins *Tursiops truncatus* using long-term photographic records. Dis Aquat Org 90:105–112

Burdett Hart L, Rotstein DS, Wells RS, Schwacke LH (2011) Lacaziosis and lacaziois-like prevalence among common bottlenose dolphins (*Tursiops truncatus*) from the west coast of Florida, USA. Dis Aquat Org 95:49–56

Cockcroft V, DeKock A, Lord D, Ross G (1989) Organochlorines in bottlenose dolphins *Tursiops truncatus* from the east coast of South Africa. S Afr J Mar Sci 8:207–217

Connor RC, Wells RS, Mann J, Read AJ (2000) The bottlenose dolphin, *Tursiops* spp.: social relationships in a fission–fusion society. In: Mann J, Connor RC, Tyack PL, Whitehead H (eds) Cetacean societies: field studies of dolphins and whales. University of Chicago Press, Chicago, pp 91–126

Cunningham-Smith P, Colbert DE, Wells RS, Speakman T (2006) Evaluation of human interactions with a wild bottlenose dolphin (*Tursiops truncatus*) near Sarasota Bay, Florida, and efforts to curtail the interactions. Aquat Mammal 32:346–356

DeLynn RE, Lovewell G, Wells RS, Early G (2011) Congenital scoliosis of a bottlenose dolphin. J Wildl Dis 47(4):979–983

Duffield DA, Wells RS (1991) The combined application of chromosome, protein and molecular data for the investigation of social unit structure and dynamics in *Tursiops truncatus*. In: Hoelzel AR (ed) Genetic ecology of whales and dolphins. Report of the International Whaling Commission, Special Issue 13. Cambridge, pp 155–169

Duffield DA, Wells RS (2002) The molecular profile of a resident community of bottlenose dolphins, *Tursiops truncatus*. In: Pfeiffer CJ (ed) Molecular and cell biology of marine mammals. Krieger, Melbourne, pp 3–11

Duignan PJ, House C, Odell DK, Wells RS, Hansen LJ, Walsh MT, St. Aubin DJ, Rima BK, Geraci JR (1996) Morbillivirus infection in bottlenose dolphins: Evidence for recurrent epizootics in the western Atlantic and Gulf of Mexico. Marine Mammal Science 12:499-515

Dunshea G, Duffield D, Gales N, Hindell M, Wells RS, Jarman SN (2011) Telomeres as age markers in animal molecular ecology. Mol Ecol Resour 11:225–235

Esch C, Sayigh L, Wells R (2009) Quantification of parameters of signature whistles of bottlenose dolphins. Mar Mamm Sci 25:976–986

Fauquier DA, Kinsel MJ, Dailey MD, Sutton GE, Stolen MK, Wells RS, Gulland FMD (2009) Prevalence and pathology of lungworm infection in bottlenose dolphins (*Tursiops truncatus*) from southwest Florida. Dis Aquat Org 88:85–90

Fazioli KL, Hofmann S, Wells RS (2006) Use of coastal Gulf of Mexico waters by distinct assemblages of bottlenose dolphins, Tursiops truncatus. Aquatic Mammals 32:212–222

Fire SE, Fauquier D, Flewelling LJ, Henry M, Naar J, Pierce R, Wells RS (2007) Brevetoxin exposure in bottlenose dolphins (*Tursiops truncatus*) associated with *Karenia brevis* blooms in Sarasota Bay, Florida. Mar Biol 152:827–834

Fire SE, Flewelling LJ, Wang Z, Naar J, Henry MS, Pierce RH, Wells RS (2008) Florida red tide and brevetoxins: association and exposure in live resident bottlenose dolphins (*Tursiops truncatus*) in the eastern Gulf of Mexico, USA. Mar Mamm Sci 24:831–844

Ford JKB, Fisher HD (1983) Group-specific dialects of killer whales (*Orcinus orca*) in British Columbia. In: Payne RS (ed) Communication and behavior of whales. Westview, Boulder, pp 129–161

Foster EA, Franks DW, Mazzi S, Darden SK, Balcomb KC, Ford JKB, Croft DP (2012) Adaptive prolonged postreproductive life span in killer whales. Science 337:1313

Fripp D, Owen C, Quintana-Rizzo, Shapiro A, Buckstaff K, Jankowski K, Wells RS, Tyack P (2005) Bottlenose dolphin (*Tursiops truncatus*) calves appear to model their signature whistles on the signature whistles of community members. Anim Cognit 8:17–26

Gannon DP, Berens EJ, Camilleri SA, Gannon JG, Brueggen MK, Barleycorn A, Palubok V, Kirkpatrick GJ, Wells RS (2009) Effects of *Karenia brevis* harmful algal blooms on nearshore fish communities in southwest Florida. Mar Ecol Prog Ser 378:171–186

Goldberg TL, Wrangham RW (1997) Genetic correlates of social behaviour in wild chimpanzees: evidence from mitochondrial DNA. Anim Behav 54:559–570

Goodall J (1983) The chimpanzees of Gombe: patterns of behavior. Belknap Press of Harvard University Press, Cambridge

Hall AJ, McConnell BJ, Rowles TK, Aguilar A, Borrell A, Schwacke L, Reijnders PJH, Wells RS (2006) An individual-based model framework to assess the population consequences of polychlorinated biphenyl exposure in bottlenose dolphins. Environ Health Perspect 114(suppl 1):60–64

Hall AJ, Wells RS, Sweeney JC, Townsend FI, Balmer BC, Hohn AA, Rhinehart HL (2007) Annual, seasonal and individual variation in hematology and clinical blood chemistry profiles in bottlenose dolphins (*Tursiops truncatus*) from Sarasota Bay, Florida. Comp Biochem Physiol A 148:266–277

Hart LB, Rotstein DS, Wells RS, Allen J, Barleycorn A, Balmer BC, Lane SM, Speakman T, Zolman ES, Stolen M, McFee W, Goldstein T, Rowles TK, Schwacke LH (2012) Skin lesions on common bottlenose dolphins (*Tursiops truncatus*) from three sites in the Northwest Atlantic, USA. PLoS One 7(3):e33081. doi:10.1371/journal.pone.0033081

Hohn AA, Scott MD, Wells RS, Sweeney JC, Irvine AB (1989) Growth layers in teeth from known-age, free-ranging bottlenose dolphins. Mar Mamm Sci 5(4):315–342

Houde M, Wells RS, Fair PA, Bossart GD, Hohn AA, Rowles TK, Sweeney JC, Solomon KR, Muir DCG (2005) Polyfluoroalkyl compounds in free-ranging bottlenose dolphins (*Tursiops truncatus*) from the Gulf of Mexico and the Atlantic Ocean. Environ Sci Technol 39:6591–6598

Houde M, Balmer BC, Brandsma S, Wells RS, Rowles TK, Solomon KR, Muir DCG (2006) Perfluorinated alkyl compounds in relation with life-history and reproductive parameters in bottlenose dolphins (*Tursiops truncatus*) from Sarasota Bay, Florida, USA. Environ Toxicol Chem 25:2405–2412

Irvine B, Wells RS (1972) Results of attempts to tag Atlantic bottlenose dolphins (*Tursiops truncatus*). Cetology 13:1–5

Irvine AB, Scott MD, Wells RS, Kaufmann JH (1981) Movements and activities of the Atlantic bottlenose dolphin, *Tursiops truncatus*, near Sarasota, Florida. Fish Bull US 79:671–688

Janik V, Sayigh LS, Wells RS (2006) Signature whistle shape conveys identity information to bottlenose dolphins. Proc Natl Acad Sci USA 103:8293–8297

Janik VM, King SL, Sayigh LS, Wells RS (2013) Identifying signature whistles from recordings of groups of unrestrained bottlenose dolphins (*Tursiops truncatus*). Mar Mamm Sci 29:109–122

Kucklick J, Schwacke L, Wells R, Hohn A, Guichard A, Yordy J, Hansen L, Zolman E, Wilson R, Litz J, Nowacek D, Rowles T, Pugh R, Balmer B, Sinclair C, Rosel P (2011) Bottlenose dolphins as indicators of persistent organic pollutants in waters along the US East and Gulf of Mexico coasts. Environ Sci Technol 45:4270–4277

Lahvis GP, Wells RS, Kuehl DW, Stewart JL, Rhinehart HL, Via CS (1995) Decreased lymphocyte responses in free-ranging bottlenose dolphins (*Tursiops truncatus*) are associated with increased concentrations of PCBs and DDT in peripheral blood. Environ Health Perspect 103:67–72

Mann J, Connor RC, Tyack PL, Whitehead H (eds) (2000) Cetacean societies: field studies of dolphins and whales. University of Chicago Press, Chicago

Mann D, Hill-Cook M, Manire CA, Greenhow D, Montie E, Powell J, Wells RS, Bauer G, Cunningham-Smith P, Lingenfelser R, DiGiovanni R, Stone A, Brodsky M, Stevens R, Kieffer G, Hoetjes P (2010) Hearing loss in stranded odontocete dolphins and whales. PLoS One 5(11):e13824. doi:10.1371/journal.pone.0013824

McHugh KA, Allen JB, Barleycorn A, Wells RS (2011a) Natal philopatry, ranging behavior, and habitat selection of juvenile bottlenose dolphins in Sarasota Bay, FL. J Mammal 92:1298–1313

McHugh KA, Allen JB, Barleycorn AA, Wells RS (2011b) Severe harmful algal bloom events influence juvenile common bottlenose dolphin behavior and sociality in Sarasota Bay, Florida. Mar Mamm Sci 27:622–643

Miller DL, Woshner V, Styer EL, Ferguson S, Knott KK, Gray MJ, Wells RS, O'Hara TM (2011) Histological findings in free-ranging Sarasota Bay bottlenose dolphin (*Tursiops truncatus*) skin: mercury, selenium and seasonal factors. J Wildl Dis 47(4):1012–1018

Moors TL (1997) Is 'menage a trois' important in dolphin mating systems? Behavioral patterns of breeding female bottlenose dolphins. M.Sc. thesis, University of California, Santa Cruz

Nollens HH, Rivera R, Palacios G, Wellehan JFX, Saliki JT, Caseltine SL, Smith CR, Jensen ED, Hui J, Lipkin WI, Yochem PK, Wells RS, St. Leger J, Venn-Watson S (2009) New recognition of *Enterovirus* infections in bottlenose dolphins (*Tursiops truncatus*). Vet Microbiol 139:170–175

Noren SR, Wells RS (2009) Postnatal blubber deposition in free-ranging common bottlenose dolphins (*Tursiops truncatus*) with considerations to buoyancy and cost of transport. J Mammal 90:629–637

Norris KS, Dohl TP (1980a) The structure and function of cetacean schools. In: Herman LM (ed) Cetacean behavior: mechanisms and functions. Wiley, New York, pp 211–261

Norris KS, Dohl TP (1980b) The behavior of the Hawaiian spinner porpoise, *Stenella longirostris*. Fish Bull US 77:821–849

Norris KS, Würsig B, Wells RS, Würsig M (1994) The Hawaiian spinner dolphin. University of California Press, Los Angeles

Nowacek SM, Wells RS, Solow AR (2001) Short-term effects of boat traffic on bottlenose dolphins, *Tursiops truncatus*, in Sarasota Bay, Florida. Mar Mamm Sci 17:673–688

Owen CFW (2001) A comparison of maternal care by primiparous and multiparous bottlenose dolphins, *Tursiops truncatus*: does parenting improve with experience? M.Sc. thesis, University of California, Santa Cruz, CA

Owen ECG (2003) The reproductive and ecological functions of the pair-bond between allied adult male bottlenose dolphins, *Tursiops truncatus*, in Sarasota Bay, Florida. Ph.D. dissertation, University of California, Santa Cruz

Owen ECG, Hofmann S, Wells RS (2002) Ranging and social association patterns of paired and unpaired adult male bottlenose dolphins, *Tursiops truncatus*, in Sarasota, Florida, provide no evidence for alternative male strategies. Can J Zool 80:2072–2089

Parsons KM, Balcomb KC III, Ford JKB, Durban JW (2009) The social dynamics of southern resident killer whales and conservation implications for this endangered population (*Orcinus orca*). Anim Behav 77:963–971

Powell JR, Wells RS (2011) Recreational fishing depredation and associated behaviors involving common bottlenose dolphins (*Tursiops truncatus*) in Sarasota Bay, Florida. Mar Mamm Sci 27:111–129

Quintana-Rizzo E, Mann DA, Wells RS (2006) Estimated communication range of social sounds used by bottlenose dolphins (*Tursiops truncatus*). J Acoust Soc Am 120:1671–1683

Read AJ, Wells RS, Hohn AA, Scott MD (1993) Patterns of growth in wild bottlenose dolphins, *Tursiops truncatus*. J Zool Lond 231:107–123

Reynolds JE III, Wells RS, Eide SD (2000) The bottlenose dolphin: biology and conservation. University Press of Florida, Gainesville, p 289

Rowles TK, Schwacke LS, Wells RS, Saliki JT, Hansen L, Hohn A, Townsend F, Sayre RA, Hall AJ (2011) Evidence of susceptibility to morbillivirus infection in cetaceans from the United States. Mar Mamm Sci 27:1–19

Ruiz C, Nollens HH, Venn-Watson S, Green LG, Wells RS, Walsh MT, Chittick E, McBain JF, Jacobson ER (2009) Baseline circulating immunoglobulin G levels in managed collection and free-ranging bottlenose dolphins (*Tursiops truncatus*). Dev Comp Immunol 33:449–455

Sayigh LS, Tyack PT, Wells RS, Scott MD (1990) Signature whistles of free-ranging bottlenose dolphins *Tursiops truncatus*: stability and mother-offspring comparisons. Behav Ecol Sociobiol 26:247–260

Sayigh LS, Tyack PL, Wells RS, Scott MD, Irvine AB (1995) Sex difference in signature whistle production of free-ranging bottlenose dolphins, *Tursiops truncatus*. Behav Ecol Sociobiol 36:171–177

Sayigh LS, Tyack PL, Wells RS, Solow AR, Scott MD, Irvine AB (1999) Individual recognition in wild bottlenose dolphins: a field test using playback experiments. Anim Behav 57:41–50

Sayigh LS, Esch HC, Wells RS, Janik VM (2007) Facts about signature whistles of bottlenose dolphins (*Tursiops truncatus*). Anim Behav 74:1631–1642

Schwacke LH, Voit EO, Hansen LJ, Wells RS, Mitchum GB, Hohn AA, Fair PA (2002) Probabilistic risk assessment of reproductive effects of polychlorinated biphenyls on bottlenose dolphins (Tursiops truncatus) from the southeast United States coast. Environmental Toxicology and Chemistry 21:2752–2764.

Schwacke LH, Hall AJ, Townsend FI, Wells RS, Hansen LJ, Hohn AA, Bossart GD, Fair PA, Rowles TK (2009) Hematologic and serum biochemical reference intervals for free-ranging common bottlenose dolphins (*Tursiops truncatus*) and variation in the distributions of clinicopathologic values related to geographic sampling site. Am J Vet Res 70:973–985

Schwacke LH, Twiner MJ, De Guise S, Balmer BC, Wells RS, Townsend FI, Rotstein DC, Varela RA, Hansen LJ, Zolman ES, Spradlin TR, Levin M, Leibrecht H, Wang Z, Rowles TK (2010) Eosinophilia and biotoxin exposure in bottlenose dolphins (*Tursiops truncatus*) from a coastal area impacted by repeated mortality events. Environ Res 110:548–555

Schwacke LH, Zolman ES, Balmer BC, De Guise S, George RC, Hoguet J, Hohn AA, Kucklick JR, Lamb S, Levin M, Litz JA, McFee WE, Place NJ, Townsend FI, Wells RS, Rowles TK (2011) Anemia, hypothyroidism, and immune suppression associated with polychlorinated biphenyl exposure in bottlenose dolphins (*Tursiops truncatus*). Proc R Soc B Biol Sci 279: 48–57

Scott MD, Wells RS, Irvine AB (1990a) A long-term study of bottlenose dolphins on the west coast of Florida. In: Leatherwood S, Reeves RR (eds) The bottlenose dolphin. Academic, San Diego, pp 235–244

Scott MD, Wells RS, Irvine AB, Mate BR (1990b) Tagging and marking studies on small cetaceans. In: Leatherwood S, Reeves RR (eds) The bottlenose dolphin. Academic, San Diego, pp 489–514, 653 pp

Sellas AB, Wells RS, Rosel PE (2005) Mitochondrial and nuclear DNA analyses reveal fine scale geographic structure in bottlenose dolphins (*Tursiops truncatus*) in the Gulf of Mexico. Conserv Genet 6:715–728

St. Aubin DJ, Forney KA, Chivers SJ, Scott MD, Danil K, Romano T, Wells RS, Gulland FMD (2013) Hematological, serum and plasma chemical constituents in pantropical spotted dolphins (*Stenella attenuata*) following chase, encirclement and tagging. Mar Mamm Sci 29:14–35

Tolley KA, Read AJ, Wells RS, Urian KW, Scott MD, Irvine AB, Hohn AA (1995) Sexual dimorphism in wild bottlenose dolphins (*Tursiops truncatus*) from Sarasota, Florida. J Mammal 76(4):1190–1198

Twiner MJ, Fire S, Schwacke L, Davidson L, Wang Z, Morton S, Roth S, Balmer B, Rowles T, Wells R (2011) Concurrent exposure of bottlenose dolphins (*Tursiops truncatus*) to multiple toxins in Sarasota Bay, Florida, USA. PLoS One 6(3):e17394. doi:10.1371/journal.pone.0017394

Urian KW, Hofmann S, Wells RS, Read AJ (2009) Fine-scale population structure of bottlenose dolphins, *Tursiops truncatus*, in Tampa Bay, Florida. Mar Mamm Sci 25:619–638

Venn-Watson S, Rivera R, Smith CR, Saliki JT, Casteline S, St. Leger J, Yochem P, Wells RS, Nollens H (2008) Exposure to novel parainfluenza virus and clinical relevance in two bottlenose dolphin (*Tursiops truncatus*) populations. Emerg Infect Dis 14:397–405

Venn-Watson S, Townsend FI, Daniels RL, Sweeney JC, McBain JW, Klatsky LJ, Hicks CL, Staggs LA, Rowles TK, Schwacke LH, Wells RS, Smith CR (2010) Hypocitraturia in common bottlenose dolphins (*Tursiops truncatus*): assessing a potential risk factor for urate nephrolithiasis. Comp Med 60:1–5

Watwood SL, Tyack PL, Wells RS (2004) Whistle sharing in paired male bottlenose dolphins, *Tursiops truncatus*. Behav Ecol Sociobiol 55:531–543

Watwood SL, Owen ECG, Tyack PL, Wells RS (2005) Signature whistle use by free-swimming and temporarily restrained bottlenose dolphins, *Tursiops truncatus*. Anim Behav 69:1373–1386

Weir JS, Duprey NMT, Würsig B (2008) Dusky dolphin (*Lagenorhynchus obscurus*) subgroup distribution: are shallow waters a refuge for nursery groups? Can J Zool 86:1225–1234

Wells RS (1986) Structural aspects of dolphin societies. Ph.D. dissertation, University of California, Santa Cruz

Wells RS (1991) The role of long-term study in understanding the social structure of a bottlenose dolphin community. In: Pryor K, Norris KS (eds) Dolphin societies: discoveries and puzzles. University of California Press, Berkeley, pp 199–225

Wells RS (1993) Why all the blubbering? BISON Brookfield Zoo 7(2):12–17

Wells RS (2000) Reproduction in wild bottlenose dolphins: Overview of patterns observed during a long-term study. In: Duffield D, Robeck T (eds) Bottlenose dolphin reproduction workshop report. AZA Marine Mammal Taxon Advisory Group, Silver Spring, pp 57–74

Wells RS (2003) Dolphin social complexity: lessons from long-term study and life history. In: de Waal FBM, Tyack PL (eds) Animal social complexity: intelligence, culture, and individualized societies. Harvard University Press, Cambridge, pp 32–56

Wells RS (2009a) Learning from nature: bottlenose dolphin care and husbandry. Zoo Biol 28:1–17

Wells RS (2009b) Identification methods. In: Perrin WF, Würsig B, Thewissen JGM (eds) Encyclopedia of marine mammals, 2nd edn. Elsevier, San Diego, pp 593–599

Wells RS (2010) Feeling the heat: potential climate change impacts on bottlenose dolphins. Whalewatcher J Am Cetacean Soc 39(2):12–17

Wells RS, Scott MD (1990) Estimating bottlenose dolphin population parameters from individual identification and capture-release techniques. In: Hammond PS, Mizroch SA, Donovan GP (eds) Individual recognition of cetaceans: use of photo-identification and other techniques to estimate population parameters. Report of the International Whaling Commission, Special Issue 12, Cambridge, pp 407–415

Wells RS, Scott MD (1994) Incidence of gear entanglement for resident inshore bottlenose dolphins near Sarasota, Florida. In: Perrin WF, Donovan GP, Barlow J (eds) Gillnets and cetaceans. Report of the International Whaling Commission, Special Issue 15, p 629

Wells RS, Scott MD (1997) Seasonal incidence of boat strikes on bottlenose dolphins near Sarasota, Florida. Mar Mamm Sci 13:475–480

Wells RS, Scott MD (1999) Bottlenose dolphin *Tursiops truncatus* (Montagu, 1821). In: Ridgway SH, Harrison R (eds) Handbook of marine mammals: the second book of dolphins and porpoises, vol 6. Academic, San Diego, pp 137–182

Wells RS, Boness DJ, Rathbun GB (1999) Behavior. Pp. 324-422 In: Reynolds JE III, Rommel SA (eds) Biology of marine mammals. Smithsonian Institution Press, Washington, DC. Pp 324–422

Wells RS, Irvine AB, Scott MD (1980) The social ecology of inshore odontocetes. In: Herman LM (ed) Cetacean behavior: mechanisms and functions. Wiley, New York, pp 263–317

Wells RS, Scott MD, Irvine AB (1987) The social structure of free-ranging bottlenose dolphins. In: Genoways H (ed) Current mammalogy, vol 1. Plenum Press, New York, pp 247–305

Wells RS, Hofmann S, Moors TL (1998) Entanglement and mortality of bottlenose dolphins (*Tursiops truncatus*) in recreational fishing gear in Florida. Fish Bull 96:647–650

Wells RS, Rhinehart HL, Hansen LJ, Sweeney JC, Townsend FI, Stone R, Casper D, Scott MD, Hohn AA, Rowles TK (2004) Bottlenose dolphins as marine ecosystem sentinels: developing a health monitoring system. EcoHealth 1:246–254

Wells RS, Tornero V, Borrell A, Aguilar A, Rowles TK, Rhinehart HL, Hofmann S, Jarman WM, Hohn AA, Sweeney JC (2005) Integrating life history and reproductive success data to examine potential relationships with organochlorine compounds for bottlenose dolphins (*Tursiops truncatus*) in Sarasota Bay, Florida. Sci Total Environ 349:106–119

Wells RS, Allen JB, Hofmann S, Bassos-Hull K, Fauquier DA, Barros NB, DeLynn RE, Sutton G, Socha V, Scott MD (2008) Consequences of injuries on survival and reproduction of common bottlenose dolphins (*Tursiops truncatus*) along the west coast of Florida. Mar Mamm Sci 24:774–794

Woshner V, Knott K, Wells R, Willetto C, Swor R, O'Hara T (2008) Mercury and selenium in blood and epidermis of bottlenose dolphins (*Tursiops truncatus*) from Sarasota Bay, Florida (USA): interaction and relevance to life history and hematologic parameters. EcoHealth 5(1):1–11. doi:10.1007/s10393-008-0164-2

Würsig B, Jefferson TA (1990) Methods of photo-identification for small cetaceans. In: Hammond PS, Mizroch SA, Donovan GP (eds) Individual recognition of cetaceans: use of photo-identification and other techniques to estimate population parameters. Report of the International Whaling Commission, Special Issue 12, Cambridge, pp 43–55

Würsig B, Würsig M (1980) Behavior and ecology of the dusky dolphin, *Lagenorhynchus obscurus*, in the South Atlantic. Fish Bull 77:871–890

Yordy JE, Mollenhauer MAM, Wilson RM, Wells RS, Hohn A, Sweeney J, Schwacke LH, Rowles TK, Kucklick JR, Peden-Adams MM (2010a) Complex contaminant exposure in cetaceans: a comparative E-SCREEN analysis of bottlenose dolphin blubber and mixtures of four persistent organic pollutants. Environ Toxicol Chem 29:2143–2153

Yordy J, Wells RS, Balmer BC, Schwacke L, Rowles T, Kucklick JR (2010b) Life history as a source of variation for persistent organic pollutant (POP) patterns in a community of common bottlenose dolphins (*Tursiops truncatus*) resident to Sarasota Bay, FL. Sci Total Environ 408:2163–2172

Yordy JE, Pabst DA, McLellan WA, Wells RS, Rowles TK, Kucklick JR (2010c) Tissue-specific distribution and whole body burden estimates of persistent organic pollutants in the bottlenose dolphin (*Tursiops truncatus*). Environ Toxicol Chem 29:1–11

Yordy JE, Wells RS, Balmer BC, Schwacke LH, Rowles TK, Kucklick JR (2010d) Partitioning of persistent organic pollutants (POPs) between blubber and blood of wild bottlenose dolphins: implications for biomonitoring and health. Environ Sci Technol 44:4789–4795

Chapter 9
Life History Tactics in Monkeys and Apes: Focus on Female-Dispersal Species

Juichi Yamagiwa, Yukiko Shimooka, and David S. Sprague

J. Yamagiwa (✉)
Laboratory of Human Evolution Studies, Graduate School of Science, Kyoto University, Sakyo, Kyoto 606-8502, Japan
e-mail: yamagiwa@jinrui.zool.ktoto-u.ac.jp

Y. Shimooka
Department of Natural and Environmental Science, Teikyo University of Science, Yatsusawa, Uenohara, Yamanashi 409-0193, Japan

D.S. Sprague
Ecosystem Informatics Division, National Institute for Agro-Environmental Sciences, Kannondai 3-1-3, Tsukuba, Ibaraki 305-8604, Japan

J. Yamagiwa and L. Karczmarski (eds.), *Primates and Cetaceans: Field Research and Conservation of Complex Mammalian Societies*, Primatology Monographs, DOI 10.1007/978-4-431-54523-1_9, © Springer Japan 2014

Abstract Primates show life history traits similar to those of cetaceans, such as small litter size, long gestation, long lactation, and long lifespan, in spite of striking contrasts in habitats, diet, mobility, and range size between them. Ecological factors (food and predation) may influence their life history traits in various ways, but social factors (social structure and reproductive strategies) may be more important for the life history of primates, in which both sexes live together even outside the breeding season. Group-living primates are classified into female-bonded species and female-dispersal species, based on the patterns of female dispersal after maturity. A comparison of life history parameters shows that female-dispersal species have a slower life history (gestation length, weaning age, age at first reproduction, and interbirth interval) than the female-bonded species, except for neonatal weight and weaning weight, which may be determined in relationship to female body weight. To elucidate factors promoting the slow life history, we focus on Atelinae and Hominidae (female-dispersal species) and examine their interspecific and intraspecific variation in social structure and male reproductive tactics in relationship to life history traits. Most Atelinae species form multimale and multifemale groups, and variation in their life history features may reflect relationships among males and their reproductive tactics. In howler monkeys, both males and females disperse, and infanticide by males may lead to a fast life history. In other Atelines, infanticide rarely occurs, although it has the effect of reducing interbirth interval. Forcible copulation by males occasionally occurs in spider monkeys. Variations in grouping among females reflecting their flexible foraging efforts according to distribution of high-quality foods may have some effects on the fast–slow continuum in the life history features of female Atelinae. Hominidae exhibit larger variations in life history features than Atelinae, probably because of their diverse social structure. Solitary nature and male reproductive tactics may have great influences on the life history of female great apes. Female orangutans, who usually live a solitary life, show the slowest life history. Maturing female orangutans need a longer time to establish their own home range and relationships with reproductive mates than female chimpanzees and gorillas, who transfer into other groups immediately after emigration. Female gorillas show the lowest age at first reproduction and the shortest interbirth interval. Intensive caretaking of the immature by male gorillas may facilitate early weaning, and infanticide by males may promote a prolonged bonding between a protector male and females to shorten the interbirth interval. Similar life history traits have been found in four long-term study sites of chimpanzees. Only females at Bossou show a fast life history, probably the result of high-quality foods and single male group composition under isolated conditions. The more frequent and stable association between females and males and more promiscuous mating in bonobos may facilitate the search for mating partners and lead to a shorter interbirth interval than chimpanzees. Frugivorous orangutans and chimpanzees may suffer more costs of female dispersal through decreased foraging efficiency than folivorous gorillas, and chimpanzees with fission–fusion grouping may suffer more social stress than gorillas in highly cohesive groups. Such differences may generally shape the fast–slow continuum of life history in female-dispersal primate species.

Keywords Age at first reproduction • Atelinae • Female-dispersal species • Hominidae • Interbirth interval • Life history

9.1 Life History Traits of Cetaceans and Primates

Life history is expressed as a scheduling of development and reproduction within a life cycle (Stearns 1992). Parameters of life history, such as gestation length, prenatal and postnatal growth rate, weaning age, age to first reproduction, interbirth interval, and lifespan, vary allometrically with body size (Read and Harvey 1989; Charnov 1991, 1993; Purvis and Harvey 1995). In mammals, large species take longer to grow to maturity, have a longer gestation, have a longer interval between births, and have fewer young per litter than smaller species. This "fast–slow continuum" in the speed of life is also found among cetaceans and primates.

A cetacean is characterized by its large body size (for the blue whale, a body length of 31 m) and large neonatal mass (more than 15 % of maternal mass) because of its aquatic habitats and advanced feeding techniques, such as the filtering observed in mysticetes (Whitehead and Mann 2000). Among cetaceans, life history traits differ between mysticetes and odontocetes (Fig. 9.1). Although no difference is found in neonatal weight relative to adult female body weight, weaning age and age at first parturition of mysticetes are lower than those of odontocetes. The mysticetes show fast life history patterns and no obvious correlation between body size and the speed of their life history processes (Martin and Rothery 1993; Kasuya 1995). By contrast, the odontocetes generally follow the fast–slow continuum, possibly caused by predation pressure and food availability being variable with body size (Whitehead and Weilgart 2000; Whitehead and Mann 2000). Duration of lactation is extremely variable among cetacean species, ranging from 6 months in baleen whales to 6 years in bottlenose dolphins (Connor et al. 2000). Mysticetes produce milk with the highest fat content (30–53 %) among mammals and wean offspring at an earlier age than most of the odontocetes, irrespective of body length (Oftedal 1997). In odontocetes, the length of lactation has a positive correlation with body length, and weaning occurs gradually, probably the result of the necessity of learning how to feed on highly mobile prey (Mann and Smuts 1999). Food supply may affect the duration of lactation, which constitutes a constraint factor on the future reproduction of mothers (Martin and Rothery 1993; Mann et al. 2000). Cetaceans are gregarious and socially diverse, but most of their social structure is matrilineal, in which males disperse from their natal groups (Kasuya 1995). As observed in bottlenose dolphins and striped dolphins, kin-related females tend to associate and cooperate in rearing calves in a school (Miyazaki and Nishiwaki 1978; Shane et al. 1986). In some matrilineal social structures (e.g., pilot whales and killer whales), menopause has been found (Whitehead 1998). Patrilineal social structure has been found in Baird's beaked whales, in which males reach maturity earlier and live longer than females (Kasuya and Jones 1984). Male Baird's beaked whales care for weaned offspring, which is exceptional among cetaceans (Kasuya et al. 1997).

Primates show life history traits similar to those of cetaceans, such as small litter size, long gestation, long lactation, and long lifespan, in spite of striking contrasts in habitats, diet, mobility, and range size between them (Harvey et al. 1987; Read and Harvey 1989; Ross 1998; Whitehead and Mann 2000). Primates lie at the slow end of the fast–slow continuum (Fig. 9.1; Table 9.1). Haplorhines have a slower life history than strepsirrhines of the same size (Purvis et al. 2003). Recent arguments

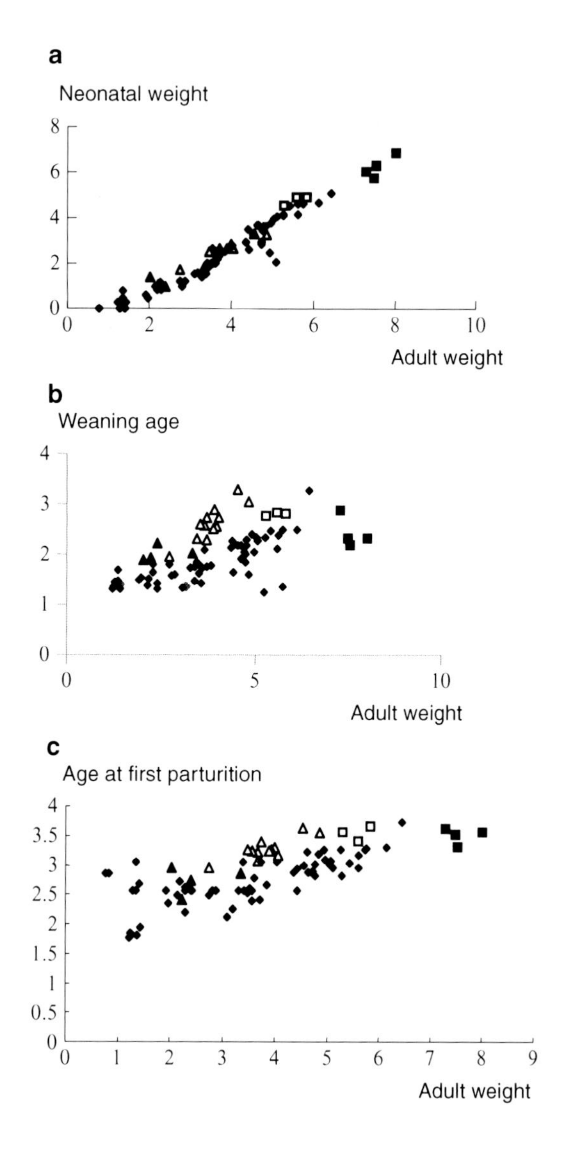

Fig. 9.1 Life history traits of cetaceans and primates among mammals relative to female body weight: neonatal weight (**a**), weaning age (**b**), and age at first parturition (**c**). *Open squares*, mysticetes; *filled squares*, odontocetes; *open triangles*, anthropoids; *filled triangles*, prosimians; *filled diamonds*, other mammals. (Data from the list by Purvis and Harvey 1995)

have proposed determinant factors of the slow life history in primates, such as large brain size (Allman et al. 1993; Martin 1996), high risk of juvenile mortality (Janson and van Schaik 1993), nutritional risk (Borries et al. 2001; Altmann and Alberts 2003; Anderson et al. 2008), and arboreal lifestyle (Eisenberg 1981; Martin 1995), but no single factor seem to fully explain this (Harvey and Purvis 1999; van Schaik and Deaner 2002).

Ecological factors may influence the life history traits of primates in various ways (Kappeler et al. 2003). The low growth rate of primates may be caused by a negative association between mortality rates and growth rates, and juvenile vulnerability to food shortage and predation may shape their life history traits (Janson and

Table 9.1 Species averages for mammalian life history variables

Common name	Scientific name	Adult weight (g)	Neonatal weight (g)	Weaning age (day)	Age at first parturition (days)
Nutria	*Myocaster coypus*	5,300	227	56	257
Uinta ground squirrel	*Spermophilus armatus*	266		21	365
Belding's ground squirrel	*S. beldingi*	257		27	365
Daurian ground squirrel	*S. dauricus*	200			365
Golden-mantled ground squirrel	*S. lateralis*	156	7	33	523
Yellow-bellied marmot	*Marmota flaviventris*	2,510	34	29	1,100
Eastern chipmink	*Tamias striatus*	95	3	35	223
Eastern grey squirrel	*Sciurus carolinensis*	568	16	63	306
Red squirrel	*Tamiasciurus hudsonicus*	200	7	62	413
East African mole rat	*Tachyoryctes splendens*	195	15	43	159
Meadow jumping mouse	*Zapus hudsonius*	19	1	28	365
Western jumping mouse	*Z. princeps*	26	1	25	478
Yellow-necked mouse	*Apodemus flavicollis*	27	2	21	89
Bank vole	*Clethrionomys glareolus*	17	2	20	58
White-footed mouse	*Peromyscus leucopus*	24	2	23	63
Deer mouse	*P. maniculatus*	18	2	23	70
Pika	*Ochotona princeps*	140	10	24	303
Brown hare	*Lepus europaeus*	3,730	115	26	243
European rabbit	*Oryctolagus cuniculus*	1,550	38	23	175
Eastern cotton-tail rabbit	*Sylvilagus floridans*	1,270	35	22	129
Greater horseshoe bat	*Rhinolophus ferrumequinum*	23	6	48	1,100
Big brown bat	*Eptesicus fuscus*	23	3	29	365
Eastern (U.S.) pipistrelle bat	*Pipistrellus subflavus*	7			730
Common pipistrelle bat	*P. pipistrellus*	6	1		730
European mole	*Talpa europaea*	85	4	31	365
European hedgehog	*Erinaceus europaeua*	771	16	40	365
Giant panda	*Ailuropoda melanoleuca*	120,000	106	183	1,170
Black bear	*Ursus americanus*	87,200	293	253	1,830
Ringed seal	*Phoca hispida*	69,700	4,400	40	1,530
Grey seal	*Halichoerus ursinus*	182,000	14,100	18	1,810
Elephant seal	*Mirounga leonina*	558,000	40,600	23	1,900
Fur seal	*Callorhinus ursinus*	42,400	4,930	82	1,700
River otter	*Lutra canadensis*	4,600	149	122	1,180
European skunk	*Mustela putorius*	646	10	38	365
Striped skunk	*Mephitis mephitis*	2,000	26	53	365
American badger	*Taxidea taxus*	4,100	104	54	593
Grey wolf	*Canis lupus*	27,000	404	43	365
Grey fox	*Urocyon cinereoargenteus*	3,300	102	42	393
Red fox	*Vulpes vulpes*	3,650	100	59	365
Arctic fox	*Alopex lagopus*	3,000	70	70	328
Domestic cat	*Felis catus*	2,620	97	56	365
Bobcat	*Lynx rufus*	7,100	317	60	460

(continued)

Table 9.1 (continued)

Common name	Scientific name	Adult weight (g)	Neonatal weight (g)	Weaning age (day)	Age at first parturition (days)
Wild boar	*Sus scrofa*	54,800	817	98	687
Warthog	*Phacochoerus aethiopicus*	54,000	720	111	776
Hippopotamus	*Hippopotamus amphibius*	1,360,000	42,800	304	2,010
Impala	*Aepyceros melampus*	44,400	4,830	154	733
Waterbuck	*Kobus ellipsiprymnus*	188,000	13,000	211	654
Kob	*K. kob*	59,200	4,470	197	641
Topi	*Damaliscus lunatus*	128,000	11,300		898
Cape buffalo	*Syncerus caffer*	547,000	40,300	304	1,800
Chamois	*Rupicapra rupicapra*	26,100	3,070	183	843
Himalayan tahr	*Hemitragus jemlahicus*	35,200	2,000	149	965
Dall's sheep	*Ovis dalli*	60,000	3,620	149	1,010
Chinese musk deer	*Moschus berezovskii*	10,900	648		1,100
Red deer	*Cervus elaphus*	109,000	8,800	216	1,030
European roe deer	*Capreolus capreolus*	23,000	866	136	733
Moose	*Alces alces*	408,000	13,200	131	897
Caribou	*Rangifer tarandus*	93,600	5,720	111	1,210
Mule deer	*Odocoileus hemionus*	56,300	2,800	69	677
Horse	*Equus caballus*	410,000	41,500	238	1,460
Plains zebra	*E. burchelli*	268,000	32,200	294	1,090
African elephant	*Loxodonta africana*	2,770,000	115,000	1,886	5,460
Blue whale	*Balaenoptera musculus*	105,000,000	7,250,000	210	3,650
Humpback whale	*Megaptera novaeangliae*	35,000,000	2,000,000	150	2,008
Grey whale	*Eschrichtius robustus*	31,466,000	500,000	210	3,285
Sperm whale	*Physeter macrocephalus*	20,000,000	1,016,000	720	4,088
False killer whale	*Pseudorca crassidens*	700,000	80,000	630	4,380
Beluga	*Delphinapterus leucas*	400,000	79,000	660	2,555
Bottlenose dolphin	*Tursiops* spp.	200,000	32,000	570	3,650
Eastern gorilla	*Gorilla beringei*	71,000	1,900	1,090	3,650
Chimpanzee	*Pan troglodytes*	34,300	1,740	1,900	4,310
White-handed gibbon	*Hylobates lar*	5,340	400	548	2,446
Gelada baboon	*Theropithecus gelada*	11,427	465	540	1,460
Guinea baboon	*Papio papio*	9,750	710	365	2,008
Rhesus macaque	*Macaca mulatta*	5,370	466	192	1,095
Toque monkey	*Macaca sinica*	3,590	446	391	1,730
Black and white colobus	*Colobus gueraza*	7,900	445	330	1,752
Savanna monkey	*Cercopithecus aethiops*	2,980	336	201	1,825
Spider monkey	*Ateles* spp.	8,440	425	760	1,825
Howler monkey	*Alouatta* spp.	4,670	295	372	1,679
Lion tamarin	*Leontopithecus rosalia*	559	50	90	876
Ring-tailed lemur	*Lemur catta*	2,210	65	105	730
Slender loris	*Loris tardigradus*	5,370	466	192	1,095
Allen's bushbaby	*Galago alleni*	255	10	170	548
Tarsiers	*Tarsius* spp.	173	12	84	256

Sources: Harvey et al. (1987), Purvis and Harvey (1995), Whitehead and Mann (2000), Kappeler and Pereira (2003)

van Schaik 1993). Primates may be adapted to the low mortality rates prevalent in their ancestral habitat (tropical forests), because other arboreal mammals, such as bats (Jones and MacLarnon 2001), also have low mortality rates. Primates living in the more unpredictable habitats have higher birthrates and earlier age at first reproduction (Ross 1998). The apes that are strictly distributed in and around the tropical forests have slow life history traits, whereas Old World monkeys living in variable habitats have relatively rapid life history traits. Forest macaques have a longer interbirth interval and a later age at first reproduction than opportunistic macaque species living in a variety of habitats, even in sympatric conditions (Ross 1992). However, the age of first reproduction in olive baboons is highly heritable (Williams-Blangero and Blangero 1995). Large intraspecific variations are also found in some life history traits. Female vervet monkeys may respond to limited access to food resources by delaying reproduction (Cheney et al. 1988). Female Japanese macaques and savanna baboons with high rank tend to mature earlier than females with low rank (Altmann et al. 1988; Gouzoules et al. 1982; Takahata et al. 1999). These observations may suggest that life history traits of primates may have evolved as a species-specific strategy as well as the immediate responses to environment changes.

Social structure and social behavior are also important for life history traits. Among mammals, primates have a unique social feature in that the two sexes live together even outside the breeding season. This lifestyle may result in diversity of social structure and may characterize the fast–slow continuum in relationship to social systems. Group size and socionomic sex ratio (the number of adult males per female within a group) can change feeding and reproductive strategies of both sexes and thereby affect life history parameters (Dunbar 1988; Sterck et al. 1997; Nunn and Pereira 2000). Female gregariousness, social relationships, or alloparental care of dependent infants may also change life history traits such as postnatal growth rate, weaning age, and interbirth interval (Fairbanks 1990; Stanford 1992; Van Noodwijk and van Schaik 2005). High infant growth rates do not appear to be correlated with environmental factors (diet, climate, or habitat) but with nonmaternal care, which allows mothers to increase birthrates by decreasing interbirth interval (Ross and MacLarnon 1995). Male reproductive strategies may constitute a strong selective force on life history traits. Infanticide by males promotes prolonged male–female association (van Schaik 2000) and complex male–infant relationships (Paul et al. 2000) and affects patterns of female movements between groups (Steenbeck 2000; Yamagiwa and Kahekwa 2001) and female reproductive biology (Watts 2000; van Noordwijk and van Schaik 2000; Yamagiwa et al. 2009). Recent findings show large variations in social structure and behavior between species and within species (Barton et al. 1996; Henzi and Barrett 2003; Doran et al. 2002; Yamagiwa et al. 2003). Life history traits are also easy to change, relatively independently, via selection (Kappeler et al. 2003). However, it is still unclear how such social variation is linked with life history variation.

In this chapter, we focus on female dispersal as the limiting factor of life history parameters. Group-living primates have been classified into female-bonded species and female-dispersal species (Wrangham 1980). Most of the macaques, cercopithecines, and *Cebus* monkeys form a group in which females remain during their entire life.

They usually associate with kin-related females and form coalitions with them in agonistic contexts (Watanabe 1979; Silk 1982; Dunbar 1988; Harcourt 1992; Henzi and Barrett 1999). Cooperation and support of kin-related females increase female reproductive success. The linear dominance rank is stable among females and between kin-groups of females. Females of the kin-groups with higher rank have higher birth-rates, younger age at first parturition, and lower infant mortality than females of kin-groups with lower rank (Drickamer 1974; Silk 1987; Itoigawa et al. 1992; Paul and Kuester 1996), although these tendencies are not consistent in some species (Cheney et al. 1988; Takahata et al. 1999). On the other hand, females of Hominidae and Atelinae usually leave their natal groups and spend their reproductive life without related females (Wrangham 1987; Yamagiwa 1999; Strier 1999a). Social relationships with males or unrelated females that they join are important for their reproductive success. The elder females or females joining earlier are dominant to younger females or those joining later (Goodall 1986; Watts 1991a; Idani 1991; Crockett and Pope 1993; Printes and Strier 1999; Nishimura 2003). However, intervention by males in conflicts (Watts 1997), sociosexual behavior among females (Kano 1992), and the fission–fusion nature of grouping (Wrangham and Smuts 1980; Goodall 1986; Strier 1992) reduce dominance effects and prevent females from having prolonged antagonistic interactions. Because of the lack of support from kin-related females, male reproductive strategies including infanticide may affect life history parameters in female-dispersal species (Strier 1999a; Harcourt and Stewart 2007). Here, we also examine inter- and intraspecific variation in social structure and male reproductive tactics in relationship to life history traits.

9.2 Life History Tactics of Female-Bonded and Female-Dispersal Species

Anthropoid primates form either female-bonded or female-dispersal groups, except for orangutans, in which both females and males spend a solitary life after maturity. Among them, we selected primate species with medium and large body size for comparisons of life history traits: 20 species for female-bonded and 15 species for female-dispersal (Table 9.2). The orangutan was included in the female-dispersal species because all females separate from their mothers before maturity. As life history parameters, gestation length, neonatal weight, weaning age, weaning weight, age at first parturition, and interbirth interval were used for interspecific comparison with reference to female body weight. We used the database constructed by Kappeler and Pereira (2003) and added to it some reliable data from recent reports. Most of the data were from observations on wild populations, but some data were from observations on provisioned or captive individuals. For statistical tests, we used the analysis of covariance (ANCOVA) in the Analyze/Fit Model section in JMPs.

Gestation length of female-dispersal species was significantly longer than that of female-bonded species ($F=69.05$, $p<0.0001$; Fig. 9.2a). Length of gestation of Atelinae (*Ateles*, *Brachyteles*) was far longer than that of Cercopithecinae

Table 9.2 Life history parameters of female-bonded and female-dispersal primate species

Species	Female-bonded (M) or female-dispersal (P)	Adult female body weight (g)	Gestation length (day)	Neonatal Weight (g)	Weaning age (day)	Weaning weight (g)	Age at first reproduction (years)	Interbirth interval (months)
Cebus abifrons	M	2,067	155	228	269		4	18
Alouatta palliata	P	4,020	186	320	325	1,100	3.6	20
Ateles fusciceps	P	9,160	226		486		4.9	
A. geoffroyi	P	7,290	225	426	750	2,000	7	37
A. paniscus	P	8,440	230	425	760	3,790	5	24
Brachyteles arachnoides	P	8,070	233		638		7.5	34
Lagothrix lagotricha	P	5,585	223	432	315		5	24
Cercopithecus ascanius	M	2,920	172	371			5	
C. cephus	M	2,805	170	339			5	
C. diana	M	3,900		460	365		5.3	12
C. pogonius	M	2,900	170	339			5	
Macaca arctoides	M	8,400	178	489	393	2,300	3.8	19
M. fascicularis	M	3,574	160	326	330	1,700	3.9	13
M. mulatta	M	5,370	165	466	192	1,454	3	12
M. nemestrina	M	4,900	167	444	234	1,417	3.9	14
M. nigra	M	4,600	170	457			5.4	18
M. radiata	M	3,700	162	388	365	2,000	4	12
M. silenus	M	5,000	180	407	365		4.9	17
Lophocebus albigena	M	6,209	175		365	2,170	4.1	
Cercocebus torquatus	M	5,500	171				4.7	
Papio anubis	M	11,700	180	915	584		4.5	25
P. cynocephalus	M	9,750	173	710	365	2,500	5.5	23
P. hamadryas	P	9,900	170	695	561	3,100	6.1	24

(continued)

Table 9.2 (continued)

Species	Female-bonded (M) or female-dispersal (P)	Adult female body weight (g)	Gestation length (day)	Neonatal Weight (g)	Weaning age (day)	Weaning weight (g)	Age at first reproduction (years)	Interbirth interval (months)
Theropithecus gelada	M	11,427	170	465	540	3,900	4	24
Colobus gereza	M	7,900	170	445	330	1,600	4.8	20
Semnopithecus entellus	M	6,910	184	500	249	2,100	3.4	17
Nasalis larvatus	M	9,593	166	450	210	2,000	4.5	18
Hylobates lar	P	5,340	205	400	548	1,070	6.7	30
H. syndactylus	P	10,568	232	513	639		5.2	50
Pongo pygmaeus	P	35,700	264	1,728	2,190	11,000	15	96
Gorilla gorilla/beringei	P	80,000	258	1,996	1,090		10	51
Pan paniscus	P	33,200	240	1,400	1,080	8,500	14	75
P. troglodytes	P	40,400	228	1,750	1,680	8,500	14	69

Sources: Anderson et al. (2008), Fedigan and Rose (1995), Furuichi and Hashimoto (2002), Harcourt and Stewart (2007), Kappeler and Pereira (2003), Nishida et al. (2003), Nishimura (2003), Shimooka et al. (2008), Wich et al. (2004)

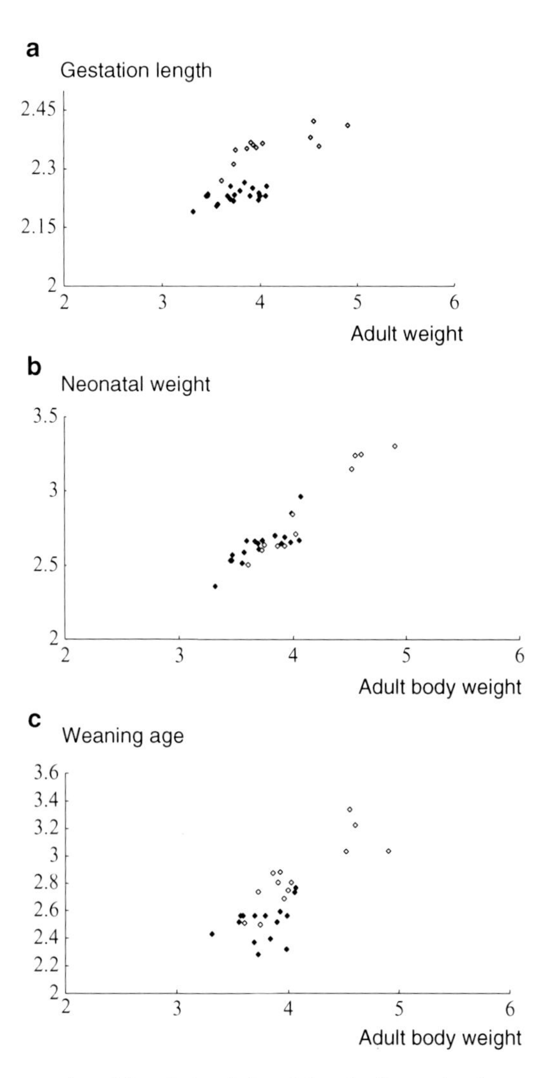

Fig. 9.2 Life history traits of female-bonded and female-dispersal primate species (relative to female body weight): gestation length (**a**), neonatal weight (**b**), weaning age (**c**), weaning weight (**d**), age at first parturition (**e**), interbirth interval (**f**). *Filled diamonds*, female-bonded species; *open diamonds*, female-dispersal species

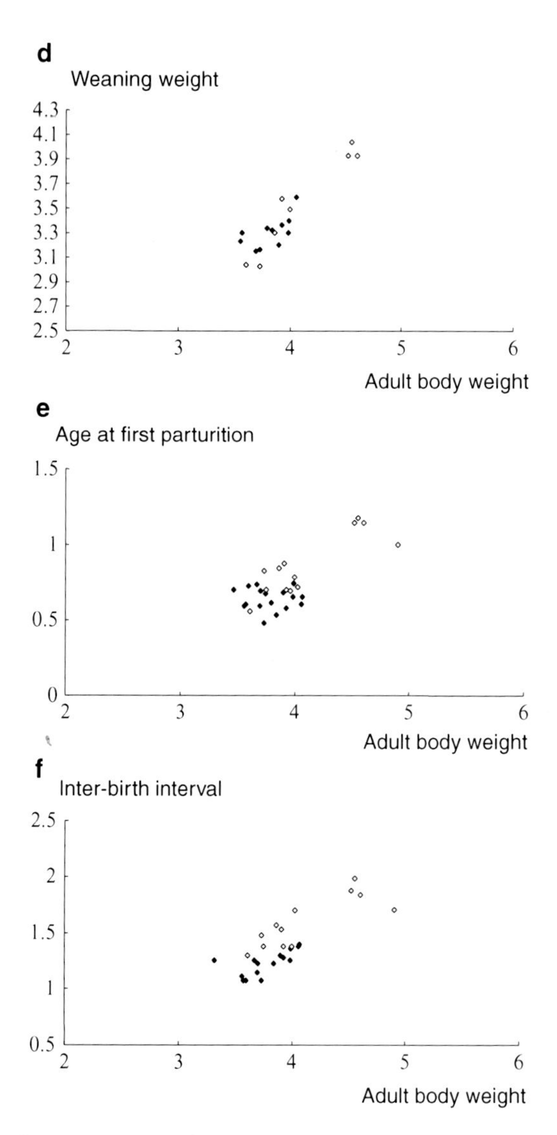

Fig. 9.2 (continued)

(*Cercopithecus*, *Macaca*, *Cercocebus*, *Lophocebus*, *Papio*, *Theropithecus*) and Colobinae (*Colobus*, *Semnopithecus*). Gestation length of Hominidae was diverse and did not correlate with body weight among the four species. Neonatal weight tended to increase relative to female body weight, and no difference was found here between the female-bonded and the female-dispersal species ($F = 106.86, p = 0.3877$; Fig. 9.2b). Although no difference was found in weaning weight ($F = 45.05$, $p = 0.9586$; Fig. 9.2c), weaning ages of female-dispersal species were significantly higher than those of female-dispersal species ($F=34.28$, $p<0.01$; Fig. 9.2d). In primate species, weaning generally occurs around the time when infants reach about one third of adult body weight (Lee et al. 1991; Lee 1996). Female-dispersal species tend to have a higher age at first parturition ($F=23.97$, $p<0.05$; Fig. 9.2e) and longer interbirth interval ($F=56.63$, $p<0.0001$; Fig. 9.2f) than female-bonded species. These results suggest that female-dispersal species may have slower life history traits than female-bonded species. Neonatal weight and weaning weight may be determined in relationship to female body weight irrespective of female movement patterns. However, female dispersal and reproduction without help from kin-related females may result in a slow life history.

Female-dispersal species form various social structures, such as solitary, monogamous, polygynous, or multimale/multifemale groups. Which aspects of female dispersal or social features lead them to slow life history? To answer this question, we compared life history traits between genera, between species, and between populations within species of Atelinae and Hominidae, which have been extensively studied at several sites.

9.3 Factors Leading to Slow Life History of Atelinae and Hominidae

Atelinae and Hominidae are typical taxa in which all genera have common social features, such as the lack of female kin bonding and female dispersal from the natal group or mother (Goodall 1986; Rosenberger and Strier 1989; Strier 1999b; Harcourt and Stewart 2007). Atelines live in the tropical forests of Central and South America and are divided into four genera: howler monkeys (*Alouatta*), spider monkeys (*Ateles*), woolly monkeys (*Lagothrix*), and muriquis (*Brachyteles*). Hominidae (great apes), except for humans, are divided into three genera and live in the tropical forests of Asia (orangutans, *Pongo*) and Africa (gorillas, *Gorilla*; chimpanzees, *Pan*). Atelines usually form groups including multiple females and males. Although male howler monkeys also disperse from their natal groups, males of the other three Atelines remain in their natal groups to associate with kin-related males, as observed for chimpanzees and bonobos (Symington 1988; Strier 1999a). The social structures of the great apes are highly differentiated. Both male and female orangutans usually live alone and partially overlap their home range with neighboring individuals of the same sex (Galdikas 1984; Delgado and van Schaik 2000; van Schaik 1999). Gorillas form a cohesive group consisting of a mature male and several females with their offspring. Both female and male gorillas tend to emigrate from their natal groups, and only females immigrate into other groups in which they start reproduction (Yamagiwa and Kahekwa 2001; Stokes et al. 2003; Robbins et al. 2009). Chimpanzees form large groups including multiple males and females, and only females emigrate from their natal groups (Nishida 1979; Goodall 1986; Boesch and Boesch-Achermann 2000).

Table 9.3 Life history parameters of atelines

Species	Body weight (kg)	Age at first reproduction (years), average (range)	Interbirth interval (years), average (range)	Source
Alouatta seniculus	5.6	5.2	(17–20)	Crockett and Pope (1993)
Lagothrix lagotricha	5.8	9	34.7	Nishimura (2003)
Ateles geoffroyi	7.5	7	34.7	Fedigan and Rose (1995)
Brachyteles sp.	9.5	(8–14)	36.0	Strier (1996)

Analogous to the great apes, Atelines are characterized by the largest body weight and the slowest life history among neotropical primates (Table 9.3). The slower reproductive traits of woolly monkeys compared to howler monkeys, which have a similar body size, may reflect their differences in female reproductive costs in relationship to social features. Female howler monkeys can have more diverse options than female woolly monkeys, such as transferring into other groups, joining males to establish a new group, or remaining in their natal group to breed (Crockett and Pope 1993; Strier 1999a). Aggressive interactions among males over mating partners are frequent, and infanticide by males occurs in howler monkeys (Crockett and Seklic 1984; Agoramoorthy and Rudran 1995; Crockett and Janson 2000). In the Venezuelan red howler monkeys (*Alouatta seniculus*), infanticide reduces the interbirth interval, and the risk of infanticide increases with the number of females within a group (Crockett and Rudran 1987; Crockett and Janson 2000). Infanticide may prompt female emigration and lead to a fast life history. In the other three Ateline genera with male philopatry, however, infanticide rarely occurs. On the other hand, coalitional aggression by males to other males, including killing of immature males and forced copulations by males, have been reported in spider monkeys (Campbell 2003, 2006; Valeo et al. 2006; Gibson et al. 2008). There are large variations in the average interval between consecutive viable births of spider monkeys among long-term study sites [32.0 months in Mexico (Ramos-Fernandez 2003); 34.5 months in Peru (Symington 1988); 43.7 months in Columbia (Y. Shimooka, unpublished data)]. Within-group competition among males and their mating strategies, including sexual coercion, may change the interbirth interval of spider monkeys (Gibson et al. 2008; Shimooka et al. 2008).

A comparison of life history parameters in female great apes indicates that orangutans have the slowest and gorillas have the fastest life history (Table 9.4). These differences are inconsistent with female body weight, and the life history parameters vary with male mating strategies. There are two types of sexually mature male orangutan: "flanged" males, with fully developed secondary sexual features, cheek pads, long hair, and a throat sack, and "non-flanged" males, looking younger without these sexual features but actually having reached the adult age. Flanged males emit loud calls and maintain antagonistic relationships with each other, competing over access to females (Galdikas 1985; Rodman and Mitani 1987; van Schaik and van Hooff 1996). Non-flanged males occasionally travel in groups to follow the same females (van Schaik et al. 2004). The strong female mating preference for the dominant flanged males facilitates their exclusive mating, but roaming non-flanged

Table 9.4 Life history parameters of the great apes

Species/subspecies	Study site	Age at first reproduction (years), average (range)	Interbirth interval (years), average (range)	Source
Pongo abelii	Ketambe	15.4 (13–18)	9.3 (6.3–11.6)	Wich et al. (2009)
Pongo pygmaeus	Tanjung Putting	15.7 (15–16)	7.7 (5.3–10.4)	Galdikas and Wood (1990), Tilson et al. (1993)
P. pygmaeus (rehabilitant)	Sepilok	10.9 (7–16)	6.2 (5.0–7.6)	Kuze et al. (2008), Kuze et al. (2012)
Pongo abelii (zoo-born, captive)		16.4 (15–17)	5.8	Anderson et al. (2008)
Pongo pygmaeus (zoo-born, captive)		15.5 (14–16)	6.3	Anderson et al. (2008)
Gorilla beringei beringei	Virunga	10.1 (9–13)[a]	3.9 (3.0–7.3)	Watts (1991a, b)
Gorilla beringei graueri	Kahuzi	10.6 (9–12)	4.6 (3.4–6.6)	Yamagiwa et al. (2003)
Gorilla gorilla gorilla (captive)		9.2 (6–19)	4.2 (2.4–6.4)	Kirschofer (1987), Sivert et al. (1991)
Pan troglodytes schweinfurthii	Mahale	13.2 (12–23)[a]	5.7 (5.3–7.3)[a]	Nishida et al. (2003)
Pan troglodytes schweinfurthii	Gombe	13.3 (11–17)	5.2 (3–8)	Goodall (1986), Wallis (1997)
Pan troglodytes verus	Bossou	10.9 (10–14)	5.3 (4–11)	Sugiyama (2004)
Pan troglodytes verus	Tai	14.3 (13–19)	5.8 (4–10)	Boesch and Boesch-Achermann (2000)
Pan troglodytes (provisioned)	River Gambia	14.3 (13–18)	5.7 (2–9)	Marsden et al. (2006)
Pan troglodytes (captive)	Japan	11.6 (5–37)	4.2	Udono et al. (1989), Yoshihara (1999)
Pan troglodytes (captive)	CIRMF (Gabon)	11.2 (8–17)	4.2 (2.8–5.2)	Tutin (1994)
Pan paniscus	Wamba	14.2 (13–15)	4.8	Kuroda (1989), Furuichi et al. (1998), Furuichi and Hashimoto (2002)

[a]Median

males occasionally force females to mate with them. Females do not form prolonged consorts with either flanged or non-flanged males. Females with dependent infants rarely associate with males, which never take care of infants. The lack of the male's care and protection may promote the female's solitary travel and preclude early weaning and reproduction.

Gorillas form a cohesive group and have no territoriality with neighboring groups (Schaller 1963; Watts 1998; Tutin 1996; Yamagiwa et al. 1996; Bermejo 2004). High cohesiveness and one-male group composition may have promoted a rapid life history compared to other apes. The leading male monopolizes most of the copulations with fertile females and takes intensive care of the offspring before and after weaning (Fossey 1979; Stewart and Harcourt 1987; Fletcher 2001; Stewart 2001). These social features may facilitate weaning at an earlier age, shorter interbirth interval, and female reproduction at an earlier age for gorillas than for chimpanzees. Furthermore, infanticide by male gorillas occurs as a mating tactic to resume female estrus and thereby shorten interbirth interval (Watts 1989; Yamagiwa et al. 2009).

Female chimpanzees and bonobos copulate with multiple males and take care of their infants by themselves (Tutin 1979; Goodall 1986; Kano 1992). Female chimpanzees tend to associate or interact with other adults less frequently than males, and mothers with dependent infants rarely join males (Wrangham 1979; Nishida 1979; Boesch and Boesch-Achermann 2000). Infanticide occurs in chimpanzees, but promiscuous mating may reduce it, as do female tactics, with paternity confusion (Hasegawa 1989; Van Noordwijk and van Schaik 2000). Fission–fusion features and promiscuous mating may prevent males from monopolizing mating and lead to a slower life history than gorillas.

Life history parameters in female great apes vary with their social features. Female orangutans, who usually live a solitary life, show the slowest life history. Maturing females need to establish their own home range and relationships with reproductive mates after separation from their mothers. They need a longer time to attain these tasks than female chimpanzees and gorillas, who transfer into other groups immediately after emigration. Solitary travel for weeks or months by female chimpanzees or gorillas has rarely been observed (Wrangham 1979; Nishida 1979; Goodall 1986; Boesch and Boesch-Achermann 2000; Watts 2003; Stokes et al. 2003; Yamagiwa et al. 2003). Female chimpanzees and gorillas may easily find mates for reproduction in the group they join and thus may not need to establish their own ranging areas. Instead, they need to establish social relationships with unrelated conspecifics within the new group. Immigrant females usually are harassed by resident females in both chimpanzees and gorillas (Goodall 1986; Idani 1991; Furuichi 1997; Watts 1991a, 1994; Harcourt and Stewart 2007). Female gorillas get support from the leading males, who frequently intervene in conflicts among females (Watts 1997; Harcourt and Stewart 2007). Immigrant female bonobos first establish affiliative relationships with resident females through sociosexual behavior (Idani 1991; Kano 1992; Furuichi 1997; Hohmann et al. 1999). Although group life may facilitate female chimpanzees in starting or resuming reproduction earlier than do female orangutans, more complex social relationships within a group and unassisted caretaking may prevent them from having a fast life history.

The costs of female transfer may also prevent Atelinae from having rapid reproduction (Strier 1999a). Female woolly monkeys transfer between groups and give birth seasonally, and immigrant females are usually accepted peacefully (Nishimura 1994, 2003). The most frequent fission–fusion in grouping is found in spider monkeys (Symington 1990; Strier 1992). Female muriquis transfer throughout the year, and immigrant females are occasionally threatened by resident females and start copulation several months after immigration (Printes and Strier 1999; Strier and Ziegler 2000). Variations in grouping and relationships among females are the driving force of the fast–slow continuum in the life history of Atelinae.

9.4 Ecological Factors Versus Social Factors Influencing Life History Parameters in Hominidae

Ecological factors, such as nutritional conditions and predation risks, may also change life history parameters. Ecological risk aversion theory predicts that seasonal fluctuation in the availability of high-quality foods, increasing intraspecific feeding competition, and low predation risk, reducing mortality rate, may lead to a slow life history (Janson and van Schaik 1993). In this respect, great apes with low predation risk have a slower life history than *Cercopithecus* monkeys, and frugivorous orangutans and chimpanzees have a slower life history than folivorous gorillas. Spider monkeys, relying most heavily on fruits with the most frequent fission–fusion grouping, have the slowest life history among Atelinae (Symington 1990; Strier 1992; Shimooka 2005). Local variation within a genus or species in relationship to different environmental conditions may elucidate the ecological role in shaping life history parameters. The great apes have been intensively studied for many years at different sites and are the best subjects for examining these local variations.

Sumatran orangutans are more frugivorous and tend to associate more frequently than Bornean orangutans (van Schaik 1999; Wich et al. 1999; Delgado and van Schaik 2000). The association rate and time tend to increase with an increase in fruit availability (Sugardjito et al. 1987). However, the association of female orangutans or an increase in association of mother–offspring units along with an increase in food availability may not be linked with fast reproduction but does extend the interbirth interval (Wich et al. 2004). Two models tried to explain these tendencies. The ecological energetic model predicts that the lower and unpredictable energy availability in Borneo may lead to reproductive output scheduled according to sufficient energy availability during the period of fruit abundance (Knott 2001; Knott et al. 2009). The ecological life history model predicts that the higher energy availability and lower mortality in Sumatra may lead to slow life history (Wich et al. 2004, 2009). Another explanation is that the difference in life history may be caused by female response to the ratio of flanged/non-flanged males and to their different mating strategies. In Sumatra, where more non-flanged males (trying forcible mating) are available, females may have slow reproduction (Delgado and van Schaik 2000). In Borneo, where more flanged males (trying monopolization of mating) are available, females may have fast reproduction. However, more detailed observations on mating and reproduction are needed to verify these interpretations.

The slower physical maturation of frugivorous western gorillas compared to folivorous mountain gorillas may support the risk-aversion hypothesis (Breuer et al. 2009). The longer interbirth interval is also suggested for western gorillas, although no difference is observed in other reproductive parameters between them (Robbins et al. 2004). Between the subspecies of *Gorilla beringei*, some differences are observed in life history parameters (Table 9.4). In the montane forest of Kahuzi, *G. b. graueri* live at a lower altitude and have a more frugivorous diet than does *G. b. beringei* in the Virungas; they show slightly longer interbirth intervals and lower infant mortality than *G. b. beringei* (Yamagiwa et al. 2003). However, this difference may be caused by the presence or absence of infanticide by males, rather than ecological factors. Killing of infants by males occurred frequently in the Virunga gorilla population as a male reproductive tactic to hasten resumption of female estrus (Fossey 1984; Watts 1989). To avoid infanticide, females tend to travel with silverbacks (fully matured males) and to join a group with multiple males to seek more reliable protection (Watts 1996; Robbins 1999). These female strategies may reduce interbirth intervals in the Virungas. By contrast, infanticide has not been observed until recently in Kahuzi, and females occasionally form all-female groups for a prolonged period after the death of the leading silverbacks (Yamagiwa and Kahekwa 2001). Infant mortality (until the second year from birth) is higher for Virunga gorillas (33.9 %) than for Kahuzi gorillas (26.1 %), and 37 % of infant mortality in the Virungas was the result of infanticide (Watts 1991b; Yamagiwa et al. 2003). The interval between the death of an infant and a next birth for Virunga gorillas (1.0 years) is shorter than that for Kahuzi gorillas (2.2 years) (Yamagiwa et al. 2003). Female Virunga gorillas may find mates and resume reproduction more rapidly after the death of infants. Three cases of infanticide have recently been observed in a group of Kahuzi gorillas, and the birthrate in the groups is very high (Yamagiwa et al. 2009, 2011). The occurrence of infanticide may promote the fast life history of gorillas.

A comparison among four study sites of chimpanzees in natural habitats shows a similarity in interbirth interval (5.2–5.8 years on average). Age at first reproduction is also relatively constant among sites (13.2–14.3 years on average), except for chimpanzees at Bossou (10.9 years). Sugiyama (2004) attributed these findings to nutritional conditions. In Bossou, high-quality foods, such as fruits and nuts, are concentrated in the study group's core area, and their tool using behavior may mitigate low nutrition during the period of fruit scarcity (Sugiyama 1997, 2004; Yamakoshi 1998). Small group size and isolated conditions may also speed the start of female reproduction in Bossou. Group size of Bossou chimpanzees is kept around 20, which is far smaller than those of Gombe, Mahale, and Tai (30–100). The study group of Bossou has been isolated from neighboring groups for 26 years, and most of the females started their first reproduction in their natal groups (Sugiyama 2004). The study group included only one adult male for more than 10 years (who stayed as an alpha male for 20 years), and the maturing males emigrated, probably because of increased competition with the leading male (Sugiyama 1999). Isolation from neighboring populations and the polygynous composition of the study group may have limited female mate choice to a single male and promoted their earlier start of reproduction. Females who gave birth in their natal group also were younger at their first birth than immigrant females in Mahale (Nishida et al. 2003). Females emigrating from their natal groups may suffer several costs for reproduction, such as reduced foraging efficiency in unknown ranges or harassment by resident females in the unfamiliar groups (Pusey 1980; Goodall 1986; Williams et al. 2002; Nishida et al. 2003).

Nutritional conditions also constitute limiting factors on the life history of great apes. Orangutans in captivity show faster reproduction than those in natural habitats (Knott and Kahlenberg 2007). Artificial feeding and grouping may permit faster growth and reproduction of orangutans than in their solitary natural habitats (Knott 2001). Rehabilitant free-ranging female orangutans also show an earlier age at first birth and a longer interbirth interval than wild females (Kuze et al. 2012). High energy intake from provisioning may enable their faster reproduction. The life history parameters of gorillas in natural habitats have been considered to be similar to those in captivity. A folivorous diet may supplement fruit scarcity, and the cohesive group formation with a one-male mating system may facilitate faster female reproduction at a level closer to their evolutionary potential (Tutin 1994; Harcourt and Stewart 2007). However, data on life history have come from folivorous mountain gorillas, and data in captivity are from frugivorous western gorillas. Recent studies on western gorillas in their natural habitats show a slower life history than captive gorillas, and regular provisioning may promote the faster life history of gorillas (Robbins et al. 2004; Breuer et al. 2009)

Female chimpanzees in captivity also tend to start reproduction earlier and have shorter interbirth intervals (Table 9.4). A rich nutritional condition from regular feeding, the limited selection of mates, and restricted movement may facilitate fast reproduction, as observed for Bossou chimpanzees. However, the interbirth interval at Bossou is longer than those in captivity and similar to those in natural habitats. Provisioned but free-ranging chimpanzees from the rehabilitation project in River Gambia National Park show a similar age at first reproduction and interbirth interval to those in natural habitats, rather than to those in captivity (Marsden et al. 2006). These observations suggest that limited movement under confined conditions may promote the shorter interbirth intervals of female chimpanzees in captivity.

9.5 Discussion

9.5.1 Costs of Female Transfer

In contrast to cetaceans, characterized by high mobility and animal diets in aquatic environments, the life history traits of primates with vegetarian diets are strongly linked with dispersal patterns of females. As primate socioecology predicts (Sterck et al. 1997), food availability and predation pressure shape female gregariousness and association between sexes, which in turn change life history parameters. Most cetaceans form matrilineal social groups, and kin-related females tend to cooperate in rearing calves in a school (Kasuya 1995; Shane et al. 1986). Food availability relative to body size may affect the duration of lactation and thereby female reproductive strategies (Martin and Rothery 1993; Whitehead and Weilgart 2000; Mann et al. 2000). By contrast, some primate species (Atelinae and Hominidae) form nonmatrilineal groups in which females transfer, and female dispersal may promote slow life history (Strier 1999a; Kappeler et al. 2003; Harcourt and Stewart 2007). Our study suggests that male reproductive tactics, adding to ecological factors, may affect the cost of female transfer and shape the fast–slow continuum in the life history traits of female-dispersal species.

In both Atelinae and Hominidae, high gregariousness among females observed in howler monkeys and gorillas is linked with faster life history than fission–fusion grouping in spider monkeys, muriquis, and chimpanzees and solitary travel in orangutans. In female-dispersal species of primates, in which females usually start reproductive life after separation from their mothers or from their natal groups, frequent association with other conspecifics and high group cohesiveness may reduce costs of female foraging and promote rapid reproduction. Solitary travel is costly in terms of the need for vigilance against predators and for finding and occupying high-quality food patches, and these costs may lead to delayed age at first reproduction and a longer interbirth interval. Association with other adults, especially with a single male that positively takes care of an infant, will decrease these costs and enable them to develop a faster life history.

In male philopatric species such as wooly monkeys, spider monkeys, muriquis, and chimpanzees, differences in life history may reflect a female's flexible foraging efforts, according to the different spatiotemporal distribution of high-quality foods, as a means to attain reproductive success. Female atelines show large variations in grouping within and between species according to the distribution of high-quality foods (Symington 1990; Strier 1992; Shimooka 2005), and female chimpanzees also show large variations in grouping between and within species (Goodall 1986; Kano 1992; Boesch and Boesch-Achermann 2000). Female great apes may promote these variations as a way of coping with strong feeding competition because of their lower digestive ability for unripe fruits and mature leaves compared with Old World monkeys (Yamagiwa 2004). In both orangutans and chimpanzees, females with dependent infants tend to travel without other adult conspecifics (van Schaik 1999; Wrangham 1979), probably because of the higher costs of feeding competition. Female bonobos form larger parties than female orangutans and chimpanzees, and even females with dependent infants usually associate with other adult females and males. Large fruit patches available throughout the year and abundant fallback foods such as terrestrial herbs may mitigate the cost of grouping among bonobos (Wrangham 1986). Frequent sociosexual behavior may also reduce social tension induced by feeding competition among them (Kuroda 1980; Kano 1989; Kitamura 1989; Parish 1994). Female gorillas do not alter their grouping patterns in response to fruit availability, although they extend their daily path length with an increasing frugivorous diet (Goldsmith 1999; Doran et al. 2002; Yamagiwa et al. 2003).

9.5.2 *Male Reproductive Tactics and Life History of Female Great Apes*

Sexual coercion of males may have strong influences on life history traits in female-dispersal species of primates. Infanticide by males promotes faster life history in howler monkeys and gorillas (Crockett and Janson 2000; Harcourt and Stewart 2007; Yamagiwa et al. 2011). To avoid infanticide, females move and choose males with which to associate, and they stay with males who have the ability to protect their infants. These movements result in stable associations between females and the protector males that lead to a fast life history.

Infanticide may shape local variation in group composition within the genus *Gorilla* through female strategies against it. In the Virungas, where infanticide by extra-group males frequently occurs, females tend to transfer into multiple male groups, seeking more protection against infanticide (Watts 1989, 1996). The female preference for multimale groups may have increased the number of females within the multimale groups and enabled maturing males to have a mating opportunity (Robbins 1995, 2001; Watts 1996, 2000; Robbins and Robbins 2005). The dominant males tend to tolerate mating by kin-related subordinate males and cooperate to defend their groups against solitary males or other groups (Robbins 2001; Watts 2003). A recent genetic analysis of paternity in four multimale groups in the Virungas indicated that both dominant and subordinate males enjoyed reproductive success, with the dominants siring an average of 85 % of group offspring (Bradley et al. 2005). Philopatric males tend to start reproduction earlier and to sire more offspring than dispersal males (Robbins and Robbins 2005). Female choice of multimale groups may have prevented male dispersal from the natal group and consequently promoted faster life histories of both males and females at Virungas (Harcourt and Stewart 2007).

By contrast, eastern lowland gorillas rarely form multimale groups and females occasionally transfer with other females and immatures at Kahuzi, where infanticide has rarely been reported (Yamagiwa et al. 1993; Yamagiwa and Kahekwa 2001). Although long-term studies have not yet been conducted on western gorillas in the lowland tropical forests, very little infanticide has been reported (Stokes et al. 2003). Frugivorous diets may prevent gorillas from forming large multimale groups in the lowland forests, and females may not have the option of transferring into multimale groups (Yamagiwa et al. 2003). Females tend to prefer to transfer into smaller groups and remain together to accept new males after the death of the leading male (Stokes 2004). Recent DNA analysis suggested a network among related males in separate but neighboring groups instead of forming multimale groups in the population of western lowland gorillas at Mondika (Bradley et al. 2004). Consequently, by forming a network of related males in neighboring one-male groups, rather than forming multimale groups, frugivorous western gorillas may be able to avoid infanticide (Fig. 9.3).

Although infanticide rarely occurs in species with male philopatry, it functions to reduce interbirth interval and may shape interspecies variation in spider monkeys

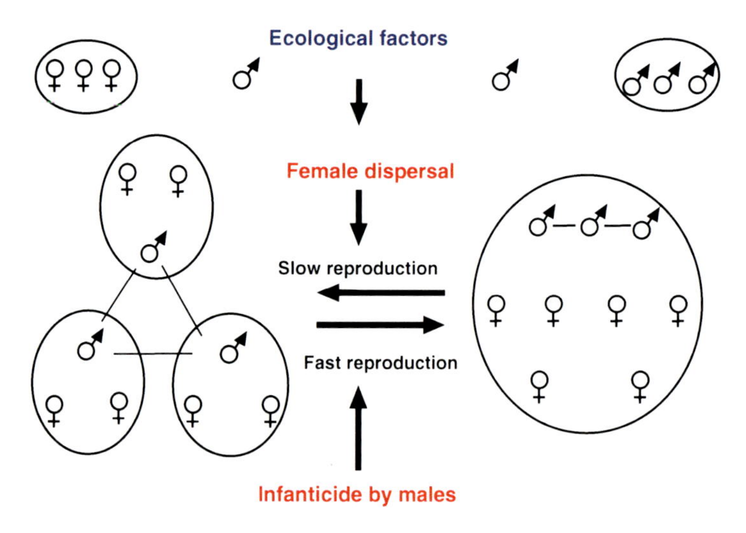

Fig. 9.3 Variation in social structure and life history of *Gorilla beringei*

(Gibson et al. 2008; Shimooka et al. 2008). Their fission–fusion grouping is characterized by frequent associations among kin-related males within a group and communal defense by these males (Symington 1990, 1988; Strier 1992; Shimooka 2005). Within-group aggression among males and forced mating by males tend to occur in the groups of spider monkeys in which operational sex ratios were highly skewed toward males (Gibson et al. 2008). These observations suggest that competition among males over mates and male coercive mating may promote faster life history traits in atelines, although long-term data on individual demography are needed for further speculation.

By contrast, sexual coercion may promote a slow life history in the genus *Pan*. As do spider monkeys, both chimpanzees and bonobos form fission–fusion grouping and males tend to remain in their natal groups after maturity (Nishida 1979; Goodall 1986; Kano 1992). Females show promiscuous mating patterns, but their gregariousness and male mating tactics vary between species and across populations (Tutin 1979; Tutin and McGinnis 1981; Furuichi 1987). Male chimpanzees are generally more gregarious than females, and the most dominant males try to monopolize mating (Tutin 1979; Hasegawa and Hiraiwa-Hasegawa 1983; Boesch and Boesch-Achermann 2000). Males form a coalition for communal defense that occasionally results in fatal communal attacks in eastern chimpanzees (Goodall et al. 1979; Nishida et al. 1985; Watts et al. 2006). Infanticide by extra-group males also occurs as an extension of their territorial aggression (Goodall 1977; Hamai et al. 1992; Arcadi and Wrangham 1999; Watts and Mitani 2000; Muller 2007). Consequently, male eastern chimpanzees sire most of their offspring within their communities (Constable et al. 2001; Inoue et al. 2008; Wroblewski et al. 2009; Newton-Fisher et al. 2010). Infanticide by females was also observed in Gombe and Budongo and is considered as a female tactic for increased conflict over resources with new immigrants (Goodall 1977; Townsend et al. 2007; Muller 2007). These aggressions may promote solitary travel of females, especially females with suckling infants, to avoid infanticide, and thus may lead to slow reproduction.

There are considerable differences in coalitional formation and intergroup attacks by males between western and eastern chimpanzees (Goodall et al. 1979; Nishida et al. 1985; Herbinger et al. 2001; Wilson and Wrangham 2003; Watts et al. 2006; Lehman and Boesch 2003). Female western chimpanzees tend to associate and groom with each other frequently at both Bossou and Tai compared to eastern chimpanzees (Sugiyama 1988; Boesch 1991; Lehmann and Boesch 2008). Nevertheless, the interbirth interval parameters of female chimpanzees at both Tai and Bossou and the age of first reproduction at Tai are similar to those of female eastern chimpanzees (Table 9.4). However, the age of first reproduction at Bossou is far earlier than those of other sites, reflecting the conditions observed in captivity. The uniqueness of Bossou chimpanzees is shown by their one-male group composition, where only one male monopolizes mating for a prolonged period (Sugiyama 2004). This social situation is similar to that in captivity, where most groups include only one adult male. Variations in social features between western and eastern chimpanzees are also explained by the distribution of defendable high-quality food resources and predation risk (Wittig and Boesch 2003; Lehmann and Boesch 2008). High nutritional conditions lead to a shorter interbirth interval in captivity than in the wild (Table 9.4). The age of first reproduction may be influenced by both nutritional conditions and male mating tactics, especially the degree of mating monopolization.

The influence of male mating tactics on the speed of life history is distinct between the two species of *Pan*. Male bonobos tend to associate with females rather than with other males, and their ranks reflect their mothers' ranks because of the mothers' strong support in agonistic conflicts between males (Kano 1992; Parish 1994; Furuichi 1997; Hohmann et al. 1999). Different groups of bonobos sometimes intermingle to stay together, and both females and males exhibit affiliative social interactions between groups (Idani 1990; Hashimoto et al. 2008). Weaker competition among male bonobos enables females to maintain stable association with males and may facilitate the search for mating partners and lead to a shorter interbirth interval than chimpanzees (Fig. 9.4).

In summary, male mating tactics may change the life history of the great apes in different ways. Female dispersal and independent reproduction from related conspecifics may enable them to form various social structures and flexible life history traits according to male mating strategies. Ecological factors basically shape the gregariousness of females in female-dispersal species, but they can choose from a wide variety of feeding strategies, from individual foraging to moving in cohesive groups. Males also take various mating tactics according to female movement and association patterns, which in turn also vary with male associations and mating strategies. Although the influences of these ecological and social factors on the life history of great apes differ between genera, between species, and between populations, the solitary nature may urge females to choose a slower life history, whereas stable associations between males and females may promote a faster life history (Fig. 9.5). Frugivorous orangutans and chimpanzees may suffer more costs of

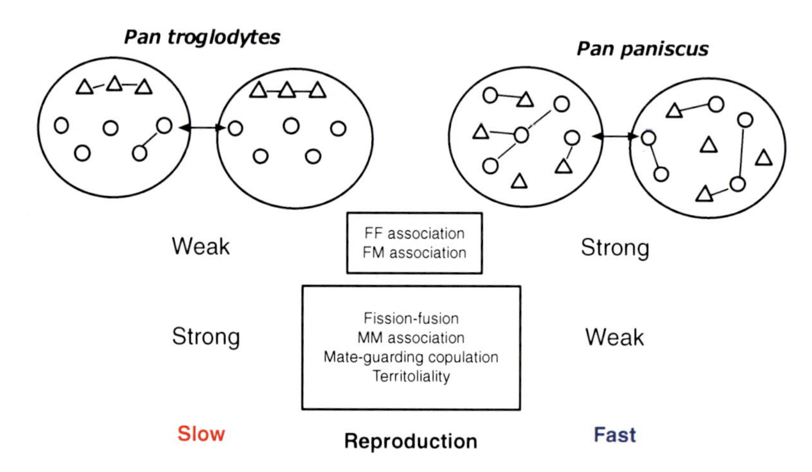

Fig. 9.4 Variation of social features and life history of *Pan*

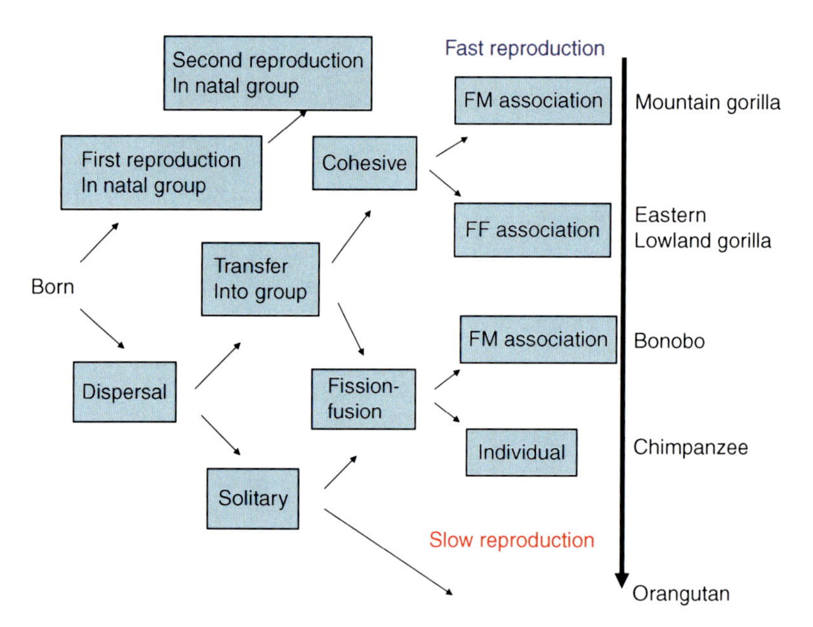

Fig. 9.5 Costs of female transfer and fast–slow continuum of life history

female movement through decreased foraging efficiency than folivorous gorillas, and chimpanzees with fission–fusion grouping may suffer more social stress than gorillas in highly cohesive groups. Such differences may generally shape the fast–slow continuum of life history in female-dispersal primate species.

Acknowledgments This study was financed in part by Grants-in-Aid for Scientific Research by the Ministry of Education, Culture, Sports, Science, and Technology, Japan (No. 162550080, No. 19107007, and No. 24255010 to J. Yamagiwa), by Grant-in-Aid for Young Scientists (B) 19770213 to Shimooka. the Global Environmental Research Fund by Japanese Ministry of Environment (F-061 to T. Nishida, Japan Monkey Centre), the Kyoto University Global COE Program "Formation of a Strategic Base for Biodiversity and Evolutionary Research," and Science and Technology Research Partnership for Sustainable Development of JST/JICA, and was conducted in cooperation with CRSN and ICCN. We appreciate to Dr. Kosei Izawa for his kind permission for the usage of a picture.

References

Agoramoorthy G, Rudran R (1995) Infanticide by adult and subadult males in free-ranging red howler monkeys of Venezuela. Ethology 99:75–88

Allman JM, McLaughlin T, Hakeem A (1993) Brain weight and life-span in primate species. Proc Natl Acad Sci USA 90:118–122

Altmann J, Alberts SC (2003) Variability in reproductive success viewed from a life-history perspective in baboons. Am J Hum Biol 15:401–409

Altmann J, Hausfater G, Altmann SA (1988) Determinants of reproductive success in savanna baboons. In: Clutton-Brock TH (ed) Reproductive success. University of Chicago Press, Chicago, pp 403–418

Anderson HB, Emery Thompson M, Knott M, Perkins CD (2008) Fertility and mortality patterns of Bornean and Sumatran orangutans: is there a species difference in life history? J Hum Evol 54:34–42

Arcadi AC, Wrangham RW (1999) Infanticide in chimpanzees: review of cases and a new within-group observation from the Kanyawara study group in Kibale National Park. Primates 40:337–351

Barton RA, Byrne RW, Whiten A (1996) Ecology, feeding competition, and social structure in baboons. Bahav Ecol Sociobiol 38:321–329

Bermejo M (2004) Home-range use and intergroup encounters in western gorillas (*Gorilla g. gorilla*) at Lossi Forest, North Congo. Am J Primatol 64:223–232

Boesch C (1991) The effects of leopard predation on grouping patterns in forest chimpanzees. Behaviour 117:220–242

Boesch C, Boesch-Achermann H (2000) The chimpanzees of Tai Forest. Cambridge University Press, Cambridge

Borries C, Koenig A, Winkler P (2001) Variation of life history traits and mating patterns in female langur monkeys (*Semnopithecus entellus*). Behav Ecol Sociobiol 50:391–402

Bradley BJ, Doran-Sheehy DM, Lukas D, Boesch C, Vigilant L (2004) Dispersed male networks in western gorillas. Curr Biol 14:510–513

Bradley BJ, Robbins MM, Williamson EA, Steklis HD, Steklis NG, Eckhardt N, Boesch C, Vigilant L (2005) Mountain gorilla tug-of-war: silverbacks have limited control over reproduction in multimale groups. Proc Natl Acad Sci USA 102:9418–9423

Breuer T, Breuer-Ndoundou Hockemba M, Olejniczak C, Parnell RJ, Stokes EJ (2009) Physical maturation, life-history classes and age estimates of free-ranging western gorillas: insights from Mbeli Bai, Republic of Congo. Am J Primatol 71:106–119

Campbell CJ (2003) Female directed aggression in free-ranging *Ateles geoffroyi*. Int J Primatol 24:223–238

Campbell (2006) Lethal intragroup aggression by adult male spider monkeys (*Ateles geoffroyi*). Am J Primatol 68:1197–1201

Charnov EL (1991) Evolution of life history variation among female mammals. Proc Natl Acad Sci USA 88:1134–1137

Charnov EL (1993) Life history invariants: some explorations of symmetry in evolutionary ecology. Oxford University Press, Oxford

Cheney DL, Seyfarth RM, Andleman S, Lee PC (1988) Reproductive success in vervet monkeys. In: Clutton-Brock TH (ed) Reproductive success. University of Chicago Press, Chicago

Connor RC, Wells R, Mann J, Read A (2000) The bottlenose dolphin: social relationships in a fission–fusion society. In: Mann J, Connor R, Tyack P, Whitehead H (eds) Cetacean societies: field studies of whales and dolphins. University of Chicago Press, Chicago, pp 91–126

Constable JL, Ashley MV, Goodall J, Pusey AE (2001) Noninvasive paternity assignment in Gombe chimpanzees. Mol Ecol 10:1279–1300

Crockett CM, Janson CH (2000) Infanticide in red howlers: female group size, male membership, and a possible link to folivory. In: van Schaik CP, Janson CH (eds) Infanticide by males and its implications. Cambridge University Press, Cambridge, pp 75–98

Crockett CM, Pope TR (1993) Consequences of sex differences in dispersal for juvenile red howler monkeys. In: Pereira ME, Fairbanks LA (eds) Juvenile primates: life history, development, and behavior. Oxford University Press, Oxford, pp 104–118

Crockett CM, Rudran R (1987) Red howler monkey birth data. II: Interannual, habitat, and sex comparisons. Am J Primatol 13:369–384

Crockett CM, Seklic R (1984) Infanticide in red howler monkeys (*Alouatta seniculus*). In: Hausfater G, Hrdy SB (eds) Infanticide: comparative and evolutionary perspectives. Aldine De Gruyter, New York, pp 173–191

Delgado R, van Schaik CP (2000) The behavioral ecology and conservation of the orangutan (*Pongo pygmaeus*): a tale of two islands. Evol Anthropol 9:201–218

Doran DM, Jungers WL, Sugiyama Y, Fleagle JG, Heesy CP (2002) Multivariate and phylogenetic approaches to understanding chimpanzee and bonobo behavioral diversity. In: Boesch C, Hohmann G, Marchant LF (eds) Behavioural diversity in chimpanzees and bonobos. Cambridge University Press, Cambridge, pp 14–34

Drickamer LC (1974) A ten-year summary of reproductive data for free ranging *Macaca mulatta*. Folia Primatol 21:61–80

Dunbar RIM (1988) Primate social system. Helm, London

Eisenberg JF (1981) The mammalian radiations: an analysis of trends in evolution, adaptation, and behavior. Chicago University Press, Chicago

Fairbanks LA (1990) Reciprocal benefits of allomothering for female vervet monkeys. Anim Behav 40:553–562

Fedigan LM, Rose LM (1995) Inter-birth interval variation in three sympatric species of Neotropical monkey. Am J Primatol 37:9–24

Fletcher A (2001) Development of infant independence from the mother in wild mountain gorillas. In: Robbins MM, Sicotte P, Stewart KJ (eds) Mountain gorillas: three decades of research at Karisoke. Cambridge University Press, London, pp 153–182

Fossey D (1979) Development of the mountain gorilla (*Gorilla gorilla beringei*): the first thirty-six months. In: Hamburg DA, McCown ER (eds) The great apes. Benjamin/Cummings, Menlo Park, pp 138–184

Fossey D (1984) Infanticide in mountain gorillas (*Gorilla gorilla beringei*) with comparative notes on chimpanzees. In: Hausfater G, Hrdy SB (eds) Infanticide: comparative and evolutionary perspectives. Aldine, Hawthorne, pp 217–236

Furuichi T (1987) Sexual swelling, receptivity and grouping of wild pygmy chimpanzee females at Wamba, Zaire. Primates 28:309–318

Furuichi T (1997) Agonistic interactions and matrifocal dominance rank of wild bonobos (*Pan paniscus*) at Wamba. Int J Primatol 18:855–875

Furuichi T, Hashimoto C (2002) Why female bonobos have a lower copulation rate during estrus than chimpanzees. In: Boesch C, Hohmann G, Marchant LF (eds) Behavioral diversity in chimpanzees and bonobos. Cambridge University Press, New York, pp 156–167

Furuichi T, Idani G, Ihobe H, Kuroda S, Kitamura K, Mori A, Enomoto T, Okayasu N, Hashimoto C, Kano T (1998) Population dynamics of wild bonobos (*Pan paniscus*) at Wamba. Int J Primatol 19:1029–1043

Galdikas BMF (1984) Adult female sociality among wild orangutans at Tanjung Putting Reserve. In: Small MF (ed) Female primates: studies by women primatologists. Liss, New York, pp 217–235
Galdikas BMF (1985) Orangutan sociality at Tanjung Putting, Central Borneo. Int J Primatol 9:1–35
Gibson KN, Vick LG, Palma AC, Carrasco FM, Taub D, Ramos-Fernandez G (2008) Intra-community infanticide and forced copulation in spider monkeys: a multi-site comparison between Cocha Cashu, Peru and Punta Laguna, Mexico. Am J Primatol 70:485–489
Goldsmith ML (1999) Ecological constraints on the foraging effort of western gorillas (*Gorilla gorilla gorilla*) at Bai Hokou, Central African Republic. Int J Primatol 20:1–23
Goodall J (1977) Infant killing and cannibalism in free-living chimpanzees. Folia Primatol (Basel) 28:259–282
Goodall J (1986) The chimpanzees of Gombe. Belknap, Cambridge
Goodall J, Bandora A, Bergmann E, Busse C, Matama H, Mpongo E, Pierce A, Riss D (1979) Intercommunity interactions in the chimpanzee population of the Gombe National Park. In: Hamburg DA, McCown ER (eds) The great apes. Benjamin/Cummings, Menlo Park, pp 13–53
Gouzoules H, Gouzoules S, Fedigan LM (1982) Behavioral dominance and reproductive success in female Japanese monkeys (*Macaca fuscata*). Anim Behav 30:1138–1190
Hamai M, Nishida T, Takasaki H (1992) New records of within-group infanticide and cannibalism in wild chimpanzees. Primates 33:151–162
Harcourt AH (1992) Coalitions and alliances: are primates more complex than non-primates? In: Harcourt AH, de Waal FBM (eds) Coalitions and alliances in human and other animals. Oxford University Press, Oxford, pp 445–472
Harcourt AH, Stewart KJ (2007) Gorilla society: conflict, compromise and cooperation between the sexes. University of Chicago Press, Chicago
Harvey PH, Purvis A (1999) Understanding the ecological and evolutionary reasons for life history variation: mammals as a case study. In: McGlade J (ed) Advanced ecological theory: principles and applications. Blackwell, Oxford, pp 232–248
Harvey PH, Martin RD, Clutton-Brock TH (1987) Life histories in comparative perspective. In: Smuts BB, Cheney DL, Seyfarth RM, Wrangham RW, Struhsaker TT (eds) Primate societies. University of Chicago Press, Chicago, pp 181–196
Hasegawa T (1989) Sexual behavior of immigrant and resident female chimpanzees at Mahale. In: Heltne PG, Marquardt LA (eds) Understanding chimpanzees. Harvard University Press, Cambridge, pp 90–103
Hasegawa T, Hiraiwa-Hasegawa M (1983) Opportunistic and restrictive mating among wild chimpanzees in the Mahale Mountains, Tanzania. J Ethol 1:75–85
Hashimoto C, Tashiro Y, Hibino E, Mulabwa M, Yangonze K, Furuichi T, Idani G, Takenaka O (2008) Longitudinal structure of a unit-group of bonobos: male philopatry and possible fusion of unit-groups. In: Furuichi T, Thompson J (eds) The bonobos: behavior, ecology, and conservation. Springer, New York, pp 107–119
Henzi P, Barrett L (1999) The value of grooming to female primates. Primates 40:47–59
Henzi P, Barrett L (2003) Evolutionary ecology, sexual conflict, and behavioral differentiation among baboon populations. Evol Anthropol 12:217–230
Herbinger L, Boesch C, Rothe H (2001) Territory characteristics among three neighboring chimpanzee communities in the Tai National Park, Ivory Coast. Int J Primatol 32:143–167
Hohmann G, Gerloff U, Tautz D, Fruth B (1999) Social bonds and genetic ties: kinship, association, and affiliation in a community of bonobos (*Pan paniscus*). Behaviour 136:1219–1235
Idani G (1990) Relations between unit-groups of bonobos at Wamba, Zaire: encounters and temporary fusions. Afr Stud Monogr 11:153–186
Idani G (1991) Social relationships between immigrant and resident bonobo (*Pan paniscus*) females at Wamba. Folia Primatol (Basel) 57:83–95
Inoue E, Inoue-Murayama M, Vigilant L, Takenaka O, Nishida T (2008) Relatedness in wild chimpanzees: influence of paternity, male philopatry, and demographic factors. Am J Phys Anthropol 137:256–262

Itoigawa N, Tanaka T, Ukai N, Fujii H, Kurokawa T, Koyama T, Ando A, Watanabe Y, Imakawa S (1992) Demography and reproductive parameters of a free-ranging group of Japanese macaques (*Macaca fuscata*) in Katsuyama. Primates 33:49–68

Janson CH, van Schaik CP (1993) Ecological risk aversion in juvenile primates: slow and steady wins the race. In: Pereira ME, Fairbanks LA (eds) Juvenile primates: development and behavior. Oxford University Press, Oxford, pp 57–74

Jones KE, MacLarnon A (2001) Bat life histories: testing models of mammalian life history evolution. Evol Ecol Res 3:465–476

Kano T (1989) The sexual behavior of pygmy chimpanzees. In: Heltne PG, Marcquardt LA (eds) Understanding chimpanzees. Harvard University Press, Cambridge, pp 176–183

Kano T (1992) The last ape: pygmy chimpanzee behavior and ecology. Stanford University Press, Stanford

Kappeler PM, Pereira ME (eds) (2003) Primate life history and socioecology. University of Chicago Press, Chicago

Kappeler PM, Pereira ME, van Schaik CP (2003) Primate life histories and socioecology. In: Kappeler PM, Pereira ME (eds) Primate life histories and socioecology. Chicago University Press, Chicago, pp 1–23

Kasuya T (1995) Overview of cetacean life histories. In: Blix AS, Walløe, Ulltang Ø (eds) Whales, seals, fish and man: proceedings of the international symposium on the biology of marine mammals in the northeast Atlantic, Tromsø, Norway, 29 November–1 December 1994. Elsevier, Amsterdam, pp 481–497

Kasuya T, Jones LL (1984) Behavior and segregation of the Dall's porpoise in the northwestern North Pacific Ocean. Sci Rep Whales Res Inst 35:107–128

Kasuya T, Balcomb K, Brownell RL Jr (1997) Life history of Baird's whales off the Pacific coast of Japan. Rep Int Whaling Comm 47:969–979

Kirschofer R (1987) International studbook of the gorilla, *Gorilla gorilla*, 1985. Frankfurt Zoological Garden, Frankfurt

Kitamura K (1989) Genito-genital contacts in the pygmy chimpanzee (*Pan paniscus*). Afr Stud Monogr 10:46–67

Knott CD (2001) Female reproductive ecology of the apes: implications for human evolution. In: Ellison PT (ed) Reproductive ecology and human evolution. de Gruyter, New York, pp 429–463

Knott CD, Kahlenberg SM (2007) Orangutans in perspective. In: Campbell CJ, Fuentes A, MacKinnon KC, Panger M, Beader SK (eds) Primates in perspective. Oxford University Press, New York, pp 290–305

Knott CD, Thompson ME, Wich SA (2009) The ecology of female reproduction in wild orangutans. In: Wich SA, Utami SS, Mitra Setia T, van Schaik CP (eds) Orangutans: geographical variation in behavioral ecology and conservation. Oxford University Press, Oxford, pp 171–188

Kuroda S (1980) Social behavior of the pygmy chimpanzees. Primates 21:181–197

Kuroda S (1989) Developmental retardation and behavioral characteristics of pygmy chimpanzees. In: Heltne PG, Marquardt AE (eds) Understanding chimpanzees. Chicago Academy of Sciences, Chicago, pp 184–193

Kuze N, Sipangkui S, Malim T, Bernard H, Ambu L, Kohshima S (2008) Reproductive parameters over a 37-year period of free-ranging female Borneo orangutans at Sepilok Orangutan Rehabilitation Centre. Primates 49:126–134

Kuze N, Dellatore D, Banes GL, Pratje P, Tajima T, Russon AE (2012) Factors affecting reproduction in rehabilitant female orangutans: young age at first birth and short inter-birth interval. Primates 53:181–192

Lee PC (1996) The meaning of weaning: growth, lactation, and life history. Evol Anthropol 5:87–96

Lee PC, Majluf P, Gordon IJ (1991) Growth, weaning and maternal investment from a comparative perspective. J Zool 225:99–114

Lehman J, Boesch C (2003) Social influences on ranging patterns among chimpanzees (*Pan troglodytes verus*) in the Tai National Park, Cote d'Ivoire. Behav Ecol 14(5):642–649
Lehmann J, Boesch C (2008) Sexual differences in chimpanzee sociality. Int J Primatol 29:65–81
Mann J, Smuts BB (1999) Behavioral development of wild bottlenose dolphin newborns. Behaviour 136:529–566
Mann J, Connor RC, Barre LM, Heithaus MR (2000) Female reproductive success in wild bottlenose dolphins (*Tursiops* sp.): life history, habitat, provisioning, and group size effects. Behav Ecol 11:210–219
Marsden BS, Marsden D, Thompson EM (2006) Demographic and female life history parameters of free-ranging chimpanzees at the chimpanzee rehabilitation project, River Gambia National Park. Int J Primatol 27:391–410
Martin RD (1995) Phylogenetic aspects of primate reproduction: the context of advanced maternal care. In: Pryce CR, Martin RD, Skuse D (eds) Motherhood in human and nonhuman primates. Karger, Basel, pp 16–26
Martin RD (1996) Scaling of the mammalian brain: the maternal energy hypothesis. News Physiol Sci 11:149–156
Martin AR, Rothery P (1993) Reproductive parameters of female long-finned pilot whales (*Globicephala melas*) around the Faroe Islands. In: Donovan GP, Lockyer CH, Martin AR (eds) Biology of northern hemisphere pilot whales: a collection of papers. Reports of the International Whaling Commission, Special Issue 14. International Whaling Commission, Cambridge, pp 263–304
Miyazaki N, Nishiwaki M (1978) School structure of the striped dolphin off the Pacific coast of Japan. Sci Rep Whales Res Inst 30:65–115
Muller MN (2007) Chimpanzee violence: femmes fatales. Curr Biol 17:365–366
Newton-Fisher NE, Thompson ME, Reynolds V, Boesch C, Vigilant L (2010) Paternity and social rank in wild chimpanzees (*Pan troglodytes*) from the Budongo Forest, Uganda. Am J Phys Anthropol 142:417–428
Nishida T (1979) The social structure of chimpanzees of Mahale Mountains. In: Hamburg DA, McCown ER (eds) The great apes. Benjamin/Cummings, Menlo Park, pp 73–121
Nishida T, Hiraiwa-Hasegawa M, Hasegawa T, Takahata Y (1985) Group extinction and female transfer in wild chimpanzees in the Mahale Mountains. Z Tierpsychol 67:284–301
Nishida T, Corp N, Hamai M, Hasegawa T, Hiraiwa-Hasegawa M, Hosaka K, Hunt KD, Itoh N, Kawanaka K, Matsumoto-Oda A, Mitani JC, Nakamura M, Norikoshi K, Sakamaki T, Turner L, Uehara S, Zamma K (2003) Demography, female life history, and reproductive profiles among the chimpanzees of Mahale. Am J Primatol 59:99–121
Nishimura A (1994) Social interaction patterns of woolly monkeys (*Lagothrix lagothricha*): a comparison among the atelines. Sci Eng Rev Doshisha Univ 35(2):235–254
Nishimura A (2003) Reproductive parameters of wild female *Lagothrix lagotricha*. Int J Primatol 24:707–722
Nunn CL, Pereira ME (2000) Group histories and offspring sex ratios in ringtailed lemurs (*Lemur catta*). Behav Ecol Sociobiol 48:18–28
Oftedal OT (1997) Lactation in whales and dolphins: evidence of divergence between baleen- and toothed-species. J Mamm Gland Biol Neoplasia 2:205–230
Parish AR (1994) Sex and food control in the "uncommon chimpanzee:" how bonobo females overcome a phylogenetic legacy of male dominance. Ethol Sociobiol 15:157–179
Paul A, Kuester J (1996) Differential reproduction in male and female Barbary macaques. In: Fa JE, Lindburg DG (eds) Evolution and ecology of macaque societies. Cambridge University Press, Cambridge, pp 293–317
Paul A, Preuchoft S, van Schaik CP (2000) The other side of the coin: infanticide and the evolution of affiliative male–infant interactions in Old World primates. In: van Schaik CP, Janson CH (eds) Infanticide by males and its implications. Cambridge University Press, Cambridge, pp 269–292
Printes RC, Strier KB (1999) Behavioral correlates of dispersal in female muriquis (*Brachyteles arachnoides*). Int J Primatol 20:941–960

Purvis A, Harvey PH (1995) Mammalian life history evolution: a comparative test for Charnov's model. J Zool 237:259–283
Purvis A, Webster AJ, Agapow P-M, Jones KE, Isaac NJB (2003) Primate life histories and phylogeny. In: Kappeler PM, Pereira ME (eds) Primate life histories and socioecology. Chicago University Press, Chicago, pp 25–40
Pusey AE (1980) Inbreeding avoidance in chimpanzees. Anim Behav 28:543–552
Read AF, Harvey PH (1989) Life history differences among the eutherian radiations. J Zool 219:329–353
Robbins MM (1995) A demographic analysis of male life history and social structure of mountain gorillas. Behaviour 132:21–47
Robbins MM (1999) Male mating patterns in wild multimale mountain gorilla groups. Anim Behav 57:1013–1020
Robbins MM (2001) Variation in the social system of mountain gorillas: the male perspective. In: Robbins MM, Sicotte P, Stewart KJ (eds) Mountain gorillas: three decades of research at Karisoke. Cambridge University Press, Cambridge, pp 29–58
Robbins AM, Robbins MM (2005) Fitness consequences of dispersal decisions for male mountain gorillas (*Gorilla gorilla beringei*). Behav Ecol Sociobiol 58:295–309
Robbins MM, Bermejo M, Cipolletta C, Magliocca F, Parnell RJ, Stokes E (2004) Social structure and life-history patterns in western gorillas (*Gorilla gorilla gorilla*). Am J Primatol 64:145–159
Robbins AM, Stoinski TS, Fawsett KA, Robbins MM (2009) Socioecological influences on the dispersal of female mountain gorillas: evidence of a second folivore paradox. Behav Ecol Sociobiol 63:477–489
Rodman PS, Mitani JC (1987) Orangutans: sexual dimorphism in a solitary species. In: Smuts BB, Cheney DL, Seyfarth RM, Wrangham RW, Struhsaker TT (eds) Primate societies. University of Chicago Press, Chicago, pp 146–154
Rosenberger AL, Strier KB (1989) Adaptive radiation of the ateline primates. J Hum Evol 18:717–750
Ross C (1992) life history patterns and ecology of macaque species. Primates 33:207–215
Ross C (1998) Primate life histories. Evol Anthropol 6:54–63
Ross C, MacLarnon A (1995) Ecological and social correlates of maternal expenditure on infant growth in haplorhine primates. In: Pryce C, Martin RD, Skuse D (eds) Motherhood in human and nonhuman primates: biological and social determinants. Karger, Basel, pp 37–46
Schaller GB (1963) The mountain gorilla: ecology and behavior. University of Chicago Press, Chicago
Shane SH, Wells RS, Wursig B (1986) Ecology, behavior and social organization of the bottlenose dolphin: a review. Mar Mamm Sci 2:34–63
Shimooka Y (2005) Sexual differences in ranging of *Ateles belzebuth belzebuth* at La Macarena, Colombia. Int J Primatol 26:385–406
Shimooka Y, Campbell CJ, Di Fiore A, Felton AM, Izawa K, Link A, Nishimura A, Ramos-Fernandez G, Wallace RB (2008) Demography and group composition of *Ateles*. In: Campbell CJ (ed) Spider monkeys: behavior, ecology, and evolution of the genus *Ateles*. Cambridge University Press, Cambridge, pp 329–348
Silk J (1982) Altruism among female *Macaca radiata*: explanations and analysis of patterns of grooming and coalition formation. Behaviour 79:162–168
Silk J (1987) Social behavior in evolutionary perspective. In: Smuts BB, Cheney DL, Seyfarth RM, Wrangham RW, Struhsaker TT (eds) Primate societies. University of Chicago Press, Chicago, pp 318–329
Sivert J, Karesh WB, Sunde V (1991) Reproductive intervals in captive female western lowland gorillas with a comparison to wild mountain gorillas. Am J Primatol 24:227–234
Stanford CB (1992) Costs and benefits of allomothering in wild capped langurs (*Presbytes pileata*). Behav Ecol Sociobiol 30:29–34
Stearns SC (1992) The evolution of life histories. Oxford University Press, Oxford

Steenbeck R (2000) Infanticide by males and female choice in wild Thomas's langurs. In: van Schaik CP, Janson CH (eds) Infanticide by males and its implications. Cambridge University Press, Cambridge, pp 153–177

Sterck EHM, Watts DP, van Schaik CP (1997) The evolution of female social relationships in nonhuman primates. Behav Ecol Sociobiol 41:291–309

Stewart KJ (2001) Social relationships of immature gorillas and silverbacks. In: Robbins MM, Sicotte P, Stewart KJ (eds) Mountain gorillas: three decades of research at Karisoke. Cambridge University Press, London, pp 183–214

Stewart KJ, Harcourt AH (1987) Gorillas: variation in female relationships. In: Smuts B, Cheney D, Seyfarth R, Wrangham RW, Struhsaker T (eds) Primate societies. University of Chicago Press, Chicago, pp 155–164

Stokes EJ (2004) Within-group social relationships among females and adult males in wild western lowland gorillas (*Gorilla gorilla gorilla*). Am J Primatol 64:233–246

Stokes EJ, Parnell RJ, Olejniczak C (2003) Female dispersal and reproductive success in wild western lowland gorillas (*Gorilla gorilla gorilla*). Behav Ecol Sociobiol 54:329–339

Strier KB (1992) Ateline adaptations: behavioral strategies and ecological constraints. Am J Phys Anthropol 88:515–524

Strier KB (1999a) The atelines. In: Dolhinow P, Fuentes A (eds) The nonhuman primates. Mayfield, Mountain View, pp 109–122

Strier KB (1999b) Why is female kin bonding so rare? Comparative sociality of neotropical primates. In: Lee PC (ed) Comparative primate socioecology. Cambridge University Press, Cambridge, pp 300–319

Strier KB, Ziegler TE (2000) Lack of pubertal influences on female dispersal in muriqui monkeys, *Brachyteles arachnoides*. Anim Behav 59:849–860

Sugardjito J, te Boekhorst IJA, van Hooff JARAM (1987) Ecological constraints on the grouping of wild orangutans (*Pongo pygmaeus*) in the Gunung Leuser National Park, Sumatra, Indonesia. Int J Primatol 8:17–41

Sugiyama Y (1988) Grooming interactions among adult chimpanzees at Bossou, Guinea, with special reference to social structure. Int J Primatol 9:393–407

Sugiyama Y (1997) Social tradition and the use of tool-composites by wild chimpanzees. Evol Anthropol 6:23–27

Sugiyama Y (1999) Socioecological factors of male chimpanzee migration at Bossou, Guinea. Primates 40:61–68

Sugiyama Y (2004) Demographic parameters and life history of chimpanzees at Bossou, Guinea. Am J Primatol 124:154–165

Symington MM (1988) Demography, ranging patterns, and activity budgets of black spider monkeys (*Ateles paniscus chamek*) in the Manu National Park, Peru. Am J Primatol 15:45–67

Symington MM (1990) Fission–fusion social organization in *Ateles* and *Pan*. Int J Primatol 11:47–61

Takahata Y, Huffmann MA, Suzuki S, Koyama N, Yamagiwa J (1999) Why dominants do not consistently attain high mating and reproductive success: a review of longitudinal Japanese macaque studies. Primates 40:143–158

Townsend SW, Slocombe KE, Emery Thompson ME, Zuberbühler K (2007) Female-led infanticide in wild chimpanzees. Curr Biol 17:355–356

Tutin CEG (1979) Mating patterns and reproductive strategies in a community of wild chimpanzees (*Pan troglodytes schweinfurthii*). Behav Ecol Sociobiol 6:29–38

Tutin CEG (1994) Reproductive success story: variability among chimpanzees and comparisons with gorillas. In: Wrangham RW, McGrew WC, de Waal FBM, Heltne PG (eds) Chimpanzee cultures. Harvard University Press, Cambridge, pp 181–194

Tutin CEG (1996) Ranging and social structure of lowland gorillas in the Lopé Reserve, Gabon. In: McGrew WC, Marchant LF, Nishida T (eds) Great ape societies. Cambridge University Press, Cambridge, pp 58–70

Tutin CEG, McGinnis PR (1981) Chimpanzee reproduction in the wild. In: Graham CE (ed) Reproductive biology of the great apes. Academic Press, New York, pp 239–264
Udono T, Sasaoka S, Inoue M, Takenaka A, Takenaka O (1989) Breeding of the chimpanzee in Sanwakagaku Kenkyusho Reichorui Center and paternity discrimination. Primate Res 5:157 (in Japanese)
Valeo A, Schaffner CM, Vick L, Aureli F, Ramos-Fernandez G (2006) Intragroup lethal aggression in wild spider monkeys. Am J Primatol 68:732–737
Van Noodwijk MA, van Schaik CP (2005) Development of ecological competence in Sumatran orangutans. Am J Phys Anthropol 127:79–94
Van Noordwijk MA, van Schaik CP (2000) Reproductive patterns in eutherian mammals: adaptations against infanticide? In: van Schaik CP, Janson CH (eds) Infanticide by males and its implications. Cambridge University Press, Cambridge, pp 322–360
van Schaik CP (1999) The socioecology of fission–fusion sociality in orangutans. Primates 40:73–90
van Schaik CP (2000) Infanticide by male primates: the sexual selection hypothesis revisited. In: van Schaik CP, Janson CH (eds) Infanticide by males and its implications. Cambridge University Press, Cambridge, pp 27–60
van Schaik CP, Deaner RO (2002) Life history and cognitive evolution in primates. In: Tyack PL, de Waal FBM (eds) Animal social complexity. Harvard University Press, Cambridge
van Schaik CP, van Hooff JARAM (1996) Toward an understanding of the orangutan's social system. In: McGrew WC, Marchant LF, Nishida T (eds) Great ape societies. Cambridge University Press, Cambridge, pp 3–15
van Schaik CP, Preuschoft S, Watts DP (2004) Great ape social systems. In: Russon AE, Begun DR (eds) The evolution of thought: evolutionary origins of great ape intelligence. Cambridge University Press, Cambridge, pp 190–209
Watanabe K (1979) Alliance formation in a free-ranging troop of Japanese macaques. Primates 20:459–474
Watts DP (1989) Infanticide in mountain gorillas: new cases and a reconsideration of the evidence. Ethology 81:1–18
Watts DP (1991a) Harassment of immigrant female mountain gorillas by resident females. Ethology 89:135–153
Watts DP (1991b) Mountain gorilla reproduction and sexual behavior. Am J Primatol 24:211–225
Watts DP (1994) Agonistic relationships between female mountain gorillas (*Gorilla gorilla beringei*). Behav Ecol Sociobiol 34:347–358
Watts DP (1996) Comparative socio-ecology of gorillas. In: McGrew WC, Marchant LF, Nishida T (eds) Great ape societies. Cambridge University Press, Cambridge, pp 16–28
Watts DP (1997) Agonistic interventions in wild mountain gorilla groups. Behaviour 134:23–57
Watts DP (1998) Long-term habitat use by mountain gorillas (*Gorilla gorilla beringei*). 1. Consistency, variation, and home range size and stability. Int J Primatol 19:651–680
Watts DP (2000) Causes and consequences of variation in male mountain gorilla life histories and group membership. In: Kappeler P (ed) Primate males. Cambridge University Press, Cambridge, pp 169–179
Watts DP (2003) Gorilla social relationships: a comparative overview. In: Taylor AB, Goldsmith ML (eds) Gorilla biology: a multidisciplinary perspective. Cambridge University Press, Cambridge, pp 302–327
Watts DP, Mitani JC (2000) Infanticide and cannibalism by male chimpanzees at Ngogo, Kibale National Park, Uganda. Primates 41:357–365
Watts DP, Muller M, Amsler S, Mbabazi G, Mitani JC (2006) Lethal intergroup aggression by chimpanzees in the Kibale National Park, Uganda. Am J Primatol 68:161–180
Whitehead H (1998) Cultural selection and genetic diversity in matrilineal whales. Science 282:1708–1711

Whitehead H, Mann J (2000) Female reproductive strategies of cetaceans: life histories and calf care. In: Mann J, Connor RC, Tyack PL, Whitehead H (eds) Cetacean societies: field studies of dolphins and whales. University of Chicago Press, Chicago, pp 219–246

Whitehead H, Weilgart L (2000) The sperm whale: social females and roving males. In: Mann J, Connor RC, Tyack PL, Whitehead H (eds) Cetacean societies: field studies of dolphins and whales. University of Chicago Press, Chicago, pp 154–172

Wich SA, Sterck EHM, Utami SS (1999) Are orangutan females as solitary as chimpanzee females? Folia Primatol (Basel) 70:23–28

Wich SA, Utami-Atmoko SS, Mitra Setia T, Rijksen HD, Schurmann C, van Hooff JARAM, van Schaik CP (2004) Life history of wild Sumatran orangutans (*Pongo abelii*). J Hum Evol 47:385–398

Wich SA, de Vries H, Ancrenaz M, Perkins L, Shumaker RW, Suzuki A, van Schaik CP (2009) Orangutan life history variation. In: Wich SA, Utami SS, Mitra Setia T, van Schaik CP (eds) Orangutans: geographical variation in behavioral ecology and conservation. Oxford University Press, Oxford, pp 65–75

Williams J, Pusey AE, Carlis JV, Farm BP, Goodall J (2002) Female competition and male territorial behaviour influence female chimpanzees' ranging patterns. Anim Behav 63:347–360

Williams-Blangero S, Blangero J (1995) Heritability of age at first birth in captive olive baboons. Am J Primatol 37:233–239

Wilson ML, Wrangham RW (2003) Intergroup relations in chimpanzees. Annu Rev Anthropol 32:363–392

Wittig RM, Boesch C (2003) Food competition and linear dominance hierarchy among female chimpanzees in the Tai National Park. Int J Primatol 24:847–867

Wrangham RW (1979) Sex differences in chimpanzee dispersion. In: Hamburg DA, McCown ER (eds) The great apes. Benjamin/Cummings, Menlo Park, pp 480–489

Wrangham RW (1980) An ecological model of female-bonded primate groups. Behaviour 75:262–300

Wrangham RW (1986) Ecology and social relationships in two species of chimpanzees. In: Rubenstein DI, Wrangham RW (eds) Ecological aspects of social evolution: birds and mammals. Princeton University Press, Princeton, pp 352–378

Wrangham RW (1987) Evolution of social structure. In: Smuts BB, Wrangham RW, Cheney DL, Struhsaker TT, Seyfarth RM (eds) Primate societies. University of Chicago Press, Chicago

Wrangham RW, Smuts B (1980) Sex differences in the behavioral ecology of chimpanzees in the Gombe National Park, Tanzania. J Reprod Fertil Suppl 28:13–31

Wroblewski EE, Murray CM, Keele BF, Schumacher-Stankey JC, Hahn BH, Pusey AE (2009) Male dominance rank and reproductive success in chimpanzees, *Pan troglodytes schweinfurthii*. Anim Behav 77:873–885

Yamagiwa J (1999) Socioecological factors influencing population structure of gorillas and chimpanzees. Primates 40:87–104

Yamagiwa J (2004) Diet and foraging of the great apes: ecological constraints on their social organizations and implications for their divergence. In: Russon AE, Begun DR (eds) The evolution of thought: evolutionary origins of great ape intelligence. Cambridge University Press, Cambridge, pp 210–233

Yamagiwa J, Kahekwa J (2001) Dispersal patterns, group structure and reproductive parameters of eastern lowland gorillas at Kahuzi in the absence of infanticide. In: Robbins MM, Sicotte P, Stewart KJ (eds) Mountain gorillas: three decades of research at Karisoke. Cambridge University Press, London, pp 89–122

Yamagiwa J, Mwanza N, Spangenberg A, Maruhashi T, Yumoto T, Fischer A, Steinhauer BB (1993) A census of the eastern lowland gorillas *Gorilla gorilla graueri* in Kahuzi-Biega National Park with reference to mountain gorillas *G. g. beringei* in the Virunga region, Zaire. Biol Conserv 64:83–89

Yamagiwa J, Maruhashi T, Yumoto T, Mwanza N (1996) Dietary and ranging overlap in sympatric gorillas and chimpanzees in Kahuzi-Biega National Park, Zaire. In: McGrew WC, Marchant LF, Nishida T (eds) Great ape societies. Cambridge University Press, Cambridge, pp 82–98

Yamagiwa J, Kahekwa J, Basabose AK (2003) Intra-specific variation in social organization of gorillas: implications for their social evolution. Primates 44:359–369

Yamagiwa J, Kahekwa J, Basabose AK (2009) Infanticide and social flexibility in the genus *Gorilla*. Primates 50:293–303

Yamagiwa J, Basabose AK, Kahekwa J, Bikaba D, Ando C, Matsubara M, Iwasaki N, Sprague DS (2011) Long-term research on Grauer's gorillas in Kahuzi-Biega National Park, DRC: life history, foraging strategies, and ecological differentiation from sympatric chimpanzees. In: Kappeler PM, Watts DP (eds) Long-term field studies of primates. Springer, New York, pp 385–412

Yamakoshi G (1998) Dietary responses to fruit scarcity of wild chimpanzees at Bossou, Guinea: possible implications for ecological importance of tool use. Am J Phys Anthropol 106:283–295

Yoshihara K (1999) Present situation of chimpanzees in Japan. Primate Res 15:267–271 (in Japanese)

Chapter 10
Social Conflict Management in Primates: Is There a Case For Dolphins?

Marina Cords and Janet Mann

A male and female juvenile dolphin pet each other. Petting is analogous to primate grooming and is often seen after conflicts. (Photograph credit: Courtesy of Ewa Krzyszscyk, Shark Bay Dolphin Research Project)

M. Cords (✉)
Department of Ecology, Evolution and Environmental Biology, Columbia University, New York, NY 10027, USA
e-mail: mc51@columbia.edu

J. Mann
Departments of Biology and Psychology, Georgetown University, Washington, DC 20057, USA

J. Yamagiwa and L. Karczmarski (eds.), *Primates and Cetaceans: Field Research and Conservation of Complex Mammalian Societies*, Primatology Monographs, DOI 10.1007/978-4-431-54523-1_10, © Springer Japan 2014

Abstract Gregarious animals face unavoidable conflicts of interest and thus therefore are likely to evolve behavioral mechanisms that allow them to manage conflict and thus maintain their social bonds. Multiple forms of conflict management characterize primates, but far less research has focused on dolphins, especially under natural conditions. Captive studies of dolphins have confirmed post-conflict reconciliation, a well-studied form of conflict management in primates. The fission–fusion nature of dolphin social systems, along with the vast home ranges of individuals, pose particular difficulties for the study of conflict management. Conflicts among male allies are likely to be a fruitful area for further research on conflict management, both because allies are valuable social partners and because they interact frequently over extended periods.

Keywords Aggression • Alliance • Conflict • Conflict management • Reconciliation • Social organization

10.1 Introduction

Conflicts of interest characterize members of any animal population but are especially acute for those living in social groups. Disputes over resources, mates, relationships, movement patterns, or other activities can compromise group integrity. Further, in species in which group living is based on individualized cooperative relationships, escalated aggressive conflicts have the potential to disrupt those relationships and thus to threaten both the benefits and the mechanisms of group living. Gregarious animals are therefore expected to have evolved a capacity to manage conflict (Aureli et al. 2002).

10.2 Conflict Management in Primates and Dolphins

Conflict management includes behavior that prevents aggressive escalation of conflicts and which mitigates or repairs the damage caused by such escalation (Cords and Killen 1998; Aureli and de Waal 2000, Appendix B). Studies of nonhuman primates provide various examples of conflict management behavior in multiple species. For example, ritualized dominance relationships, the development of routines and social conventions (such as respect for possession), and displays of reassurance that precede situations in which conflict is likely to erupt are types of behavior that reduce the likelihood of escalated aggression in nonhuman primates. In addition, animals with a conflict of interest may simply avoid each other, at least temporarily. Should aggressive conflict nevertheless erupt, primates often use various tactics to keep aggression relatively mild and brief. For example, they may adhere to ritualized forms of aggression that are less physically dangerous, redirect received aggression onto a third party to end the original aggressive interaction, or heed the "policing" interventions of powerful individuals that quickly bring escalated fighting to an end. After aggression is over, nonhuman primates have been shown to engage in several kinds of "post-conflict" interactions, which both reduce anxiety triggered by the previous aggressive conflict and reestablish a cooperative relationship with a former opponent, either directly or through its relatives (Wittig and Boesch 2003).

Best studied among primates are patterns of post-conflict friendly reunion, or "reconciliation" (Arnold et al. 2010). In a typical case, former opponents interact in an affiliative way within a few minutes after their aggression has ceased. They are selectively attracted to each other (although attraction to one another's kin has also been documented). Some studies have demonstrated that such post-conflict reunions reduce the chance of subsequent aggression, that individual opponents reduce self-directed behavior associated with anxiety, and that they restore levels of tolerance to pre-conflict levels (Aureli et al. 2002). Because approaching an individual who may still be aggressively motivated is risky, we expect reconciliation to be strategically targeted. It should occur only when aggression causes anxiety or disrupts cooperative relationships, and particularly when the opponent is a valuable social partner (likely to interact in a way that benefits the subject) but unpredictable, and when a prior history of generally friendly interaction patterns facilitates affiliation after aggression (Cords and Aureli 2000; Aureli et al. 2002). There is much evidence that partner value influences the tendency to reconcile, although it is often indirect (Watts 2006; Arnold et al. 2010).

Of the approximately 35 species of delphinids, all are highly social, living in stable (e.g., killer whale, false killer whale, pilot whale) or temporary (e.g., bottlenose dolphin, spotted dolphin) groups. Some species show heavy scarring (e.g., Risso's dolphin, *Grampus griseus*; MacLeod 1998) or tooth rake marks (Scott et al. 2005; MacLeod 1998) and clearly must engage in frequent battle. These scars and marks are likely to be good indicators of intraspecific aggression in delphinids and reveal which individuals are most vulnerable to attack. Species with extensive markings would, in general, be good candidates for studying aggression and conflict management. Although the highly social nature of these animals coupled with battle scars suggests that conflict management mechanisms should be part of their social life, little research has addressed this topic to date. Three studies of reconciliation in captive bottlenose dolphins involved two to seven dolphins of mixed sex (Weaver 2003; Tamaki et al. 2006; Holobinko and Waring 2010). These studies revealed high rates of post-conflict affiliation, and one study found some evidence that affiliation (flipper rubbing) reduced the likelihood of subsequent conflict (Tamaki et al. 2006). Although these results suggest parallels with primates, the captive environment—where continuous observation is possible—is likely to have influenced the dolphins' behavior: particularly, captive dolphins are unable to avoid each other, unlike their wild counterparts. Confirmation of these patterns of behavior in wild populations, as in primates, is therefore important.

Logistic difficulties are undoubtedly a major reason why the study of conflict management in delphinids is still in its infancy. The open fission–fusion nature of many delphinid societies presents particular challenges, because individuals may not encounter each other for weeks, months, and even years. Avoidance or reduced levels of association may be especially important ways of managing conflict in these spatially dispersed societies, but they are probably the hardest behavioral patterns to study. In addition, the difficulties inherent in observing cetaceans mean that observers not only miss some proportion of agonistic and affiliative (or conciliatory) interactions but often may have difficulty tracking association and avoidance following such interactions. Post-conflict behavior is especially hard to study in wild populations.

10.3 The Nature of Conflict in Primates and Dolphins

Mammalian conflicts are often over resources, mates, or status. Even if finding or feeding on prey is conducted socially (in groups), most delphinids catch individual prey items (fish or squid) that are swallowed quickly. Occasionally dolphins "display" their catch to others, who approach the fish closely for apparent inspection, but never challenge the owner or attempt to steal prey (Mann et al. 2007). Thus, direct feeding competition is unlikely to lead to aggressive conflicts. Rarely do dolphins chase the same individual prey item, and doing so would probably result in failure for both. An exception might be mammal-eating killer whales, which not only hunt cooperatively but also share prey, typically with kin (Baird and Dill 1996). Food-sharing with kin has also been documented in fish-eating killer whales, although cooperative hunting has not been documented (Ford and Ellis 2006). Although much primate aggression occurs in the context of feeding, and involves contests over enduring feeding sites, primates rarely reconcile when the conflict involves food, probably because the stakes are small (Aureli et al. 2002). Cooperative hunting in killer whales (and carnivores such as spotted hyenas; Wahaj et al. 2001) may raise the stakes, however, because the risk of injury and resource value are high. For the same reason, maintaining close cooperative bonds and conflict management would be critical, regardless of the source of conflict, when group members are highly interdependent.

For most delphinids, however, conflict over mating, both within and between the sexes, might be a more fruitful context in which to examine conflict resolution. Males form enduring alliances in bottlenose dolphins and perhaps other delphinid species (Connor et al. 2000). In Shark Bay bottlenose dolphins, alliances of two or three males consort with and show aggression toward individual females (Connor et al. 1996, 2000; Owen et al. 2002; Scott et al. 2005). Cycling females experience much more aggression than noncycling females, and conflicts between females are exceedingly rare (Scott et al. 2005). The majority of Shark Bay bottlenose dolphins have tooth rake markings from conspecifics, suggesting that most individuals regularly receive attacks from others. Fresh wounds are more commonly observed on cycling females than on females in other reproductive states (Scott et al. 2005). Watson-Capps and Mann (unpublished data), studying male–female interactions during consortships of Shark Bay bottlenose dolphins, recently found that affiliation rates were significantly higher within 10 min post conflict than at any other time. This affiliation may placate aggressive male alliances or repair intersexual relationships. Because consortships can last for weeks, or even months, females may be highly motivated to placate aggressive males and reduce the costs of prolonged association with males.

Well-developed conflict resolution mechanisms should also occur between male allies, who are in direct reproductive competition, and yet must cooperate against other alliances competing for the same female. Studies of nonhuman primates have provided some evidence that frequent allies are more likely to reconcile aggressive conflicts (Watts 2006), even in cases in which the alliance is not directly linked to acquiring a mate.

10.4 Conclusion

The study of conflict management in dolphins is still in its infancy, but would provide a valuable context in which to confirm or extend general patterns that have emerged from studies of primates. Conflict between allies is likely to be the most fruitful context for exploring reconciliation in delphinids, not only because allies are valuable partners, but also because male allies stay together and post-conflict observations are possible. Future research in this area will help identify the forces that shape group living in delphinids.

References

Arnold K, Fraser ON, Aureli F (2010) Postconflict reconciliation. In: Campbell CJ, Fuentes A, MacKinnon KC, Bearder SK, Stumpf R (eds) Primates in perspective, 2nd edn. Oxford University Press, Oxford, pp 608–625

Aureli F, de Waal F (2000) Natural conflict resolution. University of California Press, Berkeley

Aureli F, Cords M, van Schaik CP (2002) Conflict resolution following aggression in gregarious animals: a predictive framework. Anim Behav 64:325–343

Baird RW, Dill LM (1996) Ecological and social determinants of group size in transient killer whales. Behav Ecol 7:408–416

Connor RC, Richards AF, Smolker RA, Mann J (1996) Patterns of female attractiveness in Indian Ocean bottlenose dolphins. Behaviour 133:37–69

Connor RC, Wells R, Mann J, Read A (2000) The bottlenose dolphin, *Tursiops* sp.: social relationships in a fission–fusion society. In: Mann J, Connor R, Tyack P, Whitehead H (eds) Cetacean societies: field studies of dolphins and whales. University of Chicago Press, Chicago, pp 91–126

Cords M, Aureli F (2000) Reconciliation and relationship qualities. In: Aureli F, de Waal F (eds) Natural conflict resolution. University of California Press, Berkeley, pp 177–198

Cords M, Killen M (1998) Conflict resolution in human and non-human primates. In: Langer J, Killen M (eds) Piaget, evolution, and development. Erlbaum Associates, Hillsdale, pp 193–217

Ford JKB, Ellis GM (2006) Selective foraging by fish-eating killer whales *Orcinus orca* in British Columbia. Mar Ecol Prog Ser 316:185–199

Holobinko A, Waring GH (2010) Conflict and reconciliation behavior trends of the bottlenose dolphin (*Tursiops truncatus*). Zoo Biol 29:567–585

MacLeod CD (1998) Intraspecific scarring in odontocete cetaceans: an indicator of male 'quality' in aggressive social interactions? J Zool 244:71–77
Mann J, Sargeant BL, Minor M (2007) Calf inspection of fish catches: opportunities for oblique social learning? Mar Mamm Sci 23:197–202
Owen ECG, Wells RS, Hofmann S (2002) Ranging and association patterns of paired and unpaired adult male Atlantic bottlenose dolphins, *Tursiops truncatus*, in Sarasota, Florida, provide no evidence for alternative male strategies. Can J Zool 80:2072–2089
Scott E, Mann J, Watson-Capps JJ, Sargeant BL, Connor RC (2005) Aggression in bottlenose dolphins: evidence for sexual coercion, male–male competition, and female tolerance through analysis of tooth-rake marks and behaviour. Behaviour 142:21–44
Tamaki N, Morisaka T, Michihiro T (2006) Does body contact contribute towards repairing relationships? The association between flipper-rubbing and aggressive behavior in captive bottlenose dolphins. Behav Processes 37:209–215
Wahaj SA, Guse KR, Holekamp KE (2001) Reconciliation in the spotted hyena (*Crocuta crocuta*). Ethology 107:1057–1074
Watts DP (2006) Conflict resolution in chimpanzees and the valuable-relationships hypothesis. Int J Primatol 27:1337–1364
Weaver A (2003) Conflict and reconciliation in captive bottlenose dolphins, *Tursiops truncatus*. Mar Mamm Sci 19:836–846
Wittig RM, Boesch C (2003) The choice of post-conflict interactions in wild chimpanzees (*Pan troglodytes*). Behaviour 140:1527–1559

Chapter 11
Evolution of Small-Group Territoriality in Gibbons

Warren Y. Brockelman, Anuttara Nathalang, David B. Greenberg, and Udomlux Suwanvecho

W.Y. Brockelman (✉)
Ecology Lab, Bioresources Technology Unit, BIOTEC, 113 Science Park, Paholyothin Road, Klong Luang, Pathum Thani 12120, Thailand

Institute of Molecular Biosciences, Mahidol University, Salaya Campus, Phutthamonthon, Nakhon Pathom 73170, Thailand
e-mail: wybrock@cscoms.com

A. Nathalang • U. Suwanvecho
Ecology Lab, Bioresources Technology Unit, BIOTEC, 113 Science Park, Paholyothin Road, Klong Luang, Pathum Thani 12120, Thailand

D.B. Greenberg
Marine Science Institute, University of California, Santa Barbara, CA 93106, USA

J. Yamagiwa and L. Karczmarski (eds.), *Primates and Cetaceans: Field Research and Conservation of Complex Mammalian Societies*, Primatology Monographs, DOI 10.1007/978-4-431-54523-1_11,

Abstract This chapter endeavors to establish the basic environmental and social factors that have enabled the evolution of territorial behavior in gibbons, and perhaps other animals, and precluded it in cetaceans. These factors are given as three basic conditions, followed by some hypotheses and testable predictions that follow. These conditions concern (a) relatively homogeneous (nonclumped) resource distribution; (b) high mobility and foraging efficiency; and (c) range use exclusivity. Evidence from a study of diet and foraging in white-handed gibbons (*Hylobates lar*) in central Thailand is brought to bear in testing predictions from conditions (a) and (b). The feeding range of the study group is relatively homogeneous and, although it changes in size seasonally, it does not shift much in location. The relatively long daily foraging path in relationship to range area suggests highly efficient foraging. Evidence is presented that the gibbons' food sources are often known and frequently revisited, although they change from month to month. Because territory defense entails costs as well as benefits, defended territory should be set at a size at which resource limitation begins to occur in the population. Seasonal changes in ranging and social behavior suggest that this is the case in the study group.

Keywords Foraging efficiency • Gibbons • *Hylobates* • Monogamy • Resource dispersion hypothesis • Resource distribution • Resource limitation • Territory

11.1 Introduction

The social system of gibbons (family Hylobatidae) is unique among the apes, typically consisting of socially monogamous and territorial pairs (Gittins and Raemaekers 1980; Brockelman and Srikosamatara 1984; Leighton 1987; Reichard 2003; Bartlett 2009a; Brockelman 2009). The great apes and humans (family Hominidae) have surprisingly diverse social systems, including solitary living (orangutans), age-graded male groups (gorillas), multimale societies with flexible subgrouping (chimpanzees and bonobos), and flexible, complex multilayered societies (humans) (McGrew et al. 1996; Grueter et al. 2012). Most hominids, in contrast to hylobatids, have female–female associations, typically among close kin (orangutans are the main exception; van Schaik and van Hooff 1996). Ape social systems may provide few clues about phylogeny, but apparent cases of convergence may teach us about selective pressures favoring one social system or another (Chapman and Rothman 2009). For example, approximately one third of human societies have been classified as socially monogamous (and others facultatively so), and proposed explanations for monogamy in humans have come from studies of birds and other animals (Low 2003). Whether we can productively pursue explanations across groups as different and distant as cetaceans and primates is a new challenge that we now take up.

No dolphin (Delphinidae) and probably no known cetacean live in small, socially monogamous, territorial groups (Gowans et al. 2008). What factors might make this impossible? We seek answers from this synthesis of ideas relating to gibbons.

We first present a framework of three conditions that we believe have shaped monogamy and territoriality in gibbons: these are (a) resource distribution and use;

(b) group mobility and foraging efficiency; and (c) range exclusivity, or factors affecting the internalization of benefits from resource defense. We then present new data that permit tests of some hypotheses under conditions (a) and (b) drawn from our study of social behavior, foraging, and ranging behavior of white-handed gibbons (*Hylobates lar*) on the Mo Singto study area and forest dynamics plot in Khao Yai National Park, Thailand. The factors under (c), range exclusivity, include several types of facilitating conditions: that the defending males enjoy relatively high paternity of offspring produced on the territory (reproductive exclusivity); that residents use resources more efficiently than do nonresidents; and that interspecific competitors do not excessively exploit, or are prevented from exploiting, resources within the territory (resource use exclusivity) (Brockelman and Srikosamatara 1984). Although we believe these factors are important in the evolution of territoriality, we do not discuss them in detail here, but rather focus on ranging and resource use.

11.2 Monogamy in Primates

Social monogamy (which does not necessarily imply reproductive or genetic monogamy) has evolved independently about ten times in primates (van Schaik and Kappeler 2003) and occurs in about 10 % of primate species. Monogamy predominates in virtually all species of gibbons (Hylobatidae); serious exceptions have been found in some species of *Nomascus*, the northernmost genus, in which most groups are polygynous (Fan and Jiang 2009; Jiang et al. 1999; Fan et al. 2010). These *Nomascus* groups tend to average slightly larger (>4 individuals) than have been found in *Hylobates* groups in Southeast Asia (3–4 individuals on average). It is not known whether this difference results in weaker territorial defense. The four currently recognized gibbon genera (*Hylobates, Hoolock, Nomascus*, and *Symphalangus*) (Brandon-Jones et al. 2004; Mootnick and Groves 2005) diverged some 6 to 10 million years (Myr) ago (Hayashi et al. 1995; Roos and Geissmann 2001; Chan et al. 2010; Matsudaira and Ishida 2010), so the gibbon social system appears to be highly conservative. Monogamy associated with territoriality also predominates, although not exclusively, in both the Lemuridae and the Callitrichinae (van Schaik and Kappeler 2003).

Two types of monogamous species have been recognized: those in which males provide direct care of young, and those species—most notably gibbons—in which males provide little or no direct care of young (Kleiman 1977; Wittenberger and Tilson 1980; Dunbar 1988; Clutton-Brock 1991; Woodroffe and Vincent 1994). Explaining monogamy presents the greater challenge in the latter group: why would a male stay with one female in whom he had no long-term investment, or if he did not provide essential care for her young? Territorial defense, however, may be considered indirect male care through resource defense (Wittenberger and Tilson 1980; Rutberg 1983; Bartlett 2009a; Brockelman 2009).

11.3 Essential Conditions Favoring Territoriality and Small Group Size in Gibbons

The gibbon social system results from a combination of interacting environmental factors and inherited predispositions. It is difficult to distinguish causes and effects, or to identify the "prime mover" that came first during evolution. We argue that the conditions just proposed influence and reinforce one another in a unified complex that favors resource defense territoriality and small group size.

11.3.1 Sufficient Resources Within a Small Range

Permanent resource defense is favored when resources are stable and evenly distributed (Emlen and Oring 1977). First, to be defendable, resources must remain within the range and not drift across the boundaries. This condition poses no problem for forest frugivores but is a critical factor for marine mammals (Steele 1985; Gowans et al. 2008).

Several factors determine whether adequate resources can be found within a small home range. Adequate or "sufficient" cannot be defined in absolute terms. "Adequate" resources, in theory, are determined by a balance between benefits and costs of defense, measured by time and energy, and related to effects on fitness. In a small territory, the benefits of increasing defense will exceed the costs. As territory size increases, however, costs of defense increase while benefits gained level off, as resource needs become saturated and exploitation efficiency declines (Ebersole 1980).

The adequacy of resources depends on their degree of spatial and temporal homogeneity, the diet of the animals, and the number of competitors sharing the resources. A related factor that we consider critical in gibbons and other primates which defend territories is intimate knowledge of resources within the range, which increases foraging efficiency (Milton 1980; Bartlett 2009a, pp. 146–147).

The spatial and temporal homogeneity of resources determine the "food security" of a territory of a given size (MacDonald and Carr 1989). Food security is the probability that a territory will satisfy the minimum nutritional requirements of the occupants for a given period of time (e.g., 1 year). Waser and Wiley (1979) also recognized that resource patchiness and home range size are related, stating that "the sufficiency of each individual's share of sites increases as the available resource becomes more evenly distributed across sites and time periods." If resources are highly heterogeneous, that is, are very patchy in space or time, then a relatively large area will be required to satisfy all nutritional requirements. With a less patchy environment, minimally sized territories satisfy nutritional needs, and the amount of resources available in the territory may suffice for only a single breeding pair. A larger area has more resources, on average, and will support a larger group. These factors comprise the resource dispersion hypothesis (RDH), which was developed largely from studies of the social systems of bats (Bradbury and Vehrencamp 1976) and canids (Kruuk and Parish 1982; Mills 1982; Carr and MacDonald 1986). Extra-group members may merely be tolerated by the primary breeding pair, or may provide benefits such as parental care, improved territorial defense, or defense against predators. In the latter case, group size may increase above that mandated by the minimum territory size as providing resource security. The RDH helps to explain how gibbons are able to survive in such small ranges as compared to most other primates.

Gibbons obtain their energy primarily from ripe succulent fruits (Gittins and Raemaekers 1980; Leighton 1987; Leighton and Leighton 1983; Bartlett 2009a; Brockelman 2011; McConkey 2009), but these are not uniformly available all year. The "suitability" of any fruit species, however, depends on what other fruits are available, because lower-quality species, and more leaves, are consumed when preferred species are scarce (Bartlett 2009a; Fan et al. 2009). The RDH model assumes implicitly that resources are limiting to the population and that the home range or defended territory represents the minimum area required to provide sufficient resources over time for the occupants.

We must consider several types of resource heterogeneity to evaluate food security, including relative abundance of the species consumed at any one time, their spatial patchiness, and their seasonal and interannual variation in availability. Within the territory of our study group (see following), preferred fruit species density ranges over at least three orders of magnitude, from very rare (<1 tree/10 ha) to common (≥10 trees/ha).

Fruiting tree size or crown size is usually considered to be an aspect of patchiness, but in this discussion we consider "patchiness" to be mainly an aspect of distribution and clumping of food sources, not food tree size. Gibbons can target food trees of all sizes, in contrast to primates that live in large groups, which may target only relatively large sources (e.g., *Alouatta palliata*; Milton 1980).

Seasonality in fruiting species occurs in all primate habitats and is usually extensive (Brockman and van Schaik 2005). In addition, some species, particularly *Ficus* spp., fruit irregularly. Primates may be stressed during the dry season, and dry season severity can affect ranging behavior (Hemingway and Bynum 2005; Bartlett 2009a). Interannual variation in phenology of individual fruit species is also widespread in tropical habitats and must affect overall food availability. We need more long-term field studies to investigate how this affects primate populations (e.g. Tutin and White 1998; Chapman et al. 2005; Marshall and Leighton 2006). Natural selection responds to food security over the long term, and so we must plan longer-term studies.

In the predictions below, we test those marked with an asterisk (*) with our data.

Hypothesis A: Relative resource patchiness affects range size and group size in gibbons.

Prediction A1*: Habitats of monogamous gibbons have fruit sources distributed more or less randomly.

Prediction A2*: Fruit source availability and feeding ranges in monogamous gibbons do not shift markedly between seasons.

Prediction A3: In habitats of polygamous gibbons, fruit sources are more patchily distributed and feeding ranges may shift seasonally.

11.3.2 Mobility and Range Size

Mitani and Rodman (1979) established that territory size, group mobility, and the ability to defend the territory are related. "Relative mobility" is measured by the average daily range length, and is related to mean distance across the territory, as estimated by the diameter of a circle equal in area to the territory. Territorial species all have average daily ranges longer than this diameter, and gibbons (*Hylobates lar*) had daily ranges approximately 2.3 times the average diameter (Mitani and Rodman 1979; Bartlett 2009b). This observation suggests that gibbons should be very capable of defending their territories.

The difficulty of defending a territory against incursions by neighboring groups should be proportional to its area or to the length of its perimeter and the benefits of defending should be proportional to its area (Ebersole 1980; Schoener 1971, 1983). As territory size increases, benefits and costs of defense increase, but not proportionally. Although resource supply increases with area, the efficiency of exploitation, and the efficiency of defense, should decline.

Hypothesis B: Small territories are more efficiently exploited than larger territories.

Prediction B1: Small territories are better defended than large territories, resulting in less incursion by, and less overlap with, neighboring groups.

Prediction B2: Small territories are more fully exploited for fruits than large ones, as rarer and smaller food sources are more easily monitored and used more efficiently.

Prediction B3: Gibbons, especially in groups with small territories, tend to forage by traveling directly to food sources out of sight, as opposed to feeding opportunistically (tested by Asensio et al. 2011); this is because in small territories gibbons can reach all parts of the area more easily, and their ability to discover and remember food sources is increased. This prediction should be tested both within and between species of gibbons.

Prediction B4*: Gibbons, having a relatively long daily range length relative to territory area, make repeated use of preferred fruit sources on successive or alternate days.

Prediction B5: Territory size in gibbons tends to be set at the area at which food limitation begins to reduce reproductive fitness. This is a consequence of the fact that territory defense, as well as foraging activities, have costs that must be offset against the benefits gained. It is also a consequence of the fact that populations tend to increase until resource limitation brings birth and death rates into alignment on average.

11.4 Study Area and Methods

Our study site was the 30-ha Mo Singto forest dynamics plot (14.4°N latitude) in Khao Yai National Park, central Thailand (Brockelman et al. 2011). The site is in seasonal evergreen forest at 720–820 m altitude and receives an average of 2,200 mm annual rainfall, mostly during May–October (Bartlett 2009a; Brockelman et al. 2011). Fruit production in the forest is highly seasonal, and many species preferred by gibbons experience high interannual variability (Brockelman 2011).

We collected data by following gibbon group "A" about the plot for 6 days (including at least 5 full days) each month in 2003–2005 and recording all trees and lianas on which the adult female (the focal animal) fed. The adult female usually led the group, which had five members, in foraging. Usually, all members fed together in the same fruit trees. We analyzed the 2004 data because in that year *Nephelium melliferum* (Sapindaceae), a common tree species preferred by gibbons in April–May, fruited heavily.

We censused (tagged, mapped, identified) all trees ≥1 cm in diameter and all lianas ≥2.5 cm in diameter during 2004–2005. A total of 203 species of trees ≥10 cm in diameter occurred on the plot at an average density of 514 stems ha^{-1}, as well as at least 120 species of lianas, many of which are important food species for gibbons. Most of the plot is covered with old-growth forest with some large gaps containing regenerating forest, reaching 1 ha in area, resulting from storms during the past 30 years or so (Brockelman et al. 2011).

We used MS Access to manage the data, and ArcView (ESRI, Redlands, CA, USA) to map trees. We estimated and mapped the monthly 6-day feeding ranges as minimum convex polygons (Jennrich and Turner 1969) around all fruit sources (trees and lianas) visited. We conducted several analyses to investigate the intensity of use of the total yearly range (Table 11.1; Figs. 11.2, 11.4). Correlations are tested with the product–moment correlation coefficient.

11.5 Results

We used our data to examine predictions A1 (that habitats of monogamous gibbons have fruit sources distributed more or less randomly), A2 (that fruit source availability and feeding ranges in monogamous gibbons do not shift markedly between seasons), and B4 (that gibbons, having relatively long daily range length relative to territory area, make repeated use of preferred fruit sources on successive or alternate days. Other predictions for which evidence exists in the literature are discussed in the next section.

A total of 61 fruit species were used in all periods, including 16 species of *Ficus*. *Ficus*, *Nephelium*, and other sources that were utilized are widely scattered over the plot, covering nearly the whole yearly feeding range (Fig. 11.1). Some large gaps in their distributions do occur (such as in the center about 400 m north), created by storms up to 20 years ago. Large gaps have fewer fruit sources and are more difficult for gibbons to cross than more mature forest. During 9 of the 12 months, 6-day feeding ranges covered most of the 22.8-ha yearly range (Fig. 11.1, Table 11.1). During the lean fruiting months of November–February, the 6-day ranges contracted considerably, to as small as ~3 ha in February. During January, the highly preferred fruit of the liana *Elaeagnus conferta* (Elaeagnaceae) became available, and the range

Table 11.1 Monthly areas of feeding ranges (% of yearly area) (as shown in Fig. 1.2) indicating number of fruit species consumed, number of fruit sources visited, number of visits to most-preferred species/number of sources of same, and including *Ficus* species

Month	Area in hectares (%)	Fruit species	Total sources	Total visits	Preferred species	Visits/number of sources for preferred species
January	9.1 (40)	11	45	74	*Elaeagnus conferta*	28/16
February	3.2 (14)	10	11	22	*Ficus stricta*	9/1
March	16.9 (74)	9	35	57	*Syzygium syzygioides*	26/12
April	15.0 (66)	10	24	58	*Prunus javanica*	14/8
May	18.5 (81)	16	47	92	*Nephelium melliferum*	45/23
June	13.6 (63)	10	29	38	*Balakata baccata*	12/5
July	18.9 (83)	23	51	72	*Miliusa lineata*	11/3
August	14.4 (63)	14	41	62	*Aidia densifolia*	16/12
September	15.8 (69)	17	45	66	*Garcinia benthamii*	20/14
October	13.4 (59)	13	37	59	*Choerospondias axillaris*	14/10
November	9.1 (40)	10	19	29	*Choerospondias axillaris*	6/5
December	6.9 (30)	8	17	30	*Elaeagnus conferta*	9/4
Total	22.8 (100)	61	401	659		
Mean ± SD	12.9 4.8	12.6 4.3	33.4 13.2	54.9 $(9.2/\text{day})^{-1}$ 21.1 $(1.5/\text{day})^{-1}$		

expanded. The size of the monthly range is correlated with the number of sources utilized ($r=0.675$, $p<0.05$, $n=12$ months. The number of species utilized per 6-day period was relatively stable over the year, usually 8–14, but increased to 16 in May, 23 in July, and 17 in September; it was only weakly correlated with range area ($r=0.415$, $p>0.05$) (Table 11.1). Fewer fruit species, and fewer trees, were used during periods of low fruit availability. Gibbons respond to fruit shortage by traveling less and eating more young leaves, shoots, flowers, and insects (Bartlett 2003, 2009a; unpublished data).

Prediction A2 appears correct, as the centroids of monthly feeding ranges (computed as averages of x and y coordinates of all visits to fruit sources in a given month) shifted little during the year (Fig. 11.2). We included all repeat visits to the same sources when calculating these centroids (see discussion of Prediction B4). All averages of range centroids lie within ~100 m on the east–west axis of the plot and ~200 m on its north–south axis. The averages for October–December shifted about 100 m eastward, but the ranges for all 12 months overlapped in the center of the whole range. Because fruiting of individual species is highly seasonal, these results imply that individual fruit species are not very clumped in distribution, although we have insufficient space here to demonstrate this quantitatively.

Further ways to view the 6-day feeding ranges appear in Figs. 11.3 and 11.4. Figure 11.3 shows spatial variation in the number of 6-day ranges that overlap different parts of the yearly range. Numerical results indicate that 63 % of the aggregate area is overlapped by at least six monthly ranges, and that 40 % is overlapped by at least nine monthly ranges. One hectare of the area in the lower center of the plot falls within all 12 monthly ranges. This appears to be a small core activity area,

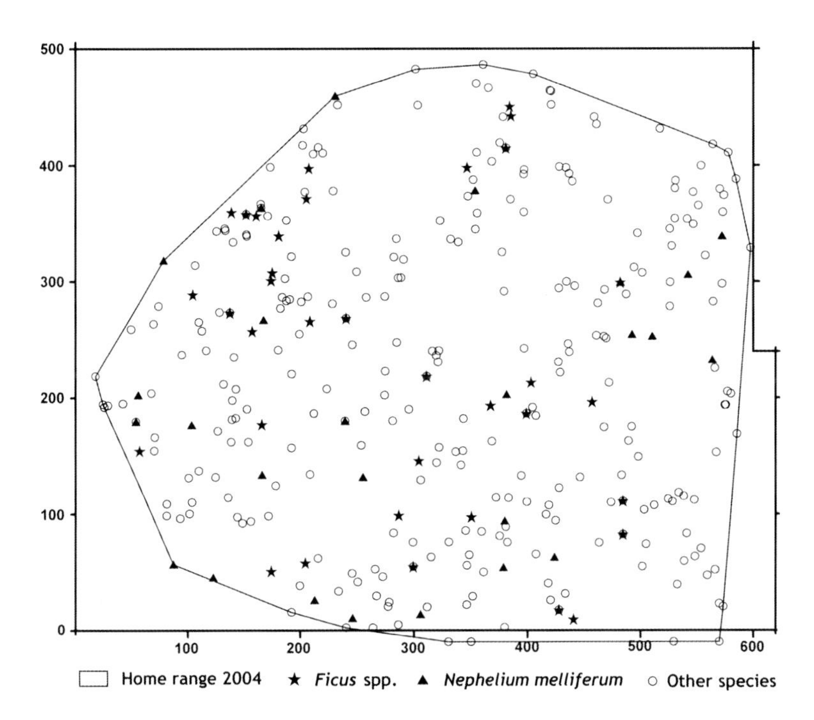

Fig 11.1 Locations of all fruit sources used by the adult female of group A during 12 6-day monthly periods in 2004, showing fig, *Nephelium melliferum*, and other fruit sources. The minimum convex polygon feeding range for the year is shown. Plot dimensions are in meters from the origin

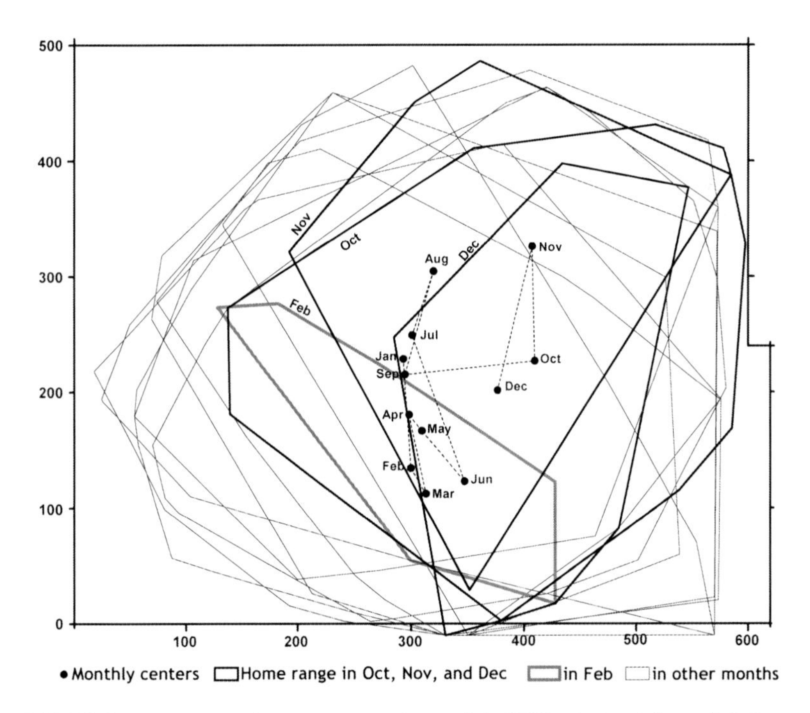

Fig. 11.2 Minimum convex polygon ranges each month in 2004 calculated from all fruit sources used during each 6-day sample period in the given month. The centroid of each 6-day range (average *x* and *y* coordinates) is shown. Gibbons had the smallest ranges in the months of October, November, December. and February (shown with *thicker lines*). Plot dimensions in meters from the origin

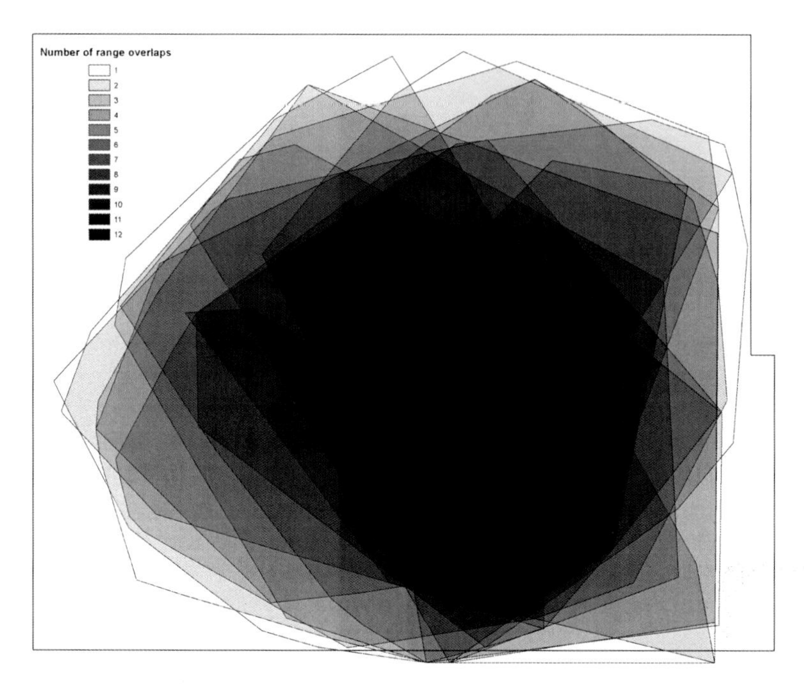

Fig. 11.3 Spatial variation in the number of overlapping 6-day activity ranges within the total yearly range

but it was not visited every day. This area coincides roughly with the square hectare area containing most frequent fruit visits (Fig. 11.4). Eight other hectares of relatively heavy use (>30 visits) are scattered throughout the yearly range. Most hectares of light use (<20 visits) contain old large canopy gaps. By comparing Figs. 11.1 and 11.4, we also see that hectares of heavy use tend to have more sources of figs, many of which were visited repeatedly (see following).

Prediction B4 is satisfied by repeated visits, on successive or alternate days, to the most preferred fruit sources. We have tested this prediction by examining the frequency distribution of visits per fruit source over all months of the year, and by identifying the species with the most visits per month (Table 11.1). Over the year, 35 % of fruit sources were visited more than once; we actually consider this an underestimate, given the short 6-day sample periods. The ratio of repeat visits to total visits made to fruit sources better estimates the role of knowledge in foraging, perhaps. We eliminated the first sampling day each month from the 6-day samples used in this calculation (many of these would be erroneously scored as first visits, and therefore bias the results). The ratio of repeat visits to total visits over the whole year was 0.43, and it varied from 0.30 to 0.67 per month. The monthly ratios were not correlated with the relative scarcity of fruits each month.

The most preferred trees in the yearly range were frequently revisited (Table 11.1). In April, nine visits were made to a single *Prunus javanica*; in May, seven visits were made to just one *Nephelium melliferum*. This pattern also held for rare species

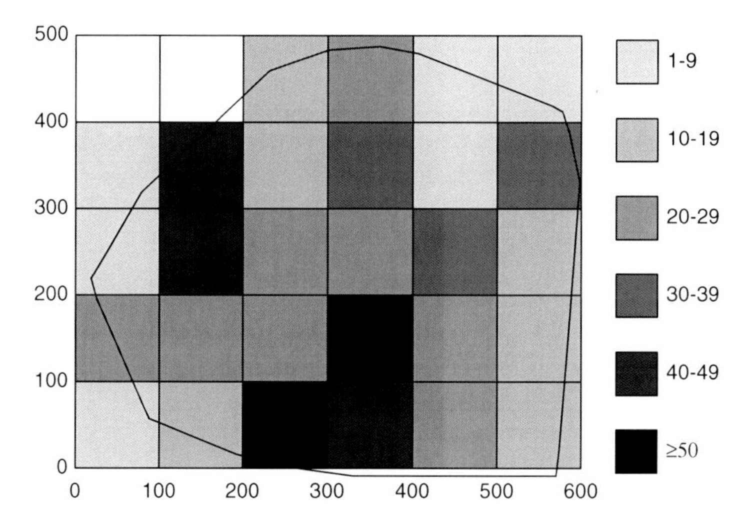

Fig. 11.4 Spatial variation in the frequency of fruit tree visits per hectare within the total yearly range

of which only one or two sources were available. In February, when fruit availability was lowest, a large *Ficus stricta* was visited nearly twice a day. In December, the gibbons largely used the relatively uncommon liana *Elaeagnus conferta*, making nine visits to four sources.

11.6 Discussion

We have discussed the conditions that we believe have shaped the evolution of the main socioecological characteristics of gibbons (i.e., very small group size with monogamous tendency and territorial resource defense). We believe these predisposing conditions include (a) resources homogeneously distributed within a relatively small range, (b) high group mobility relative to the area covered, and (c) the ability to internalize benefits of resource defense. Here, we mainly discuss predictions that have been addressed with data from the Mo Singto site.

11.6.1 Resource Distribution

Gibbon home ranges are usually within 15–60 ha, and their defended territories are usually within 10–45 ha (Chivers 1984; Leighton 1987; but see Fan and Jiang 2008). Mo Singto gibbons are typical in this regard: the monthly range of group A has varied from 23–26 ha (depending on the year and how it was measured). These gibbons are approximately at carrying capacity, with virtually no vacant space between the defended territories of different groups (Brockelman et al. 1998; Reichard 2009). The relative homogeneity of resources in the territory of group A appears typical of tropical forest gibbons, but comparative quantitative data are lacking. Preliminary evidence, however, suggests that fruit trees utilized by a group of *Nomascus concolor* in the subtropical Wuliang Mountain Reserve, China, are more clumped than at Mo Singto (Fan and Jiang 2008). That study group had a relatively large range (>1 km^2) with three core areas of intensive use, corresponding to the fruit species used in different seasons. Most interesting is that this group and its neighbors are reportedly polygynous (Jiang et al. 1999 and personal communication). Further detailed comparisons with species of *Nomascus*, which occur in southern China, Laos, and Vietnam, might be extremely productive. Savini et al. (2008) have also found considerable variation in territory size in the Mo Singto gibbon population, and this could be studied more fully.

It is beyond the scope of this chapter to test our hypotheses on all other territorial species of primates. We note, however, one early comparison of five species of monkeys in the Kibale Forest by Waser and Wiley (1979). The two species that experienced greatest variation in resource availability had the most overlapping group activity ranges. However, the many differences in diet, social structure, and ranging ability among the five species made comparisons difficult.

11.6.2 Ranging and Foraging

Bartlett (2009b) found that daily ranging path lengths of group A at Mo Singto averaged from 720 m in November to 1,660 m in April. Daily variation was considerable, however, and path lengths often exceeded 3,000 m from April to July (Brockelman, unpublished data). Average daily path lengths of *Nomascus concolor* on Mt. Wuliang, Yunnan, varied somewhat more, from 584 m to 2,356 m (Fan and Jiang 2008). At both Wuliang and Mo Singto, path length varied directly with the proportion of fruit in the diet. At Wuliang, the *N. concolor* group used only 19–50 % of the yearly range in monthly samples whereas at Mo Singto, *H. lar* used more than 50 % of the total range during 8 months of the year, and more than 80 % in May and July. Fan and Jiang (2008) attributed the large total home range of *N. concolor* to the relatively low density of fruit trees (figs, especially), as well as to shifting of the monthly range resulting from the patchy distribution of the fruit species.

Larger territories incur additional costs beyond greater difficulty defending the boundary. Learning and monitoring resource locations should also be more difficult. The value of a small, well-defended territory seems clear: gibbons with small ranges should be able to exploit resources more quickly than primates with larger ranges (Whitington 1992). This theory has not been tested quantitatively because we need suitable methods for measuring foraging efficiency. Further analysis of the frequency distribution of visits to particular food sources might be useful.

Finally, our data support Prediction B5, that territory size in gibbons tends to be set at the area at which food limitation begins to reduce reproductive fitness. Gibbons at Mo Singto use fewer fruit species, visit fewer fruit sources, and have shorter daily ranging paths in the early dry season months than at other times of the year (Bartlett 2003, 2009a; Brockelman 2011). However, that gibbons can forage efficiently in small ranges does not allow them to escape the stresses of seasonal food shortage (see also Fan et al. 2009; Leighton and Leighton 1983; Marshall and Leighton 2006; Marshall et al. 2009). Gibbons eat preferred fruit species whenever possible, but during lean months they ingest more nonpreferred fruits (which have less sugary pulp), as well as more leaves and flowers (e.g., McConkey et al. 2003; Fan et al. 2009). Social behaviors such as play and intergroup conflicts are also reduced at these times Bartlett (2003). Given the high interannual variation in fruiting of many preferred species at Mo Singto (Brockelman 2011), the degree of food shortage must vary substantially from year to year.

Marshall et al. (2009) monitored a large sample of trees for several years in Gunung Palung National Park, Kalimantan, Indonesia, and found that the number of fruit species consumed was highest in the season of low fruit abundance in *Hylobates albibarbis*. This finding is contrary to our results (Brockelman 2011; this study) in which the number of species used per 6-day sample was lowest during lean fruiting months. The difference between our respective results appears to result from the very different sampling methods we used. Marshall et al. (2009) pooled feeding records for different months within seasons over several years, whereas we monitored feeding over short time periods during which a study group was followed continuously. Their data show that the potential range of foods consumed during lean fruiting seasons is high while ours show that actual dietary diversity during short lean periods is low. Thus, gibbons in general are able to take advantage of a large variety of nonpreferred or "fallback" foods (Marshall and Wrangham 2007), as available, to survive in seasons and in years of low or unpredictable fruit availability. It should also be borne in mind that floristic diversity in Borneo is several times greater than in the more seasonal forests of central Thailand.

The positive correlations between monthly range size, food sources, and food visits suggests that gibbons are "energy maximizers", and that increasing food intake in months of high food availability leads to higher fitness (Schoener 1971, 1983; Hixon 1982). This finding is in accord with the findings in some territorial animals, and with theoretical considerations, that increasing food density leads to larger optimal territory size (Ebersole 1980). If gibbons were only seeking an "optimal" level of food input, then food use would level off, and not continue to increase, with increasing range use. Months of low numbers of species and sources used (February, November, December) are the same months determined by Bartlett (2009a) to have the lowest food availability in the group A territory 10 years earlier. These findings suggest a new interpretation of territory size in gibbons: the total annual territory size is not that which guarantees an adequate food supply in lean months (as that would likely be too costly to defend in lean months), but is close to that area which a group can defend in months of most plentiful food supply.

11.6.3 Conclusions

There are multiple explanations for territoriality and monogamy in gibbons, and we must not attribute their social system to any single factor. The list of important conditions is long, and failing to consider any one factor, such as repulsion among females or high mobility, might change our predictions.

The factors we have outlined must be complicated further by the likelihood of positive feedback on selection that reduces group size and range size in gibbons (Fig. 11.5). For example, selection for smaller territories should lead to more efficient foraging and easier range defense, which must generate further reductions in territory size. Smaller groups should then lead to increased genetic cohesion among members, which should increase the defending male's genetic paternity. Primate species that live in larger groups also use knowledge to their advantage, but they tend to rely on fewer and larger fruit sources than do gibbons (Milton 1980; Goodall 1986; Wrangham 1977).

All conditions that may lead to the evolution of small groups and territoriality, as outlined in this chapter, relate to food distribution and foraging behavior. We therefore echo the conclusions that Clutton-Brock and Harvey (1977) made in the early days of primate socioecology. Gibbons have evolved highly specialized limbs and body proportions that facilitate rapid locomotion and efficient feeding in tree crowns. These adaptations preclude running on four limbs and regular use of the ground, but they facilitate defense of resources within the forest canopy. The need to exploit relatively homogeneous, stationary, and defendable resources has led to divergence of gibbons from primates that live in larger groups within larger ranges (Wright 1986; Oates 1987; MacDonald and Carr 1989; Bartlett 2009a). The reduced need to defend against predators may also have been an important factor (van Schaik 1983; Terborgh 1986).

That cetaceans lack a similar social system is probably also rooted in resource distribution. A fission–fusion system of foraging without resource defense is more

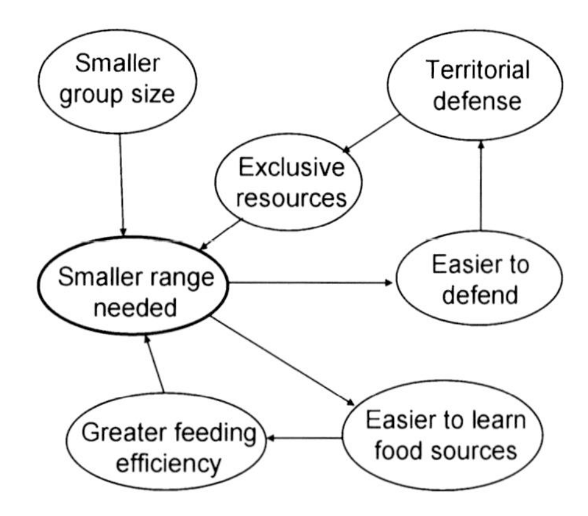

Fig. 11.5 Causal scheme postulating positive feedback mechanisms among components of natural selection that may have led to the evolution of small territory size and easier resource defense in gibbons

typical of dolphins (Delphinidae) (e.g., see Würsig et al., this volume), and probably compares more closely with foraging by canopy primates such as spider monkeys, which have more flexible grouping and (usually) larger ranges than do gibbons (Robbins et al. 1991).

Acknowledgments We thank our field workers, Amnart Boonkongchart, Saiwaroon Chongko, Jantima Saentorn, Ratasart Somnuek, Umaporn Martmoon, and Wisanu Chongko, for their hard work following gibbons and other tasks, students Chanpen Wongsriphuek, Wirong Chanthorn, and Petchprakai Wongsorn for their field help, and Onuma Petrmitr for herbarium support. J.F. Maxwell is thanked for his invaluable botanical help. We thank Mr. Prawat Wohandee and Mr. Narong Mahunnop, past directors of Khao Yai National Park, for their constant support. Thad Bartlett, Susan Lappan, Kim McConkey, and several anonymous reviewers provided valuable comments on the manuscript. The photograph of gibbons in a fig tree was taken by Kulpat Saralamba. This research was funded by Biodiversity and Training Program grants BRT 239001, BRT 242001, BRT R_346005 and BRT R349009.

References

Asensio N, Brockelman WY, Malaivijitnond S, Reichard UH (2011) Gibbon travel paths are goal oriented. Anim Cognit 14:395–405

Bartlett TQ (2003) Intragroup and intergroup social interactions in white-handed gibbons. Int J Primatol 24:239–259

Bartlett TQ (2009a) The gibbons of Khao Yai: seasonal variation in behavior and ecology. Pearson Prentice Hall, Upper Saddle River

Bartlett TQ (2009b) Seasonal home range use and defendability in white-handed gibbons (*Hylobates lar*) in Khao Yai National Park, Thailand. In: Lappan S, Whittaker DJ (eds) The gibbons: new perspectives on small ape socioecology and population biology. Springer, New York, pp 265–275

Bradbury JW, Vehrencamp SL (1976) Social organization and foraging in emballonurid bats. II. A model for the determination of group size. Behav Ecol Sociobiol 2:383–404

Brandon-Jones D, Eudey AA, Geissmann T, Groves CP, Melnick DJ, Morales JC, Shekelle M, Stewart C-B (2004) Asian primate classification. Int J Primatol 25:97–164

Brockelman WY (2009) Ecology and the social system of gibbons. In: Lappan S, Whittaker DJ (eds) The gibbons: new perspectives on small ape socioecology and population biology. Springer, New York, pp 211–239

Brockelman WY (2011) Rainfall patterns and unpredictable fruit production in seasonally dry evergreen forest, and their effects on gibbons. In: McShea WJ, Davies SJ, Phumpakphan N (eds) The ecology and conservation of seasonally dry forests in Asia. Smithsonian Institution Scholarly Press, Washington, DC, pp 195–216

Brockelman WY, Srikosamatara S (1984) Maintenance and evolution of social structure in gibbons. In: Preuschoft H, Chivers DJ, Brockelman WY, Creel N (eds) The lesser apes: evolutionary and behavioural biology. Edinburgh University Press, Edinburgh, pp 298–323

Brockelman WY, Reichard U, Treesucon U, Raemaekers JJ (1998) Dispersal, pair formation and social structure in gibbons (*Hylobates lar*). Behav Ecol Sociobiol 42:329–339

Brockelman WY, Nathalang A, Gale GA (2011) The Mo Singto forest dynamics plot, Khao Yai National Park, Thailand. Nat Hist Bull Siam Soc 57:35–56

Brockman DK, van Schaik CP (eds) (2005) Seasonality in primates: studies of living and extinct human and non-human primates. Cambridge University Press, Cambridge

Carr GM, MacDonald DW (1986) The sociality of solitary foragers: a model based on resource dispersion. Anim Behav 34:1540–1549

Chan Y-C, Roos C, Inoue-Murayana M, Inoue E, Shih C-C, Pei KJ-C, Vigilant L (2010) Mitochondrial genome sequences effectively reveal the phylogeny of *Hylobates* gibbons. PLoS One 5(12):e14419. doi:10.1371/journal.pone.0014419
Chapman CA, Rothman JM (2009) Within-species differences in primate social structure: evolution of plasticity and phylogenetic constraints. Primates 50:12–22
Chapman CA, Chapman LJ, Struhsaker TT, Zanne AE, Clark CJ, Poulsen JR (2005) A long-term evaluation of fruiting phenology: importance of climate change. J Trop Ecol 21:1–14
Chivers DJ (1984) Feeding and ranging in gibbons: a summary. In: Preuschoft H, Chivers DJ, Brockelman WY, Creel N (eds) The lesser apes: evolutionary and behavioural biology. Edinburgh University Press, Edinburgh, pp 267–281
Clutton-Brock TH (1991) The evolution of parental care. Princeton University Press, Princeton
Clutton-Brock TH, Harvey PH (1977) Species differences in feeding and ranging behaviour in primates. In: Clutton-Brock TH (ed) Primate ecology: studies of feeding and ranging behaviour in lemurs, monkeys and apes. Academic, London, pp 557–579
Dunbar RIM (1988) Primate social systems. Chapman & Hall, London
Ebersole JP (1980) Food density and territory size: an alternative model and a test on the reef fish *Eupomacentrus leucostictus*. Am Nat 115:492–509
Emlen ST, Oring LW (1977) Ecology, sexual selection, and the evolution of mating systems. Science 197:215–223
Fan P-F, Jiang X-L (2008) Effects of food and topography on ranging behavior of black crested gibbon (*Nomascus concolor jingdongensis*). Am J Primatol 70:871–878
Fan P-F, Jiang X-L (2009) Maintenance of multifemale social organization in a group of *Nomascus concolor* at Wuliang Mountain, Yunnan, China. Int J Primatol 31:1–13
Fan P-F, Ni Q-Y, Sun G-Z, Huang B, Jiang X-L (2009) Gibbons under seasonal stress: the diet of the black crested gibbons (*Nomascus concolor*) on Mt. Wuliang, central Yunnan. Primates 50:37–44
Fan P-F, Fei H-L, Xiang Z-F, Zhang W, Ma C-Y, Huang T (2010) Social structure and group dynamics of the Cao Vit gibbon (*Nomascus nasutus*) in Bangliang, Jingxi, China. Folia Primatol (Basel) 81:245–253
Gittins SP, Raemaekers JJ (1980) Siamang, lar and agile gibbons. In: Chivers DJ (ed) Malayan forest primates: ten years' study in tropical rain forest. Plenum, New York, pp 63–105
Goodall J (1986) The chimpanzees of Gombe: patterns of behavior. Harvard University Press, Cambridge
Gowans S, Würsig B, Karczmarski L (2008) The social structure and strategies of delphenids: predictions based on and ecological framework. Adv Mar Ecol 53:195–294
Grueter CC, Chapais B, Zinner D (2012) Evolution of multilevel social systems in nonhuman primates and humans. Int J Primatol 33:1002–1037
Hayashi S, Hayasaka K, Takenaka O, Hori S (1995) Molecular phylogeny of gibbons inferred from mitochondrial sequences: preliminary report. J Mol Evol 41:351–365
Hemingway CA, Bynum N (2005) The influence of seasonality on primate diet and ranging. In: Brockman DK, van Schaik CP (eds) Seasonality in primates: studies of living and extinct human and non-human primates. Cambridge University Press, Cambridge, pp 57–104
Hixon MA (1982) Energy maximizers and time minimizers: theory and reality. Am Nat 119:596–599
Jennrich RI, Turner FB (1969) Measurement of non-circular home range. J Theor Biol 22:227–237
Jiang X-L, Wang Y-X, Wang Q (1999) Coexistence of monogamy and polygamy in black-crested gibbons (*Hylobates concolor*). Primates 40:607–611
Kleiman DG (1977) Monogamy in mammals. Q Rev Biol 52:39–69
Kruuk H, Parish J (1982) Factors affecting population density, group size and territory size of the European badger, *Meles meles*. J Zool (Lond) 196:31–39
Leighton DR (1987) Gibbons: territoriality and monogamy. In: Smuts BB, Cheney DL, Seyfarth RM, Wrangham RW, Struhsaker TT (eds) Primate societies. University of Chicago Press, Chicago, pp 135–145
Leighton M, Leighton D (1983) Vertebrate responses to fruiting seasonality within a Bornean rain forest. In: Sutton SL, Whitmore TC, Chadwick AC (eds) Tropical rain forest: ecology and management. Blackwell, Oxford, pp 181–196

Low BS (2003) Ecological and social complexities in human monogamy. In: Reichard UH, Boesch C (eds) Monogamy: mating strategies and partnerships in birds, humans and other mammals. Cambridge University Press, Cambridge, pp 161–176
MacDonald PW, Carr GM (1989) Food security and the rewards of tolerance. In: Standen V, Foley RA (eds) Comparative socioecology: the behavioural ecology of humans and other animals. Blackwell, Oxford, pp 75–99
Marshall AJ, Leighton M (2006) How does food availability limit the population density of white-bearded gibbons? In: Hohmann G, Robbins MM, Boesch C (eds) Feeding ecology in apes and other primates: ecological, physical, and behavioral aspects. Cambridge University Press, Cambridge, pp 313–335
Marshall AJ, Wrangham RW (2007) Evolutionary consequences of fallback foods. Int J Primatol 28:1219–1235
Marshall AJ, Cannon CH, Leighton M (2009) Competition and niche overlap between gibbons (*Hylobates albibarbis*) and other frugivorous vertebrates in Gunung Palung National Park, West Kalimantan, Indonesia. In: Lappan S, Whittaker DJ (eds) The gibbons: new perspectives on small ape socioecology and population biology. Springer, New York, pp 161–188
Matsudaira K, Ishida T (2010) Phylogenetic relationships and divergence dates of the whole mitochondrial genome sequences among three gibbon genera. Mol Phylogenet Evol 55:454–459
McConkey KR (2009) The seed dispersal niche of gibbons in Bornean dipterocarp forests. In: Lappan S, Whittaker DJ (eds) The gibbons: new perspectives on small ape socioecology and population biology. Springer, New York, pp 189–207
McConkey KR, Ario A, Aldy F, Chivers DJ (2003) Influence of forest seasonality on gibbon food choice in the rainforests on Barito Ulu, central Kalimantan. Int J Primatol 24:19–33
McGrew WC, Marchant LF, Nishida T (eds) (1996) Great ape societies. Cambridge University Press, Cambridge
Mills MGM (1982) Factors affecting group size and territory size in the brown hyaena, *Hyaena brunnea*, in the southern Kalahari. J Zool (Lond) 198:39–51
Milton K (1980) The foraging strategy of howler monkeys. Columbia University Press, New York
Mitani JC, Rodman PS (1979) Territoriality: the relation of ranging pattern and home range size to defendability, with an analysis of territoriality among primate species. Behav Ecol Sociobiol 5:241–251
Mootnick AR, Groves CP (2005) A new generic name for the Hoolock gibbon (Hylobatidae). Int J Primatol 26:971–976
Oates JF (1987) Food distribution and foraging behavior. In: Smuts BB, Cheney DL, Seyfarth RM, Wrangham RW, Struhsaker TT (eds) Primate societies. University of Chicago Press, Chicago, pp 197–209
Reichard U (2003) Social monogamy in gibbons: the male perspective. In: Reichard UH, Boesch C (eds) Monogamy: mating strategies and partnerships in birds, humans and other mammals. Cambridge University Press, Cambridge, pp 190–213
Reichard UH (2009) The social organization and mating system of Khao Yai white-handed gibbons: 1992–2006. In: Lappan S, Whittaker DJ (eds) The gibbons: new perspectives on small ape socioecology and population biology. Springer, New York, pp 347–384
Robbins D, Chapman CA, Wrangham RW (1991) Group-size and stability—why do gibbons and spider monkeys differ? Primates 32:301–305
Roos CI, Geissmann T (2001) Molecular phylogeny of the major hylobatid divisions. Mol Phylogenet Evol 19:486–494
Rutberg AT (1983) The evolution of monogamy in primates. J Theor Biol 104:93–112
Savini T, Boesch C, Reichard UH (2008) Home-range characteristics and the influence of seasonality on female reproduction in white-handed gibbons (*Hylobates lar*) at Khao Yai National Park, Thailand. Am J Phys Anthropol 135:1–12
Schoener TW (1971) Theory of feeding strategies. Annu Rev Ecol Syst 2:369–404
Schoener TW (1983) Simple models of optimal feeding-territory size. Am Nat 121:608–629
Steele JH (1985) A comparison of terrestrial and marine ecological systems. Nature (Lond) 313:355–358

Terborgh J (1986) The social systems of New World primates: an adaptionist view. In: Else JG, Lee PC (eds) Primate ecology and conservation. Cambridge University Press, Cambridge, pp 199–211
Tutin CEG, White LJT (1998) Primates, phenology and frugivory: present, past and future patterns in the Lopé Reserve, Gabon. In: Newbery DM, Prins HHT, Brown ND (eds) Dynamics of tropical communities. Blackwell, London, pp 309–337
van Schaik C (1983) Why are diurnal primates living in groups? Behaviour 87:120–144
van Schaik CP, Kappeler PM (2003) The evolution of social monogamy in primates. In: Reichard UH, Boesch C (eds) Monogamy: mating strategies and partnerships in birds, humans and other mammals. Cambridge University Press, Cambridge, pp 59–80
van Schaik CP, van Hooff JARAM (1996) Towards an understanding of the orangutan's social system. In: Marchant LF, Nishida T, McGrew WC (eds) Great ape societies. Cambridge University Press, Cambridge, pp 3–15
Waser PM, Wiley RH (1979) Mechanisms in the evolution of spacing in animals. In: Marler P, Vandenberg JG (eds) Handbook of behavioral neurobiology, vol 3. Plenum, New York, pp 159–223
Whitington C (1992) Interactions between lar gibbons and pig-tailed macaques at fruit sources. Am J Primatol 26:61–64
Wittenberger JF, Tilson RL (1980) The evolution of monogamy: hypotheses and evidence. Annu Rev Ecol Syst 11:197–232
Woodroffe R, Vincent A (1994) Mothers' little helpers: patterns of male care in mammals. Trends Ecol Evol 9:294–297
Wrangham RW (1977) Feeding behavior of chimpanzees in Gombe National Park, Tanzania. In: Clutton-Brock TH (ed) Primate ecology: studies of feeding and ranging behavior in lemurs, monkeys and apes. Academic, London, pp 504–538
Wright PC (1986) Ecological correlates of monogamy in *Aotus* and *Callicebus*. In: Else JG, Lee PC (eds) Primate ecology and conservation. Cambridge University Press, Cambridge, pp 159–167

Part III
Demography, Genetics, and Issues in Conservation

Chapter 12
Northern Muriqui Monkeys: Behavior, Demography, and Conservation

Karen B. Strier

Northern muriqui (*Brachyteles hypoxanthus*) mother and infant at the RPPN Feliciano Miguel Abdala. (Photograph by Fernanda P. Tabacow)

K.B. Strier (✉)
Department of Anthropology, University of Wisconsin-Madison, 1180 Observatory Drive Madison, Madison, WI 53706, USA
e-mail: kbstrier@wisc.edu

J. Yamagiwa and L. Karczmarski (eds.), *Primates and Cetaceans: Field Research and Conservation of Complex Mammalian Societies*, Primatology Monographs, DOI 10.1007/978-4-431-54523-1_12,

Abstract The northern muriqui (*Brachyteles hypoxanthus*) is a critically endangered primate species, endemic to the Atlantic forest of southeastern Brazil. Long-term data from one of the largest populations, which inhabits the 957-ha forest at the RPPN Feliciano Miguel Abdala in Caratinga, Minas Gerais, provides insights into the dynamics between behavior, demography, and conservation. With some 328 individuals as of June 2012, this population has quintupled since systematic monitoring began in 1982, and it now represents roughly 33 % of the species. This rapid expansion can be attributed to female-biased infant sex ratios and high survivorship during the first two decades of research. However, male-biased infant sex ratios, compounded by reduced survivorship among dispersing females compared to patrilocal males, have been documented in recent years. The adult sex ratio is projected to become increasingly male biased as these cohorts mature, and the population growth rate is expected to decline. Moreover, demographic fluctuations are expected to affect males and females differently because of sex differences in dispersal patterns and social dynamics. Specifically, the strong affiliative relationships that have persisted among philopatric males are predicted to be sensitive to increases in both the absolute and relative number of adult males, resulting in higher levels of male competition and possible disruptions in male social networks. This case study illustrates the importance of long-term behavioral data for estimating population viabilities and identifying conservation priorities for this and other critically endangered species.

Keywords Behavioral plasticity • *Brachyteles hypoxanthus* • Conservation Demography • Dispersal • Life history • Northern muriqui • Population viability • Reproductive rates • Sex ratios

12.1 Introduction

Long-term field studies provide unique sources of data on the demography and life histories of wild primates (Strier and Mendes 2009, 2012). These data are necessary for realistic assessments of the viability of endangered populations (Coulson et al. 2001), and they also provide important insights into population fluctuations over time (Durant 2000; Metcalf and Pavard 2007; Strier and Ives 2012). Fluctuations in the demographic and life history variables that influence population viabilities, such as reproductive rates, sex ratios, and survival probabilities, are also known to have corresponding effects on levels of competition within and between primate groups, and therefore can affect the composition of groups and the dynamics of male and female social relationships (Altmann and Altmann 1979; Dunbar 1979). Long-term studies are critical for documenting both the ways in which primates adjust their behavior in response to local demographic conditions and the ways in which behavior feeds back to affect demography and life histories (Charpentier et al. 2008). The value of incorporating the synergistic interactions between demographic and behavioral variables into conservation assessments cannot be overstated (Curio 1996;

Cowlishaw and Dunbar 2000; Sutherland and Gosling 2000; Bradbury et al. 2001; Caro 2007; Strier and Ives 2012).

Documenting the ways in which demography and behavior influence each other requires long-term data because of the time lags involved in these interactions. For example, the impact of infant sex ratios on adult sex ratios and levels of reproductive competition will only become evident after surviving infants of current birth cohorts reach sexual maturity and begin to compete with older adults and with one another for access to mates. Age at sexual maturation can vary from about 2 to 12 years in small-bodied (e.g., callitrichins) and large-bodied (e.g., chimpanzees) primates, respectively (Ross and Jones 1999), and the time required to document the reproductive and behavioral consequences of skewed infant sex ratios will vary accordingly. Moreover, the effects of infant sex ratios on adult sex ratios are filtered by sex-specific patterns of mortality and dispersal, which determine the extent to which the recruitment of breeding males or females through intergroup transfers affect the demographic conditions of their natal groups relative to other groups in the population (Strier 2000).

Demographic conditions can also shift during the course of an individual's lifespan, and despite the extended maturational time lags in many primates, the pace of individual behavioral responses to demographic conditions is still faster than the generations required for evolution to act. Indeed, the high levels of phenotypic plasticity displayed by most primates may reflect their histories with fluctuating selection pressures, which can result in behavioral polymorphisms that functionally resemble genetic polymorphisms despite the different processes by which they are established and maintained (Lee and Kappeler 2003; Strier 2003a).

Identifying the ways in which primates adjust their behavior in response to demographic fluctuations over their lifespans can simultaneously advance our understanding of behavioral plasticity and provide insights into the adaptive potentials of primates (Strier 2009). Anthropogenic disturbances at local, regional, and global scales are altering primate habitats and their communities at unprecedented rates (Boyce et al. 2006). The rapidity of these alterations, which include climate change, and their effects on the conservation status of primates, makes investigations into demographic and behavioral interactions an urgent priority for research and conservation alike because both can change within an individual's lifetime, instead of across the generations involved in evolutionary adaptations. Although there is no dispute that increasing available habitat through the establishment of corridors and reforestation efforts may be the only way to improve the prospects for isolated populations of critically endangered primates, understanding the interactions between demographic fluctuations and behavioral responses can nonetheless provide valuable insights into the potential of populations to recover to viable sizes when the primates and their habitats are protected (Strier and Ives 2012).

In this chapter, I review the implications of some of the major demographic changes that have occurred during a 29-year field study of one population of northern muriqui monkeys (*Brachyteles hypoxanthus*; Fig. 12.1). After briefly describing the species and study population, I review the fluctuating demographic conditions that affect both group and population-wide sex ratios. I then consider the associated behavioral changes that are predicted to occur in response to projected shifts in

Fig. 12.1 Northern muriqui (*Brachyteles hypoxanthus*) mothers and infants at the RPPN Feliciano Miguel Abdala. (Photograph courtesy of Carla B. Possamai)

adult sex ratios, and evaluate the implications of these changes for the long-term persistence and conservation status of northern muriquis. Many of the data reviewed here have also been recently discussed elsewhere, although exact sample sizes may vary depending on the years included in the analyses (e.g., Strier and Mendes 2012; Strier and Ives 2012).

12.2 Background

The northern muriqui is endemic to the Atlantic forest of southeastern Brazil and, as are other species endemic to this ecosystem, it is critically endangered. Fewer than 1,000 northern muriquis are now known to occur in a dozen isolated populations in the states of Minas Gerais and Espírito Santos, and only three of these populations include more than 200 individuals (Mendes et al. 2005). The largest of these populations inhabits the 957-ha forest at the Reserva Particular Patrimônio Natural-Feliciano Miguel Abdala (RPPN-FMA; previously known as the Estação Biológica de Caratinga, 19°44′S, 41°49′W), where our studies during the past 29 years have been based. With 328 individuals as of June 2012, the RPPN-FMA muriqui population has more than quintupled in size since 1982, and it now represents roughly one third of the entire population of the species (updated from Strier et al. 2006).

The muriquis at the RPPN-FMA have never been provisioned, and in contrast to the sympatric capuchin monkeys, they have never been observed or reported to engage in crop raiding. The forest had been subjected to some selective logging in the past, but removal of forest products ceased entirely by 2001, when the forest was officially designated as a privately owned nature reserve (Castro 2001). Even before its official conversion to a nature reserve, some of the abandoned pastures and coffee fields within and surrounding the forest had already been left to regenerate to the point that the muriquis and other primates were actively ranging through and feeding on the vegetation in these areas. In contrast to the ongoing disturbances that threaten many other northern muriqui populations, the area of suitable habitat available to the RPPN-FMA muriquis has increased since the onset of the study. Habitat regeneration and expansion, together with effective hunting prohibitions, have contributed to the ongoing protection and recovery of this population (Strier and Boubli 2006; Strier and Mendes 2012).

12.2.1 Demographic Context

When observations were initiated in the early1980s, there appeared to be two mixed-sex groups and an estimated total population of some 40–45 individuals (Valle et al. 1984). Systematic studies were initiated in one of the original groups (Matão group) in July 1983. The other original group (Jaó) is known to have fissioned in the late 1980s (Strier et al. 1993) to form a third group (M2), and again in 2002, to form a fourth group (Nadir group). Both the Jaó and Nadir groups have been monitored since 2002, and the M2 group has been monitored since 2003 (Strier and Boubli 2006; Strier et al. 2006). The Matão group has been monitored on a near-daily basis during July 1983–1984, and from June 1986 through the present; the other three groups in the population have been monitored roughly every 3 weeks since the population-wide monitoring began.

In July 1982, the 22 individuals in the Matão group included 6 adult males, 8 adult females, of which 6 were carrying infants less than 6 months of age, and 2 juvenile males. The absence of a larger class of juveniles of different age-sex classes was most peculiar, but it was not likely to have been a consequence of recent hunting pressure, because the ease with which the muriquis habituated to the presence of human observers corroborated local reports that neither the muriquis nor the other three species of sympatric primates (brown howler monkeys, tufted capuchin monkeys, and buffy-headed marmosets) had been hunted for many years (Strier 1999a). Among the other possible causes for the initial skewed age structure (e.g., disease, climatic conditions that negatively affected female reproduction, and predation), predation seems to be the most likely reason by default. Although disease cannot be ruled out, the local farmers who routinely passed through the main part of Matão group's home range and occasionally entered the forest did not report encounters with visibly ailing monkeys or corpses or skeletal materials. Similarly, there were no indications or reports of extreme climatic conditions that might have resulted in reproductive failure during the previous years. Predators are known to prey on muriquis in this forest (Bianchi and Mendes 2007), and 2 infants less than 2 years of age were suspected of having been preyed on by an avian and a terrestrial predator, respectively (Printes et al. 1996). More recently, body parts suspected of belonging to a third missing infant were discovered in feline scats (Possamai et al. 2007).

12.2.2 Behavioral Context

Muriqui society is characterized by unusually peaceful, egalitarian relationships among males and females (Strier 1990, 1992). Males are philopatric, and females typically disperse from their natal groups before the onset of puberty (Printes and Strier 1999; Strier and Ziegler 2000). For more than a decade, the rate of natal female emigrations from the Matão group was comparable to that of female immigrations into the Matão group. In the mid-1990s, however, the number of Matão female emigrants began to exceed the number of immigrants, raising questions about source-sink dynamics and whether the infant sex ratios of the other groups differed from that of the Matão group (Strier 2005). Indeed, the need to investigate these possibilities was an important stimulus behind the decision to expand the demographic monitoring of the Matão group to include all the other groups in this population.

The most striking behavioral change to occur as the Matão group began to grow was the shift from cohesive to fluid grouping patterns and an expansion in the size of the Matão group's home range (Strier et al. 1993). The group home range has continued to shift and expand into the southern part of the forest as first the M2 group, and then the Nadir group, became established and began using the northern parts of the Matão group's original home range (Dias and Strier 2003; Boubli et al. 2005). However, the muriquis' day ranges have not increased with group size, implying that their fluid grouping patterns, coupled with their ability to consume substantial quantities of leaves, have offset the increases in competition over food that have been correlated with group size in other primates (Dias and Strier 2003). More recently, we have also documented an increase in the group's terrestrial activities (Mourthé et al. 2007), suggesting that they may now be engaged in vertical niche expansion, despite the potentially higher predation risks to which they are exposed when on the ground (Tabacow et al. 2009; Strier and Ives 2012).

12.2.3 Life History Context

Muriquis have slow life histories that resemble more closely those of apes than other monkeys similar to them in body size (Strier 1999b, 2003b). Male age at sexual maturity has been documented only in the Matão group to date (Strier 1997). Males in this group become sexually active at a median age of 6.19 years ($n=30$; updated from Strier and Mendes 2012), but do not achieve their first complete copulations, which terminate with ejaculation (Possamai et al. 2005), until a median age of 6.74 years ($n=30$; updated from Strier and Mendes 2012). Females typically disperse from their natal groups, and median age at first parturition for females of known age is 9.00 years ($n=21$; updated from Strier et al. 2006). Only 4 of 49 females that survived to 6 years (the minimum dispersal age) have reproduced in the same group as their mother, with their first parturition occurring at 7.0, 7.3, 7.5, and 8.6 years of age, respectively (updated from Martins and Strier 2004; Strier and Mendes 2012). With the exception of two group fissioning events (Strier et al. 2006), no cases of female secondary dispersal subsequent to giving birth have occurred (Strier 2008).

Most births occur in the peak dry season months (June–August), but births have been documented in all months of the year (updated from Strier et al. 2001). The median birth interval following infants that survive to 2 years is 35.74 months ($n=73$; updated from Strier 2005), and does not appear to vary significantly with infant sex (Strier 1999c; Strier and Ives 2012). As with other primates, infant deaths before weaning can result in shorter subsequent birth intervals (Strier 2004). Nonetheless, birth intervals have declined unexpectedly over time (Strier and Ives 2012).

The upper limit of muriqui lifespans is still unresolved. Although the last of the adult males that were present in the Matão group in 1983 died in September 2005, both the 1982 male infants, as well as one of the 1982 female infants that remained and reproduced in her natal group, were still alive as of June 2012, and the female is still reproductively active at 30 years of age. Two of the six original adult females that were carrying new infants in 1982 are still alive (as of August 2012) and reproductively active. Extrapolating from the youngest age at first reproduction known for nulliparous females, these females would have been at least 7.0 years of age if they were carrying their first infants in 1982, and are therefore at least 35 years of age as of 2012. A fourth adult female, who was visibly nulliparous in 1982 and gave birth to her first infant in 1983, is also still alive and reproductively active in 2012.

Although female northern muriquis can evidently survive and reproduce into their thirties, if not beyond, there is some indication of declining fertility as a consequence of low estradiol peaks during the ovarian cycles of one of the oldest females in the Matão group to date (Strier and Ziegler 2005). This female was carrying an infant in 1982 and is still alive in 2012, but has not reproduced since July 2001. A second mother that was carrying an infant when observations were initiated in 1982 disappeared from the population in October 2007 and is presumed to have died; nonetheless, she survived 14.9 years after her last known parturition, and 1.7 years after her last observed copulation.

12.3 Demographic Fluctuations

Muriqui population growth rates, similar to those of other mammals, are limited by the number of reproducing females and by female reproductive rates (Ross 1998). Although female reproductive rates are known to vary with local population densities and food availability and with individual age, health, and rank in other species (Lee and

Kappeler 2003), the number of reproductive females in a population is a product of stochastic fluctuations in infant sex ratios and survivorship. In matrilocal societies, infant sex ratios and survivorship also determine the number of reproductive females in groups, but when females disperse, as is the case for northern muriquis, the distribution of females across groups, and the sex ratios of these groups, is also mediated by local demographic and ecological conditions that affect female dispersal decisions (Strier 2000).

Population viability analyses based on demographic conditions in the Matão group during the first 10 years (Strier 1993/1994, 2000) and 16 years (Rylands et al. 1998) of the study were consistent in implicating female-biased sex ratios and low mortality rates as the main factors responsible for the increase in the group's size and the potential of the group, and by inference, the population, to continue to expand. More recently, however, infant sex ratios have increased and immature survivorship has decreased in the Matão group, similar to the male-biased sex ratios and immature survivorship that have characterized the other groups in the population since their monitoring was initiated in 2002 (Strier 2005; Strier et al. 2006).

12.3.1 Sex Ratios at Birth

Infant sex ratios in the Matão group were consistently female biased (median = 0.50), with nearly twice as many females as males born between 1982 and 1999. From 2000 to 2007, this pattern reversed, with sex ratios at birth more than twice as high (median = 1.58) as those during the previous 18 years. Indeed, analyses through 2010 indicate a significant increase in the male-biased birth sex ratios (Strier and Ives 2012). Although 12 of the males born in the Matão group since 2000 had reached sexual maturity by 2012, it will still be some time before the reproductive maturation of either males or females from the more recent male-biased infant cohorts begins to impact adult sex ratios, which are still female biased. Moreover, there are indications that recent cohorts of infants of both sexes may have lower probabilities of surviving to adulthood than their predecessors.

12.3.2 Infant and Juvenile Mortality

Immature survival in the Matão group was high for both males and females from 1982 to 2001. Male infants had a 95.4 % probability of surviving to year 1, a 90.9 % chance of surviving to year 2, and an 87.9 % chance of surviving to year 3. Female infants had similarly high probabilities of surviving to year 1 (95.3 %), and slightly lower survival probabilities to year 2 (88.7 %) and year 3 (81.1 %). Between 2002 and 2007, however, the probability of surviving to year 3 had declined to 63.5 % for males and 59.5 % for females. Survivorship to year 3 in the other groups was slightly higher for both males (68.4 %) and females (65.5 %) during this time period. During these years, only 66.97 % of males and 64.0 % of females in the population have survived to age 3, and even fewer (60.55 % of males and 58.67 % of females) have survived to age 5. Increases in mortality across all age-sex classes have occurred through 2010 (Strier and Ives 2012). No immatures younger than 3 years of age have been known to survive without their mothers, and no cases of male dispersal at any age have been observed (see following).

Combining the data on female mortality associated with dispersal across all the groups yields a population-wide estimate of 71.6 % survival, or 28.4 % of females that die during or within a few months of dispersal. Mortality for dispersing adolescent females is more than 20 % higher than for philopatric males (4.55 %) in the same 5- to 7-year-old age class in this population, further reducing the number of reproductive females entering the breeding population in both absolute and relative terms. The male bias in sex ratios at birth, sustained by comparable mortality rates for both sexes to year 5, will be even more pronounced among adults because of the loss of additional females relative to males as a consequence of female dispersal.

12.4 Behavioral Responses to Fluctuating Adult Sex Ratios

Projected increases in adult sex ratios are expected to affect males and females in different ways. Dispersal decisions by either or both sexes could alter the sex ratios of their groups and potentially impact population-wide sex ratios through the impact of dispersal on mortality. For example, if females remain and reproduce in their natal groups, as three of the Matão females have done to date, their improved survivorship compared to dispersing females would result in more a greater proportion of females surviving to enter adulthood as breeders. All three of the females that remained in this natal group were younger when they gave birth to their first offspring than the median age of dispersing Matão group females were at their first reproductions. Thus, in addition to increased survivorship, natal females may also gain a potentially longer reproductive lifespan because they begin reproducing earlier (Strier 2008). Females are not expelled from their natal groups through overt or targeted aggression (Printes and Strier 1999), but whether variation in the onset of female puberty is a consequence or cause of their dispersal decisions is not clear (Martins and Strier 2004).

Despite the advantages to females that reproduce in their natal groups, there are disadvantages associated with the risks of inbreeding with their fathers or other males that are closely related paternally or maternally so long as related males are philopatric. In fact, only one female has been observed to copulate on only two occasions with one of her adult sons, suggesting that mechanisms of mother–son kin recognition and inbreeding avoidance may exist in this species (Strier 1997; Possamai et al. 2007; Tolentino et al. 2008). However, two of the three natal females have copulated with the same maternal brother, implying that inbreeding avoidance mechanisms, at least among siblings, may be less effective. Nonetheless, recent analyses of genetic paternity in a cohort of 22 infants showed that none was sired by close biological kin (Strier et al. 2011).

Female dispersal decisions affect the adult sex ratios of groups through their choices of which nonnatal groups to join. From 2002 to July 2012, 50 % of the 28 dispersing females from the other three groups in the population have joined the Matão group, which is more than the 33 % expected if they were distributing themselves equally among all three nonnatal groups. Curiously, only 2 of the 16 Matão females that have dispersed during this period have joined the M2 group (updated from Strier et al. 2006). Moreover, only 2 females have transferred between the Nadir and Jaó groups (1 from each group), perhaps because the fissioning of the Jaó group in 2002 is still so recent. Whether these dispersal choices reflect a balance between preferences for associating with other females that are already familiar from their natal groups (and in some cases, related) and avoidance of familiar males or are caused by other factors is not known, but whatever their basis, they affect the adult sex ratios, and therefore improve reproductive opportunities for patrilocal males in the groups that they join, while negatively impacting those in the groups that females leave or fail to join.

The immigration of 14 females into the Matão group between 2002 and 2012 has helped to offset the number of natal Matão female emigrants ($n=16$). Indeed, the influx of females into the Matão group may be responsible, at least in part, for the persistence of this group instead of its fissioning. Although we do not have comparable records on the number of copulating males and females in the other groups, it seems likely that unfavorable adult sex ratios may have contributed to the decisions of some males to leave their natal Jaó group on both occasions when subsets of Jaó females fissioned (Strier et al. 2006). In both cases, these transient males initially maintained associations with at least two different mixed-sex groups before ultimately joining the newly established Nadir group (Tokuda et al. 2012).

Whether the result of group fissioning or a paucity of female immigrants, unfavorable adult sex ratios should stimulate typically patrilocal males to transfer out of their natal groups in search of more favorable reproductive opportunities. The demographic threshold for males to leave their natal groups is predicted to be higher in patrilocal societies than it is for males in matrilocal societies (Strier 2009), but the advantages of doing so are similar. In both cases, dispersing males would gain by increasing their reproductive opportunities, and if they disperse with only a subset of related males, they also gain by avoiding reproductive competition with the other male kin remaining in their natal groups. However, if adult sex ratios become similarly male biased across groups in the population, then the tolerant, egalitarian relationships that have characterized male muriqui intragroup dynamics may begin to change in more dramatic ways.

The dynamics of male social relationships, at least in the Matão group, are based on strong spatial associations and affiliative interactions (Strier et al. 2002). Male associations and interactions are typically polyadic, by contrast to those of females, which tend to be more targeted, with each female maintaining strong affiliations with only a few other females (Strier 2011). The extensive networks that underlie male sociality may make male muriquis more sensitive than the females to the cognitive challenges of tracking one another as the absolute number of males in their groups increases (sensu Dunbar 1992). Thus, just as an increase in the relative number of adult males to females will elevate levels of reproductive competition among males, an increase in the absolute number of adult males may undermine the social mechanisms that facilitate their tolerance toward one another's mating activities. The consequences of increased competition and a disruption in male tolerance are likely to be manifested in greater reproductive skew, with consequences for the genetic composition of the population.

12.5 Implications for Conservation

The male-biased infant sex ratios, compounded by increasing mortality across all age-sex classes, could result in a corresponding decline in this population's growth during the next decade if these recent conditions persist (Strier and Ives 2012). The muriquis may respond to these changing demographic conditions by adjusting their dispersal and mating patterns, with consequences for the demography, genetics, and viability of the population.

Episodic deviations from the normative pattern of female dispersal and male philopatry have already been documented, and therefore the mechanisms and consequences of these deviations are easy to envision. By contrast, the tolerant, egalitarian relationships among males have persisted without any indications of change. This persistence is striking considering that adult sex ratios in the Matão group have increased from 0.75 in 1983 to 1.10 in 2011. Yet, there is no evidence of any corresponding increase in the levels of male–male competition to date. However, if current low levels of reproductive skew (Strier et al. 2011) were to increase with both the relative and absolute numbers of males, then genetic variation in the subsequent generations could decline. Increased competition, compounded by weaker bonds among males as their networks are disrupted, could also stimulate males to disperse instead of remaining in their natal groups, further reducing the genetic variation between groups in the population.

The hypothesized effects of both female and male responses to current and projected demographic conditions are derived from comparative analyses of the behavior of other primates, such as those describing the effects of group size or male numbers on levels of competition (Struhsaker 2008) and maturation (Charpentier et al. 2008). Continued demographic analyses using updated demographic parameters can provide more accurate estimates of the likely consequences (Strier and Ives 2012). Similarly, whether the benefits of increased survivorship and earlier reproduction outweigh the greater risks of inbreeding for matrilocal females can be assessed with sensitivity analyses that take genetics, as well as demography, into account. Specifically, the consequences of varying levels of male reproductive skew could be modeled to estimate the effects of mating patterns on paternity and population genetics. However, whether the muriquis will respond as predicted remains an empirical question, which can only be resolved by continued monitoring of their demography and behavior into the future.

Despite the recent documented decline in the population's growth rate (Strier and Ives 2012), there is no way of knowing whether the present demographic conditions will persist long enough to have lasting effects. Successive years of female-biased births could dilute the effects of the increasingly male-biased birth sex ratio and thus lower adult sex ratios in the future.

Moreover, despite recent indications of density-dependent effects, the shorter birth intervals have resulted in an unexpected increase in fertility (Strier and Ives 2012). Thus, current management plans to increase the available habitat through the establishment of corridors and reforestation efforts may be the only way to restore the continuing high rate of growth of this population.

The demographic shifts that have occurred over this long-term study of one northern muriqui population provide important cautionary perspectives about extrapolating from demographic conditions at any particular point in time into the future. Stochasticity in infant sex ratios and the survivorship of infants to adulthood, even when mediated by behavioral responses, can have significant effects in small populations of critically endangered species. Assumptions that small, disturbed populations are demographically stable, or that the primates are behaving in evolutionarily adaptive, normative ways, can be seriously misleading. Long-term studies that document the interactions between demography and behavior are necessary to identify the processes that affect population persistence and conservation priorities in northern muriquis and other critically endangered species (Strier and Mendes 2012; Strier and Ives 2012).

Acknowledgments I thank the editors, J. Yamagiwa and L. Karczmarski, for inviting me to contribute to their volume, CNPq and the Abdalla family for permission to conduct research at the RPPN-FMA, Preserve Muriqui and CI-Brasil for logistical support in the field, and S.L. Mendes for his long-term collaboration on the project. The field study has been supported by a variety of sources, including the National Science Foundation (BNS 8305322, BCS 8619442, BCS 8958298, BCS 9414129, BCS 0621788, BCS 0921013), National Geographic Society, the Liz Claiborne and Art Ortenberg Foundation, Fulbright Foundation, Sigma Xi Grants-in-Aid, Grant #213 from the Joseph Henry Fund of the NAS, World Wildlife Fund, L.S.B. Leakey Foundation, Chicago Zoological Society, Lincoln Park Zoo Neotropic Fund, Center for Research on Endangered Species (CRES), Margot Marsh Biodiversity Foundation, Conservation International, the University of Wisconsin-Madison, and CNPq-Brazilian National Research Council. This research has complied with all U.S. and Brazilian regulations. I also thank the many people who have contributed to the long-term demographic data records from 1983 to 2012 (in alphabetical order): L. Arnedo, M.L. Assunção, N. Bejar, J.P. Boubli, P.S. Campos, T. Cardoso, A. Carvalho, D. Carvalho, C. Cäsar, A.Z. Coli, C.G. Costa, P. Coutinho, L. Dib, Leonardo G. Dias, Luiz G. Dias, D.S. Ferraz, A. Ferreira, F. Fernandez, J. Fidelis, A.R.G. Freire Filho, J. Gomes, D. Guedes, V.O. Guimarães, R. Hack, M.F. Iurck, M. Kaizer, M. Lima, M. Maciel, I.I. Martins, W.P. Martins, F.D.C. Mendes, I.M. Mourthé, F. Neri, M. Nery, S. Neto, C.P. Nogueria, A. Odalia Rímoli, A. Oliva, L. Oliveira, F.P. Paim, C.B. Possamai, R.C. Printes, J. Rímoli, S.S. Rocha, R.C. Romanini, R.R. dos Santos, M. Schultz, B.G.M. da Silva, J.C. da Silva, A.B. Siqueira de Morais, V. Souza, D.V. Slomp, F.P. Tabacow, W. Teixeira, M. Tokudo, K. Tolentino, and E.M. Veado.

References

Altmann SA, Altmann J (1979) Demographic constraints on behavior and social organization. In: Bernstein IS, Smith EO (eds) Primate ecology and human origins. Garland, New York, pp 47–64

Bianchi RC, Mendes SL (2007) Ocelot (*Leopardus pardalis*) predation on primates in Caratinga Biological Station, southeast Brazil. Am J Primatol 69:1173–1178

Boubli JP, Tokuda M, Possamai C, Fidelis J, Guedes D, Strier KB (2005) Dinâmica intergrupal de muriquis-do-norte, *Brachyteles hypoxanthus*, na Estação Biológica de Caratinga, MG: O comportamento de uma unidade de machos (all male band) no vale do Jaó. In: Resumos LD (ed) XI congresso Brasileiro de primatologia, Porto Alegre, p 41

Boyce MS, Haridas CV, Lee CT, NCEAS Stochastic Demography Working Group (2006) Demography in an increasingly variable world. Trends Ecol Evol 21:141–148

Bradbury RB, Payne RJH, Wilson JD, Krebs JR (2001) Predicting population responses to resource management. Trends Ecol Evol 16:440–445

Caro T (2007) Behavior and conservation: a bridge too far? Trends Ecol Evol 22:394–400

Castro MI (2001) RPPN Feliciano Miguel Abdala-a protected area for the northern muriqui. Neotropical Primates 9:128–129

Charpentier MJE, Tung J, Altmann J, Alberts SC (2008) Age at maturity in wild baboons: genetic, environmental and demographic influences. Mol Ecol 17:2026–2040

Coulson T, Maces GM, Hudson E, Possingham H (2001) The use and abuse of population viability analysis. Trends Ecol Evol 16:219–221

Cowlishaw G, Dunbar RIM (2000) Primate conservation biology. University of Chicago Press, Chicago

Curio E (1996) Conservation needs ethology. Trends Ecol Evol 11:260–263

Dias LG, Strier KB (2003) Effects of group size on ranging patterns in *Brachyteles arachnoides hypoxanthus*. Int J Primatol 24:209–221

Dunbar RIM (1979) Population demography, social organization, and mating strategies. In: Bernstein IS, Smith EO (eds) Primate ecology and human origins. Garland, New York, pp 67–88

Dunbar RIM (1992) Neocortex size as a constraint on group size in primates. J Hum Evol 22:469–493

Durant S (2000) Dispersal patterns, social organization and population viability. In: Gosling LM, Sutherland WJ (eds) Behaviour and conservation. Cambridge University Press, Cambridge, pp 172–197

Lee PC, Kappeler PM (2003) Socioecological correlates of phenotypic plasticity of primate life histories. In: Kappeler PM, Pereira ME (eds) Primate life histories and socioecology. Cambridge University Press, Cambridge, pp 41–65

Martins WP, Strier KB (2004) Age at first reproduction in philopatric female muriquis (*Brachyteles arachnoides hypoxanthus*). Primates 45:63–67

Mendes SL, Melo FR, Boubli JP, Dias LG, Strier KB, Pinto LPS, Fagundes V, Cosenza B, De Marco PJ (2005) Directives for the conservation of the northern muriqui, *Brachyteles hypoxanthus* (Primates, Atelidae). Neotrop Primates 13(Suppl):7–18

Metcalf CJE, Pavard S (2007) Why evolutionary biologists should be demographers. Trends Ecol Evol 22:205–212

Mourthe IMC, Guedes D, Fidelis J, Boubli JP, Mendes SL, Strier KB (2007) Ground use by northern muriquis (*Brachyteles hypoxanthus*). Am J Primatol 69:706–712

Possamai CB, Young RJ, Oliveira RCR, Mendes SL, Strier KB (2005) Age-related variation in copulations of male northern muriquis (*Brachyteles hypoxanthus*). Folia Primatol (Basel) 76:33–36

Possamai CB, Young RJ, Mendes SL, Strier KB (2007) Socio-sexual behavior of female northern muriquis (*Brachyteles hypoxanthus*). Am J Primatol 69:766–776

Printes RC, Strier KB (1999) Behavioral correlates of dispersal in female muriquis (*Brachyteles arachnoides*). Int J Primatol 20:941–960

Printes RC, Costa CG, Strier KB (1996) Possible predation on two infant muriquis, *Brachyteles arachnoides*, at the Estação Biologica de Caratinga, Minas Gerais, Brasil. Neotrop Primates 4:85–86

Ross C (1998) Primate life histories. Evol Anthropol 6:54–63

Ross C, Jones KE (1999) Socioecology and the evolution of primate reproductive rates. In: Lee PC (ed) Comparative primate socioecology. Cambridge University Press, Cambridge, pp 73–110

Rylands AB, Strier KB, Mittermier RA, Borovansky J, Seal US (1998) Population and habitat viability assessment workshop for the muriqui (*Brachyteles arachnoides*). CBSG, Apple Valley

Strier KB (1990) New World primates, new frontiers: insights from the woolly spider monkey, or muriqui (*Brachyteles arachnoides*). Int J Primatol 11:7–19

Strier KB (1992) Causes and consequences of nonaggression in woolly spider monkeys. In: Silverberg J, Gray JP (eds) Aggression and peacefulness in humans and other primates. Oxford University Press, New York, pp 100–116

Strier KB (1993/1994) Viability analyses of an isolated population of muriqui monkeys (*Brachyteles arachnoides*): implications for primate conservation and demography. Primate Conserv 14-15(1993–1994):43–52

Strier KB (1997) Mate preferences in wild muriqui monkeys (*Brachyteles arachnoides*): reproductive and social correlates. Folia Primatol (Basel) 68:120–133

Strier KB (1999a) Faces in the forest: the endangered muriqui monkeys of Brazil. Harvard University Press, Cambridge

Strier KB (1999b) The atelines. In: Fuentes A, Dolhinow P (eds) Comparative primate behavior. McGraw-Hill, New York, pp 109–114

Strier KB (1999c) Predicting primate responses to "stochastic" demographic events. Primates 40:131–142

Strier KB (2000) From binding brotherhoods to short-term sovereignty: the dilemma of male Cebidae. In: Kappeler PM (ed) Primate males: causes and consequences of variation in group composition. Cambridge University Press, Cambridge, pp 72–83

Strier KB (2003a) Demography and the temporal scale of sexual selection. In: Jones CB (ed) Sexual selection and reproductive competition in primates: new perspectives and directions. American Society of Primatologists, Norman, pp 45–63

Strier KB (2003b) Primatology comes of age: 2002 AAPA luncheon address. Yearbk Phys Anthropol 122:2–13

Strier KB (2004) Reproductive strategies of New World primates: interbirth intervals and reproductive rates. In: Mendes SL, Chiarello AG (eds) A Primatologia no Brasil-8. IPEMA/ Sociedade Brasileiro de Primatologia, Vitoria, pp 53–63

Strier KB (2005) Reproductive biology and conservation of muriquis. Neotropical Primates 13(suppl):41–47

Strier KB (2008) The effects of kin on primate life histories. Annu Rev Anthropol 37:21–36

Strier KB (2009) Seeing the forest through the seeds: mechanisms of primate behavioral diversity from individuals to populations and beyond. Curr Anthropol 50:213–228

Strier KB (2011) Social plasticity and demographic variation in primates. In: Sussman RW, Cloninger CR (eds) The origins and nature of cooperation and altruism in non-human and human primates. Springer, New York, pp 179–192

Strier KB, Boubli JP (2006) A history of long-term research and conservation of northern muriquis (*Brachyteles hypoxanthus*) at the Estação Biológica de Caratinga/RPPN-FMA. Primate Conserv 20:53–63

Strier KB, Ives AR (2012) Unexpected demography in the recovery of an endangered primate population. PLoS One 7(9):e44407. doi:10.1371/journal.pone.0044407

Strier KB, Mendes SL (2009) Long-term field studies of South American primates. In: Garber PA, Estrada A, Bicca-Marques JC, Heymann E, Strier KB (eds) South American primates. Springer, New York, pp 139–155

Strier KB, Mendes SL (2012) The northern muriqui (*Brachyteles hypoxanthus*): lessons on behavioral plasticity and population dynamics from a critically endangered primate. In: Kappeler P, Watts D (eds) Long-term studies of primates. Springer, Heidelberg, pp 125–140

Strier KB, Ziegler TE (2000) Lack of pubertal influences on female dispersal in muriqui monkeys (*Brachyteles arachnoides*). Anim Behav 59:849–860

Strier KB, Ziegler TE (2005) Variation in the resumption of cycling and conception by fecal androgen and estradiol levels in female muriquis (*Brachyteles hypoxanthus*). Am J Primatol 67:69–81

Strier KB, Mendes FDC, Rímoli J, Rímoli AO (1993) Demography and social structure in one group of muriquis (*Brachyteles arachnoides*). Int J Primatol 14:513–526

Strier KB, Mendes SL, Santos RR (2001) The timing of births in sympatric brown howler monkeys (*Alouatta fusca clamitans*) and northern muriquis (*Brachyteles arachnoides hypoxanthus*). Am J Primatol 55:87–100

Strier KB, Dib LT, Figueira JEC (2002) Social dynamics of male muriquis (*Brachyteles arachnoides hypoxanthus*). Behaviour 139:315–342

Strier KB, Boubli JP, Possamai CB, Mendes SL (2006) Population demography of northern muriquis (*Brachyteles hypoxanthus*) at the Estação Biológica de Caratinga/Reserva Particular do Patrimônio Natural-Feliciano Miguel Abdala, Minas Gerais, Brazil. Am J Phys Anthropol 130:227–237

Strier KB, Chaves PB, Mendes SL, Fagundes V, Di Fiore A (2011) Low paternity skew and the influence of maternal kin in an egalitarian, patrilocal primate. Proc Natl Acad Sci USA 108:18915–18919

Struhsaker TT (2008) Demographic variability in monkeys: implications for theory and conservation. Int J Primatol 29:19–34

Sutherland WJ, Gosling LM (2000) Advances in the study of behaviour and their role in conservation. In: Gosling LM, Sutherland WJ (eds) Behaviour and conservation. Cambridge University Press, Cambridge, pp 3–9

Tabacow FP, Mendes SL, Strier KB (2009) Spread of a terrestrial tradition in an arboreal primate. Am Anthropol 111:238–249

Tokuda M, Boubli JP, Izar P, Strier KB (2012) Social cliques in male northern muriquis (*Brachyteles hypoxanthus*). Curr Zool 58:342–352

Tolentino K, Roper JJ, Passos FC, Strier KB (2008) Mother–offspring associations in northern muriquis, *Brachyteles hypoxanthus*. Am J Primatol 70:301–305

Valle CMC, Santos IB, Santos IB, Alves MC, Pinto CA, Mittermeier RA (1984) Algumas observações sobre o comportamento do mono (Brachyteles arachnoides) em ambient natural environment (Fazenda Montes Claros, Municipio de Caratinga, Minas Gerais, Brasil). In: Mello M (ed) Primatologia no Brasil. Sociedade Brasileiro de Primatologia, Brasilia, pp 271–283

Chapter 13
Indo-Pacific Humpback Dolphins: A Demographic Perspective of a Threatened Species

Shiang-Lin Huang and Leszek Karczmarski

S.-L. Huang
Department of Environmental Biology and Fishery Science, Center for Marine Bioenvironment and Biotechnology, National Taiwan Ocean University, Keelung, Taiwan
e-mail: slhuang@mail.ntou.edu.tw

L. Karczmarski (✉)
The Swire Institute of Marine Science, School of Biological Sciences, The University of Hong Kong, Hong Kong
e-mail: leszek@hku.hk

J. Yamagiwa and L. Karczmarski (eds.), *Primates and Cetaceans: Field Research and Conservation of Complex Mammalian Societies*, Primatology Monographs, DOI 10.1007/978-4-431-54523-1_13,

Abstract Indo-Pacific humpback dolphins inhabit shallow coastal waters within the tropics and subtropics of the Indian and Western Pacific Oceans. Their taxonomy remains unresolved, between a single widespread and highly variable species, two species, and three species being currently proposed. Their inshore distribution renders them highly susceptible to the adverse effects of many human activities; for most of the known remaining populations their continuous survival is a subject of major conservation concern. In this chapter, we describe the use of demographic analysis to quantify population trend and, more informatively, predict the risk (probabilities) of extinction. The results of demographic analyses provide valuable means of assessing conservation status. Using the population of humpback dolphins from the Pearl River Estuary as an example, we show the power of demographic analyses, predicting a significant population decline before it is directly documented by other standard techniques. Comparing our findings with known, albeit limited data from southeast Africa, and considering the current ambiguity of the taxonomic classification adopted by IUCN, we question the current listing of humpback dolphins under the IUCN Red List of Threatened Species. We urge that their conservation status classification be reconsidered as it likely understates, perhaps severely, the threats faced by many fragmented populations off Southeast Asia and the western Indian Ocean.

Keywords Conservation status • Demographic analyses • Pearl River Estuary • Population trend • Probability of extinction • *Sousa chinensis* • *Sousa plumbea*

13.1 Introduction

Humpback dolphins, the genus *Sousa*, inhabit coastal waters of tropical West Africa, the Indian Ocean, and the western Pacific Ocean. They generally associate with shallow-water coastal habitats (Karczmarski et al. 2000). Although the choice of specific habitats might differ between locations and regions in response to varying coastal environments, the overall pattern frequently reoccurs (Karczmarski 2000; Jefferson and Karczmarski 2001; Parra and Ross 2009). In many parts of the world, however, inshore coastal habitats are becoming increasingly degraded through overharvesting and habitat destruction, which, along with incidental and deliberate kills in fishing gear, represent the greatest threat to humpback dolphins. With habitats fast diminishing, populations become increasingly fragmented and more susceptible to further anthropogenic pressure and environmental stochasticity. For most of the known remaining populations of humpback dolphins, their continuous survival is a subject of major conservation concern (Karczmarski 2000; Jefferson and Karczmarski 2001; Reeves et al. 2008; Jefferson et al. 2009; Ross et al. 2010).

The taxonomy of the genus *Sousa* remains unresolved: from one to five species have been proposed (Ross et al. 1994). Currently, most researchers recognize either two (Jefferson and Karczmarski 2001) or three (Rice 1998) species of *Sousa*. The three-species taxonomy distinguishes *S. teuszi* off West Africa, *S. plumbea* in the western Indian Ocean, and *S. chinensis* off southeast Asia and in the western Pacific

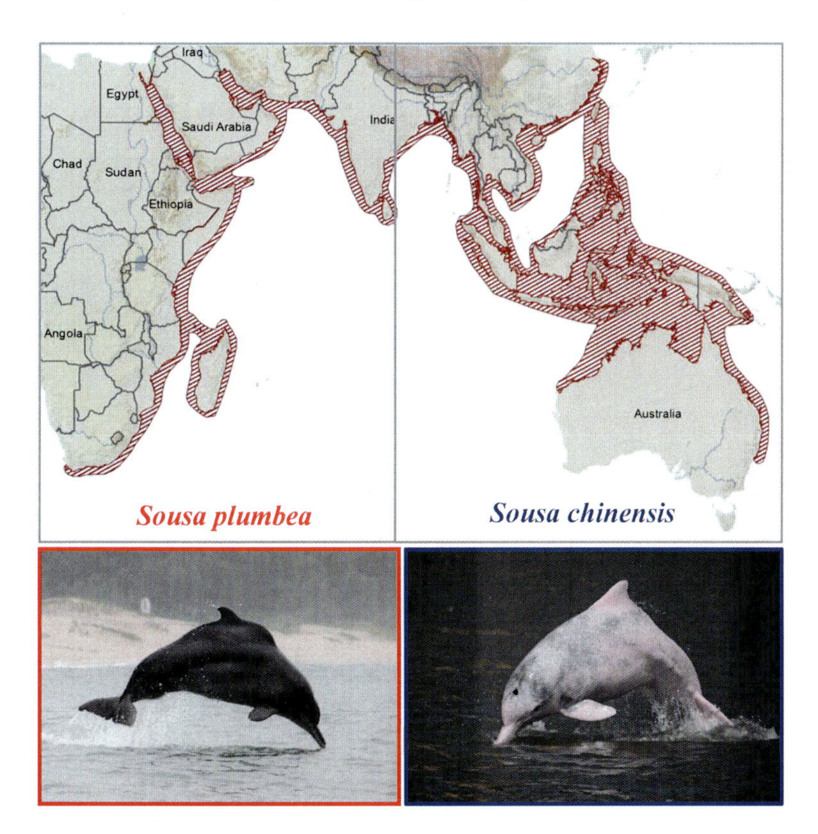

Fig. 13.1 The taxonomy of the Indo-Pacific humpback dolphins remains unresolved. Some researchers recognize a single widespread and highly variable species, *Sousa chinensis*. Others, including both authors of this chapter, consider humpback dolphins in the Indo-Pacific to consist of two species: *S. plumbea* in the western Indian Ocean, from South Africa to the east coast of India, and *S. chinensis*, from the east coast of India to China and Australia. The International Union for Conservation of Nature (IUCN) recognizes them as two geographic forms that differ in their external morphology. The *plumbea* form is uniformly gray in color and has a well-pronounced dorsal hump. Newborns of the *chinensis* form are dark gray, but become light gray as juveniles and white to light pink as adults and lack the prominent dorsal hump, with only a slight dorsal ridge instead. [Map source: IUCN 2008. Photography credit: S. Atkins (Endangered Wildlife Trust, South Africa), *S. plumbea*, and R. Tang (Cetacean Ecology Lab, The Swire Institute of Marine Science, The University of Hong Kong), *S. chinensis*. See also the title page photograph]

Ocean. When two-species taxonomy is considered (e.g., current IUCN classification), *S. chinensis* and *S. plumbea* are combined into one species—the Indo-Pacific humpback dolphin *S. chinensis*—ranging from South Africa in the west to southeast China and northeast Australia in the east, with dolphins in the central and western Indian Ocean referred to as the *plumbea* form and those in the eastern Indian Ocean and western Pacific Ocean referred to as the *chinensis* form (Fig. 13.1).

This taxonomic classification and resulting IUCN conservation status have been the subject of intense debates (e.g., Global Mammal Assessment: Cetacean Red List

Assessment. IUCN Species Survival Commission, Cetacean Specialist Group, 22–26 January 2007, La Jolla, CA, USA). Although both authors of this chapter support the three-species taxonomy (recognizing that even further revision to a four-species taxonomy might be needed; e.g. Frère et al. 2008), we refer to the two-species taxonomy here to emphasize important conservation implications. We focus here primarily on the *chinensis* form, known locally in China and Taiwan as the Chinese white dolphin, and more specifically on the animals inhabiting the Pearl River Estuary (PRE), which is one of the most ecologically degraded coastal habitats in Southeast Asia (MacKinnon et al. 2012), yet harbors the world's largest population of humpback dolphins. Our focus here on species and locations have further range-wide implications, beyond the PRE and Southeast Asia, and potentially useful to researchers working with other mammals with conservation concerns, both marine and terrestrial.

13.2 Why Are Demographic Analyses Important?

Demographic parameters, including population size, age-specific survivorship, generation length, and instantaneous rate of increase, provide a baseline foundation for the management of species and populations (Wade 1998; Stolen and Barlow 2003; Moore and Read 2008; Currey et al. 2009a; Huang et al. 2012a, b; Mei et al. 2012). Estimates of demographic parameters allow population dynamics models to quantitatively predict population trends and risk of extinction, a powerful approach that can facilitate stepping up conservation efforts and moving on from precautionary management to informed conservation strategies (Lacy 1993; Harwood 2000; Fujiwara and Caswell 2001).

For most cetaceans, reliable demographic estimates and population trend models are rare, primarily because of limited life history data, which leaves policy makers and conservation managers with a considerable challenge when making management decisions. Incomplete evidence can easily lead to misguided judgments of conservation status, which in turn can misguide and sometimes delay the implementation of appropriate conservation strategies. Consequences of such a chain of events can have severe, even catastrophic, implications for the survival of species and populations (Taylor et al. 2000; Thompson et al. 2000).

When addressing conservation challenges of threatened or endangered species or populations, the one question of paramount importance that frequently comes up is this: *How many years do we still have to reverse population decline?* The answer often depends on the actual rate of decline that can be estimated through a set of demographic analyses. For example, the Yangtze finless porpoise (*Neophocaena asiaeorientalis asiaeorientalis*), the only freshwater subspecies of porpoise, has been reported to be in decline based on evidence from early life-table analysis (Yang et al. 1998) and long-term census data (Zhang et al. 1993; Zhao et al. 2008). Recent comparative demographic analyses, however, provide further insights into this process, indicating an accelerating decline, from an average 1.66 % abundance per year before 1993 to 6.17 % abundance per year after 1993 (Mei et al. 2012). Under the present scenario, the Yangtze finless porpoise is likely to become extinct within the next 30 years, or possibly sooner (Mei et al. 2012). Such estimates of how rapidly a population is declining and how soon a population is likely to become extinct quantify the urgency of conservation actions and can provide a powerful tool in monitoring the effectiveness of management initiatives.

Under Criteria A and C1 of the IUCN Red List Categories and Criteria Version 3.1, the population status, either *NT* (*Near Threatened*), *VU* (*Vulnerable*), *EN* (*Endangered*), or *CR* (*Critically Endangered*), is classified by the percentage of decline within one, two, or three generations (IUCN 2001). When the exact generation length is not known, it is often accepted that three generations can be substituted with 10 years to facilitate assessment procedures (IUCN 2001). In cetaceans, however, where one generation length is longer than 10 years for many species (Taylor et al. 2007a), such an approach can lead to an underestimation of the risk of extinction, producing a status assessment that might not truly reflect the real conservation threat. On the other hand, estimates of the percentage of population decline (or change) require the accumulation of long-term census data, which is frequently challenged by the high costs and demanding logistics of collecting cetacean sighting data at sea.

13.3 Long-Term Data: How Long Is Long Enough?

Long-term abundance estimates can be used to reflect population trends (Chaloupka et al. 1999; Wilson et al. 1999; Stevick et al. 2003). However, a simple application of statistical methods to historic abundance estimates should be viewed cautiously as they may yield statistically "significant" yet biased and potentially misguided results (Gerrodette 1987; Thompson et al. 2000; Taylor et al. 2007b). Such arbitrary conclusions are of particular concern when the abundance estimates have high variation. For example, Jefferson and Hung (2004) applied the least-squares method to abundance estimates of humpback dolphins in Hong Kong waters between 1995 and 2002 by a cubic polynomial regression to describe the abundance fluctuations in those years:

$$\left(N(t) = -0.210t^3 + 5.867t^2 - 46.312t + 166.5, r^2 = 0.515\right)$$

When extrapolating this regression into the future, a "declining" trend was suggested. Previously, Jefferson (2000) had reported a trend of Hong Kong humpback dolphins ($N(t) = 134.989 \times 10^{-0.072t}$, $r_2 = 0.463$) based on abundance estimates between 1995 and 1998. With this rate of decline ($\lambda = 10^{-0.072} = 0.847$), demographic models predict more than 85 % loss of original abundance in just 10 years since

2000, which clearly did not happen and was recently challenged by the latest abundance estimate (see further; Chen et al. 2010).

The power analysis shows that the ability to detect population trends from periodical census data is highly dependent on the estimate of instantaneous rate of increase and the variation of abundance estimates (Gerrodette 1987; Taylor and Gerrodette 1993; Thompson et al. 2000; Taylor et al. 2007b; Huang et al. 2012a). The variation in traditional census data from transect sampling correlates highly with the number of abundance estimates (Taylor and Gerrodette 1993) and intrinsically restricts the statistical resolution of census data to reveal population trends within relatively short timeframes (Taylor and Gerrodette 1993; Taylor et al. 2007b; Huang et al. 2012a), effectively restricting the accurate classification of population status (Fig. 13.2). Decline of a threatened population often results from excessive anthropogenic influences that outweigh the capability of intrinsic recovery. Systematic census surveys, however, are seldom implemented across the entire geographic range of a population until the population decline, or anthropogenic impacts, become explicitly apparent. It is at this point that conservation awareness usually increases, but unfortunately it may already be too late to preserve the species of concern, as tragically exemplified by the recent history of the decline and extinction of the baiji (*Lipotes vexillifer*) and associated research efforts (Wang et al. 2006).

The importance of long-term studies cannot be overemphasized, especially when working with long-lived, slow-reproducing mammals with complex social structures. Much of our current knowledge of cetaceans stems from multigenerational studies that extend over many years, sometimes decades. However, in the case of threatened species and populations, status assessment based on long-term census data and resulting management decisions, if ever implemented, frequently represent "crisis management" rather than "crisis prevention"; such decisions are often delayed and come too late.

13.4 The Demographic Approach

Demographic analyses provide an alternative way to detect past and, particularly, predict future trends (Fujiwara and Caswell 2001; Koschinski 2002), which can be achieved by long-term mark–recapture analysis (Chaloupka et al. 1999; Fujiwara and Caswell 2001; Stevick et al. 2003; Currey et al. 2009a, b; Verborgh et al. 2009), or life-table analysis (Barlow and Boveng 1991; Caswell et al. 1998; Dans et al. 2003; Stolen and Barlow 2003; Moore and Read 2008; Huang et al. 2012b; Mei et al. 2012). Both methods are sensitive to sample size and require long-term datasets (mark–recapture) or extensive specimen collection from strandings or bycatch (life table) that may not be suitable for small and regionally confined populations.

In this approach, the demographic rates, including generation time (T_0) and instantaneous rate of increase (r), can be calculated using the standard method (Krebs 1989):

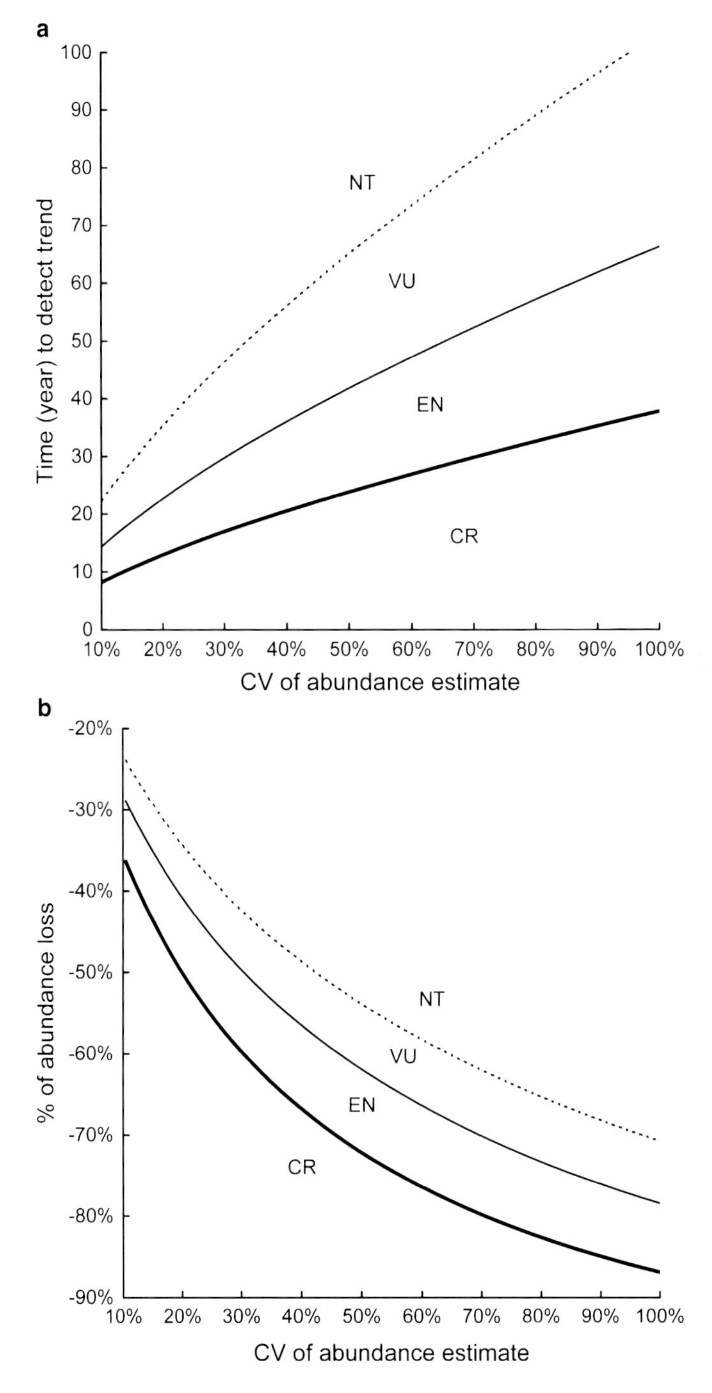

Fig. 13.2 (**a**) Time needed to detect population decline by periodic census investigation. (**b**) Percentage of abundance loss when the abundance decline becomes detectable at different levels of abundance estimate coefficient of variation (*CV*). *Cutoff lines* represent the boundary conditions meeting the classifications of *VU* (Vulnerable), *EN* (Endangered), or *CR* (Critically Endangered) status under IUCN Criterion A2-4, where population decline is higher than 30 % (*VU*), 50 % (*EN*), or 80 % (*CR*) of initial abundance within three generations for freshwater cetaceans (IUCN 2001). (Reproduced from Huang et al. 2012a)

$$T_0 = \frac{\sum x \times l(x) \times m(x)}{\sum l(x) \times m(x)} \tag{13.1}$$

where $l(x)$ and $m(x)$ are age-specific survivorship and the age-specific reproduction rate, respectively. The value of r can be estimated as follows:

$$r = \frac{\ln\left(\sum l(x) \times m(x)\right)}{T_0} \tag{13.2}$$

$m(x)$ can be approximately defined as follows:
$m(x) = 0$, when $0 \leq x < Am$, and

$$m(x) = \frac{\rho}{RI} \tag{13.3}$$

when $Am \leq x < Ax$, where Am represents female age at maturation, Ax represents female expected lifespan, RI is the reproduction (calving) interval, and ρ represents the expected proportion of female calves (often assumed to be 0.50) (Huang et al. 2012a, b; Mei et al. 2012). The age-specific survivorships $l(x)$, on the other hand, can be constructed by life-table analysis or mark–recapture analysis. In life-table analysis, the life-table parameters, including number alive at age x (n_x), proportion surviving to age x (l_x), and mortality rate at age x (q_x), can be calculated using a standard method (Krebs 1989; Barlow and Boveng 1991; Stolen and Barlow 2003; Moore and Read 2008; Huang et al. 2012b; Mei et al. 2012) and used to build a age-specific survivorship model ($l(x)$) with quantifiable range of uncertainties (Stolen and Barlow 2003; Huang et al. 2012b; Mei et al. 2012). In the mark–recapture analysis, $l(x)$ can be constructed by

$$l(x) = Sc \times Sa^{x-1} \tag{13.4}$$

where Sc and Sa are the survival rates of calf ($x \leq 1$) and non-calf animals (Huang et al. 2012a).

The usefulness of the demographic approach comes from the fact that it not only estimates current/past population trends and status but also forecasts future population change under an assumed (and usually the most optimistic) scenario with factored-in stochasticity (Box 1). This trend prediction forecasts the risk and measures the probability of extinction (PE) within a specific timeframe, for example, three or five generations. The PE estimate itself becomes another quantitative assessment of population status under Criterion E in the IUCN Red List Categories and Criteria Version 3.1 (IUCN 2001).

Box 1

The process to project dynamic change of population size by an individual-based Leslie matrix model (Slooten et al. 2000; Currey et al. 2009a; Huang et al. 2012a, b; Mei et al. 2012):

1. An individual survived from age x at year t to age $x+1$ at year $t+1$ whenever the random number σ that ranges between 0 and 1 exceeded the mortality rate at age x, $q(x)=1-S$, where S is survival rate, either Sc (for $x \leq 1$) or Sa (for $x > 1$). In case when $\sigma \leq q(x)$, the individual dolphin was considered a victim of mortality; otherwise, it survived.
2. A female that survived to the next year was determined to give birth by comparing σ with $1/RI_i$, with the presence of newborn, when $\sigma \leq 1/RI_i$.
3. The sex of the newborn was male when σ exceeded the sex ratio ρ (default=0.50); otherwise, the calf was a female.
4. For each of the foregoing simulations, a new random number σ was generated.

Recent studies of population trends of the Yangtze finless porpoise provide strong evidence validating the usefulness of demographic analyses based on age-at-death data. Systematic boat surveys in the mainstream of the Yangtze River have shown a drastic decline in abundance from more than 2,550 animals (Zhang et al. 1993) to fewer than 1,225 animals (Zhao et al. 2008) in 15 years. Meta-analyses of data collected between 1990 and 2007 estimate an average of 6.4 % decline of abundance per annum (Xiujiang Zhao, unpublished data). By comparison, independent life-table analysis based on age-at-death data from incidentally killed porpoises estimates the current rate of change of the Yangtze finless porpoise population at a negative value of −0.0637 (Mei et al. 2012). The consistency of these results from two independent studies applying different methodological approaches demonstrates the valuable applicability of demographic analyses.

Despite the utility of demographic analysis in estimating population trends and forecasting the risk (probability) of extinction, especially when applied to threatened populations, the interpretation of demographic analysis, especially projections of population trends, should be cautious. Demographic analyses based on age-at-death data usually assume a stable age distribution across the population structure (Caughley 1966), and departure from this assumption may bias the demographic rate estimates (Gaillard et al. 1998). For most cetacean species, the validity in meeting this assumption is difficult to assess, unless there is a long-term photo-identification (photo-ID) monitoring of individual animals (as in Hamilton et al. 1998; see also Wells, this volume). However, the life history patterns of many delphinids, with their relatively long calving interval, long life expectancy of females, and high adult survival rates (for humpback dolphins, see

Jefferson 2000; Taylor et al. 2007a; Huang et al. 2008; Jefferson et al. 2011), may buffer the bias caused by violation of the stable age distribution assumption (Stolen and Barlow 2003).

Most demographic models assume that current environmental conditions are constant and do not change much over time (Lacy 1993; Caswell et al. 1999; Winship and Trites 2006). In reality, environmental conditions may and often do change with time, sometimes substantially so, as is the case in Chinese waters, thus altering the risk of local extinction. This change may be either positive or negative but is unlikely to remain constant under ever-increasing human pressures.

Species or populations usually begin attracting conservation attention only after they are exposed to severe environmental degradation or habitat loss, or when once-common animals are no longer seen frequently. In such instances, the animals may indeed be struggling for their long-term biological survival in the face of environmental uncertainty of increasingly deteriorating habitats. Although some components of habitat deterioration may be reversible in the short-to-medium term (e.g., environmentally degradable pollutants; Chevé 2000), other components can only be addressed over the long term or may be irreversible (e.g., bioaccumulative pollutants and habitat loss from land reclamation). Moreover, many of these factors are closely related to the exponentially increasing economic and human population growth (Chevé 2000; Bearzi et al. 2004, 2010; Piroddi et al. 2011; Wang et al. 2011). Therefore, it is more realistic to assume that the survivorship of threatened species or populations predicted by demographic analyses is higher than the true survivorship; the parameter estimates and projections should be therefore viewed as optimistic estimates that have an unknown degree of uncertainty, especially as the predictive future estimates deviate further from actual data.

13.5 The Humpback Dolphins of the Pearl River Estuary

It has been suggested that there are 2,517 to 2,555 humpback dolphins in the waters of the PRE (Chen et al. 2010), which is substantially more than in any other area where humpback dolphins are known to occur (Table 13.1). One could assume, therefore, that this population might be strong enough to resist demographic stochasticity and environmental pressures. However, because of their proximity to the world's busiest seaport and airport, several large densely populated urban centers, and a fast developing economy with the fastest and ever-increasing rate of major infrastructural development in the entire PRE region (Fig. 13.3), the dolphins inhabiting PRE waters are exposed to many adverse effects of human activities. In fact, there are few other small cetacean populations that face the range and intensity of human-induced pressures which exist within the PRE (Wilson et al. 2008).

In Hong Kong waters, these animals have received scientific and conservation attention since the mid-1990s (Jefferson 2000; Jefferson and Hung 2004; Parsons 2004; Hung et al. 2006; Jefferson et al. 2011), instigated at first by large-scale anthropogenic impacts resulting from the massive construction of the Hong Kong International

Table 13.1 Known population size and abundance estimates for Indo-Pacific humpback dolphins

Populations	Areas	*N*	Source
Algoa Bay	South Africa	466	Karczmarski et al. (1999), Karczmarski (2000)
Richards Bay	South Africa	166 170	Durham (1994) Atkins and Atkins (2002)
Zanzibar (south coast)	East Africa	63	Stensland et al. (2006)
Maputo Bay	Mozambique	105	Guissamulo and Cockcroft (2004)
Great Sandy Strait	Australia	150	Cagnazzi et al. (2011)
Cleveland Bay	Australia	54	Parra et al. (2006)
Moreton Bay	Australia	163	Parra et al. (2004)
Moreton Bay	Australia	119	Corkeron et al. (1997)
Dafengjiang River	China	114	Chen et al. (2009)
Hepu	China	39	Chen et al. (2009)
Hong Kong + adjacent area	China	1,028	Jefferson (2000)
Leizhou Bay	China	237	Zhou et al. (2007)
Xiamen	China	76	Chen et al. (2009)
Xiamen	China	86	Chen et al. (2008a, b)
Goa Bay	India	842	Sutaria and Jefferson (2004)
Gulf of Kachch	India	174	Sutaria and Jefferson (2004)
Eastern Taiwan Strait	Taiwan	99 85	Wang et al. (2007) Yu et al. (2010)

Airport at Chek Lap Kok. Despite these commendable research efforts, however, considerable information gaps remain; much of the population vital parameters, dynamics and structure, and various aspect of their behavioral ecology remain poorly understood. The lack of analyses assessing cumulative effects of the multitude of threats faced by the animals inhabiting the PRE should be of major concern. Although the Agriculture, Fisheries and Conservation Department of the Hong Kong Government has been monitoring the abundance of humpback dolphins in Hong Kong waters since early investigations in the mid-1990s, the current available data are insufficient to reliably estimate the population demographic processes that ultimately determine the species biological persistence. Without such basic knowledge, there is very little in the way of educated guidelines that could lead authorities toward informed management decisions. Consequently, as presented further in this chapter, conservation measures remain ineffective, as they have evidently been so far in Hong Kong and much of the PRE. At present, the forecast for humpback dolphins inhabiting the Pearl River Estuary is grim, and it will remain so unless cumulative effects of anthropogenic impacts and the dolphin population trends and structure are assessed and addressed in a timely manner and properly incorporated into environmental management strategies. Such an approach is the only reasonable and responsible way toward conscientious and effective management planning.

In a recent study, using data collected from humpback dolphins stranded on the mainland China coast of the PRE, Huang and colleagues (2012b) applied Siler's

Fig. 13.3 Humpback dolphins in the Pearl River Estuary (PRE) inhabit waters flanked by the large and densely populated urban centers of Hong Kong, Shenzhen, Zhuhai, and Macau, with the world's busiest seaport and airport, and major infrastructural development across the entire PRE region. [Map source: Landsat data, USGS (http://www.usgs.gov/pubprod/aerial.html#satellite)]

competitive risk model of survivorship (Siler 1979) to empirical life-table parameters to construct a modeled life table (Fig. 13.4), which was used to calculate demographic rates (Table 13.2). A continuous rate of population decline of 2.46 % per annum was estimated. It was projected that if the estimated rate of decline remains constant, the current population will be diminished by ~74 % after only three generations (approximately 51 years; Fig. 13.5) and ~58 % of model simulations meet the criteria for conservation status classification as Endangered under Criterion A3b (Fig. 13.6), applying IUCN Red List Categories and Criteria Version 3.1. Under a more pessimistic scenario, with ~40 % probability (SD 1.25 %, CI 37.7–41.3 %), the model projection suggests that the PRE humpback dolphins will decline by more than 80 % of current population numbers within 51 years (three generations).

One might argue that the PRE humpback dolphins may not be at such risk as they are likely to represent an open population with a continuous influx of individuals from peripheral areas. However, being the largest population in the region, the PRE

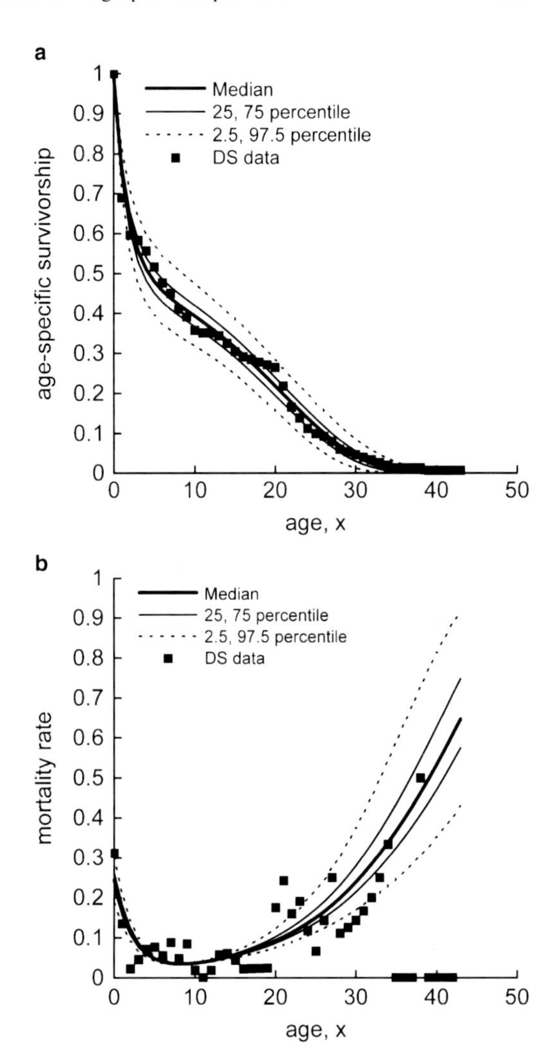

Fig. 13.4 Age-specific survivorship $l(x)$ (**a**) and mortality rates $q(x)$ (**b**) of humpback dolphins from the Pearl River Estuary (PRE) based on empirical (*dotted*) and modeled fitted (*line*) calculations. *DS data* represents data acquired from aged stranded dolphins. *Dashed lines* represent the 2.5 and 97.5 confidence intervals. (Reproduced from Huang et al. 2012b)

Table 13.2 Demographic parameter estimates for humpback dolphins from the Pearl River Estuary, where T_0=generation time, and r=instantaneous rate of increase, calculated as $r = \frac{\ln\left(\sum l(x)\times m(x)\right)}{T_0}$, where $l(x)$ and $m(x)$ are age-specific survivorships and reproductive rates, respectively

	Mean	SD	Median	CI
T_0	17.01	0.76	16.98	15.81–18.32
r	−0.0249	0.0091	−0.0245	−0.0405 to −0.0107

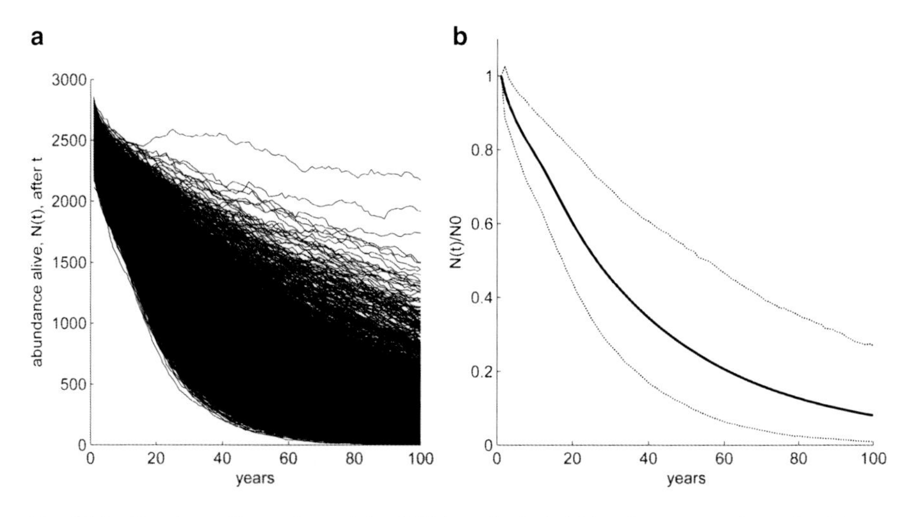

Fig. 13.5 Abundance fluctuations of the PRE humpback dolphin after *t* years, shown by (**a**) stochastic plots that illustrate variation in prediction and (**b**) deterministic plot of median (*solid lines*) and CI (*dashed lines*) of percentage of population alive. (Reproduced from Huang et al. 2012b)

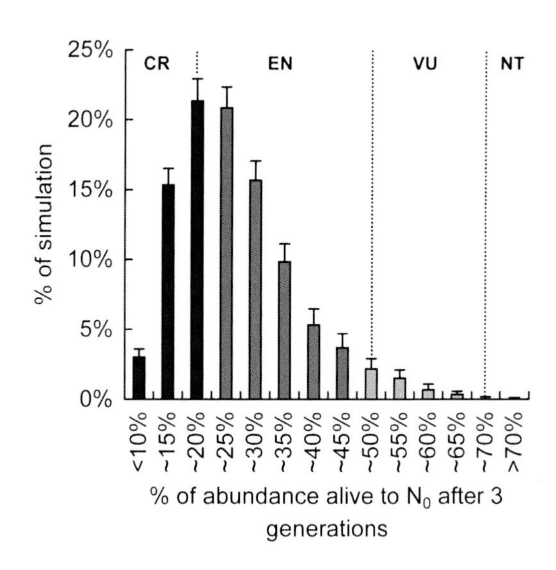

Fig. 13.6 Percent distribution (% +SD) of abundance alive after three generations, indicating percentage of simulations meeting the criteria for classification as critically endangered (CR, 39.33 %), endangered (EN, 57.60 %), vulnerable (VU, 2.89 %), or near-threatened (NT, 0.05 %) for rate of decline (Criterion A3b: IUCN 2001). (Reproduced from Huang et al. 2012b)

population may actually act more as a "source" rather than "sink," with many individuals moving to neighboring areas and populations and seemingly persisting on a larger spatiotemporal scale. Under such a meta-population scenario, the numbers of the PRE humpback dolphins may in fact decline more rapidly than our current predictions.

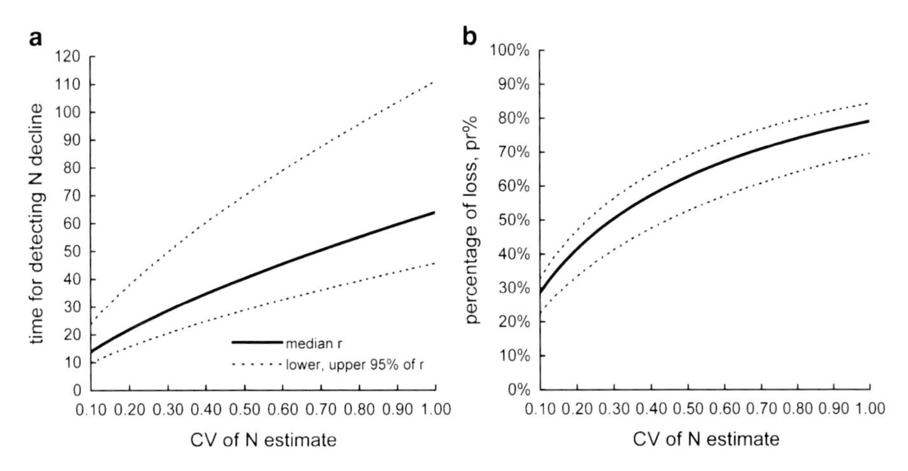

Fig. 13.7 Power analysis showing years needed to detect population trends through annual abundance estimates (**a**) and the percentage (*pr%*) of abundance change (decline) (**b**) at T_D at different scales of CV of abundance estimates using the estimated *r*. (Reproduced from Huang et al. 2012b)

Another fallacy that underestimates the risk of decline of the PRE humpback dolphins may derive from the lack of direct evidence of a decline in abundance at present (i.e., observed versus predicted model validation), especially from periodic abundance estimates. The population status usually trends downward far before the signs of population decline become evident in a deteriorating environment (Drake and Griffen 2010; Huang et al. 2012a). The power of detecting the signs of abundance decline by traditional transect techniques used for cetaceans depends on the variation (CV) of abundance estimates and rate of decline (Gerrodette 1987; Taylor and Gerrodette 1993; Thompson et al. 2000), and it is low and typically requires decades of abundance estimates, or a rapid population decline to detect trends in a shorter time span. The detectable rate of population decline within a reasonable time span, not longer than 10 years of investigation, would have to exceed 5 % of abundance annually for populations with an abundance comparable to PRE humpback dolphins (Taylor and Gerrodette 1993). For cetacean species such as humpback dolphins that have a generation length exceeding 17 years, or longer (Taylor et al. 2007a; Huang et al. 2012b), a 5 % decline in abundance annually would result in an abundance loss greater than 92.6 % after three generations, exceeding the criteria for classification as Critically Endangered status (Criterion A3b: IUCN 2001). For the PRE humpback dolphins, the calculated number of years needed to detect a population trend through periodic abundance estimates (T_D) ranged from 13.8 (when CV of abundance estimate was 10 %) to 63.8 years (when CV = 100 %) with median *r* estimates (Fig. 13.7). The percentage of decline after T_D ranged from 28.6 % (CV = 10 %) to 79 % (CV = 100 %) of the current abundance. Analyses using traditional transect sampling techniques of abundance estimations, even numerically "long term," are very unlikely to detect the current declining trend within one generation (17 years) as the current census dataset itself has a CV higher than 30 % (Chen et al. 2010)[1]

The Pearl River Estuary region is one of the fastest growing economic regions in the world, a process that is likely accompanied by increasing anthropogenic pressures on a wide variety of biota. Large-scale projects that result in land reclamation, dredging, intense boat traffic, and other impacts increase the incidental mortality of the PRE humpback dolphins directly and indirectly (Jefferson 2000; Reeves et al. 2008; Jefferson et al. 2009). The efflux of persistent organic pollutants into PRE waters, such as PCBs (polychlorinated biphenyls), OCPs (organochlorine pesticides), PBDEs (polybrominated diphenyl ethers), and heavy metals, shows a worrying trend (Minh et al. 1999; Jefferson 2000; Parsons 2004; Leung et al. 2005; Xing et al. 2005; Hung et al. 2006) that likely increases the vulnerability of this population. Many of these persistent organic pollutants are thought to be endocrine disruptive with further demographic and developmental consequences (Birnbaum 1994; Guillette et al. 1994; Cheek and McLachlan 1998; Danzo 1998; Vartiainen et al. 1999; Crews et al. 2000). The cumulative effect of these impacts can gradually decrease population survival rates.

The technique described here, the life-table analysis based on the age-structure of collected carcasses, cannot detect recent trends in survival or mortality because the samples were collected over a period of more than 10 years; producing estimates of survival in which temporal changes are conflated with sampling error. Although a comparative study of life-table structure in a temporal scale could contrast demographic consequences on population survival with increasing anthropogenic pressures (Currey et al. 2009b; Mei et al. 2012), comparable methods cannot be applied here because of the lack of long-term life-table data (as in Mei et al. 2012).

If the current environmental conditions of the PRE continue to worsen without effective mitigation and management measures, the rate of decline of humpback dolphins in the region is likely to accelerate. Direct evidence supporting such concerns comes from the accelerating decline of the Yangtze finless porpoise mentioned earlier. The rate of decline was slow before the mid-1990s (Yang et al. 1998) but accelerated rapidly within a decade (Wei et al. 2002; Wang et al. 2005; Zhao et al. 2008; Mei et al. 2012), which directly corresponds with the fast economic growth and large-scale developments along the middle and lower reaches of the Yangtze River (Wang et al. 2011). The PRE humpback dolphins are facing a similar increase in anthropogenic disturbance from rapid human population growth and economic growth in nearby regions, including resource depletion, accumulation of pollutants, habitat destruction, and alteration of hydrological patterns (Jefferson and Hung 2004; Dudgeon et al. 2006; Reeves et al. 2008; Jefferson et al. 2009; Kreb et al. 2010). The history of population decline on the Yangtze finless porpoise, chronicled by a long-term (36 years) census investigation, provides a precautionary

[1] At the time of going to print, the Agriculture, Fisheries and Conservation Department of the Hong Kong Government announced that first signs of decline have been detected from the periodic line transect surveys; suggesting that a population decline of a considerable magnitude must have been taking place for a long time, as our demographic model predicts.

warning for the persistence of PRE humpback dolphins under the deteriorating state of the environment.

Recent questionnaire surveys conducted among fisherman residing along the 500-km coast between the PRE and the coastal city of Xiamen indicate that historically humpback dolphins occurred continuously throughout this range and apparently in larger numbers (Wang et al. 2010). However, much of the coastal habitat between the PRE and Xiamen has been heavily degraded in recent decades, making much of the area unsuitable for humpback dolphins, which effectively fragments the previously continuous distribution to relict populations in the PRE and, considerably smaller, in the coastal region of Xiamen. Sightings of humpback dolphins between the PRE and Xiamen have been very infrequent in recent years (Wang et al. 2010). Comparable pressure from habitat degradation and fragmentation can also be expected to occur in waters neighboring western reaches of the PRE as the result of similarly high economic and population growth (Zhou et al. 2007; Chen et al. 2009).

13.6 Status and Risk Assessment Under Scarcity of Data: The Likely Risk of Underestimating the Extinction Risk

The current IUCN Red List of Threatened Species lists humpback dolphins globally as "Near Threatened (NT)," with the Taiwan population regionally sublisted as "Critically Endangered (CR)" (Reeves et al. 2008). This classification understates the threats faced by the PRE humpback dolphins, and likely to a greater extent those faced by the fragmented and small populations off southeast Asia and western Indian Ocean, especially off the east coast of Africa.

In the Pearl River Estuary region, even though genetic evidence suggests historical exchange between populations in the PRE and off Xiamen (Chen et al. 2008a, b), this is unlikely to persist to date because of drastic habitat degradation (Wang et al. 2010), which defines the eastern boundary of the PRE humpback dolphin population to within Hong Kong waters (Hung 2008). The range of the PRE humpback dolphins is unlikely to exceed an area of 20,000 km^2 and might even be confined to less than 5,000 km^2 (Chen et al. 2010). The western boundary of this population, however, remains undetermined (Chen et al. 2010, 2011). With the current evidence of abundance decline, this population should be classified as Vulnerable (extent of occurrence $<20,000$ km^2) or even Endangered (extent of occurrence $<5,000$ km^2) under Criterion B1v (IUCN 2001). The number of mature individuals among the PRE humpback dolphins is unlikely to exceed 2,500; as the estimate of total N approximates 2,500 animals, and with the projected rate of decline of 2.5 % annually, the estimated 59 % (SD 12.6 %) decline after two generations meets the classification criteria as Endangered under Criterion C1 (IUCN 2001). Consequently, the classification of the largest humpback dolphin population as Endangered (58 % of simulations) appears to be the most appropriate representation of its current conservation status under IUCN Criteria A3b, B1v, and C1 (IUCN 2001), whereas a classification as Critically Endangered (39 % of simulations) might also be considered.

In Maputo Bay, Mozambique, the estimate of calf survival rate (0.53; Guissamulo and Cockcroft 2004) is lower than that of the PRE humpback dolphins (0.61; Jefferson et al. 2011). Considering the similarity of life history patterns in both areas, the population decline of humpback dolphins in Maputo Bay is likely to occur faster than in the PRE, possibly exceeding 80 % within three generations. In another location, Richards Bay, an estuarine habitat on the subtropical KwaZulu-Natal coast of South Africa, 370 km south of Maputo Bay, survival rates of neither calves (*Sc*) nor non-calf animals (*Sa*) are known, but early studies indicate a high incidental mortality rate approximating 4.5 % per annum, caused by shark net entanglement (Cockcroft 1990; Durham 1994). A simple comparison of this incidental mortality rate with an undisturbed *Sa* estimate (0.95; Karczmarski 2000; Jefferson and Karczmarski 2001; Taylor et al. 2007a) suggests that the rate of population decline in Richards Bay could be extremely high.

Both Maputo Bay and Richards Bay humpback dolphins can temporarily "maintain" their numbers by influx from other populations/areas (Guissamulo 2008; S. Atkins and L. Karczmarski, unpublished data). In the long term, however, the overall population figures are alarmingly low (Durham 1994; Atkins and Atkins 2002; Keith et al. 2002; Guissamulo and Cockcroft 2004; see Table 13.1). Other known populations in the region, such as in Algoa Bay, 880 km south of Richards Bay (Karczmarski et al. 1999) or Inhambane Bay and Bazaruto archipelago, 420 and 640 km north of Maputo Bay, respectively (V.G. Cockcroft, A.T. Guissamulo, and L. Karczmarski, unpublished data) are geographically distant and neither large enough nor sufficiently demographically productive (Karczmarski 2000; Taylor et al. 2007a) to compensate for any rapid decline in the sink populations. A similar situation is being repeated throughout the coastal region off east Africa. With few remaining populations, ranging in size from a few tens to the low few hundred individuals (see Table 13.1), separated by distances as great as several hundred kilometers (Reeves et al. 2008; V.G. Cockcroft, A.T. Guissamulo and L. Karczmarski, unpublished data), the regional trend of humpback dolphins in the western Indian Ocean is likely to be declining much faster than currently recognized (Reeves et al. 2008). The forecast for the *plumbea* humpback dolphins in the western Indian Ocean may be even worse than for the *chinensis* form in the PRE and off Southeast Asia. Relevant population models would be informative in assessing the status and predicting trends and should be applied whenever existing data allow. In the absence of reliable estimates, the precautionary conservation principle needs to be applied and should be urgently implemented (Thompson et al. 2000; Huang et al. 2012a). Inaction can result in substantial population loss before any conclusive evidence from the data currently collected becomes available. A revision of the current IUCN conservation status of humpback dolphins should be a first such step toward an appropriate conservation strategy.

There are a number of considerable concerns in regard to humpback dolphin conservation, both in the industrialized world such as Hong Kong and the Pearl River Delta, where substantial funding has been unsuccessful in generating scientifically sound management schemes, and in developing parts of the world such as Southern and East Africa, where scarcity of funding seldom allows long-term sophisticated research. In both cases, effective conservation measures are long overdue.

It is imperative that action be taken immediately. We urge that the current IUCN conservation status of humpback dolphins be reconsidered, with more discriminative treatment of the *chinensis* form and the *plumbea* form (as proposed to IUCN by L. Karczmarski, 2007[2]), because the current status underestimates the local threats to humpback dolphins across their severely fragmented range in both the Indo-Pacific (*chinensis* form) and western Indian Ocean (*plumbea* form). An ongoing large-scale genetic study (M. Mendez and H. Rosenbaum, personal communication) may soon provide evidence for taxonomic revisions. In the meantime, in lieu of other evidence to the contrary, we believe the precautionary principle toward risk assessment should be applied and act as a pragmatic approach to conservation politics. As the remaining largest population of humpback dolphins currently faces a major risk of local extirpation, such risk is likely to be only magnified in many of the much smaller populations elsewhere. We suggest that reconsidering the individual status of the *plumbea* and *chinensis* forms and updating the current IUCN listing is urgently needed as it would far more appropriately reflect their current conservation status and threats to the future survival of local populations.

Acknowledgments S.-L. Huang was funded by the National Science Council of Taiwan (grant NSC 101-2311-B-019-002). L. Karczmarski was supported by the Research Grant Council of Hong Kong (GRF grant HKU768110M), the University Professorial Sponsorship Programme of the Ocean Park Conservation Foundation Hong Kong (OPCFHK), and The University of Hong Kong (HKU) Funding Programme for Basic Research. We kindly thank Cynthia Yau and Glenn Gailey for valuable comments on the early draft of the manuscript.

References

Atkins S, Atkins BL (2002) Abundance and site fidelity of Indo-Pacific humpback dolphins (*Sousa chinensis*) at Richards Bay, South Africa. Working paper SC/54/SM25 of the 54th meeting of the International Whaling Commission, Shimonoseki, April–May 2002

Barlow J, Boveng P (1991) Modeling age-specific mortality for marine mammal populations. Mar Mamm Sci 7:50–65

Bearzi G, Holcer D, di Sciara GN (2004) The role of historical dolphin takes and habitat degradation in shaping the present status of northern Adriatic cetaceans. Aquat Conserv Mar Freshw Ecosyst 14:363–379

Bearzi G, Agazzi S, Gonzalvo J, Bonizzoni S, Costa M, Petroselli A (2010) Biomass removal by dolphins and fisheries in a Mediterranean Sea coastal area: do dolphins have an ecological impact on fisheries? Aquat Conserv Mar Freshw Ecosyst 20:549–559

Birnbaum LS (1994) Endocrine effects of prenatal exposure to PCBs, dioxins, and other xenobiotics: implications for policy and future research. Environ Health Perspect 102:676–679

Cagnazzi DDB, Harrison PL, Ross GJB, Lynch P (2011) Abundance and site fidelity of Indo-Pacific humpback dolphins in the Great Sandy Strait, Queensland, Australia. Mar Mamm Sci 27:255–281

Caswell H, Brault S, Read AJ, Smith TD (1998) Harbor porpoise and fisheries: an uncertainty analysis of incidental mortality. Ecol Appl 8:1226–1238

[2]Global Mammal Assessment: Cetacean Red List Assessment. IUCN Species Survival Commission, Cetacean Specialist Group, 22–26 January 2007, La Jolla, California, USA.

Caswell H, Fujiwara M, Brault S (1999) Declining survival probability threatens the North Atlantic right whale. Proc Natl Acad Sci USA 96:3308–3313

Caughley G (1966) Mortality patterns in mammals. Ecology 47:906–918

Chaloupka M, Osmond M, Kaufman G (1999) Estimating seasonal abundance trends and survival probabilities of humpback whales in Hervey Bay (east coast Australia). Mar Ecol Prog Ser 184:291–301

Cheek AO, McLachlan JA (1998) Environmental hormones and the male reproductive system. J Androl 19:5–10

Chen B, Zheng D, Zhai F, Xu X, Sun P, Wang Q, Yang G (2008a) Abundance, distribution and conservation of Chinese White Dolphins (*Sousa chinensis*) in Xiamen, China. Mamm Biol 73:156–164

Chen H, Khai K, Chen J, Wen H, Chen S, Wu Y (2008b) A preliminary investigation on genetic diversity of *Sousa chinensis* in the Pearl River Estuary and Xiamen of Chinese waters. J Genet Genomics 35:491–497

Chen B, Zheng D, Yang G, Xu X, Zhou K (2009) Distribution and conservation of the Indo-Pacific humpback dolphin in China. Integr Zool 4:240–247

Chen T, Hung SK, Qiu Y, Jia X, Jefferson TA (2010) Distribution, abundance, and individual movements of Indo-Pacific humpback dolphins (*Sousa chinensis*) in the Pearl River Estuary, China. Mammalia 74:117–125

Chen T, Qiu Y, Jia X, Hung SK, Liu W (2011) Distribution and group dynamics of Indo-Pacific humpback dolphins (*Sousa chinensis*) in the western Pearl River Estuary, China. Mamm Biol 76:93–96

Chevé M (2000) Irreversibility of pollution accumulation. Environ Resour Econ 16:93–104

Cockcroft VG (1990) Dolphin catches in the Natal shark nets, 1980 to 1988. S Afr J Wildl Res 20:44–51

Corkeron PJ, Morissette NM, Porter L, Marsh H (1997) Distribution and status of humpback dolphins, *Sousa chinensis*, in Australian waters. Asian Mar Biol 14:49–59

Crews D, Willingham E, Skipper JK (2000) Endocrine disruptors: present issues, future directions. Q Rev Biol 75:243–260

Currey RJC, Dawson SM, Slooten E (2009a) An approach for regional threat assessment under IUCN Red List criteria that is robust to uncertainty: the Fiordland bottlenose dolphins are critically endangered. Biol Conserv 142:1570–1579

Currey RJC, Dawson SM, Slooten E, Schneider K, Lusseau D, Boisseau OJ, Hasse P, Williams JA (2009b) Survival rates for a declining population of bottlenose dolphins in Doubtful Sound, New Zealand: an information theoretic approach to assessing the role of human impacts. Aquat Conserv Mar Freshw Ecosyst 19:658–670

Dans SL, Alonso MK, Pedraza SN, Crespo EA (2003) Incidental catch of dolphins in trawling fisheries off Patagonia, Argentina: can populations persist? Ecol Appl 13:754–762

Danzo BJ (1998) The effects of environmental hormones on reproduction. Cell Mol Life Sci 54:1249–1264

Drake JM, Griffen BD (2010) Early warning signals of extinction in deteriorating environments. Nature (Lond) 467:456–459

Dudgeon D, Arthington AH, Gessner MO, Kawabata Z-I, Knowler DJ, Lévêque C, Naiman RJ, Prieur-Richard A-H, Soto D, Stiassny MLJ, Sullivan CA (2006) Freshwater biodiversity: importance, threats, status and conservation challenges. Biol Rev 81:163–182

Durham B (1994) The distribution and abundance of the humpback dolphin (*Sousa chinensis*) along the Natal coast, South Africa. M.Sc. thesis, University of Natal, South Africa

Frère CH, Hale PT, Porter L, Cockcroft VG, Dalebout ML (2008) Phylogenetic analysis of mtDNA sequences suggests revision of humpback dolphin (*Sousa* spp.) taxonomy is needed. Mar Freshw Res 59:259–268

Fujiwara M, Caswell H (2001) Demography of the endangered North Atlantic right whale. Nature (Lond) 414:537–541

Gaillard J-M, Festa-Bianchet M, Yoccoz NG (1998) Population dynamics of large herbivores: variable recruitment with constant adult survival. Trends Ecol Evol 13:58–63

Gerrodette T (1987) A power analysis for detecting trends. Ecology 68:1364–1372

Guillette LJJ, Gross TS, Masson GR, Matter JM, Percival HF, Woodward AR (1994) Developmental abnormalities of the gonad and abnormal sex hormone concentrations in juvenile alligators from contaminated and control lakes in Florida. Environ Health Perspect 102:680–688

Guissamulo A (2008) Ecological studies of bottlenose and humpback dolphins in Maputo Bay, southern Mozambique. Ph.D. thesis, University of Kwazulu-Natal, South Africa

Guissamulo A, Cockcroft VG (2004) Ecology and population estimates of Indo-Pacific humpback dolphins (*Sousa chinensis*) in Maputo Bay, Mozambique. Aquat Mamm 30:94–102

Hamilton PK, Knowlton AR, Marx MK, Kraus SD (1998) Age structure and longevity in north Atlantic right whales, *Eubalaena glacialis*, and their relation to reproduction. Mar Ecol Prog Ser 171:285–293

Harwood J (2000) Risk assessment and decision analysis in conservation. Conserv Biol 95:219–226

Huang S-L, Ni I-H, Chou L-S (2008) Correlations in cetacean life history traits. Raffles Bull Zool (Suppl) 19:285–292

Huang S-L, Hao Y, Mei Z, Turvey ST, Wang D (2012a) Common pattern of population decline for freshwater cetacean species in deteriorating habitats. Freshw Biol 57:1266–1276

Huang S-L, Karczmarski L, Chen J, Zhou R, Lin W, Zhang H, Li H, Wu Y-P (2012b) Demography and population trend of the largest population of Indo-Pacific humpback dolphin (*Sousa chinensis*). Biol Conserv 147:234–242

Hung SK (2008) Habitat use of Indo-Pacific humpback dolphins (*Sousa chinensis*) in Hong Kong. Ph.D. dissertation, University of Hong Kong, Hong Kong

Hung CLH, Xu Y, Lam JCW, Jefferson TA, Hung SK, Yeung LWY, Lam MHW, O'Toole DK, Lam PKS (2006) An assessment of the risks associated with polychlorinated biphenyls found in the stomach contents of stranded Indo-Pacific humpback dolphins (*Sousa chinensis*) and finless porpoises (*Neophocaena phocaenoides*) from Hong Kong waters. Chemosphere 63:845–852

IUCN (2001) The Red List categories and criteria, version 3.1. http://www.iucnredlist.org/technical-documents/categories-and-criteria/2001-categories-criteria

Jefferson TA (2000) Population biology of the Indo-Pacific hump-backed dolphin in Hong Kong waters. Wildl Monogr 144:1–65

Jefferson TA, Hung SK (2004) A review of the status of the Indo-Pacific humpback dolphin (*Sousa chinensis*) in Chinese waters. Aquat Mamm 30:149–158

Jefferson TA, Karczmarski L (2001) *Sousa chinensis*. Mamm Species 655:1–9

Jefferson TA, Hung SK, Würsig B (2009) Protecting small cetaceans from coastal development: impact assessment and mitigation experience in Hong Kong. Mar Policy 33:305–311

Jefferson TA, Hung SK, Robertson KM, Archer FI (2011) Life history of the Indo-Pacific humpback dolphin in the Pearl River Estuary, southern China. Mar Mamm Sci. doi:10.1111/j.1748-7692.2010.00462.x

Karczmarski L (2000) Conservation and management of humpback dolphins: the South African perspective. Oryx 34:207–216

Karczmarski L, Winter PED, Cockroft VG, McLachlan A (1999) Population analysis of Indo-Pacific humpback dolphins *Sousa chinensis* in Algoa Bay, Eastern Cape, South Africa. Mar Mamm Sci 15:1115–1123

Karczmarski L, Cockroft VG, McLachlan A (2000) Habitat use and preferences of Indo-Pacific humpback dolphins *Sousa chinensis* in Algoa Bay, South Africa. Mar Mamm Sci 16:65–79

Keith M, Peddemors VM, Bester MN, Ferguson JWH (2002) Population characteristics of Indo-Pacific humpback dolphins at Richards Bay, South Africa: implications for incidental capture in shark nets. S Afr J Wildl Res 32:153–162

Koschinski S (2002) Current knowledge on harbour porpoises (*Phocoena phocoena*) in the Baltic Sea. Ophelia 55:167–197

Kreb D, Reeves RR, Thomas PO, Braulik GT, Smith BD (2010) Establishing protected areas for Asian freshwater cetaceans: freshwater cetaceans as flagship species for integrated river conservation management, Samarinda, 19–24 October 2009. Yayasan Konservasi RASI, Samarinda

Krebs CJ (1989) Ecological methodology. Harper Collins, New York

Lacy RC (1993) VORTEX: a computer simulation model for population viability analysis. Wildl Res 20:45–65

Leung CCM, Jefferson TA, Hung SK, Zheng GJ, Yeung LWY, Richardson BJ, Lam PKS (2005) Petroleum hydrocarbons, polycyclic aromatic hydrocarbons, organochlorine pesticides and polychlorinated biphenyls in tissues of Indo-Pacific humpback dolphins from south China waters. Mar Pollut Bull 50:1713–1719

MacKinnon J, Verkuil YI, Murray N (2012) IUCN situation analysis on East and Southeast Asian intertidal habitats, with particular reference to the Yellow Sea (including the Bohai Sea). IUCN Species Survival Commission, occasional paper no. 47. IUCN, Gland

Mei Z, Huang S-L, Hao Y, Turvey ST, Gong W, Wang D (2012) Accelerating population decline of Yangtze finless porpoise (*Neophocaena asiaeorientalis asiaeorientalis*). Biol Conserv 153:192–200

Minh TB, Watanabe M, Nakata H, Tanabe S, Jefferson TA (1999) Contamination by persistent organochlorines in small cetaceans from Hong Kong coastal waters. Mar Pollut Bull 39:383–392

Moore JE, Read AJ (2008) A Bayesian uncertainty analysis of cetacean demography and bycatch mortality using age-at-death data. Ecol Appl 18:1914–1931

Parra GJ, Ross GJB (2009) Humpback dolphins *S. chinensis* and *S. teuszii*. In: Perrin WF, Würsig B, Thewissen JGM (eds) Encyclopedia of marine mammals, 2nd edn. Academic/Elsevier, San Diego, pp 576–582

Parra GJ, Corkeron PJ, Marsh H (2004) The Indo-Pacific humpback dolphin, *Sousa chinensis* (Osbeck, 1765), in Australian waters: a summary of current knowledge. Aquat Mamm 30:197–206

Parra GJ, Corkeron PJ, Marsh H (2006) Population sizes, site fidelity and residence patterns of Australian snubfin and Indo-Pacific humpback dolphins: implications for conservation. Biol Conserv 129:167–180

Parsons ECM (2004) The potential impacts of pollution on humpback dolphins, with a case study on the Hong Kong population. Aquat Mamm 30:18–37

Piroddi C, Bearzi G, Gonzalvo J, Christensen V (2011) From common to rare: the case of the Mediterranean common dolphin. Biol Conserv 144:2490–2498

Reeves RR, Dalebout ML, Jefferson TA, Karczmarski L, Laidre K, O'Corry-Crowe G, Rojas-Bracho L, Secchi ER, Slooten E, Smith BD, Wang JY, Zhou K (2008) *Sousa chinensis*. In: IUCN 2013. IUCN red list of threatened species. Version 2013.1. http://www.iucnredlist.org/details/20424. Downloaded on 05 September 2013

Rice DW (1998) Marine mammals of the world: systematics and distribution. Special publication no. 4. Society for Marine Mammalogy, New Zealand

Ross GJB, Heinsohn GE, Cockcroft VG (1994) Humpback dolphins *Sousa chinensis* (Osbeck, 1765), *Sousa plumbea* (G. Cuvier, 1829) and *Sousa teuszii* (Kukenthal, 1892). In: Ridgway SH, Harrison R (eds) Handbook of marine mammals, vol 5. The first book of dolphins. Academic, London, pp 23–42

Ross PS, Dungan SZ, Hung SK, Jefferson TA, Macfarquhar C, Perrin WF, Riehl KN, Slooten E, Tsai J, Wang JY, White BN, Würsig B, Yang SC, Reeves RR (2010) Averting the baiji syndrome: conserving habitat for critically endangered dolphins in Eastern Taiwan Strait. Aquat Conserv Mar Freshw Ecosyst 20:685–694

Siler W (1979) A competing risk model for animal mortality. Ecology 60:750–757

Slooten E, Fletcher D, Taylor BL (2000) Accounting for uncertainty in risk assessment: case study of Hector's dolphin mortality due to gillnet entanglement. Conserv Biol 14:1264–1270

Stensland E, Carlén I, Särnblad A, Bignert A, Berggren P (2006) Population size, distribution, and behavior of Indo-Pacific bottlenose (*Tursiops aduncus*) and humpback (*Sousa chinensis*) dolphins off the south coast of Zanzibar. Mar Mamm Sci 22:667–682

Stevick PT, Allen J, Clapham PJ, Friday N, Katona SK, Larsen F, Lien J, Mattila D, Palsboll PJ, Sigurjonsson J, Smith TD, Oien N, Hammond PS (2003) North Atlantic humpback whale abundance and rate of increase four decades after protection from whaling. Mar Ecol Prog Ser 258:263–273

Stolen MK, Barlow J (2003) A model life table for bottlenose dolphins (*Tursiops truncatus*) from the Indian River Lagoon System, Florida, U.S.A. Mar Mamm Sci 19:630–649

Sutaria D, Jefferson TA (2004) Records of indo-pacific humpback dolphins (*Sousa chinensis*, Osbeck, 1765) along the coasts of India and Sri Lanka: an overview. Aquat Mamm 30:125–136
Taylor BL, Gerrodette T (1993) The uses of statistical power in conservation biology: the vaquita and northern spotted owl. Conserv Biol 7:489–500
Taylor BL, Wade PR, de Master DP, Barlow J (2000) Incorporating uncertainty into management models for marine mammals. Conserv Biol 14:1243–1252
Taylor BL, Chivers SJ, Larese J, Perrin WF (2007a) Generation length and percent mature estimates for IUCN assessments of cetaceans. Administrative report LJ-07-01. Southwest Fisheries Science Center, La Jolla
Taylor BL, Martinez M, Gerrodette T, Barlow J, Hrovat YN (2007b) Lessons from monitoring trends in abundance of marine mammals. Mar Mamm Sci 23:157–175
Thompson PM, Wilson B, Grellier K, Hammond PS (2000) Combining power analysis and population viability analysis to compare traditional and precautionary approaches to conservation of coastal cetaceans. Conserv Biol 14:1253–1263
Vartiainen T, Kartovaara L, Tuomisto J (1999) Environmental chemicals and changes in sex ratio: analysis over 250 years in Finland. Environ Health Perspect 107:813–815
Verborgh P, de Stephanis R, Pérez S, Jaget Y, Barbraud C, Guinet C (2009) Survival rate, abundance, and residency of long-finned pilot whales in the Strait of Gibraltar. Mar Mamm Sci 25:523–536
Wade PR (1998) Calculating limits to the allowable human-caused mortality of cetaceans and pinnipeds. Mar Mamm Sci 14:1–37
Wang D, Hao Y, Wang K, Zhao Q, Chen D, Wei Z, Zhang X (2005) Aquatic resource conservation: the first Yangtze finless porpoise successfully born in captivity. Environ Sci Pollut Res Int 12:247–250
Wang K, Wang D, Zhang X, Pfluger A, Barrett L (2006) Range-wide Yangtze freshwater dolphin expedition: the last chance to see Baiji? Environ Sci Pollut Res 13:418–424
Wang JY, Yang SC, Hung SK, Jefferson TA (2007) Distribution, abundance and conservation status of the eastern Taiwan Strait population of Indo-Pacific humpback dolphins, *Sousa chinensis*. Mammalia 71:157–165
Wang X, Yan C, Zhu Q (2010) Distribution and historical decline process of Chinese white dolphin from Xiamen to the Pearl River Estuary. In: International workshop on population connectivity and conservation of *Sousa chinensis* off Chinese coast, Nanjing, 4–7 June
Wang JH, Wei QW, Zou YC (2011) Conservation strategies for the Chinese sturgeon, *Acipenser sinensis*: an overview on 30 years of practices and future needs. J Appl Ichthyol 27:176–180
Wei Z, Wang D, Kuang X, Wang K, Wang X, Xiao J, Zhao Q, Zhang X (2002) Observations on behavior and ecology of the Yangtze finless porpoise (*Neophocoena phocaenoides asiaeorientalis*) group at Tian-e-Zhou Oxbow of the Yangtze River. Raffles Bull Zool (Suppl 10):97–103
Wilson B, Hammond PS, Thompson PM (1999) Estimating size and assessing trends in a coastal bottlenose dolphin population. Ecol Appl 9:288–300
Wilson B, Porter L, Gordon J, Hammond PS, Hodgins N, Wei L, Lin J, Lusseau D, Tsang A, van Vaerebeek K, Wu YP (2008) A decade of management plans, conservation initiatives and protective legislations for Chinese white dolphin (*Sousa chinensis*): an assessment of progress and recommendations for future management strategies in the Pearl River Estuary, China. Report. WWF, Hong Kong, 7–11 April
Winship AJ, Trites AW (2006) Risk of extirpation of Steller sea lions in the Gulf of Alaska and Aleutian Islands: a population viability analysis based on alternative hypotheses for why sea lions declined in western Alaska. Mar Mamm Sci 22:124–155
Xing Y, Lu Y, Dawson RW, Shi Y, Zhang H, Wang T, Liu W, Ren H (2005) A spatial temporal assessment of pollution from PCBs in China. Chemosphere 60:731–739
Yang G, Zhou K, Gao A, Chang Q (1998) A study on the life table and dynamics of three finless porpoise populations in the Chinese waters. Acta Theriol Sin 18:1–7
Yu H-Y, Lin T-H, Chang W-L, Chou L-S (2010) Using the mark-recapture method to estimate the population size of Sousa chinensis in Taiwan. In: Workshop on population connectivity and conservation of Sousa chinensis off Chinese Coast, Nanjing, p 13

Zhang X, Liu R, Zhao Q, Zhang G, Wei Z, Wang X, Yang J (1993) Populations of the finless porpoise in the middle and lower reaches of the Yangtze River, China. Acta Theriol Sin 13:260–270

Zhao X, Barlow J, Taylor BL, Pitman RL, Wang K, Wei Z, Stewart BS, Turvey ST, Akamatsu T, Reeves RR, Wang D (2008) Abundance and conservation status of the Yangtze finless porpoise in the Yangtze River, China. Biol Conserv 141:3006–3018

Zhou K, Xu X, Tian C (2007) Distribution and abundance of Indo-Pacific humpback dolphins in Leizhou Bay, China. N Z J Zool 34:35–42

Chapter 14
Mountain Gorillas: A Shifting Demographic Landscape

Elizabeth A. Williamson

Adult male "silverback" mountain gorilla with infant. (© David Pluth)

Abstract Large-scale habitat destruction and poaching in the 1950s, 1960s, and 1970s had major impacts on the population size and demography of mountain gorillas (*Gorilla beringei beringei*) in Rwanda. In those three decades, the population of the Virunga Volcanoes was halved: groups became unstable, and infanticide was relatively common. Intensive conservation efforts began in the 1980s and have

E.A. Williamson (✉)
Psychology, School of Natural Sciences, University of Stirling, Scotland, UK
e-mail: e.a.williamson@stir.ac.uk

J. Yamagiwa and L. Karczmarski (eds.), *Primates and Cetaceans: Field Research and Conservation of Complex Mammalian Societies*, Primatology Monographs,
DOI 10.1007/978-4-431-54523-1_14, © Springer Japan 2014

enabled the gorilla population to recover. The present study took place during a period of social stability in the lives of three Karisoke gorilla groups. Characterized by few female transfers, no known infanticide, and only one silverback male departure from the research groups, there were striking increases in both group size and the number of adult males per group. I consider how these changes have occurred and implications for the management of this Critically Endangered primate. Despite encouraging growth, this population is so small that it remains extremely vulnerable to human disturbance. If mountain gorillas are to survive in this volatile region, a hands-on approach to their conservation may be justified.

Keywords Demography • *Gorilla beringei* • Group formation • Karisoke • Rwanda • Socioecology • Solitary silverback • Virunga

14.1 Mountain Gorilla Social Systems

Mountain gorillas (*Gorilla beringei beringei*) are characterized as living in stable, cohesive, polygynous groups typically composed of one adult male "silverback" and a median of five adult females, together with their offspring (Harcourt and Stewart 2007). Silverback is the term used for mature male gorillas from the age of 12 years, when hairs on the saddle of the back start to turn grey, and some individuals are already capable of siring an infant at that age (Bradley et al. 2005). Silverbacks are considered to be fully grown by age 15 (Watts 1991; Watts and Pusey 1993), by which time 50 % of them are likely to have left their natal group and become solitary (Robbins 1995; Watts 2000). The other half of the silverback population does not leave the natal group, and thus groups can become multimale (Watts 2000; Robbins 2001, 2003). Although gorillas are considered to have evolved in a one-male mating system, a significant proportion of mountain gorilla groups contain two or more silverbacks. In their age-graded social system, a hierarchy exists amongst the adult male gorillas, and competition among the males can be intense, particularly when females are in estrus (Harcourt et al. 1980; Robbins 2003). Males that had a strong affiliative relationship with a dominant silverback when they were infants are more likely to remain in their natal group (Harcourt and Stewart 1981), and a few eventually inherit leadership of that group. The accepted model of new group formation is that females transfer out of a group to join a solitary male, or that a breeding group splits permanently into two units.

Nearly 500 mountain gorillas live in the 455-km^2 Virunga Volcanoes of the Democratic Republic of Congo, Rwanda, and Uganda (Gray et al. 2013). The groups studied by the Karisoke Research Centre, in Rwanda, have been habituated to the presence of human observers and can be individually identified. In just one generation, these groups have shown striking increases in both size and the number of adult males per group, with few males emigrating. At the beginning of this study, in January 1996, the research population comprised 73 gorillas in three groups of 19, 23, and 31 individuals, each with one or two adult male silverbacks. Six years

Table 14.1 Composition of three Karisoke mountain gorilla groups at beginning and end of this study

	1 January 1996				31 December 2001			
Gorilla group name	BEE	PAB	SHI	Total	BEE	PAB	SHI	Total
No. individuals	23	31	19	73	27	47	24	98
No. silverback males (>12 years)	2	2	1	5	3	3	6	12
No. blackback males (8–12 years)	2	2	4	8	5	3	3	11
No. adult females (>8 years)	7	14	5	26	7	17	7	31
No. subadults (6–8 years)	2	3	1	6	4	4	2	10
No. juveniles (3.5–6 years)	4	1	5	10	1	6	2	9
No. infants (0–3.5 years)	6	9	3	18	7	14	4	25
Adult female:adult male ratio	3.5	7.0	5.0		2.3	5.7	1.2	

later, the population had grown to 98 gorillas with 24, 27, and 47 individuals in the same three groups, and three to six adult males per group (Williamson and Gerald–Steklis 2001) (Table 14.1). I describe changes in the structure of the Karisoke population, consider how they have occurred, and present some implications for the management of this Critically Endangered primate.

14.2 The Karisoke Mountain Gorilla Population 1996–2001

The 1994 genocide in Rwanda brought a halt to intensive behavioral research on the Karisoke mountain gorillas. During the periods of insecurity and military conflict that followed, great efforts went into monitoring the gorillas and maintaining demographic records, while behavioral data were collected within the constraints of limited access to the gorillas. During this study, researchers were unable to enter the Volcanoes National Park from June 1997 to September 1998 (the park remained closed to tourists until July 1999). Observations were interrupted again between May and August 2001. Despite these difficulties, we consider the demographic records of these groups to be complete, as records of all births, deaths, and transfers have been maintained since 1967. The only demographic events that might have been missed would have been infants that were born during the 16 months when we had no access to the gorillas but which did not survive.

During this study, ten young male gorillas in three research groups matured into silverbacks. Based on our prior knowledge of gorilla demography, we expected four or five of these males to emigrate, but only one left his group to become solitary (at age 13.2 years). One died (aged 14.9 years) following infection of bite wounds inflicted during an aggressive encounter with another gorilla group, and three attained full physical maturity (i.e., reached 15 years of age) in their natal groups (Table 14.2).

This low rate of emigration (10 % in this study compared to 36 % in the preceding three decades; Robbins 2001) was not a consequence of subordinate males being permitted to stay with impunity, as they were subject to aggressive attacks by

Table 14.2 Ages of adult male mountain gorillas in this study

Name	Group	Date of birth	Age 31 December 2001 (years)	Social context
Beetsme	BEE	1 January 1966	(35.4)	Died in June 2001
Titus	BEE	24 August 1974	27.4	Group (dominant)
Pablo	PAB	31 August 1974	27.3	Group (former dominant)
Shinda	SHI	28 February 1977	24.8	Group (dominant)
Cantsbee	PAB	14 November 1978	23.1	Group (dominant)
Ndatwa	BEE	16 February 1985	(14.9)	Died in January 2000
Umurava	PAB	4 January 1986	16.0	Group
Ineza	Formerly PAB	28 January 1986	15.9	Solitary from April 1999
Amahoro	SHI	3 May 1986	15.7	Group
Kuryama	BEE	16 August 1986	15.4	Group
Ntambara	SHI	21 February 1987	14.9	Group
Ugenda	SHI	20 October 1987	14.2	Group
Gwiza	SHI	20 October 1987	14.2	Group
Inshuti	SHI	4 February 1988	13.9	Group
Joli Ami	BEE	23 February 1988	13.9	Group

the dominant male. Behavioral observations were conducted on two of the three Karisoke research groups, one of which (the SHI group) included three to six silverbacks, whereas the other (the PAB group) was unusually large (31–47 individuals during this study), with three silverbacks. Hierarchies among the males were determined using pair-wise displacements recorded between July 1999 and June 2001 (cf. Robbins 1996), a period when access to the park was consistent and data collection was regular. Rank correlated with age for all but two males: the oldest, a post-prime silverback in PAB group, and a young silverback in SHI group who was physically inferior to the rest of his cohort (undersized with a sway back).

14.3 Intragroup Aggression

Analyses of aggression focused on the young competitors over the same 2-year period, excluding a former dominant male (PAB) who had been deposed but not evicted from the group and no longer posed a challenge for leadership. Two levels of physical aggression were distinguished: (1) medium—when a single bite was observed, or one or two wounds recorded (e.g., Fig. 14.1), and (2) severe—when biting was repeated, or visible wounding was extensive. Attacks by the dominant males (CAN and SHI) were almost always directed at one particular individual in their group (UMR in PAB group and NTA in SHI group; NTA was the highest ranking and most physically developed of five young silverbacks). The level of visible wounding of these two silverbacks was not observed with other individuals, and attacks intensified when they reached the ages of 13.5 years (UMR) and 12.4 (NTA), respectively. Attacks classed as "severe" lasted up to 3 min and often resulted in

Fig. 14.1 Subordinate silverback mountain gorilla with a bloody nose after being attacked by the dominant silverback. (© DFGFI)

serious wounding. Gashes 6–15 cm long were seen on top of the head, side of face, back of neck, shoulder, back, inner arm, and hands. During attacks, both protagonists were generally silent, although onlooking females screamed. Victims were often pinned down and did not retaliate or defend themselves. They filled the air with a characteristic pungent male odor (silverbacks emit a musky odor from axillary glands in the armpit in situations of fear or excitement) and sometimes passed diarrheic dung during or immediately after the attack. Adult females occasionally tried to intervene during the dominant silverback's attacks, as did the former dominant male in PAB group on one occasion. In SHI group, the victim's mother once attempted to bite the attacking dominant silverback, but on two occasions other group members joined in the assault.

Intragroup aggression was scored from bite wounds, even if an attack had not been observed, but only when no intergroup interaction had been recorded in the preceding days. Watts (1996) noted that silverbacks receive progressively more aggression from older males as they mature, and severe aggression was recorded during 16 of the last 20 months of this study. SHI made 80 % of these attacks on NTA, despite the presence of four other young silverbacks maturing in the cohort. As the other males aged, SHI's attacks were also directed at the fourth ranking (from the age of 13.5 years) and fifth ranking (from age 13.2 years); however, his aggression toward NTA did not diminish. Interestingly, the third-ranking male (UGE) was the only silverback in the cohort observed to display any affiliative behavior with the dominant male, and he was rarely attacked.

Most attacks seemed to occur without provocation other than proximity. We might expect aggression to increase at times when competition for females was highest, that is, when females were in estrus and would typically copulate repeatedly over a 3- to 4-day period (Czekala and Sicotte 2000). Using dates of parturition, I estimated the number of females in each group that could have been cycling each month during the 2 years. On average, 2 females per group were potentially cycling for a few days each month, although in the large PAB group with 17 adult females, it is possible that 8 females were cycling during the second quarter of 1999. No significant correlation was detected between the presence of cycling females and levels of aggression recorded; however, Watts and Pusey (1993) noted that rates of male–male aggression were influenced not only by the presence and number but also the identity of estrous females. Thus, the proxy used in this study may have been too simplistic.

Despite enduring frequent and severe attacks by the dominant silverback, some maturing males stayed with their group rather than becoming solitary. Thus, there was an obvious cost to remaining in their natal groups, and the question arises: why would these males tolerate such aggression?

14.4 Mating Opportunities and Companionship

Mountain gorilla females reach sexual maturity at about 8 years old and generally have their first infant between the ages of 9 and 10 years. While infants are suckling, their mothers experience lactational anestrus and are unable to conceive again for 3–4 years after giving birth. Therefore, opportunities to mate are rare, even though females may copulate with more than one male (Robbins 2003), and a male's emigration decisions will be influenced by the number of adult females and the number and age of male competitors in his group (Robbins 1995; Watts 2000).

Dominant males do not readily tolerate copulations by subordinate males, and harassment by the dominant of another male or toward the female is well documented (Robbins 1999). Although dominant males participate in the majority of copulations, in this study nine of the ten subordinate silverbacks were seen to copulate at least once with an adult female. On one occasion when NTA was copulating, the dominant male ran at and bit him severely. However, almost all observed copulations by subordinates were surreptitious—copulatory vocalizations were subdued or suppressed—and did not attract the attention of the dominant silverback ($n=32$). Genetic studies have since revealed that subordinate silverbacks occasionally sire offspring (Bradley et al. 2005).

The high ratio of 5.6 adult females per adult male in PAB group reveals a factor likely to have had a strong influence on the three silverbacks remaining together; however, this would not explain the situation in SHI group with six resident males and only 1.2 females per male. Reproductive competition is a key determinant of reproductive strategy, but in this highly gregarious species another factor must also play a role: life in the company of others. Subordinate silverbacks who stay in their natal group have at least an occasional opportunity to mate; they also continue to engage in social interactions such as grooming. To support this hypothesis, we can cite the former existence of two all-male groups at Karisoke: male gorillas that chose to live in a group environment where individuals could interact with one another and further their social skills, but which clearly did not contribute to their reproductive success (Yamagiwa 1987a; Robbins 1996; see also Stoinski et al. 2001, 2004).

14.5 Risk of Injury

Attempting to obtain females from another silverback is a high-risk venture, which can lead to physical injury and even death; thus, opting for a solitary strategy is also costly. During this study, three adult male gorillas were fatally wounded: we found the bodies of two lone silverbacks with trauma to the head and genitalia, consistent with injury inflicted during confrontation with another silverback. The third fatality was a group-living subordinate silverback who died of septicemia following an injury sustained during an intergroup encounter.

If the chances of acquiring a female were high enough, the risks involved in pursuing a group might be worth incurring, but it seems they are not. During the first 35 years of study of the Karisoke mountain gorillas, few solitary males succeeded in forming a new group or taking over an established group (Watts 1989, 2000; Robbins 1995, 1996, 2001; Harcourt and Stewart 2007). Group takeovers by a solitary silverback can occur when the dominant male in a one-male group dies or is killed, but these events have been rare. We can find examples of group takeovers among the other gorilla subspecies (Tutin 1996; Yamagiwa and Kahekwa 2001); however, in mountain gorillas, group acquisition has not occurred as a natural phenomenon, but as a consequence of poachers killing the dominant silverback.

During the present study, only one of the ten maturing silverbacks (INZ) became solitary. We observed several subsequent interactions between this individual and his natal group, and each time he was chased away by the subordinate males, his former playmates. Although in his prime and in excellent physical condition, he remained alone throughout (and beyond) this study. A significant disadvantage for this solitary male was that his home range overlapped with multimale but no one-male groups, and he would have had to lure females away from two or more silverbacks.

14.6 Multimale Groups

Most female mountain gorillas transfer from one group to another at least once during their lifetime (Watts 1996), and transfers usually take place during encounters between groups. Gorilla groups often exchange chestbeats and vocalizations from a distance, but when encounters escalate to aggression, fighting between adult males can be intense (Harcourt 1981; Watts 1991). Intergroup interactions are contests for access to adult females and occur about once per month in the Karisoke population (Sicotte 1993, 2001). Multimale groups are better able to retain their females because two or more silverbacks can cooperate during such encounters. Typically, one silverback herds females away from the frontline, which is effective in preventing female transfer, while the other rebuffs challengers (Sicotte 1993). Therefore, so long as the dominant male loses few matings or gains compensating payoffs in inclusive fitness, he may benefit by tolerating a younger silverback as an ally against extra-group males (Watts 1996). Subordinate males in the present study were indeed seen to play active roles in intergroup interactions.

The incidence of multimale groups in the wider mountain gorilla population (groups habituated for research and tourism) has increased from 40 % in 1981 to 61 % in 2010 (Gray et al. 2013). What is more extraordinary is that there are not just two but many silverbacks in some groups; each of the three research groups has included six or more silverbacks at some point in time. It is also notable that during this study, the rate of female transfer was low, both into ($n=2$) and out of ($n=2$) the research population, and between the three research groups ($n=5$) (total number of adult females = 26–31).

Robbins (2001) noted that multimale groups perpetuate multimale groups, and a likely mechanism for this was elucidated by Parnell (2002): "If the presence of more than one silverback confers an advantage when acquiring new females or defending residents, a virtual 'arms-race' can be envisioned in which increasing numbers of silverbacks are required for a group to remain competitive. The dominant male may thus become more tolerant of subordinate males within his group. Ever greater numbers of adult males per group can be predicted with an inevitable decrease in numbers of solitary males, both as young males are tolerated in their natal group, and as the solitary route to group acquisition becomes increasingly unrewarding and potentially hazardous. Such a mechanism for the increase in multimale groups will have a 'feedback' effect on other groups. A group silverback unable to rely on coalition support may be more likely to lose females and encounter more difficulty in acquiring them. Creating conditions such that maturing males delay their emigration will be a powerful strategy for maintaining viable groups".

Males that do not emigrate contribute to groups becoming larger not simply by their own presence but also because mountain gorilla groups with more than one male tend to attract and retain more females (Robbins 1995; Yamagiwa et al. 2009). Adult females associate with adult males as a means to avoid infanticide by extra-group males (Watts 1989). Infanticide is a reproductive tactic that shortens the time which elapses before lactating females become fertile again and has accounted for 26 % of infant deaths in the Karisoke population (Robbins and Robbins 2004). Apparently females favor groups with more than one adult male because they provide better protection against infanticide in the event of the death of the dominant male (Watts 2000; Robbins 2003). Interestingly, there were no cases of infanticide by adult males in the research groups during this study. This absence of infanticide was likely a consequence of the social stability of these groups, which would be consistent with Robbins' (1995) prediction that infants in multimale groups should suffer fewer infanticidal attacks. [Two infant deaths occurred in 1996, but these were attributed to competition between new mothers and pregnant nulliparous females over the newborns, not deliberate killing by a potential mate (Warren and Williamson 2004)].

Sicotte (1993) observed that groups with more than two adult males did not exist in the Karisoke population and questioned why multimale groups were not more common at that time. However, as the research groups have grown in size and the number of resident females has increased, the likelihood of a number of male offspring growing up as a cohort has also increased, and this was the case in 1986–1987, when six male infants were born into Group 5 (the precursor to the PAB and SHI groups). By the end of the present study, the 25-year-old leader of SHI group was accompanied by five young silverbacks, 14–16 years old, and at least three of them shared the same father (Bradley et al. 2005). Watts (2000) suggested that closely related males (father and son or half-brothers) are likely to be more tolerant of one another than unrelated males, and this seems to be true of the SHI group silverbacks.

14.7 Social Stability and a Supergroup

The factors discussed above have combined to produce larger groups. Consider that in 1972 there were 96 gorillas in the Karisoke sector living in eight groups (Fossey 1983) and that in 2001 the same number could be found in just three groups. Group 5 grew in size to 35 members, five times the norm, before fissioning in 1993. The catalyst for this split was the death of the dominant male. We therefore expected the PAB group to fission when it surpassed 35 members; however, there was no catastrophic event to perturb the social balance and this group continued to grow, peaking at 65 individuals in 2006 (Vecellio 2008).

Various factors contribute to the formation and maintenance of large, multimale groups, and these have been discussed at length (e.g. Watts 1989, 2000; Robbins 1995, 2001, 2003; Robbins and Robbins 2005). Typically, a group will disintegrate when the dominant male of a single-male group dies, but when this happens in a multimale setting one of the subordinate males can take over leadership and the group remains intact, enabling offspring to mature in a stable social setting (Yamagiwa 1987b; Robbins 1995). Thus, group stability is largely assured by a multimale structure, and males who remain in breeding groups seem to have substantially higher fitness payoffs than males who emigrate. The principal advantages gained by residents are increased opportunities to mate and enhanced infant survival. It seems that solitary silverbacks have little chance of establishing a new group and reproducing successfully in this "arms race".

So, are these changes typical of the population as a whole? Surveys of the Virunga population show that the median size of gorilla groups has not changed over 30 years (median 7.5, $n=32$), but that mean group size has increased from 7.9 to 11.4 individuals, reflecting an increasing proportion of large groups (Gray et al. 2009). Habituated groups, and the Karisoke groups in particular, are significantly larger than unhabituated groups (mean, 16.8 vs. 5.9). The above-average size of gorilla groups in the Karisoke sector has been attributed to two principal factors: first, that habitat quality in this sector is better than in other sectors of the Virunga Volcanoes (McNeilage 2001), and second, the much higher level of protection afforded to the Karisoke groups through daily monitoring by gorilla trackers and researchers, which not only deters poachers but also facilitates rapid intervention by a veterinary team when needed (Kalpers et al. 2003).

14.8 Active Conservation and Management Issues

Since 1902, when mountain gorillas were first brought to international attention, human impacts on the Virunga population have been devastating, from their slaughter by museum collectors and trophy hunters to the clearing of more than half of their entire habitat in Rwanda in the late 1960s (Plumptre and Williamson 2001). Compounding the loss of habitat, targeted killing by poachers in the 1970s and early 1980s had a major impact on the gorillas' demography, leading to group breakups or takeovers and further losses through subsequent infanticide (Fossey 1983). However, by the end of the 1980s the conservation status of the Virunga gorillas had improved dramatically, brought about by daily monitoring of the habituated groups and the expansion of antipoaching patrols.

The Virunga gorillas' survival was again threatened throughout the 1990s, this time by civil conflict. Less well known than the 1994 genocide is that during 1997 and 1998 civilians fleeing armed conflict in Rwanda took refuge in the forests of the Virunga Volcanoes, building shelters and cultivating crops in the park, while armed militia controlled access to the region. Remarkably, these events left the Karisoke gorillas visibly unscathed. Active conservation has allowed the Virunga gorilla population to recover from an estimated all-time low of 254 and to attain 480 individuals in just 30 years, one and a half generations. Groups that are well protected do not suffer the same degree of human-induced mortality and habitat degradation that nonhabituated groups do and, with veterinary intervention, habituated gorillas no longer die of snare injuries. This is not the case for the entire population, and the unhabituated subpopulation is in decline (minus 0.7 % annual growth rate; Robbins et al. 2011). Consequently, mountain gorillas continue to be classified as Critically Endangered on the IUCN Red List of Threatened Species (Robbins and Williamson 2008), meaning that they face a high risk of extinction.

With a return to peace and stability in some sectors of the Virunga Volcanoes, military escorts have become a necessary accompaniment to research and tourism activities. The increased number of people in the park—tourists, trackers, researchers, and soldiers—is likely to have affected the gorillas' demographic dynamics. Unhabituated groups rarely approach monitored groups and usually flee if human observers are present. Even if interactions between habituated and unhabituated gorillas still take place, they are infrequent and thus normal social dynamics may be impeded. Incest has perhaps become inevitable and is known to have occurred in one Karisoke group, in which the dominant silverback fathered an infant with one of his sisters and mated with at least one of his aunts. The mother and four other close relatives of this male were among the 17 reproductive females in his group. Many individuals in the research groups are closely related (Bradley et al. 2005), and inbreeding is manifested by strabismus and syndactyly (e.g. Routh and Sleeman 1997). Fewer transfers between groups would reduce gene flow and may increase the level of inbreeding. If the close and sustained presence of human observers is potentially compromising normal social interactions, behavioral disturbance must be minimized by observance of strict limits to the numbers of people tracking, the distance to which they approach gorillas (no closer than 7 m), and the duration of visits (see Macfie and Williamson 2010).

From a management perspective, it is clear that active conservation—anti-poaching activities, constant monitoring of the gorillas, and maintenance of the size and integrity of the habitat—have been paramount to the gorillas' survival. It has long been noted that the Virunga gorillas are a relict population (Watts 1983:25), one so small that it is extremely vulnerable to human disturbance, and it is perhaps time to acknowledge that this is no longer a wild population but a highly managed one. We should accept this fact when assessing the feasibility of interventions. Population growth rate is more affected by changes in survivorship than by fertility, so efforts to conserve mountain gorillas are best focused on improving survivorship (Robbins et al. 2011). In reality, besides treating injuries and managing disease, few interventions are possible, but if mountain gorillas are to continue to survive in such a volatile region, a hands-on approach that incorporates knowledge of socioecology and demographic processes is justified. The increasing human pressures on mountain gorillas have already prompted a reevaluation of veterinary intervention policy: for nearly 20 years, the policy had been to intervene only if an injury or disease was human induced or life threatening; in recent years, some potentially life-threatening cases have been treated even if they were not caused by humans (Cranfield et al. 2006). With respiratory disease outbreaks among the habituated groups becoming more frequent and leading to fatalities (Palacios et al. 2011; Ryan and Walsh 2011), perhaps this reevaluation should go further. After careful analysis of human pathogen spillover, safety and cost of possible interventions, and efficiency of mitigation measures, Ryan and Walsh (2011) concluded that the conservation community should pursue proactive vaccination of great apes as a conservation strategy. Too few of these great apes remain for us to be passive about disease and natural selection "running their course".

14.9 Postscript

This study took place during a period of stability in the Karisoke gorilla population, ironically at a time when the lives of people in the region were in turmoil. In the 6 years from 1996 to 2001, only 1 silverback left one of the three Karisoke research groups; in a subsequent 6-year period, 2003–2008, 14 males dispersed from the same three groups, 11 of them to become solitary (Stoinski et al. 2009). There have also been group splits, new groups formed, infant mortality through infanticide, and a high number of female transfers (Vecellio 2008), marking the return to a social dynamic reminiscent of earlier decades. The Karisoke gorillas seem to have been able to adapt to the prevailing social conditions, showing behavioral flexibility with their shifting reproductive strategies.

14.10 Parallels Between Gorillas and Cetaceans

All whales are listed by the Convention on International Trade in Endangered Species (CITES), meaning that legal trade and products derived from them are highly controlled. In contrast, there is no legal exploitation of mountain gorillas. They are totally protected under national and international laws, on Appendix I of CITES, and it is forbidden to kill, capture, or harm them.

Both gorillas and cetaceans have low rates of reproduction and population growth. Slow demographic recovery stems from their life histories, such as a relatively long delay before first reproduction and long interbirth intervals. Actual growth rate in the Virunga gorilla population has been 1.15 % per year, well below the projected 3.8 % under ideal conditions (Gray et al. 2009).

There are similarities in population responses to legal (some cetaceans) and illegal (gorillas) hunting, as social balance in these complex mammals is easily disrupted. Their population structure is sensitive to the removal of older individuals, which may lead to fragmentation of social units. Lack of resilience to exploitation in cetaceans equates with the gorillas' vulnerability to illegal killing. Trophy hunting that selectively removed silverback gorillas—key group members—and the killing of adults to capture infant gorillas have led to group disintegrations and subsequent deaths by infanticide (Kalpers et al. 2003). Similarly, in some cetaceans, direct removals have disrupted social structure.

To conclude, management interventions must take demographic processes into account and be fully cognizant of the disproportionately large impact of removing key individuals from these socially complex mammalian populations.

Acknowledgments I thank the Rwandan Office of Tourism and National Parks for permission to work in the Volcanoes National Park and the Dian Fossey Gorilla Fund International for making this research possible. I am very grateful to J.R. Anderson, R.W. Byrne, M. Klailova, W.C. McGrew, R.J. Parnell, M.M. Robbins, M.E. Rogers, and C.E.G. Tutin for providing valuable comments on the manuscript. I am indebted to the staff of the Karisoke Research Centre, in particular Jean Damascene Hategekimana, Emmanuel Hitayezu, and the late Mathias Mpiranya, for sharing with me their insights to the gorillas' world. But most especially, I am deeply grateful to Jean Bosco Bizumuremyi, without whose endless dedication, energy, and personal sacrifice, we would have been much less effective in protecting the gorillas during very difficult times when so many people lost their lives.

References

Bradley BJ, Robbins MM, Williamson EA, Steklis HD, Gerald-Steklis N, Eckhardt N, Boesch C, Vigilant L (2005) Mountain gorilla tug-of-war: silverbacks have limited control over reproduction in multi-male groups. Proc Natl Acad Sci USA 102:9418–9423

Cranfield M, Gaffikin L, Minnis R, Nutter F, Rwego I, Travis D, Whittier C (2006) Clinical response decision tree for the mountain gorilla (*Gorilla beringei*) as a model for great apes. Am J Primatol 68:909–927

Czekala N, Sicotte P (2000) Reproductive monitoring of free-ranging female mountain gorillas by urinary hormone analysis. Am J Primatol 51:209–215

Fossey D (1983) Gorillas in the mist. Houghton-Mifflin, Boston

Gray M, McNeilage A, Fawcett K, Robbins MM, Ssebide B, Mbula D, Uwingeli P (2009) Censusing the mountain gorillas in the Virunga Volcanoes: complete sweep method vs. monitoring. Afr J Ecol 48:588–599

Gray M, Roy J, Vigilant L, Fawcett K, Basabose A, Cranfield M, Uwingeli P, Mburunumwe I, Kagoda E, Robbins MM (2013) Genetic census reveals increased but uneven growth of a critically endangered mountain gorilla population. Biol Conserv 158:230–238

Harcourt AH (1981) Intermale competition and the reproductive biology of the great apes. In: Graham CE (ed) Reproductive biology of the great apes, comparative and biomedical perspectives. Academic, New York, pp 301–318

Harcourt AH, Stewart KJ (1981) Gorilla male relationships: can differences during immaturity lead to contrasting reproductive tactics in adulthood? Anim Behav 29:206–210

Harcourt AH, Stewart KJ (2007) Gorilla society: conflict, compromise, and cooperation between the sexes. University of Chicago Press, Chicago

Harcourt AH, Fossey D, Stewart KJ, Watts DP (1980) Reproduction in wild gorillas and some comparisons with chimpanzees. J Reprod Fertil Suppl 28:59–70

Kalpers J, Williamson EA, Robbins MM, McNeilage A, Nzamurambaho A, Lola N, Mugiri G (2003) Gorillas in the crossfire: assessment of population dynamics of the Virunga mountain gorillas over the past three decades. Oryx 37:326–337

Macfie EJ, Williamson EA (2010) Best practice guidelines for great ape tourism. IUCN/SSC Primate Specialist Group, Gland

McNeilage A (2001) Diet and habitat use of two mountain gorilla groups in contrasting habitats in the Virungas. In: Robbins MM, Sicotte P, Stewart KJ (eds) Mountain gorilla: three decades of research at Karisoke. Cambridge University Press, Cambridge, pp 265–292

Palacios G, Lowenstine LJ, Cranfield MR, Gilardi KV, Spelman L, Lukasik-Braum M, Kinani J-F, Mudakikwa A, Nyirakaragire E, Bussetti AV, Savji N, Hutchison S, Egholm M, Lipkin WI (2011) Human metapneumovirus infection in wild mountain gorillas, Rwanda. Emerg Infect Dis 17:711–713

Parnell RJ (2002) Group size and structure in western lowland gorillas (*Gorilla gorilla gorilla*) at Mbeli Bai, Republic of Congo. Am J Primatol 56:193–206

Plumptre AJ, Williamson EA (2001) Conservation oriented research in the Virunga region. In: Robbins MM, Sicotte P, Stewart KJ (eds) Mountain gorilla: three decades of research at Karisoke. Cambridge University Press, Cambridge, pp 361–390

Robbins MM (1995) A demographic analysis of male life history and social structure of mountain gorillas. Behaviour 132:21–47

Robbins MM (1996) Male–male interactions in heterosexual and all-male wild mountain gorilla groups. Ethology 102:942–965

Robbins MM (1999) Male mating patterns in wild multimale mountain gorilla groups. Anim Behav 57:1013–1020

Robbins MM (2001) Variation in the social system of mountain gorillas: the male perspective. In: Robbins MM, Sicotte P, Stewart KJ (eds) Mountain gorilla: three decades of research at Karisoke. Cambridge University Press, Cambridge, pp 29–58

Robbins MM (2003) Behavioural aspects of sexual selection in mountain gorillas. In: Jones CB (ed) Sexual selection and reproductive competition in primates: new perspectives and directions. American Society of Primatologists, Norman, pp 477–501

Robbins MM, Robbins AM (2004) Simulation of the population dynamics and social structure of the Virunga mountain gorillas. Am J Primatol 63:201–223

Robbins AM, Robbins MM (2005) Fitness consequences of dispersal decisions for male mountain gorillas (*Gorilla beringei beringei*). Behav Ecol Sociobiol 58:295–309

Robbins M, Williamson EA (2008) *Gorilla beringei*. In: IUCN 2012. IUCN red list of threatened species. Version 2012.2. <www.iucnredlist.org>

Robbins MM, Gray M, Fawcett KA, Nutter FB, Uwingeli P, Mburanumwe I, Kagoda E, Basabose A, Stoinski TS, Cranfield MR, Byamukama J, Spelman LH, Robbins AM (2011) Extreme conservation leads to recovery of the Virunga mountain gorillas. PLoS One 6:e19788. doi:10.1371/journal.pone.0019788

Routh A, Sleeman J (1997) A preliminary survey of syndactyly in the mountain gorilla (*Gorilla gorilla beringei*). In: British Veterinary Zoological Society (ed) Proceedings of the medical conditions and veterinary considerations of zoo animals. Tamurlane, Canterbury, pp 22–25

Ryan SJ, Walsh PD (2011) Consequences of non-intervention for infectious disease in African great apes. PLoS One 6:e29030. doi:10.1371/journal.pone.0029030

Sicotte P (1993) Inter-group encounters and female transfer in mountain gorillas: influence of group composition on male behavior. Am J Primatol 30:21–36

Sicotte P (2001) Female mate choice in mountain gorillas. In: Robbins MM, Sicotte P, Stewart KJ (eds) Mountain gorilla: three decades of research at Karisoke. Cambridge University Press, Cambridge, pp 59–87

Stoinski TS, Hoff MP, Lukas KE, Maple TL (2001) A preliminary behavioral comparison of two captive all-male gorilla groups. Zoo Biol 20:27–40

Stoinski TS, Lukas KE, Kuhar CW, Maple TL (2004) Factors influencing the formation and maintenance of all-male gorilla groups in captivity. Zoo Biol 23:189–203

Stoinski TS, Vecellio V, Ngaboyamahina T, Ndagijimana F, Rosenbaum S, Fawcett KA (2009) Proximate factors influencing dispersal decisions in male mountain gorillas, *Gorilla beringei beringei*. Anim Behav 77:1155–1164

Tutin CEG (1996) Ranging and social structure of lowland gorillas in the Lopé Reserve, Gabon. In: McGrew WC, Marchant LF, Nishida T (eds) Great ape societies. Cambridge University Press, Cambridge, pp 58–70

Vecellio V (2008) Rapid decline in the largest group of mountain gorillas. Gorilla J 37:6–7

Warren Y, Williamson EA (2004) Transport of dead infant mountain gorillas by mothers and unrelated females. Zoo Biol 23:375–378

Watts DP (1983) Foraging strategy and socioecology of mountain gorillas (*Pan gorilla beringei*). Ph.D. Thesis, University of Chicago, IL

Watts DP (1989) Infanticide in mountain gorillas: new cases and a reconsideration of the evidence. Ethology 81:1–18

Watts DP (1991) Mountain gorilla reproduction and sexual behavior. Am J Primatol 24:211–218

Watts DP (1996) Comparative socioecology of gorillas. In: McGrew WC, Marchant LF, Nishida T (eds) Great ape societies. Cambridge University Press, Cambridge, pp 16–28

Watts DP (2000) Causes and consequences of variation in male mountain gorilla life histories and group membership. In: Kappeler P (ed) Primate males. Cambridge University Press, Cambridge, pp 169–179

Watts DP, Pusey AE (1993) Behaviour of juvenile and adolescent great apes. In: Pereira ME, Fairbanks LA (eds) Juvenile primates; life history, development and behavior. Oxford University Press, New York, pp 148–167

Williamson EA, Gerald-Steklis N (2001) Composition of gorilla groups monitored by Karisoke Research Center, 2001. Afr Primates 5:48–51

Yamagiwa J (1987a) Intra- and inter-group interactions of an all-male group of Virunga mountain gorillas (*Gorilla gorilla beringei*). Primates 28:1–30

Yamagiwa J (1987b) Male life history and the social structure of wild mountain gorillas (*Gorilla gorilla beringei*). In: Kawano S, Connell JH, Hidaka T (eds) Evolution and coadaptation in biotic communities. University of Tokyo Press, Tokyo, pp 31–51

Yamagiwa J, Kahekwa J (2001) Dispersal patterns, group structure, and reproductive parameters of eastern lowland gorillas at Kahuzi in the absence of infanticide. In: Robbins MM, Sicotte P, Stewart KJ (eds) Mountain gorilla: three decades of research at Karisoke. Cambridge University Press, Cambridge, pp 90–122

Yamagiwa J, Kahekwa J, Basabose AK (2009) Infanticide and social flexibility in the genus *Gorilla*. Primates 50:293–303

Chapter 15
Population Genetics in the Conservation of Cetaceans and Primates

Kimberly Andrews

K. Andrews (✉)
School of Biological Sciences, Durham University, South Road, Durham DH1 3LE, UK
e-mail: kimandrews@gmail.com

J. Yamagiwa and L. Karczmarski (eds.), *Primates and Cetaceans: Field Research and Conservation of Complex Mammalian Societies*, Primatology Monographs,
DOI 10.1007/978-4-431-54523-1_15, © Springer Japan 2014

Abstract Understanding the factors driving spatial and temporal variation in genetic diversity in cetaceans and primates is crucial in directing conservation and management decisions for these taxa. Spatial variation in genetic diversity can be driven by geographic barriers to dispersal, such as rivers for primates or land masses for cetaceans. Spatial variation in diversity can also be driven by population size differences across habitat patches that vary in resource abundance, with smaller populations typically exhibiting lower genetic diversity than larger populations. However, cetaceans and primates often exhibit complex genetic structure that cannot be explained by simple geographic barriers or variation in habitat abundance. In many cases this complex structure is attributed to behavioral philopatry or social structure. Many cetaceans and primates exhibit philopatry to natal ranges or philopatry to particular habitat types, and many species also exhibit complex social structure features that can influence genetic structure, such as strong group stability and sex-biased dispersal. Finally, genetic diversity can vary temporally; genetic coalescent analyses indicate that diversity has declined in the recent past for several cetacean and primate species, often with evidence for population bottlenecks during times of historical anthropogenic impact, such as habitat fragmentation or whaling. The complexity of factors influencing genetic structure over space and time for cetacean and primate species illustrates that multidisciplinary studies are required to truly understand genetic structure for these species. These studies are becoming increasingly important as cetaceans and primates face severe anthropogenic threats around the world.

Keywords Bottleneck • Genetic diversity • Habitat degradation • Habitat preference • Philopatry • Sex-biased dispersal • Social structure • Spatial variation • Temporal variation • Whaling

15.1 Introduction

Understanding spatial and temporal variation in genetic diversity within species can provide insight into numerous conservation issues, such as the identification and assessment of management units, development of optimal breeding strategies for captive populations, or identification of the geographic origin of invasive species (reviewed in Frankham et al. 2002; Allendorf and Luikart 2007). The utility of information on genetic diversity has led to a growing interest within the field of conservation biology in understanding the factors driving variation in intraspecific genetic diversity across space and time. Spatial variation in genetic diversity is often driven by the presence of barriers that restrict gene flow between groups, thus leading to the accumulation of genetic differences between groups. A variety of factors can act as barriers to gene flow, including geographic factors, such as a mountain range or an ocean separating regions; ecological factors, such as habitat differences between adjacent regions; or behavioral factors, such as differences in mating behavior between groups (reviewed in Coyne and Orr 2004). Spatial variation in genetic diversity can also be influenced by the effective population sizes of genetically distinct groups; populations with smaller effective sizes are expected to have lower genetic diversity (Wright 1931), and effective size can be influenced by a number of environmental, demographic, or behavioral factors including abundance and isolation of habitat, unequal numbers of males and females, and reproductive skew (reviewed in Allendorf and Luikart 2007).

Spatial variation in genetic diversity can also change over time as the result of stochastic processes or environmental change. Global climate shifts throughout the history of life on Earth have influenced population sizes and barriers to gene flow over time by causing range expansions, localized population extinctions, and bottlenecks. In the more recent past, humans have caused extensive environmental changes that have resulted in the loss of biodiversity at a pace more rapid than at any other time in Earth's history since the extinction of the dinosaurs 65 million years ago (Wilson 2002).

Cetaceans and primates are taxonomic groups that exhibit particularly complex patterns of intraspecific genetic diversity. Species within both these groups have high potential mobility compared to other mammalian taxa and therefore should experience few physical geographic barriers to gene flow, with this pattern expected to be particularly strong for cetaceans because of the relatively continuous nature of their marine habitat. However, genetically distinct populations are often present within both primate and cetacean species, even in the absence of obvious geographic barriers. In these cases, the complex behaviors of these species are often implicated as responsible for barriers to gene flow. For example, habitat preferences or social structure can act as barriers to gene flow and can lead to genetic structure over short geographic distances. Understanding patterns of genetic diversity is particularly important for cetaceans and primates given the severe anthropogenic threats faced by species within these taxa, such as poaching, directed fisheries or fisheries bycatch, and habitat degradation. Here, we review the geographic, ecological, behavioral, and demographic factors that have influenced the genetic diversity of cetaceans and primates across space and time.

15.2 Spatial Variation in Genetic Diversity: Barriers to Gene Flow

15.2.1 Geographic Barriers

In primates, rivers are one of the geographic features most commonly identified as a barrier to dispersal and gene flow. For example, rivers separate genetically distinct populations in bonobos (*Pan paniscus*; Eriksson et al. 2004), chimpanzees (*Pan troglodytes*; Gonder et al. 2006), mandrills (*Mandrillus leucophaeus*; Telfer et al. 2003), orangutans (*Pongo pygmaeus*; Jalil et al. 2008), gorillas (*Gorilla* spp.; Anthony et al. 2007), and mouse lemurs (*Microcebus ravelobensis*; Guschanski et al. 2007). This strong effect on genetic structure is probably a result of the poor swimming abilities of many primates (Ayres and Cluttonbrock 1992). In other cases, anthropogenic alterations of the environment have been implicated as geographic barriers to gene flow. For example, human-mediated fragmentation of forest habitats is thought to be a primary factor influencing genetic structure within the Yunnan snub-nosed monkey in Tibet (Liu et al. 2009) and the mouse lemur (*M. ravelobensis*) in northwestern Madagascar (Guschanski et al. 2007; Radespiel et al. 2008).

In cetaceans, there are relatively few physical barriers to dispersal and gene flow because of the high dispersal capabilities of these species, combined with the relatively continuous nature of the marine environment. However, more subtle environmental factors do appear to act as barriers to dispersal and gene flow in some cetaceans. The distributions of many cetacean species are restricted to climatic regions, indicating that water temperature can directly or indirectly prevent dispersal between regions. For example, a number of odontocete species are restricted to tropical or warm- temperate waters despite having distributions across multiple ocean basins, including pygmy killer whales (*Feresa attenuate*), melon-headed whales (*Peponocephala electra*), rough-toothed dolphins (*Steno bredanensis*), pantropical spotted dolphins (*Stenella attenuate*), and spinner dolphins (*Stenella longirostris*; Jefferson et al. 1993). Water temperature may also act as a barrier to gene flow on a smaller scale; for example, warm oligotrophic waters in the Bay of Biscay and the Mediterranean Sea appear to act as barriers to gene flow in harbour porpoises (*Phocoena phocoena*) (Fontaine et al. 2007). However in other cetacean species, water temperature appears to have little influence on genetic structure; for example, killer whales travel long distances across thermal boundaries (Matthews et al. 2011) and exhibit fine-scale population structure that appears unrelated to these thermal boundaries (Hoelzel et al. 2007).

For both primates and cetaceans, historic geographic barriers may also contribute to present-day genetic structure. For example, genetically distinct parapatric or sympatric populations may have originally accumulated their genetic differences while trapped in allopatric glacial refugia during glacial maxima of the Quaternary (Haffer 1969). As the earth warmed and the glaciers retreated, the ranges of these populations may have expanded to their present-day distribution, with populations coming into contact but remaining genetically distinct. The predicted genetic signature of these historical events includes population differentiation corresponding with the locations of hypothesized refugia, and evidence for recent population expansions within these locations (Haffer 1969; Hewitt 2000; Lessa et al. 2003). Quaternary hypotheses have been put forth to explain genetic structure for a number of cetacean and primate species, including harbour porpoises in the North Atlantic (Tolley et al. 2001), bottlenose dolphins (*Tursiops* spp.) across their global range (Natoli et al. 2004), gorillas across their range in central Africa (Anthony et al. 2007), and hamadryas baboons (*Papio hamadryas hamadryas*) across their range in East Africa and western Arabia (Winney et al. 2004).

Genetically distinct populations may alternatively have originally accumulated their genetic differences during historic periods of global warming. For example, species boundaries of present-day sympatric minke whales (*Balaenoptera *bonaerensis* and *B. acutorostrata*) in Antarctica are thought to have developed as a result of food limitation during a period of global warming in the Pliocene (Pastene et al. 2007). Minke whales rely upon regions of upwelling as a food source, and these regions would have become reduced and more localized during this time period as a result of warm temperatures. This restriction in food resources may have led to population fragmentation and speciation, followed by secondary contact when the climate cooled (Pastene et al. 2007).

15.2.2 Habitat Preferences

In both cetaceans and primates, genetic structure is often present even in the absence of geographic barriers. Although in some cases this genetic structure is thought to result from historic allopatry followed by range expansions, in other cases it is thought to result from present-day ecological and behavioral factors. Among other means, genetic structure can develop in the absence of geographic barriers through habitat preferences. For example, natal philopatry can promote the formation of genetically distinct population segments, even across short geographic distances, between apparently similar habitats, and without the influence of strong social bonding within habitats. For example, in spinner dolphins, genetic distinctions between islands separated by as little as 17 km in French Polynesia and 47 km in the Hawaiian Islands are likely shaped by natal philopatry to island habitats, despite an absence of strong social bonds between individuals at many of these islands (Norris et al. 1994; Östman 1994; Poole 1995; Oremus et al. 2007; Andrews et al. 2010). For coastal bottlenose dolphins, genetic structure over short geographic distances has also been attributed to natal philopatry for populations in Australia (Möller and Beheregaray 2004) and the western North Atlantic (Sellas et al. 2005; Rosel et al. 2009). For humpback whales (*Megaptera novaeangliae*), maternally directed philopatry to migratory destinations is likely responsible for genetic structure between feeding and breeding grounds (Baker et al. 1990, 1994); this philopatry occurs despite the high realized dispersal abilities of humpback whales and despite a lack of social bonding within this species, except between mother–calf pairs (Valsecchi et al. 2002; Pomilla and Rosenbaum 2006) (Fig. 15.1).

In other cases, natal philopatry can result in a pattern of genetic isolation by distance rather than (or in addition to) the formation of genetically distinct subpopulations. For example, isolation by distance has been documented for mouse lemurs in northwestern Madagascar (Radespiel et al. 2008), gorillas in central Africa (Anthony et al. 2007), the piscivorous ecotype of killer whales (*Orcinus orca*) in the North Pacific (Hoelzel et al 2007), and bottlenose dolphins in Western Australia (Krützen et al. 2004).

Genetic structure can also develop if individuals choose to remain in habitats similar to that of their natal range, even if that habitat does not necessarily include their natal range. In at least four dolphin species, genetic structure appears to depend on preferences for adjacent nearshore versus offshore habitats (common dolphins, *Delphinus*

Fig. 15.1 Two humpback whales (*Megaptera novaeangliae*) feeding in Stellwagen Bank National Marine Sanctuary. Population genetic structure of humpbacks indicates natal fidelity to feeding and breeding grounds (Baker et al. 1990, 1994). This natal fidelity is likely driven by maternally directed philopatry to migratory destinations, despite a lack of social bonding within this species except between mother–calf pairs (Valsecchi et al. 2002; Pomilla and Rosenbaum 2006). (Published with kind permission of © Alison Stimpert 2011, NMFS Permit #605-1904. All rights reserved)

spp.; Natoli et al. 2006; bottlenose dolphins, *Tursiops* spp.; Hoelzel et al. 1998b; Natoli et al. 2004; Tezanos-Pinto et al. 2009; pantropical spotted dolphins, *Stenella attenuata*; Escorza-Treviño et al. 2005; and killer whales; Hoelzel et al. 2007). In some cases, morphological differences have evolved between these genetically differentiated coastal versus pelagic populations, indicating ecological specializations that correspond with habitat preferences (Natoli et al. 2004, 2006; Tezanos-Pinto et al. 2009). Genetic differentiation between populations inhabiting parapatric, ecologically divergent habitats has been found in other cetacean species and other regions as well; oceanographic features that vary in association with population genetic differentiation over short geographic distances include water temperature, salinity, primary productivity, current patterns, and topography (e.g., pilot whales, *Globicephala* spp.; Kasuya et al. 1988; Fullard et al. 2000; and bottlenose dolphins, *Tursiops* spp.; Natoli et al. 2005; Dowling and Brown 1993; Bilgmann et al. 2007). In most of these studies it is unknown whether the oceanographic features are responsible for the genetic differentiation, or whether these features correspond to another parameter responsible for the differentiation. However, the association of genetic structure with environmental variability does indicate habitat preferences are driving genetic differentiation.

Fewer studies have investigated the influence of ecological preferences on dispersal and population genetic structure within primate species. However, a study of mountain gorillas (*Gorilla beringei beringei*) in Uganda provided evidence that female (but not male) dispersal and genetic structure are influenced by individual preferences for habitat characteristics of altitude and plant composition (Guschanski

Fig. 15.2 The genetic structure of female (but not male) mountain gorillas (*Gorilla beringei beringei*) in Uganda is associated with geographic distance, altitude, and plant composition, indicating that habitat preferences drive female genetic structure. (Published with kind permission of © Stuart Ibsen 2010. All rights reserved)

et al. 2008) (Fig. 15.2). Another study indicated that dispersal and gene flow in golden-brown mouse lemurs (*Microcebus ravelobensis*) are influenced by habitat preferences within their forest habitat, and especially by preferences for altitude (Radespiel et al. 2008). Future research may provide more evidence for the influence of habitat on population genetic structure in primate species (Vigilant and Guschanski 2009).

15.2.3 Social Structure

Social structure can add another layer of complexity to the population structure of primates and cetaceans, potentially leading to the formation of genetically distinct populations in the absence of geographic or ecological barriers to dispersal. For example, group philopatry can promote genetic divergence between regions (Sugg et al. 1996; Storz 1999; Wakeley 2000). In mammalian species, one sex is usually more dispersive than the other (Greenwood 1980), and this sex-biased dispersal is expected to result in predictable patterns of genetic structure (reviewed by Avise 1995; Avise 2004). In most mammals, males are more likely to disperse than females, and this male-biased dispersal is expected to result in greater levels of genetic structure between groups or between regions for females than males, and greater genetic structure for mitochondrial DNA (mtDNA) markers (from maternal inheritance) than autosomal markers or Y-chromosome markers (from biparental or paternal inheritance). In contrast, female-biased dispersal is expected to result in greater genetic structure for males than females, and greater genetic structure for

Fig. 15.3 Several macaque species (*Macaca* spp.) have greater genetic differences between local populations and between social units for maternally inherited mitochondrial DNA (mtDNA) than for bi-parentally inherited autosomal DNA (reviewed in Melnick and Hoelzer 1996). This difference in mtDNA versus autosomal DNA structure is likely driven by the matrilineal social structure of these species, in which females exhibit strong fidelity to natal groups throughout their lives, whereas males usually migrate from their natal group before reaching sexual maturity. (Published with kind permission of)

Y-chromosome markers than mtDNA or autosomal markers. Notably, however, the expected differences in genetic structure across different marker types could be confounded by various factors including differences in effective population sizes of markers, differences in mutation rate of markers (Hedrick 1999), or differences in effective population sizes between the sexes. For example, lower genetic divergence at mtDNA or Y-chromosome markers than autosomal markers could actually result from the greater influence of genetic drift on uniparentally inherited markers because of their fourfold lower effective population sizes (Avise 2004).

Behavioral studies indicate that primates exhibit a wide variety of dispersal patterns for both sexes, and the genetic structure of these species at different types of genetic markers generally follows expectations based on these patterns (reviewed by Di Fiore 2003). For example, female macaques (*Macaca* spp.) exhibit strong fidelity to natal groups throughout their lives, whereas males usually migrate from their natal group before reaching sexual maturity (Melnick and Pearl 1987) (Fig. 15.3). When group fission occurs, it usually occurs along matrilineal groups. As expected based on these dispersal patterns, several macaque species exhibit greater genetic differences between local populations and between social units for mtDNA than for autosomal DNA (e.g., rhesus macaques, *Macaca mulatta*; long-tailed macaques, *M. fascicularis*; toque macaques, *M. sinica*; Japanese macaques, *M. fuscata*; and pig-tailed macaques, *M. nemestrina*; reviewed in Melnick and

Hoelzer 1996). Similarly, gray mouse lemurs (*Microcebus murinus*) exhibit evidence for matrilineal social groups and male-biased dispersal. These lemurs are solitary while foraging at night, but during the day females form sleeping groups of two to four individuals while males sleep alone or in pairs (Martin 1973). Trapping data indicate that males disperse more than females (Fredsted et al. 2004), and genetic relatedness studies using microsatellite markers indicate that females in sleeping groups or with similar home ranges are closely related (Radespiel et al. 2001; Wimmer et al. 2002). Population genetic analyses of this species in the Kirindy forest in western Madagascar revealed greater spatial clustering of mtDNA haplotypes for females than males, providing evidence that genetic structure is influenced by matrilineal social structure and male-biased dispersal within this region (Wimmer et al. 2002; Fredsted et al. 2004). This spatial clustering of female haplotypes occurred despite an absence of evident geographic or ecological barriers to gene flow (Wimmer et al. 2002; Fredsted et al. 2004).

In contrast, other primate species exhibit female-biased dispersal, including chimpanzees and bonobos (*Pan* spp.), hamadryas baboons (*P. hamadryas hamadryas*), red colobus (*Piliocolobus* spp.), and species within the ateline tribe of New World monkeys (Struhsaker 1980; Kano 1992; Mitani et al. 2002; Hammond et al. 2006; Di Fiore and Campbell 2007). In these species, females generally leave their natal group before reaching sexual maturity. As expected, these species generally exhibit low genetic structure at mtDNA loci, but higher genetic structure for Y-chromosome loci. For example, mtDNA studies of chimpanzees (*Pan troglodytes*) revealed shared haplotypes and low phylogeographic structure across large geographic regions (Morin et al. 1994; Goldberg and Ruvolo 1997). On a smaller geographic scale, a study comparing genetic differentiation at the Y-chromosome versus mtDNA markers for four groups of chimpanzees in Uganda revealed higher genetic differentiation between groups for the Y chromosome than the mtDNA marker, with extensive sharing of haplotypes between groups for mtDNA, but no shared haplotypes between groups for Y-chromosome markers (Langergraber et al. 2007) (Fig. 15.4). Studies of bonobos (*Pan paniscus*) across their range in the Democratic Republic of Congo have also revealed lower genetic structure for mtDNA than Y-chromosome markers (Gerloff et al. 1999; Eriksson et al. 2006). Similarly, studies of two atelines in Amazonian Ecuador, the white-bellied spider monkeys (*Ateles belzebuth*) and woolly monkeys (*Lagothrix poeppigii*), revealed greater genetic structure for males than for females at autosomal microsatellite loci (Di Fiore et al. 2009).

Fewer studies investigating both social and genetic structure have been conducted for cetacean species than for primate species, probably because of the challenges of conducting long-term studies on species living in the marine environment and the very large home ranges of many cetacean species. One cetacean that has been studied in detail is the killer whale (*Orcinus orca*; Fig. 15.5). Long-term behavioral studies of killer whales in the North Pacific revealed high group stability for both males and females (Bigg et al. 1990; Baird and Whitehead 2000; Ford et al. 2000). This stability is thought to be responsible for the evolution of sympatric foraging specialists ("ecotypes") within this region (Hoelzel et al. 1998a). Genetic analyses of parentage within and between pods of the different ecotypes revealed that males and females remain in their natal pods, but that most matings occur

Fig. 15.4 Chimpanzees (*Pan troglodytes*) generally exhibit weak genetic structure for maternally inherited mitochondrial DNA (mtDNA) markers, but stronger genetic structure for paternally inherited Y-chromosome markers (Morin et al. 1994; Goldberg and Ruvolo 1997; Langergraber et al. 2007). This contrasting pattern for different marker types is likely driven by female-biased dispersal in chimpanzees; females generally leave their natal group before reaching sexual maturity for this species. (Published with kind permission of © Kathryn Shutt 2006. All rights reserved)

Fig. 15.5 Killer whales (*Orcinus orca*) have strong genetic divergence between populations, in some cases even between populations that have overlapping geographic ranges (Hoelzel et al. 2007). Behavioral observations and genetic parentage analyses indicate that this genetic structure is likely driven by strong group fidelity for both males and females (Bigg et al. 1990; Baird and Whitehead 2000; Ford et al. 2000; Pilot et al. 2010). (Published with kind permission of © Michael Richlen 2003. All rights reserved)

between pods rather than within pods, probably during temporary associations of pods or temporary dispersal by males (Pilot et al. 2010). Analyses of population genetic structure revealed that the social structure and mating system of killer whales had a strong influence on genetic structure (Hoelzel et al. 2007). Each population of killer whales had only one mtDNA control region haplotype, with no shared haplotypes between ecotypes. Significant genetic distinctions between each population and each ecotype were also found for nuclear microsatellite data, although, in contrast to the mtDNA control region, there were shared alleles between populations and ecotypes. Given that these ecotypes occur in sympatry, these results provide evidence that social structure is a stronger force than geographic isolation in driving population genetic structure. The stronger genetic structure for mtDNA than nuclear DNA likely reflects the male-biased genetic dispersal revealed through genetic parentage analyses.

Behavioral and genetic evidence indicates that pilot whales (*Globicephala* spp.) may have a matrilineal social structure similar to that of killer whales. Photographic identification studies have revealed long-term group fidelity for both long-finned pilot whales (*G. melas*) and short-finned pilot whales (*G. macrorhynchus*) (Heimlich-Boran 1993; Ottensmeyer and Whitehead 2003). Additionally, genetic relatedness analyses of long-finned pilot whales from the Faeroese drive-fishery indicated that both sexes remain in their natal pod, but that matings occur outside of the pod (Amos et al. 1993). As in killer whales, pilot whales exhibit population genetic divergence across short geographic distances at a number of locations across their ranges, and the stable social structures of these species have been implicated as factors responsible for this genetic structure. For example, social structure is thought to contribute to genetic structure between long-finned pilot whales in Tasmania versus New Zealand (Oremus et al. 2009) and Greenland versus the Faeroe Islands and Cape Cod (Fullard et al. 2000); and between short-finned pilot whales in Northern versus Southern Japan (Oremus et al. 2009). Based on evidence for male-biased dispersal from paternity analyses, we would expect greater genetic structure at mtDNA than autosomal DNA for these species; however, none of these pilot whale studies directly compared genetic structure at these two types of loci. Notably, however, social structure is not the only factor that has been implicated as a driver of genetic structure in pilot whales; sea surface temperature is thought to influence population genetic divergence within these species in Greenland and Japan, perhaps because of the distribution of their prey across different temperature regimes (Kasuya et al. 1988; Fullard et al. 2000).

Sperm whales (*Physeter macrocephalus*) also exhibit evidence for a matrilineal social structure, although genetic data indicate that the matrilineal structure of sperm whales is weaker than that of killer whales or pilot whales, with both related and unrelated individuals present in many "social units" (Richard et al. 1996; Lyrholm and Gyllensten 1998; Mesnick 2001). Sperm whale social structure also differs from that of killer whales and pilot whales in that social units are composed only of females and immatures; males disperse from their natal group and then live solitary or aggregated in temporary "bachelor groups" (Best 1979; Lettevall et al.

Fig. 15.6 Spinner dolphins (*Stenella longirostris*) in the northwestern Hawaiian Islands have little genetic divergence between atolls, despite the presence of strong social group stability that might be expected to deter dispersal between atolls (Andrews et al. 2010; Karczmarski et al. 2005a, b). Photographic identification data indicate genetic connectivity is driven by periodic movement of large groups between atolls (Karczmarski et al. 2005a). (Published with kind permission of © Susan Rickards 2001, NMFS Permit# GA-LOC-10021622. All rights reserved)

2002). Males also have a wider dispersal range, moving into polar waters to feed and returning to tropical regions to breed, whereas females and immatures remain in lower latitudes (Rice 1989), and thus a pattern of genetic structure reflecting male-biased dispersal between oceans might be predicted. Population genetic analyses indicate that sperm whale social structure does have a strong influence on genetic structure within this species, as indicated by the presence of greater genetic structure between social groups than between geographic regions across ocean basins (Lyrholm and Gyllensten 1998; Lyrholm et al. 1999) and within the Azores (Pinela et al. 2009). Genetic evidence also indicates that males disperse more frequently than females, with greater differentiation within and between ocean basins found for mtDNA than microsatellite loci (Lyrholm and Gyllensten 1998; Lyrholm et al. 1999; Engelhaupt et al. 2009).

The presence of strong group philopatry does not always result in genetic differentiation between social groups, however. For example, photographic identification data indicate high group stability for spinner dolphins in the atolls of the Northwestern Hawaiian Islands, with each atoll containing a single large group of 110–260 dolphins (Karczmarski et al. 2005a, b) (Fig. 15.6). Despite this high group stability, however, little or no genetic structure was found between atolls at mtDNA or autosomal loci (Andrews et al. 2010). These genetic results were supported by photographic identification data, which revealed that large groups (30–60 dolphins) periodically moved between Midway and Kure Atolls (Karczmarski et al. 2005a).

Dispersal in large groups, or "parallel dispersal," is relatively rare among mammals, and is generally thought to function in maintaining social relationships that are important for predator avoidance, resource acquisition, intraspecific competition, or the raising of young (van Hooff 2000; Handley and Perrin 2007). Parallel dispersal has also been observed in a number of primate species (reviewed by Schoof et al. 2009). However, parallel dispersal in primates usually involves groups of males, whereas the dispersing groups of spinner dolphins in the Northwestern Hawaiian atolls are composed of approximately equal numbers of males and females. This dispersal mechanism may have evolved among spinner dolphins in the Northwestern Hawaiian atolls to maintain strong social group stability while still permitting relatively frequent dispersal between atolls (Andrews et al. 2010).

15.3 Spatial Variation in Genetic Diversity: Population Size

Population genetic theory indicates that small effective population size is correlated with low neutral genetic diversity and increased risk of extinction caused by inbreeding depression, disease, environmental change, and stochastic events, and growing empirical evidence supports these ideas (Lande 1988; Frankham et al. 2002; Keller and Waller 2002; Hughes et al. 2008). Therefore, estimation of neutral genetic diversity can provide vital information regarding the vulnerability of populations. For example, mtDNA control region diversity of golden-brown mouse lemurs (*M. ravelobensis*) in northwestern Madagascar was severely reduced in small forest fragments compared to continuous forests, providing evidence that anthropogenic forest fragmentation has decreased population sizes, driven down genetic diversity, and increased population vulnerability in this species (Guschanski et al 2007).

In contrast, low genetic diversity in some populations is thought to have evolved without anthropogenic causes, and these populations may be adapted to low genetic diversity and therefore may not experience inbreeding depression. For example, low mtDNA control region and microsatellite diversity of spinner dolphins in the atolls of the Northwestern Hawaiian Islands as compared with the Main Hawaiian Islands is thought to result from historically small effective population sizes because of the small amount of resting habitat naturally available in the atolls of the Northwestern Hawaiian Islands (Andrews et al. 2010). Similarly, arguments have been put forth that the lack of mtDNA control region genetic diversity observed in the critically endangered vaquita (*Phocoena sinus*; Rosel and Rojas-Bracho 1999) results from a historically low population size rather than recent population declines (Rojas-Bracho and Taylor 1999; Taylor and Rojas-Bracho 1999), and that this species may therefore be adapted to low genetic diversity. This idea was put forth to counter arguments that the vaquita is "doomed" to extinction as a result of inbreeding depression resulting from recent human-mediated population declines.

15.4 Temporal Variation in Genetic Diversity

The spinner dolphin and vaquita examples described here illustrate some of the benefits that could be gained from an understanding of historical changes in genetic diversity over time. In these examples, knowing whether populations have always been small or whether recent declines have occurred is important in assessing population vulnerability. Recent advances in population genetics statistical methods using coalescent theory can address these and many other types of questions relevant to conservation biology by providing information on effective population sizes over time. These statistics use DNA sequence data or microsatellite allele frequency data combined with estimates of mutation rates for the genetic markers and generation length for the species (Rogers and Harpending 1992; Beaumont 1999; Beerli and Felsenstein 2001; Storz and Beaumont 2002; Drummond et al. 2005). For example, a microsatellite study of orangutans (*Pongo pygmaeus*) in northeastern Borneo using coalescent methods indicated that a population bottleneck occurred within the past few centuries, providing evidence that declines within this species were caused by recent anthropogenic destruction of forests rather than by farming or climate change in the more distant past (Goossens et al. 2006). Similarly, microsatellite studies of the Milne-Edwards' sportive lemur (*Lepilemur edwardsi*) and three mouse lemur species (*Microcebus* spp.) in northwestern Madagascar indicated major population declines for each of these species within the past few hundred years, a time of intensified human population growth and anthropogenic forest fragmentation (Olivieri et al. 2008; Craul et al. 2009). In contrast, a study of savannah baboons (*Papio cynocephalus*) in East Africa indicated that this species underwent a long-term population decline beginning in the late Pleistocene or early Holocene, with populations remaining relatively stable in more recent history, suggesting that population declines in this species were not likely caused by human impact (Storz et al. 2002).

The use of coalescent genetic methods to estimate historic population sizes has led to controversy with regard to the conservation and management of baleen whales. During the past two centuries, many populations of baleen whales were hunted to near extinction, leading to a ban on commercial whaling by the International Whaling Commission in 1986. Since that time, recovery assessments for each species have relied primarily upon comparisons of present-day population abundance estimates with pre-whaling abundance estimates based on whaling records. However, genetic estimates of pre-whaling abundance based on coalescent methods have been found to differ substantially from estimates based on whaling records. For example, genetic estimates of pre-whaling abundance for humpback whales, fin whales (*Balaenoptera physalus*), and minke whales (*Balaenoptera acutorostrata*) in the North Atlantic were as much as ten times higher than estimates based on whaling records (Roman and Palumbi 2003). These results provide controversial evidence that whale populations may not have recovered to the extent that was previously thought, and that management goals for these species should be reevaluated (Lubick 2003).

15.5 Conclusion

The examples described here illustrate the complex patterns of spatial and temporal variation in genetic diversity that commonly occur within cetacean and primate species, despite the high dispersal capabilities of these species which might be expected to homogenize genetic diversity across space and time. Furthermore, these examples illustrate the wide variety of environmental, behavioral, and demographic factors that drive the patterns of genetic diversity within these taxa, as well as the ways in which humans have influenced these patterns over time. Clearly, a thorough understanding of genetic diversity within cetacean and primate species requires multidisciplinary studies incorporating research on habitat, social structure, and demography for each species. As cetacean and primate species continue to face extinction threats caused by human impact, this multidisciplinary approach to research and management will become increasingly important if we are to preserve the diversity of cetacean and primate species on our planet.

References

Allendorf FW, Luikart G (2007) Conservation and the genetics of populations. Blackwell, Oxford

Amos B, Schlötterer C, Tautz D (1993) Social structure of pilot whales revealed by analytical DNA profiling. Science 30:670–672

Andrews KR, Karczmarski L, Au WWL et al (2010) Rolling stones and stable homes: social structure, habitat diversity, and population genetics of the Hawaiian spinner dolphin (*Stenella longirostris*). Mol Ecol 19:732–748

Anthony NM, Johnson-Bawe M, Jeffery K et al (2007) The role of Pleistocene refugia and rivers in shaping gorilla genetic diversity in central Africa. Proc Natl Acad Sci USA 104:20432–20436

Avise JC (1995) Mitochondrial DNA polymorphism and a connection between genetics and demography of relevance to conservation. Conserv Biol 9:686–690

Avise JC (2004) Molecular markers, natural history, and evolution. Sinauer Associates, Sunderland

Ayres JM, Clutton-Brock TH (1992) River boundaries and species range size in Amazonian primates. Am Nat 140:531–537

Baird RW, Whitehead H (2000) Social organization of mammal-eating killer whales: group stability and dispersal patterns. Can J Zool 78:2096–2105

Baker CS, Palumbi SR, Lambertsen RH et al (1990) Influence of seasonal migration on geographic distribution of mitochondrial DNA haplotypes in humpback whales. Nature (Lond) 344:238–240

Baker CS, Slade RW, Bannister JL et al (1994) Hierarchical structure of mitochondrial DNA gene flow among humpback whales *Megaptera novaeangliae*, worldwide. Mol Ecol 3:313–327

Beaumont MA (1999) Detecting population expansion and decline using microsatellites. Genetics 153:2013–2029

Beerli P, Felsenstein J (2001) Maximum likelihood estimation of a migration matrix and effective population sizes in n subpopulations by using a coalescent approach. Proc Natl Acad Sci USA 98:4563–4568

Best PB (1979) Social organization in sperm whales, *Physeter macrocephalus*. In: Winn HE, Olla BL (eds) Behavior of marine animals. Plenum, New York, pp 227–289

Bigg MA, Olesiuk PF, Ellis GM, Ford JKB, Balcomb KC (1990) Social organization and genealogy of resident killer whales (*Orcinus orca*) in the coastal waters of British Columbia and Washington State. Rep Int Whaling Comm 12:383–406, Special Issue

Bilgmann K, Moller LM, Harcourt RG, Gibbs SE, Beheregaray LB (2007) Genetic differentiation in bottlenose dolphins from South Australia: association with local oceanography and coastal geography. Mar Ecol Prog Ser 341:265–276

Coyne JA, Orr HA (2004) Speciation. Sinauer, Sunderland

Craul M, Chikhi L, Sousa V et al (2009) Influence of forest fragmentation on an endangered large-bodied lemur in northwestern Madagascar. Biol Conserv 142:2862–2871

Di Fiore A (2003) Molecular genetic approaches to the study of primate behavior, social organization, and reproduction. Yearb Phys Anthropol 46:62–99

Di Fiore A, Campbell CJ (2007) The atelines: variation in ecology, behavior, and social organization. In: Campbell CJ, Fuentes A, MacKinnon KC, Panger M, Beader SK (eds) Primates in perspective. Oxford University Press, New York, pp 155–185

Di Fiore A, Link A, Schmitt CA, Spehar SN (2009) Dispersal patterns in sympatric woolly and spider monkeys: integrating molecular and observational data. Behaviour 146:437–470

Dowling TE, Brown WM (1993) Population structure of the bottlenose dolphin (*Tursiops truncatus*) as determined by restriction endonuclease analysis of mitochondrial DNA. Mar Mamm Sci 9:138–155

Drummond AJ, Rambaut A, Shapiro B, Pybus OG (2005) Bayesian coalescent inference of past population dynamics from molecular sequences. Mol Biol Evol 22:1185–1192

Engelhaupt D, Hoelzel AR, Nicholson C, Frantzis A, Mesnick S, Gero S, Whitehead H, Rendell L, Miller P, De Stefanis R, Canadas A, Airoldi S, Mignucci-Giannoni AA (2009) Female philopatry in coastal basins and male dispersion across the North Atlantic in a highly mobile marine species, the sperm whale (*Physeter macrocephalus*). Mol Ecol 18:4193–4205

Eriksson J, Hohmann G, Boesch C, Vigilant L (2004) Rivers influence the population genetic structure of bonobos (*Pan paniscus*). Mol Ecol 13:3425–3435

Eriksson J, Siedel H, Lukas D et al (2006) Y-chromosome analysis confirms highly sex-biased dispersal and suggests a low male effective population size in bonobos (*Pan paniscus*). Mol Ecol 15:939–949

Escorza-Treviño S, Archer FI, Rosales M, Lang AM, Dizon AE (2005) Genetic differentiation and intraspecific structure of Eastern Tropical Pacific spotted dolphins, *Stenella attenuata*, revealed by DNA analyses. Conserv Genet 6:587–600

Fontaine MC, Baird SJE, Piry S et al (2007) Rise of oceanographic barriers in continuous populations of a cetacean: the genetic structure of harbour porpoises in Old World waters. BMC Biol 5:30

Ford J, Ellis G, Balcomb K (2000) Killer whales: the natural history and genealogy of *Orcinus orca* in British Columbia and Washington State, 2nd edn. UBC, Vancouver

Frankham R, Ballou JD, Briscoe DA (2002) Introduction to conservation genetics. Cambridge University Press, Cambridge

Fredsted T, Pertoldi C, Olesen JM, Eberle M, Kappeler PM (2004) Microgeographic heterogeneity in spatial distribution and mtDNA variability of gray mouse lemurs (*Microcebus murinus*, Primates: Cheirogaleidae). Behav Ecol Sociobiol 56:393–403

Fullard KJ, Early G, Heide-Jørgensen MP et al (2000) Population structure of long-finned pilot whales in the North Atlantic: a correlation with sea surface temperature? Mol Ecol 9:949–958

Gerloff U, Hartung B, Fruth B, Hohmann G, Tautz D (1999) Intracommunity relationships, dispersal pattern and paternity success in a wild living community of Bonobos (*Pan paniscus*) determined from DNA analysis of faecal samples. Proc R Soc Lond B Biol Sci 266:1189–1195

Goldberg TL, Ruvolo M (1997) The geographic apportionment of mitochondrial genetic diversity in East African chimpanzees, *Pan troglodytes schweinfurthii*. Mol Biol Evol 14:976–984

Gonder MK, Disotell TR, Oates JF (2006) New genetic evidence on the evolution of chimpanzee populations and implications for taxonomy. Int J Primatol 27:1103–1127

Goossens B, Chikhi L, Ancrenaz M et al (2006) Genetic signature of anthropogenic population collapse in orangutans. PLoS Biol 4:285–291

Greenwood PJ (1980) Mating systems, philopatry and dispersal in birds and mammals. Anim Behav 28:1140–1162
Guschanski K, Olivieri G, Funk SM, Radespiel U (2007) MtDNA reveals strong genetic differentiation among geographically isolated populations of the golden brown mouse lemur, *Microcebus ravelobensis*. Conserv Genet 8:809–821
Guschanski K, Caillaud D, Robbins MM, Vigilant L (2008) Females shape the genetic structure of a gorilla population. Curr Biol 18:1809–1814
Haffer J (1969) Speciation in Amazonian forest birds. Science 165:131–137
Hammond RL, Handley LJL, Winney BJ, Bruford MW, Perrin N (2006) Genetic evidence for female-biased dispersal and gene flow in a polygynous primate. Proc Biol Sci 273:479–484
Handley LJL, Perrin N (2007) Advances in our understanding of mammalian sex-biased dispersal. Mol Ecol 16:1559–1578
Hedrick PW (1999) Highly variable loci and their interpretation in evolution and conservation. Evolution 53:313–318
Heimlich-Boran JR (1993) Social organization of the short-finned pilot whale, *Globicephala macrorhynchus*, with special reference to the social ecology of delphinids. Ph.D. thesis, Cambridge University, Cambridge
Hewitt G (2000) The genetic legacy of the Quaternary ice ages. Nature (Lond) 405:907–913
Hoelzel AR, Dahlheim M, Stern SJ (1998a) Low genetic variation among killer whales (*Orcinus orca*) in the Eastern North Pacific and genetic differentiation between foraging specialists. J Hered 89:121–128
Hoelzel AR, Potter CW, Best PB (1998b) Genetic differentiation between parapatric 'nearshore' and 'offshore' populations of the bottlenose dolphin. Proc R Soc Lond B Biol Sci 265: 1177–1183
Hoelzel AR, Hey J, Dahlheim ME et al (2007) Evolution of population structure in a highly social top predator, the killer whale. Mol Biol Evol 24:1407–1415
Hughes AR, Inouye BD, Johnson MTJ, Underwood N, Vellend M (2008) Ecological consequences of genetic diversity. Ecol Lett 11:609–623
Jalil MF, Cable J, Inyor JS et al (2008) Riverine effects on mitochondrial structure of Bornean orangutans (*Pongo pygmaeus*) at two spatial scales. Mol Ecol 17:2898–2909
Jefferson TA, Leatherwood S, Webber MA (1993) FAO species identification guide. Marine mammals of the world. FAO, Rome
Kano T (1992) The last ape. Stanford University Press, Stanford
Karczmarski L, Rickards SH, Gowans S, et al (2005a) 'One for all and all for one': intra-group dynamics of an insular spinner dolphin population. In: Abstracts of the 16th biennial conference on the biology of marine mammals, San Diego, 12–17 December 2005
Karczmarski L, Wursig B, Gailey G, Larson KW, Vanderlip C (2005b) Spinner dolphins in a remote Hawaiian atoll: social grouping and population structure. Behav Ecol 16:675–685
Kasuya T, Miyashita T, Kasamatsu F (1988) Segregation of two forms of short-finned pilot whales off the Pacific coast of Japan. Sci Rep Whales Res Inst 39:77–90
Keller LF, Waller DM (2002) Inbreeding effects in wild populations. Trends Ecol Evol 17: 230–241
Krützen M, Sherwin WB, Berggren P, Gales N (2004) Population structure in an inshore cetacean revealed by microsatellite and mtDNA analysis: bottlenose dolphins (*Tursiops* sp.) in Shark Bay, Western Australia. Mar Mamm Sci 20:28–47
Lande R (1988) Genetics and demography in biological conservation. Science 241:1455–1460
Langergraber KE, Siedel H, Mitani JC et al (2007) The genetic signature of sex-biased migration in patrilocal chimpanzees and humans. PLoS One 2(10):e973
Lessa EP, Cook JA, Patton JL (2003) Genetic footprints of demographic expansion in North America, but not Amazonia, during the Late Quaternary. Proc Natl Acad Sci USA 100:10331–10334
Lettevall E, Richter C, Jaquet N et al (2002) Social structure and residency in aggregations of male sperm whales. Can J Zool 80:1189–1196

Liu ZJ, Ren BP, Wu RD et al (2009) The effect of landscape features on population genetic structure in Yunnan snub-nosed monkeys (*Rhinopithecus bieti*) implies an anthropogenic genetic discontinuity. Mol Ecol 18:3831–3846

Lubick N (2003) Ecology: new count of old whales adds up to big debate. Science 301:451

Lyrholm T, Gyllensten U (1998) Global matrilineal population structure in sperm whales as indicated by mitochondrial DNA sequences. Proc R Soc Lond B Biol Sci 265:1679–1684

Lyrholm T, Leimar O, Johanneson B, Gyllensten U (1999) Sex-biased dispersal in sperm whales: contrasting mitochondrial and nuclear genetic structure of global populations. Proc R Soc Lond B Biol Sci 266:347–354

Martin R (1973) A review of the behaviour and ecology of the lesser mouse lemur (*Microcebus murinus* J.F. Miller 1777). In: Michael R, Crook JH (eds) Comparative ecology and behaviour of primates. Academic, London, pp 1–68

Matthews CJD, Luque SP, Petersen SD, Andrews RD, Ferguson SH (2011) Satellite tracking of a killer whale (*Orcinus orca*) in the eastern Canadian Arctic documents ice avoidance and rapid, long-distance movement into the North Atlantic. Polar Biol 34:1091–1096. doi:10.1007/s00300-0100-0958

Melnick DJ, Hoelzer GA (1996) The population genetic consequences of macaque social organization and behaviour. In: Fa JE, Lindburg DG (eds) Evolution and ecology of macaque societies. Cambridge University Press, Cambridge, pp 413–443

Melnick DJ, Pearl MC (1987) Cercopithecines in multi-male groups: genetic diversity and population structure. In: Smuts BB, Cheney DL, Seyfarth RM, Wrangham RW, Struhsaker TT (eds) Primate societies. University of Chicago Press, Chicago, pp 121–134

Mesnick SL (2001) Genetic relatedness in sperm whales: evidence and cultural implications. Behav Brain Sci 24:346

Mitani JC, Watts DP, Muller MN (2002) Recent developments in the study of wild chimpanzee behavior. Evol Anthropol 11:9–25

Möller LM, Beheregaray LB (2004) Genetic evidence for sex-biased dispersal in resident bottlenose dolphins (*Tursiops aduncus*). Mol Ecol 13:1607–1612

Morin PA, Moore JJ, Chakraborty R et al (1994) Kin selection, social structure, gene flow, and the evolution of chimpanzees. Science 265:1193–1201

Natoli A, Peddemors VM, Hoelzel AR (2004) Population structure and speciation in the genus *Tursiops* based on microsatellite and mitochondrial DNA analyses. J Evol Biol 17:363–375

Natoli A, Birkun A, Aguilar A, Lopez A, Hoelzel AR (2005) Habitat structure and the dispersal of male and female bottlenose dolphins (*Tursiops truncatus*). Proc Roy Soc B Biol Sci 272:1217–1226

Natoli A, Canadas A, Peddemors VM et al (2006) Phylogeography and alpha taxonomy of the common dolphin (*Delphinus* sp.). J Evol Biol 19:943–954

Norris KS, Würsig B, Wells RS et al (1994) The Hawaiian spinner dolphin. University of California Press, Berkeley

Olivieri GL, Sousa V, Chikhi L, Radespiel U (2008) From genetic diversity and structure to conservation: genetic signature of recent population declines in three mouse lemur species (*Microcebus* spp.). Biol Conserv 141:1257–1271

Oremus M, Poole MM, Steel D, Baker CS (2007) Isolation and interchange among insular spinner dolphin communities in the South Pacific revealed by individual identification and genetic diversity. Mar Ecol Prog Ser 336:275–289

Oremus M, Gales R, Dalebout ML et al (2009) Worldwide mitochondrial DNA diversity and phylogeography of pilot whales (*Globicephala* spp.). Biol J Linn Soc 98:729–744

Östman JSO (1994) Social organization and social behavior of Hawai'ian spinner dolphins (*Stenella longirostris*). Ph.D. dissertation, University of California, Santa Cruz

Ottensmeyer CA, Whitehead H (2003) Behavioural evidence for social units in long-finned pilot whales. Can J Zool 81:1327–1338

Pastene LA, Goto M, Kanda N et al (2007) Radiation and speciation of pelagic organisms during periods of global warming: the case of the common minke whale, *Balaenoptera acutorostrata*. Mol Ecol 16:1481–1495

Pilot M, Dahlheim ME, Hoelzel AR (2010) Social cohesion among kin, gene flow without dispersal and the evolution of population genetic structure in the killer whale (*Orcinus orca*). J Evol Biol 23:20–31

Pinela AM, Querouil S, Magalhaes S et al (2009) Population genetics and social organization of the sperm whale (*Physeter macrocephalus*) in the Azores inferred by microsatellite analyses. Can J Zool 87:802–813

Pomilla C, Rosenbaum HC (2006) Estimates of relatedness in groups of humpback whales (*Megaptera novaeangliae*) on two wintering grounds of the Southern Hemisphere. Mol Ecol 15:2541–2555

Poole MM (1995) Aspects of behavioral ecology of spinner dolphins (*Stenella longirostris*) in the nearshore waters of Mo'orea, French Polynesia. Ph.D. thesis, University of California, Santa Cruz

Radespiel U, Sarikaya Z, Zimmermann E, Bruford MW (2001) Sociogenetic structure in a free-living nocturnal primate population: sex-specific differences in the grey mouse lemur (*Microcebus murinus*). Behav Ecol Sociobiol 50:493–502

Radespiel U, Rakotondravony R, Chikhi L (2008) Natural and anthropogenic determinants of genetic structure in the largest remaining population of the endangered golden-brown mouse lemur, *Microcebus ravelobensis*. Am J Primatol 70:860–870

Rice DW (1989) Sperm whale. *Physeter macrocephalus* Linnaeus, 1758. In: Ridgway SH, Harrison R (eds) Handbook of marine mammals. Academic, London, pp 177–233

Richard KR, Dillon MC, Whitehead H, Wright JM (1996) Patterns of kinship in groups of free-living sperm whales (*Physeter macrocephalus*) revealed by multiple molecular genetic analyses. Proc Natl Acad Sci USA 93:8792–8795

Rogers AR, Harpending H (1992) Population growth makes waves in the distribution of pairwise genetic differences. Mol Biol Evol 9:552–569

Rojas-Bracho L, Taylor BL (1999) Risk factors affecting the vaquita (*Phocoena sinus*). Mar Mamm Sci 15:974–989

Roman J, Palumbi SR (2003) Whales before whaling in the North Atlantic. Science 301:508–510

Rosel PE, Rojas-Bracho L (1999) Mitochondrial DNA variation in the critically endangered vaquita *Phocoena sinus* Norris and MacFarland, 1958. Mar Mamm Sci 15:990–1003

Rosel PE, Hansen L, Hohn AA (2009) Restricted dispersal in a continuously distributed marine species: common bottlenose dolphins *Tursiops truncatus* in coastal waters of the western North Atlantic. Mol Ecol 18:5030–5045

Schoof VAM, Jack KM, Isbell LA (2009) What traits promote male parallel dispersal in primates? Behaviour 146:701–726

Sellas AB, Wells RS, Rosel PE (2005) Mitochondrial and nuclear DNA analyses reveal fine scale geographic structure in bottlenose dolphins (*Tursiops truncatus*) in the Gulf of Mexico. Conserv Genet 6:715–728

Storz JF (1999) Genetic consequences of mammalian social structure. J Mammal 80:553–569

Storz JF, Beaumont MA (2002) Testing for genetic evidence of population expansion and contraction: an empirical analysis of microsatellite DNA variation using a hierarchical Bayesian model. Evolution 56:154–166

Storz JF, Beaumont MA, Alberts SC (2002) Genetic evidence for long-term population decline in a savannah-dwelling primate: inferences from a hierarchical Bayesian model. Mol Biol Evol 19:1981–1990

Struhsaker TT (1980) The red colobus monkey. University of Chicago Press, Chicago

Sugg DW, Chesser RK, Dobson FS, Hoogland JL (1996) Population genetics meets behavioral ecology. Trends Ecol Evol 11:338–342

Taylor BL, Rojas-Bracho L (1999) Examining the risk of inbreeding depression in a naturally rare cetacean, the vaquita (*Phocoena sinus*). Mar Mamm Sci 15:1004–1028

Telfer PT, Souquiere S, Clifford SL et al (2003) Molecular evidence for deep phylogenetic divergence in *Mandrillus sphinx*. Mol Ecol 12:2019–2024

Tezanos-Pinto G, Baker CS, Russell K et al (2009) A worldwide perspective on the population structure and genetic diversity of bottlenose dolphins (*Tursiops truncatus*) in New Zealand. J Hered 100:11–24

Tolley KA, Víkingsson GA, Rosel PE (2001) Mitochondrial DNA sequence variation and phylogeographic patterns in harbour porpoises (*Phocoena phocoena*) from the North Atlantic. Conserv Genet 2:349–361

Valsecchi E, Hale P, Corkeron P, Amos W (2002) Social structure in migrating humpback whales (*Megaptera novaeangliae*). Mol Ecol 11:507–518

van Hooff JARAM (2000) Relationships among non-human primate males: a deductive framework. In: Kappeler PM (ed) Primate males: causes and consequences of variation in group composition. Cambridge University Press, Cambridge, pp 183–191

Vigilant L, Guschanski K (2009) Using genetics to understand the dynamics of wild primate populations. Primates 50:105–120

Wakeley J (2000) The effects of subdivision on the genetic divergence of populations and species. Evolution 54:1092–1101

Wilson EO (2002) The future of life. Knopf, New York

Wimmer B, Tautz D, Kappeler PM (2002) The genetic population structure of the gray mouse lemur (*Microcebus murinus*), a basal primate from Madagascar. Behav Ecol Sociobiol 52:166–175

Winney BJ, Hammond RL, Macasero W et al (2004) Crossing the Red Sea: phylogeography of the hamadryas baboon, *Papio hamadryas hamadryas*. Mol Ecol 13:2819–2827

Wright S (1931) Evolution in Mendelian populations. Genetics 16:97–159

Chapter 16
Eco-toxicants: A Growing Global Threat

Victoria Tornero, Teresa J. Sylvina, Randall S. Wells, and Jatinder Singh

Human–dolphin interactions. (Photograph provided by Randal S. Wells)

The author name Teresa J. Sylvina was also published as Taranjit Kaur

V. Tornero (✉)
Stazione Zoologica Anton Dhorn di Napoli, Villa Comunale, 80121, Naples, Italy
e-mail: Victoriatornero@ub.edu; Victoria.tornero@szn.it

T.J. Sylvina • J. Singh
Virginia-Maryland Regional College of Veterinary Medicine, Virginia Polytechnic Institute and State University, Blacksburg, VA, USA

R.S. Wells
Chicago Zoological Society, c/o Mote Marine Laboratory, Sarasota, FL, USA

J. Yamagiwa and L. Karczmarski (eds.), *Primates and Cetaceans: Field Research and Conservation of Complex Mammalian Societies*, Primatology Monographs, DOI 10.1007/978-4-431-54523-1_16,

Abstract Ecotoxicology is a constantly evolving discipline, and ecological risk assessments are required to estimate and predict threats and exposures before they occur. There is a critical need to obtain species- and site-specific data to determine toxicological sensitivities under prevailing environmental conditions and so mitigate the potential effects of pollutants on wildlife, particularly on species that are already endangered. Cetaceans are considered one of the most vulnerable organisms with respect to long-term toxicity of chemicals such as organochlorines and metals. Many studies have investigated the factors affecting the accumulation, metabolization, and potential harmful effects of such pollutants, but the question of whether chemical pollution is changing the dynamics of the populations is not yet resolved. In the case of primates, although extrapolations can be made to some extent with humans, the pollution levels and their effects in wild populations are largely unknown. The present chapter examines the main cetacean and primate features that might lead to differential vulnerability among individuals, populations, and species, and evaluates how behavioral patterns can influence exposure potential and outcomes. Furthermore, the role of pollutants as stressors of the ecological health and social systems of cetaceans and primates is explored by two case studies: the first case describes site fidelity of common bottlenose dolphins (*Tursiops truncatus*) in Sarasota Bay (Florida, USA), despite the cumulative impacts that they face resulting from human activities, and the second case discusses manganese-induced neurotoxicity as a causal agent of cognitive, motor, and behavioral dysfunctions in cynomolgus monkeys (*Macaca fascicularis*).

Keywords Behavioral patterns • Case of study • Cetaceans • Chemical pollution • Metals • Organochlorines • Primates

16.1 Introduction

The reliable assessment of the nature and magnitude of the impact of environmental contaminants on ecosystems has become essential for the successful management and conservation of wildlife populations. Among pollutants, organochlorine compounds (OCs) and toxic metals have been a concern during the past decades because they have become widespread and have reached almost all natural environments. OCs were introduced in the 1930s and their production peaked in the 1960s and 1970s. Toward the end of this period, they were identified as hazardous substances and, as a result, their use was banned in most industrialized countries. However, the recalcitrant nature of some forms, such as DDTs (dichlorodiphenyltrichloroethane and its metabolites), and PCBs (polychlorinated biphenyls), and their consequent long half-life has favored their trapping in terrestrial sediments and landfills and their global transport to the marine environment (Voldner and Yi-Fan 1995). These compounds bioaccumulate in the fatty tissues of organisms and biomagnify through food webs, reaching extremely high concentrations in species located at the highest trophic levels (Tanabe 2002). Small odontocete cetaceans, as top predators, accumulate lipophilic pollutants efficiently in their blubber (Aguilar et al. 1999) and are, in effect, the vertebrates in which the highest OC concentrations have been so far reported (O'Shea and Aguilar 2001). Trace element contaminants also bioaccumulate in the organs and tissues of apex predators, reaching alarming levels in those areas where natural or anthropogenic accumulation of metals occur in elevated concentrations (Bustamante et al. 2003).

OCs have been used worldwide in industry and agriculture, and have also been used in public health as vector control agents, specifically, DDT for the control of malaria, leishmaniasis, and typhus (Azeredo et al. 2008; Gahan et al. 1945; Oliveira 1997; Smith 1991; Turusov et al. 2002; WHO 1989). OCs and heavy metals can be transported to the marine environment and also can remain locally in freshwater systems (Manirakiza et al. 2002; Simonich and Hites 1995). They can cycle through evaporation from soils and surface waters or incineration of organochlorine wastes into the atmosphere and subsequent deposition back to the Earth's surface. OCs and toxic metals released into the environment are distributed globally, in urban areas as well as once remote inaccessible areas including tropical and subtropical regions of Africa, Asia, and the Americas. As such, the role of eco-toxicants as one of the stressors that may affect primate populations is an emerging issue.

Although there is a wealth of information on concentrations of these pollutants in cetaceans, few studies have explored the relationship between contaminant exposure and toxicity under natural conditions. There is no evidence in the literature of any acute chemical poisoning episode among cetacean species. The major threat for these animals seems to be through the food web, which may provoke sublethal effects over time. High OC concentrations have been associated with adverse effects on reproduction and immune function (Béland et al. 1993; Lahvis et al. 1995; Martineau et al. 1987; Schwacke et al. 2002), and as a result of the latter, with disease outbreaks that have led to large-scale unusual mortality events (Aguilar and Borrell 1994; Geraci 1989). Severe and prolonged metal toxicity in dolphins has been linked to renal pathology and bone malformations (Lavery et al. 2009; Long et al. 1997). In recent years, the occurrence of cetacean unusual mortality events has increased and, as a consequence, some dolphin stocks around the world have declined considerably. This realization has prompted interest in assessing the potential involvement of anthropogenic chemicals in such events (Kuehl and Haebler 1995).

Pollutant concentrations in the tissues of wild primates have not been reported. In contrast to carcasses of aquatic, avian, and marine mammal species that will surface and wash ashore, wild primate carcasses are generally not found, limiting opportunistic postmortem sampling and analysis. Comparatively, their terrestrial nature, geographic range, and diet limit the amount of direct exposure to these pollutants and contribute to the paucity of information available on their ecological vulnerability. However, studies in human and herbivorous terrestrial mammals have reported OC and toxic metal exposures and differential accumulation in contaminated environments (Beyer et al. 2007; Carrizo et al. 2007; Röllin et al. 2005; Röllin et al. 2009; Suutari et al. 2009).

Examination of possible health risks from contaminant exposure is a complicated task in any organism, but particularly in those animals whose protected status does not permit experimentation under controlled laboratory conditions: this is the case of cetaceans where the lack of dose–response information hinders the identification of cause-effect relationships. Other difficulties arise from the large natural variability in exposures among species, the complexity of the chemical mixtures to which individuals are exposed, the quality of the samples examined (e.g., tissues that experience postmortem autolysis), and the impossibility to determine essential biological variables without invasively collecting internal tissues.

As with cetaceans, many primate species are protected. However, others, such as macaques, are used as models for controlled toxicology studies (Chen et al. 1983; Erikson et al. 2008; Schneider et al. 2006; Schoeffner and Thorgeirsson 2000; Thorgeirsson et al 1994). For example, long-term oral administration of DDT in the diet of cynomolgus and rhesus monkeys for 130 months provided evidence of hepatic and central nervous system toxicity in these primate species (Schoeffner and Thorgeirsson 2000; Takayama et al. 1999; Zühlke and Weinbauer 2003). Laboratory animal studies have also shown differential sensitivities among primate species, for example, with the toxic metal arsenic (Vahter 1999). However, the relevance of controlled studies to wild primates and their ecological vulnerability to eco-toxicants is unknown. Therefore, the potential effects of differential exposures to chemical pollutants on animal behavior have been certainly very little investigated in either cetaceans or primates. In addition, it is often not possible to establish unequivocally that an observed effect is caused by a particular environmental factor. Most factors co-vary and are thus significant in influencing responses. For example, other human activities not directly related to chemical releases, such as habitat disruption, directed hunts, fishery interactions, and overfishing, may interact with pollution to produce negative impacts on the populations. This chapter aims to assess the potential impact of pollution on primates and cetaceans by estimating the likelihood that health risks may be the result of the interactive effects of contaminant exposure and animal behavior. To approach this task, we examine the main factors that might account for variations in vulnerability across populations and species. Furthermore, we evaluate the potential of pollutants as stressors that affect the ecological health and social system of cetaceans and primates using two case studies: site fidelity and common bottlenose dolphins (*Tursiops truncatus*) from Sarasota Bay, Florida, and manganese-induced neurotoxicity and cynomolgus monkeys (*Macaca fascicularis*).

16.2 Behaviorally Related Factors Likely Affecting Vulnerability to Pollution

Several traits of individuals may give information on their potential sensitivity to pollutants. At the individual scale, it is well established that a number of biological variables, such as age, sex, and reproductive status, affect organochlorine and trace metal concentrations in the tissues of marine mammals, producing substantial variations among individuals (Aguilar et al. 1999; Lahaye et al. 2006; Wells et al. 2005). Research conducted on humans, and laboratory toxicology studies on rodent models and nonhuman primates, indicate that such differences may also cause differential vulnerabilities in wild primate species. As a result, information about these variables is critical before bioaccumulation of pollutants can be correctly understood and to allow further comparison between species and areas. Other factors associated with animal behavior have been relatively less explored, although they are also likely to lead to differences in the availability, uptake, persistence, and toxic effects of chemicals and, therefore, in the vulnerability to pollutants at the population scale. We discuss and compare here the implications of these factors for differences in exposure to contaminants on cetacean and primate populations. When no information is available, we refer to extrapolations based on findings in other species.

16.2.1 Habitat

Cetacean and primate populations, as well as many other mammals, usually exhibit patterns of distribution associated with specific habitats or concentrations of food. Therefore, different populations may be subject to different human impacts. Coastal marine cetaceans are at risk from chemical contaminants because they inhabit areas that suffer dense human populations and hence intense industrial and agricultural activities. It is widely recognized that dolphins living in nearshore waters tend to show the highest concentrations of chemical pollutants (O'Shea 1999). As an example, Indo-Pacific bottlenose dolphins (*Tursiops aduncus*) from South Australia have been reported to carry the highest metal levels, probably because this dolphin inhabits shallow, coastal areas more impacted by pollutants than the pelagic offshore habitats of other species of the region, such as common bottlenose dolphins or common dolphins (*Delphinus delphis*) (Lavery et al. 2008). Similarly, it is reasonable to expect higher pollutant loads in species occurring off industrialized areas than in those off less-developed regions (Aguilar et al. 2002). Nevertheless, global trends in contamination show a diminution of the concentrations in the areas where pollution was originally elevated and, nowadays, the major sources of organochlorines are present in developing countries. Moreover, the volatilization and dispersion of pollutants as a result of atmospheric transport have led to a considerable contamination in regions located further away from direct pollution sources. Now, high contaminant levels have been found in cetaceans inhabiting remote areas without large industrial activity, such as the tropical oceans and Arctic waters (Aguilar et al. 2002; Tanabe et al. 1994).

Primates inhabit a wide range of habitats including rainforest, montane forest, savanna, and urban landscapes (Higham et al. 2009; Nishida 1990; Ogawa et al. 2007; Sha et al. 2009). Nonhuman primates are being subjected to a rapidly increasing human population and anthropogenic activities that are impacting these habitats (e.g., increased urbanization, expansion of industrial activities, and the use of gasoline additives). In addition to air and water contamination in urban areas in the developing world, OCs and toxic metals have even been detected in low concentrations in rainforest, terrestrial areas, and freshwater bodies in remote areas of the world (Simonich and Hites 1995; Vandelannoote et al. 1996). Forest habitat fragmentation and loss is occurring at an alarming rate from a number of activities, including commercial logging, use of forest biomass for making charcoal, gathering of wood for fuel, and clearing of land for agricultural plots. Mining and smelting, the use of DDT for vector control, and the extension of modern agriculture practices into less-developed areas are other potential sources of pollutants in areas adjacent to primate habitats (Akagi et al. 2000; Banza et al. 2009; Campbell et al. 2008; Kishimba et al. 2004; Manirakiza et al. 2002; Sing et al. 2003; van Straaten 2000). Effects are associated with a wide range of factors in the habitat, including biological diversity, biogeochemistry, climate, and the amount, degree, and types of habitat destruction from other anthropogenic activities. In general, because of the terrestrial nature of primates, their potential for exposure, as well as the routes and degree of exposure to eco-toxicants, differs from that of cetaceans.

Information on habitat usage and patterns of occupancy is important in predicting potential impacts of chemical pollutants on wild mammal populations. Strong site fidelity, for example, can place individuals at higher risk from point-sources of pollution or hotspots (Brager et al. 2002). Long-term site fidelity and restricted home ranges have been described for a number of coastal delphinids (McSweeney et al. 2007; Wells 1991, 2003; Zolman 2002). These animals do not, or cannot, relocate in response to environmental changes, so any alteration of their habitat may have greater negative effects on them. This impact becomes particularly significant in coastal areas influenced by other anthropogenic pressures, such as increasing urbanization, fishing activities, and vessel traffic. Less is known about site fidelity for oceanic cetaceans, although some are known to move over large distances (Wells et al. 1999; Whitehead 2003). Thus, their wider offshore home range seems to make them much less susceptible to inshore human activities. However, some oceanic species have been found to carry and accumulate similar or higher burdens of environmental contaminants as inshore dolphins (Borrell and Aguilar 2005; Stockin et al. 2007). Although this might be attributable to the characteristics inherent to the species, particularly feeding behavior, it can also be reflective of some incursions of those open water dolphins into more contaminated coastal areas (Herman et al. 2005; Lahaye et al. 2006; Stockin et al. 2007).

Understanding the patterns of residency and range of movements of individuals is fundamental for the assessment of long-term population viability, although there are some limitations. Although the detection of critical zones may be reasonably attainable for inshore resident cetaceans, the identification of protection areas for highly mobile species or those with undefined patterns of migration presents many more difficulties. Through a gradual habituation process, wild primate populations accept proximity to humans, and studies show that wild primates also exhibit site fidelity (Janmaat et al. 2009; Jolly and Pride 1999; Mertl-Millhollen 2000; Murray et al. 2008).

The potential detection of any significant variation in habitat use by these animals could be a suitable indicator of significant habitat degradation or disturbance. We know of no evidence of changes in ranging patterns in cetacean species or wild primates as a function of exposure to contaminants. Moreover, in such a change, it would not be easy to associate it directly with pollution because, as already mentioned, many human activities may interact and are likely to produce effects. Similarly, oceanological features such as water temperature, depth, bottom topography, surface salinity, and sediment type have been also found to affect the distribution of cetaceans (Baumgartner 1997; Doniol-Valcroze et al. 2007; Natoli et al. 2005), so changes in patterns of distribution of the individuals could result from changes in environmental conditions. On the other hand, displacement away from the potential source of disturbance could also have detrimental effects on the individuals. In particular, diversion to lower-quality habitats has been related to decreased reproductive success (Lusseau 2003; Mann et al. 2000), which might have repercussions at the population level.

16.2.2 Feeding Behavior

A factor very likely to lead to differential vulnerability between individuals is feeding behavior. In cetaceans, the bioaccumulation of chemical pollutants occurs mostly through dietary ingestion (Borga et al. 2004; Law et al. 1991, 1992), so concentrations in preferred prey determine to a large extent the levels found in the organisms. Additionally, organochlorine compounds and many trace metals biomagnify in marine ecosystems and reach the highest concentrations in cetaceans feeding at the top of marine food webs (Aguilar et al. 1999).

The influence of diet on the accumulation of chemical contaminants can have implications at both the intra- and interspecific scale. Within a population of the same cetacean species, differences in dietary preferences among individuals have been reported and related to age (Meynier et al. 2008), sex (Barber et al. 2001; Burger et al. 2007), body size (Borga et al. 2004; Burger et al. 2007), or reproductive status (Bernard and Hohn 1989). These factors can affect the prey type and sizes eaten and, consequently, the contaminant intake. Moreover, despite foraging behavior being very species specific, dissimilarities in diet composition have been described between geographic areas, mainly associated with the most abundant prey type at each locality (Herman et al. 2005). Seasonal changes in prey distribution and abundance patterns could also lead to temporal shifts in diet from one key prey species to another (Meynier et al. 2008). Whatever the reasons for food type change, such variations would have effects on the contaminant loads found in the organisms and, therefore, would account for potential differential risks among the components of the population.

We may also expect significant interspecific differences in pollutant concentrations linked to dietary differences. Thus, as already mentioned, contaminant levels are frequently higher in toothed whales than in baleen whales as a result of the process of biomagnification. Moreover, even species sharing the same waters or feeding on the same trophic level can exploit separate food resources and hence be exposed to different risks from pollutant exposure (Borrell and Aguilar 2005; Borrell et al. 2006; Metcalfe et al. 2004). For example, it is assumed that a diet based on cephalopods constitutes a major source of cadmium for small cetaceans

(Bustamante et al. 1998). As a result, concentrations for this metal have been found to be higher in dolphin species feeding mostly on squids than in those with a diet composed mainly of fish (Das et al. 2003; Lahaye et al. 2006). The effect of diet on tissue concentration can, indeed, outweigh that of the differences in anthropogenic influences. Harbour porpoises (*Phocoena phocoena*) from temperate regions have been reported to carry far lower cadmium loads than those from Arctic waters, in spite of higher exposure to human activities. Porpoises from polar and subpolar areas can amplify much greater amounts of cadmium through feeding because food webs in those regions present a natural enrichment of this metal (Lahaye et al. 2007).

Information on dietary habits of the organisms is, therefore, essential to assess the differential susceptibility of the individuals to pollutant exposure. Such information is necessary not only for risk estimation, but also for detecting possible changes in feeding behavior. Chemical pollution may cause negative effects of an indirect nature, such as reduction of food availability through habitat degradation or changes in prey behavior. Any response from the individuals, either adjusting their diet to opportunity or moving to another area, could mean that pollution is indirectly interfering with foraging by dolphins. Although dietary plasticity could be regarded as an adaptive benefit, it also has the potential to alter contaminant exposure and resilience in those apex feeders and therefore pose long-term impacts at the population level. It must be taken into account, however, that prey modification might be the consequence of other human activities, such as overfishing or other vessel operations, or natural events, such as harmful algal blooms, that greatly impact prey availability (Gannon et al. 2009).

Pollution exposure levels depend on a variety of environmental factors, and the distribution and extent of eco-toxicants in the watershed, as well as the dietary preferences of the primate species and their proximity to polluted areas. The classic example, Minamata disease in humans, is mercury poisoning from the ingestion of contaminated fish (Eto 2000). Dietary ingestion of contaminated food items, such as soil, roots, flowers, fruits, seeds, wild and cultivated plants, insects, and small mammals, as well as polluted freshwater, are possible sources of toxicants for wild primates (Aufreiter et al. 2001; Eisler 2004; Ericksen et al. 2003; Hernandez-Aguilar et al. 2007; Hockings et al. 2009). Studies have shown that toxic metals (e.g., arsenic, mercury, and manganese) can accumulate in cultivated and wild plants (Amonoo-Neizer et al. 1996; Egler et al. 2006). In herbivorous species, however, mercury levels have been found to be low even in mercury-contaminated areas, and gastrointestinal absorption of inorganic mercury can reduce its adverse effects (Egler et al. 2006; Gardner et al. 1978). Primate populations that crop-raid in agricultural land are at risk of ingestion of toxic pesticides being used on those crops. The dose and duration of exposures, as well as individual vulnerabilities and intra- and interspecific differences, including dietary composition, feeding habits, and habituation levels, have to be taken into account when looking into the potential effects of eco-toxicants on primates. Overall, their dietary composition differs greatly from that of cetaceans, piscivorous and carnivorous terrestrial mammals, and humans who consume fish and shellfish. Hence, exposure, bioaccumulation, and biomagnification differ. Studies have not been conducted, however, to determine the level of threat posed by the ingestion of eco-toxicants by wild primates.

16.2.3 Biology and Physiology

As already mentioned, chemical pollutant concentrations exhibit substantial variability in cetacean tissues as a function of biological and ecological influences. Therefore, the estimation of individual and population-scale risks from those contaminants requires knowledge of their accumulation and excretion, principally relative to age, sex, and life history. The usual pattern of organochlorine accumulation observed in most cetaceans is that concentrations increase with age in males and tend to decrease in females after the onset of reproduction, presumably through transfer across the placenta and via lactation (Aguilar et al. 1999; Cockcroft et al. 1989). This transfer may pose a primary threat, especially to first-born calves or to calves at a decisive stage of their development (Schwacke et al. 2002; Wells et al. 2005). The patterns of variation of trace elements are less consistent. For example, concentrations of cadmium, mercury, lead, and selenium have also been found to increase with age (Bryan et al. 2007; Honda et al. 1986; Lahaye et al. 2006), which has been associated with processes of bioaccumulation and biomagnification in marine food webs (Law 1996). However, other metals such as copper and zinc do not accumulate with age. Indeed, copper concentrations have shown higher concentrations in fetuses than in adults, which may be a result of either the lower excretion rates in newborns or a specific requirement for development (Lahaye et al. 2007). On the other hand, the influence of sex on metal levels is not clear. Some studies have shown no differences between males and females, indicating that transfer of metals to offspring during pregnancy and lactation is not an important route of excretion (Lahaye et al. 2006). Other studies have reported higher concentrations in females as a result of a higher consumption of fish necessary to sustain the energy requirements of their reproductive activity (Bryan et al. 2007). Conversely, Cannella and Kitchener (1992) found lower mercury levels in pregnant and lactating sperm whales, suggesting that this may be linked to hormonal cycles or different metabolic pathways causing the redistribution of mercury in body tissues.

Body size could also play an important role in the accumulation of contaminants in cetaceans. Some differences of metals between sexes have been attributed to their sexual dimorphism (Caurant et al. 1994). It has been recognized that small animals usually build up higher concentrations of contaminants relative to their body weight than those of larger body mass (Aguilar et al. 1999).

As has been said, small odontocetes are particularly vulnerable to organochlorine compounds because they are top predators and their food intake is high, but also because a large proportion of their body is composed of fatty tissues that efficiently accumulate lipophilic compounds. Fluctuations in blubber lipid levels and lipid profile affect, presumably, the dynamics of lipophilic contaminants and thus are likely to account for variations in the tissue concentrations of pollutants.

Such fluctuations have been related to reproductive activity, as it often involves changes in behavioral traits and diet, but also to seasonal changes in body fat condition (i.e., alterations in ambient conditions) (Lockyer et al. 2003; Learmonth 2006; Samuel and Worthy 2004). As trace metals are not lipophilic, lipid mobilization is not expected to affect their concentrations. However, seasonal pattern differences in dolphins have also been reported for several trace elements and associated with changes in water conditions, dietary modifications, or prey migration (Bryan et al. 2007; Lahaye et al. 2007).

The incorporation of organochlorines in cetaceans is yet increased by their limited ability to eliminate lipophilic compounds through ventilation (via gills), which is the principal excretory function in other aquatic organisms such as fish (Boon et al. 1992). Moreover, most cetacean species show a very low metabolic capacity to decompose these toxicants as compared with terrestrial mammals, because of the specific mode of cytochrome P-450 enzyme systems, and therefore retain them in their bodies (Tanabe et al. 1994). In some cases, concentrations of PCBs have been found to be higher in cetaceans from the open seas, far from areas of high industrial activity, than in coastal and terrestrial mammals (Tanabe 2002).

Regarding trace elements, it has been proposed that cetaceans have developed efficient metabolic pathways to support elevated exposure to some metals (Das et al. 2000). The potential toxicity of these elements can be alleviated by different detoxification processes, for example, by the demethylation of organic mercury and its coprecipitation with selenium to produce compounds with little toxicity (Cardellicchio et al. 2002; Nigro and Leonzio 1996), or by the binding of metals to metallothioneins, reducing, then, their bioavailability (Caurant et al. 1996; Das et al. 2000). It must be pointed out that some essential metals such as copper and zinc are strongly regulated and therefore present little variability within and between cetacean species around the world (Mackey et al. 2003). Metallothioneins seem also to play a central role in the homeostasis of these elements (Webb and Cain 1982).

Chemical pollutant concentrations have not been reported in wild primates. However, studies in humans indicate that age, sex, diet, reproductive status, and other factors do influence toxic effects. For example, impaired pregnancy rates have been reported in women with high serum concentrations of OCs. However, in men OC levels in adipose tissue or serum have not been associated with infertility or reduction in sperm quality (Cok et al. 2009; Weiss et al. 2006). In pregnant women, OCs have been reported to be mobilized from adipose tissue to the circulation and reach the placenta. A higher body mass index of pregnant women was found to be associated with higher endosulfan concentrations in the placenta, and greater maternal weight gain with higher DDT metabolite concentrations (Lopez-Espinosa et al. 2007). Neurodevelopmental deficits have been reported in infants and children in which exposures occurred in utero, as well as postnatally (Carrizo et al. 2007; Jacobson and Jacobson 1997; Ribas-Fito et al. 2003). In utero and postnatal exposures reportedly have occurred through chronic airborne contamination as well as diet via high maternal intake of fish and marine mammal fat and breastfeeding (Azeredo et al. 2008; Dallaire et al. 2004; Ribas-Fito et al. 2003). PCB exposure in utero has shown marked individual differences in vulnerability, as well as gender effects, with females being more affected than males (Guo et al. 1994; Jacobson and Jacobson 1997). In tropical and subtropical areas where OCs are used for agriculture and vector control, metabolite concentrations in human breast milk surpass the WHO acceptable daily intake (Azeredo et al. 2008; Chikuni et al. 1991). In addition, neurocognitive changes have been reported to occur in elderly people, and evidence of dose-dependent effects in women, but not men, has also been provided (Haase et al. 2009; Lin et al. 2008).

Variations in the effects of heavy metals on primates have been reported. In the case of arsenic, a potent human carcinogen, studies on people and laboratory primates have shown marked variation in metabolism, as well as carcinogenicity. For example, the methylation of inorganic arsenic for more rapid urinary excretion differs widely. Children have a lower degree of methylation than adults, and studies also report a lower degree in men than in women, especially during pregnancy (Vahter 1999). Large interindividual differences in methylation of arsenic in people have also been reported. In contrast to humans, the marmoset and chimpanzee do not methylate inorganic arsenic at all (Vahter 1999). In macaques and squirrel monkeys given daily oral doses of methylmercury, the burden of mercury in major organs (e.g., brain, intestine, liver, and kidney) appeared to be related to dose and duration of exposure (Chen et al. 1983; Evans et al. 1977). At exposure levels considered to be safe in people, neuropsychological dysfunctions have been detected in children at 7 years of age following in utero exposure to methylmercury from maternal consumption of pilot whale meat (Grandjean et al. 1997). Frequent fruit consumption has been shown to reduce mercury exposure in persons who consumed the same number of fish meals (Passos et al. 2007). In the case of lead toxicity, a relationship between very low level lead exposure in cord blood and male-specific cognitive deficiencies in children 3 years of age has been reported (Jedrychowski et al. 2009). The urinary cobalt levels in people living very close to a mining area or smelting plant in the Democratic Republic of Congo are the highest ever reported, especially in children (Banza et al. 2009).

16.2.4 Health Condition

This factor is also crucial in understanding the true extent of accumulation of contaminants in cetaceans. For example, animals affected with disease may undergo abnormal rates of metabolization or excretion of lipophilic contaminants, which can involve changes in organochlorine body burdens (Borrell and Aguilar 1990). Similarly, animals in poor nutritional status can mobilize their lipid reserves to meet energy requirements and therefore influence their lipid-associated pollutants (O'Shea 1999). Poor health status has also been related to the elevated concentrations of elements such as mercury and zinc in some porpoise species (Lahaye et al. 2007).

The health status of the individuals has an influence on contaminant dynamics, but also on the strength and complexity of potential effects in exposed organisms.

For example, reduced immune responses associated with high contaminant levels have been proposed as the primary cause for the increasing vulnerability to disease in various populations of dolphins (Aguilar and Borrell 1994; Kuehl and Haebler 1995). Similarly, reduced reproductive capacity caused by pollutant exposure can hamper the recovery of the affected populations (Martineau et al. 1987).

In humans, prenatal exposure to immunotoxic OCs resulting from the maternal consumption of fish and marine mammal fat has shown a possible association between exposure and acute infections during the first 12 months of life (Dallaire et al. 2004). It is not known if eco-toxicants being ingested or inhaled by wild primates are in sufficient quantity to be toxic alone, or if co-exposure to different eco-toxicants would have additive or synergistic effects. In primates with long periods of gestation and lactation, there is a greater risk of exposure to eco-toxicants during critical developmental stages. With slower development to sexual maturity and longer lifespans, chronic exposure, even at very low levels, poses a greater risk for cumulative effects over the lifespan. As such, gradually eco-toxicants could take a toll on reproductive fitness and population dynamics, greatly affecting the ability of a population to sustain itself and ultimately to recover from long-term exposures (De Lange et al. 2009; Rowe 2008). Other environmental stressors and their interactions are at play in primate habitats with eco-toxicants further stressing the ecosystem, with the potential to overwhelm primate populations to the point where they are no longer able to adapt and sustain the behavioral, cognitive, and motor functioning necessary for complex social interactions that drive their communications, proliferation, and viability.

16.3 Case Studies

16.3.1 Case of Study 1: There's No Place Like Home: Ecological Health and Common Bottlenose Dolphin Site Fidelity as a Case Study

Common bottlenose dolphins have been found to exhibit strong *site fidelity* to bays, sounds, and estuaries in much of the species range (Wells and Scott 1999, 2009). Residency over decades and across multiple generations has been demonstrated, providing an apparent geographic basis to the stable social systems that have been described from a number of sites (Connor et al. 2000; Wells et al. 1987). At one study site along the west coast of Florida, Sarasota Bay, resident dolphins spanning five concurrent generations have been observed since 1970 (Irvine et al. 1981; Scott et al. 1990; Wells 1991, 2003). The strength of the resident dolphins attachment to their long-term community home range has been demonstrated through prolonged periods of extreme environmental perturbations, such as harmful algal blooms involving the red tide dinoflagellate, *Karenia brevis*, which have led to dramatic changes in prey availability (Gannon et al. 2009) and threats from direct exposure to brevetoxins (Fire et al. 2008). Long lifespans in conjunction with long-term site fidelity provide opportunities for dolphins to establish and maintain long-term social relationships, become familiar with temporal and spatial prey distribution patterns and specialized capture techniques, learn to identify and avoid predatory sharks and other threats, and engage in observational learning and cultural transmission of knowledge (Wells 2003).

In contrast to these presumed benefits of site fidelity, changes to the local environment can have adverse impacts on dolphins that do not, or cannot, exercise an option to relocate in response to ecological changes. In Sarasota Bay, increasing human activities are correlated with changes in dolphin behavior, survivorship, and reproductive success, which can be disruptive to long-established societies. Boat collisions occasionally result in mortalities, and high levels of local boat traffic lead to changes in dolphin communication, dive, and grouping patterns (Buckstaff 2004; Nowacek et al. 2001; Wells and Scott 1997). Recreational fishing gear injuries and mortalities have increased in recent years, especially when prey stocks have declined from red tides (Wells et al. 1998, 2008). Declines in large coastal shark stocks from overfishing may be responsible for increased dolphin exposure to, and mortality from, stingray barbs.

Environmental contaminants such as persistent organic pollutants (POPs) accumulate in tissues of males at levels more than an order of magnitude greater than hypothesized threshold levels for health and reproductive impacts, whereas concentrations in females drop below this threshold once they reproduce (Schwacke et al. 2002; Wells et al. 2005). The oldest documented male in the resident community (as of 2009) is 50 years old, whereas the oldest female is 59 years of age. It could be hypothesized that differences in contaminant accumulation patterns may contribute to the 15 % shorter maximum lifespan for males as compared to females. Paternity analyses nearing completion indicate that males continue to reproduce into their forties; an artificially shortened lifespan could reduce a male's reproductive contributions to the community (Wells 2003). This impact could be exacerbated by the premature death of one member of a strongly bonded male pair by reducing the apparent breeding advantage derived from pair-bonding (Owen et al. 2002; Wells et al. 1987). First-born calves exhibit higher concentrations of POPs and lower survivorship than subsequent calves (Wells et al. 2005). The loss of a first-born calf before it can reproduce comes at a cost to the overall reproductive success of the mother. Surprisingly, in the face of these cumulative impacts from human activities, emigration rates remain low, and residents born to the area tend to continue to reside in the region, generation after generation (Wells and Scott 1990; Wells 2003).

In coming years, these resident dolphins are likely to face a new and additional, complex, and interacting set of threats from global climate change. The precise nature of future threats is impossible to predict at this time. Warmer waters will provide more favorable conditions for the survival of existing pathogens (Buck et al. 2006), as well as supporting new pathogens arriving in the region. Immunosuppression from POPs could lead to higher rates of infection by these pathogens (Lahvis et al. 1995). Mortality rates for residents are significantly higher in summer than in winter. Metabolic rate studies suggest that Sarasota Bay dolphins may already be approaching the upper limits of their thermal tolerance, as summer water temperatures in the bay approach body temperature (Costa et al. 1993). The next few years and decades will likely provide opportunities to test hypotheses regarding the importance of site fidelity to the social structure and lives of these coastal dolphin communities.

16.3.2 Case of Study 2: Manganese-induced Neurotoxicity and Cynomolgus Monkeys (Macaca fascicularis)

Manganese (Mn) is an essential element and micronutrient needed to support life, with the level of Mn intake being important to good health. Too much or too little Mn is toxic to humans. Among other uses, Mn is important in the manufacturing of steel. Occupational exposure in at-risk workers (e.g., miners, welders, and chemical and industrial workers) is well established with manganese poisoning occurring from the inhalation of toxic fumes, which have neurotoxic effects. Symptoms resemble those seen with Parkinsonian disease, and include tremors, loss of balance, and slowed movement, as well as impotency and psychiatric disturbances.

With the phasing out of lead as a gasoline additive and its replacement with methylcyclopentadienyl manganese tricarbonyl (MMT), there is increasing concern about environmental contamination with manganese and the effects of chronic low-level exposures. More than 99 % of Mn particles emitted during MMT combustion are particle sizes in the respirable fraction of the aerosol (Ardeleanu et al. 1999). Only about 24 months after the partial introduction of MMT to South African petrol in the Johannesburg region, and with an eye on possible future increases in the use of MMT in the rest of the country, a community-based study on environmental exposure to manganese in first-grade schoolchildren was performed (Röllin et al. 2005). Manganese levels in blood, as well as in water supplies, soil, and classroom dust, were measured in schools in Johannesburg and Cape Town where MMT had been introduced. Higher levels of manganese were found in soil and dust samples from Johannesburg, and mean blood Mn concentrations were significantly higher in children in Johannesburg. Manganese levels in blood were found to be significantly related to classroom dust at schools. Further, a pilot study in South Africa found that levels in maternal and umbilical cord bloods of delivering women were above that considered to be a normal level (>14 μg/l) in several different geographic areas (Röllin et al. 2009).

Most studies on understanding the mechanisms of Mn neurotoxicity have been conducted in cell culture, or in rodents. However, rodents do not develop the behavioral alterations as seen in humans. Studies in primates are necessary to elucidate the spectrum of effects of Mn. In addition to transport of inhaled Mn to the blood, it has been shown in rodents and in monkeys that Mn can also be transported directly into the brain through the olfactory tract (Dorman et al. 2002, 2006). Complex, multidisciplinary studies of chronic Mn exposure on cynomolgus monkeys (*M. fascicularis*) have shown cognitive, motor, and behavioral alterations (Guilarte et al. 2006; Schneider et al. 2006). Among other effects, neurodegeneration has been detected in cortical areas of the brain, areas not previously suspected to be involved in Mn neurotoxicity, and even though they accumulate relatively low levels of Mn (Guilarte et al. 2006). Hence, Mn neurotoxicity is not only determined by the amount of Mn that a brain region accumulates, but also by the intrinsic vulnerability of a region to Mn injury (Burton and Guilarte 2009).

In general, the insidious nature of chronic, low-level environmental exposures causing only subtle changes in behavioral, cognitive, and motor functioning or reproduction are difficult to detect in wildlife. Further, sick animals often hide or cannot or do not stay with the group, and the carcasses of deceased wild animals are rapidly scavenged and not found. Thus, unless an acute, lethal exposure resulting in mass mortality occurs, contaminant poisoning would go undetected (Wren 1986). Sublethal effects of eco-toxicants in wild primates could have long-term consequences. Latent effects may manifest as impaired reproduction or loss of reproductive fitness and, ultimately, decrease the survival of individuals, groups or populations. Such declines might be difficult to separate from natural population fluctuations and are further compounded by lack of reliable population data under normal conditions. Exposure levels of wild primates to environmental manganese contamination are unknown. Experimental toxicology studies have provided evidence that chronic low-level Mn exposure can lead to cognitive, motor, and behavioral abnormalities in cynomolgus monkeys. Studies should be conducted on wild primates, particularly where habitats are adjacent to urban and industrialized areas, to determine the risk posed by prevailing levels of environmental Mn. Under natural conditions, neurotoxic effects could potentially result in deficits that may influence the ability of primates to function in complex social networks and perform motor tasks necessary for their survival.

16.4 Conclusions and Recommendations

Cetaceans are one of the most vulnerable groups of organisms to chemical pollutants and, therefore, they are likely to be the target animals of long-term toxicity of hazardous contaminants in the future. However, the question of whether chemical pollution is changing the dynamics of the populations is not yet resolved. Behavioral patterns of individuals result from an intricate compromise between requirements associated with life history traits, feeding ecology, health status, seasonal changes, reproduction, or socializing. All these factors are likely to influence the obtaining, accumulation, metabolization, and potential effects of pollutants and, hence, to lead to differential vulnerability among individuals. Awareness of possible alterations in normal behavioral patterns is therefore crucial in predicting potential risks from manmade contaminants on exposed populations. However, cetaceans are affected simultaneously by a variety of pressures that can compromise their health, so a limited approach could mean that the effects of other environmental impacts are being overlooked. There is a need for coordinated and deeper research on how those potential threats interact to be able to associate a change in distribution or behavior direct or indirectly with pollutant exposure. This imperative includes the study of other contaminant substances such as oil, polybrominated biphenyls, organotins, and polychlorinated naphthalenes, which can also influence the status of the populations and have been very little investigated. Also, identifying population boundaries, even for those widely distributed species, could be a powerful tool for the implementation of appropriate conservation and management measures that can reduce the risks or adverse effects for a given population.

Wild primates coexist in a human-dominated landscape and endure environmental conditions and risks similar to those that humans are facing. However, in general, human risk assessments do not adequately protect other biota (Rattner 2009). The order Primates is highly diverse and, although extrapolations can be made to some extent with humans, differing sensitivities and biological and behavioral traits will influence exposure potentials and outcomes. Pollution levels and their effects in wild primates are largely unknown. There is a critical need to obtain species- and site-specific data to determine toxicological sensitivities and exposure potentials under prevailing environmental conditions. Primate life history traits, such as long pre- and postpartum development periods, large per capita parental investment in few offspring, and long lifespan increase their vulnerability to even very low levels of eco-toxicants over the long term.

Mortalities or impairments of individuals from environmental contaminants can have much broader effects as these individuals are removed from the functional matrix of cetacean or primate societies. Ecotoxicology is an evolving discipline, and ecological risk assessments will aid in estimating and predicting threats (De Lange et al. 2009; Chapman 2002; Hall et al. 2006; Munns 2006; Schwacke et al. 2002). The goal is to estimate and predict exposures before they occur to mitigate the potential effects of pollutants on wildlife and their supporting habitats. These assessments should be timely, include the potential effects of climate change (e.g., temperature increases and changes in regional precipitation patterns), and identify areas and species that may be more vulnerable to climate–pollutant interactions (Noyes et al. 2009). There is an urgent need to conduct ecological risk assessments, particularly for species that are already endangered and at risk of extinction.

Acknowledgments This research has been possible thanks to the assistance of the Mammal Research Institute of the University of Pretoria. While doing this work, V.T. was supported by the UP postdoctoral fellowship program. T.K. (T.J.S.) was supported by the National Science Foundation (NSF) Award #0238069. Any opinions, findings, and conclusions or recommendations expressed in this material are those of the author(s) and do not necessarily reflect the views of NSF.

References

Aguilar A, Borrell A (1994) Abnormally high polychlorinated biphenyl levels in striped dolphins (*Stenella coeruleoalba*) affected by the 1990–92 Mediterranean epizootic. Sci Total Environ 154:237–247

Aguilar A, Borrell A, Pastor T (1999) Biological factors affecting variability of persistent pollutant levels in cetaceans. In: Reijnders PJH, Aguilar A, Donovan P (eds) Chemical Pollutants and Cetaceans. The Journal of Cetacean Research and Management (Special Issue 1). IWC

Aguilar A, Borrell A, Reijnders PJH (2002) Geographical and temporal variation in levels of organochlorine contaminants in marine mammals. Mar Environ Res 53:425–452

Akagi H, Castillo ES, Cortes-Maramba N, Francisco-Rivera AT, Timbang TD (2000) Health assessment for mercury exposure among schoolchildren residing near a gold processing and refining plant in Apokon, Tagum, Davao del Norte, Philippines. Sci Total Environ 259(1-3):31–43

Amonoo-Neizer EH, Nyamah D, Bakiamoh SB (1996) Mercury and arsenic pollution in soil and biological samples around the mining town of Obuasi, Ghana. Water Air Soil Pollut 91(3–4):363–373

Ardeleanu A, Loranger S, Kennedy G, L'Espérance G, Zayed Z (1999) Emission rates and physico-chemical characteristics of Mn particles emitted by vehicles using methylcyclopentadienyl manganese tricarbonyl (MMT) as an octane improver. Water Air Soil Pollut 115:411–427

Aufreiter S, Mahaney WC, Milner MW, Huffman MA, Hancock RG, Wink M, Reich M (2001) Mineralogical and chemical interactions of soils eaten by chimpanzees of the Mahale Mountains and Gombe Stream National Parks, Tanzania. J Chem Ecol 27(2):285–311

Azeredo A, Torres JP, de Freitas FM, Britto JL, Bastos WR, Azevedo ESCE, Cavalcanti G, Meire RO, Sarcinelli PN, Claudio L et al (2008) DDT and its metabolites in breast milk from the Madeira River basin in the Amazon, Brazil. Chemosphere 73(1):246–251

Banza CL, Nawrot TS, Haufroid V, Decree S, De Putter T, Smolders E, Kabyla BI, Luboya ON, Ilunga AN, Mutombo AM et al (2009) High human exposure to cobalt and other metals in Katanga, a mining area of the Democratic Republic of Congo. Environ Res 109(6):745–752

Barber DG, Saczuk E, Richard P (2001) Examination of beluga–habitat relationships through use of telemetry and geographic information system. Arctic 54:305–316

Baumgartner MF (1997) The distribution of Risso's dolphin (*Grampus griseus*) with respect to the physiography of the northern gulf of Mexico. Mar Mamm Sci 13(4):614–638

Béland P, De Guise S, Girard C, Lagac A, Martineau D, Michaud R, Muir DCG, Norstrom RJ, Pelletier E, Ray S, Shugart LR (1993) Toxic compounds and health and reproductive effects in St. Lawrence beluga whales. J Great Lakes Res 19:766–775

Bernard HJ, Hohn AA (1989) Differences in feeding habits between pregnant and lactating spotted dolphins (*Stenella attenuata*). J Mammal 70:211–215

Beyer WN, Gaston G, Brazzle R, O'Connell AF Jr, Audet DJ (2007) Deer exposed to exceptionally high concentrations of lead near the Continental Mine in Idaho, USA. Environ Toxicol Chem 26(5):1040–1046

Boon JP, Arnhem EV, Jansen S, Kannan N, Petrick G, Schulz D, Duinker JC, Reijnders PJH, Goksoyr A (1992) The toxicokinetics of PCBs in marine mammals with special reference to possible interactions of individual congeners with the cytochrome P450-dependent monooxygenase system: an overview. In: Walker CH, Livingstone DR (eds) Persistent pollutants in marine ecosystems. Pergamon, Tarrytown

Borga K, Fisk AT, Hoekstra PF, Muir DCG (2004) Biological and chemical factors of importance in the bioaccumulation and trophic transfer of persistent organochlorine contaminants in Arctic marine food webs. Environ Toxicol Chem 23:2367–2385

Borrell A, Aguilar A (1990) Loss of organochlorine compounds in the tissues of a decomposing stranded dolphin. Bull Environ Contam Toxicol 45:46–53

Borrell A, Aguilar A (2005) Differences in DDT and PCB residues between common and striped dolphins from the Southwestern Mediterranean. Arch Environ Contam Toxicol 48:501–508

Borrell A, Aguilar A, Tornero V, Sequeira M, Fernandez G, Alis S (2006) Organochlorine compounds and stable isotopes indicate bottlenose dolphin subpopulation structure around the Iberian Peninsula. Environ Int 32:516–523

Brager S, Dawson SM, Slooten E, Smith S, Stone GS, Yoshinaga A (2002) Site fidelity and alongshore range in Hector's dolphin, an endangered marine dolphin from New Zealand. Biol Conserv 108:281–287

Bryan CE, Christopher SJ, Balmer BC, Wells RS (2007) Establishing baseline levels of trace elements in blood and skin of bottlenose dolphins in Sarasota Bay, Florida: implications for noninvasive monitoring. Sci Total Environ 388:325–342

Buck JD, Wells RS, Rhinehart HL, Hansen LJ (2006) Aerobic microorganisms associated with free-ranging bottlenose dolphins in coastal Gulf of Mexico and Atlantic Ocean waters. J Wildl Dis 42:536–544

Buckstaff KC (2004) Effects of watercraft noise on the acoustic behavior of bottlenose dolphins, *Tursiops truncatus*, in Sarasota Bay, Florida. Mar Mamm Sci 20:709–725

Burger J, Fossi C, McClellan-Green P, Orlando EF (2007) Methodologies, bioindicators, and biomarkers for assessing gender-related differences in wildlife exposed to environmental chemicals. Environ Res 104:135–152

Burton NC, Guilarte TR (2009) Manganese neurotoxicity: lessons learned from longitudinal studies in nonhuman primates. Environ Health Perspect 117(3):325–332

Bustamante P, Caurant F, Fowler SW, Miramand P (1998) Cephalopods as a vector for the transfer of cadmium to top marine predators in the north-east Atlantic Ocean. Sci Total Environ 220:71–80

Bustamante P, Garrigue C, Breau L, Caurant F, Dabin W, Greaves J, Dodemont R (2003) Trace elements in two odontocete species (*Kogia breviceps* and *Globicephala macrorhynchus*) stranded in New Caledonia (South Pacific). Environ Pollut 124:263–271

Campbell L, Verburg P, Dixon DG, Hecky RE (2008) Mercury biomagnification in the food web of Lake Tanganyika (Tanzania, East Africa). Sci Total Environ 402(2–3):184–191

Cannella EJ, Kitchener DJ (1992) Differences in mercury levels in female sperm whale, *Physeter macrocephalus* (Cetacea: Odontoceti). Austr Mammalogy 15:121–123

Cardellicchio N, Decataldo A, Di Leo A, Misino A (2002) Accumulation and tissue distribution of mercury and selenium in striped dolphins (*Stenella coeruleoalba*) from the Mediterranean Sea (southern Italy). Environ Pollut 116:265–271

Carrizo D, Grimalt JO, Ribas-Fito N, Torrent M, Sunyer J (2007) In utero and post-natal accumulation of organochlorine compounds in children under different environmental conditions. J Environ Monit 9(6):523–529

Caurant F, Amiard JC, Amiard-Triquet C, Sauriau PG (1994) Ecological and biological factors controlling the concentrations of trace elements (As, Cd, Cu, Hg, Se, Zn) in delphinids *Globicephala melas* from the North Atlantic Ocean. Mar Ecol Prog Ser 103:207–219

Caurant F, Navarro M, Amiard JC (1996) Mercury in pilot whales: possible limits to the detoxification process. Sci Total Environ 186:95–104

Chapman PM (2002) Integrating toxicology and ecology: putting the "eco" into ecotoxicology. Mar Pollut Bull 44(1):7–15

Chen WJ, Body RL, Mottet NK (1983) Biochemical and morphological studies of monkeys chronically exposed to methylmercury. J Toxicol Environ Health 12(2–3):407–416

Chikuni O, Skare JU, Nyazema N, Polder A (1991) Residues of organochlorine pesticides in human milk from mothers living in the greater Harare area of Zimbabwe. Cent Afr J Med 37(5):136–141

Cockcroft V, DeKock A, Lord D, Ross G (1989) Organochlorines in bottlenose dolphins *Tursiops truncatus* from the east coast of South Africa. S Afr J Mar Sci 8:207–217

Cok I, Durmaz TC, Durmaz E, Satiroglu MH, Kabukcu C (2009) Determination of organochlorine pesticide and polychlorinated biphenyl levels in adipose tissue of infertile men. Environ Monit Assess 118:383–391

Connor RC, Wells RS, Mann J, Read AJ (2000) The bottlenose dolphin, *Tursiops* spp.: social relationships in a fission–fusion society. In: Mann J, Connor RC, Tyack PL, Whitehead H (eds) Cetacean societies: field studies of dolphins and whales. University of Chicago Press, Chicago

Costa DP, Worthy GAJ, Wells RS, Read AJ, Scott MD, Irvine AB, Waples DM (1993) Seasonal changes in the field metabolic rate of bottlenose dolphins, *Tursiops truncatus*. In: Tenth biennial conference on the biology of marine mammals, 11–15 November, Galveston

Dallaire F, Dewailly E, Muckle G, Vezina C, Jacobson SW, Jacobson JL, Ayotte P (2004) Acute infections and environmental exposure to organochlorines in Inuit infants from Nunavik. Environ Health Perspect 112(14):1359–1365

Das K, Debacker V, Bouquegneau JM (2000) Metallothioneins in marine mammals. Cell Mol Biol 46:283–294

Das K, Beans C, Holsbeek L, Mauger G, Berrowd SD, Rogan E, Bouquegneau JM (2003) Marine mammals from northeast Atlantic: relationship between their trophic status as determined by $\delta^{13}C$ and $\delta^{15}N$ measurements and their trace metal concentrations. Mar Environ Res 56:349–365

De Lange HJ, Lahr J, Van der Pol JJC, Wessels Y, Faber JH (2009) Ecological vulnerability in wildlife. An expert judgment and multi-criteria analysis tool using ecological traits to assess relative impact of pollutants. Environ Toxicol Chem 28(10):2233–2240

Doniol-Valcroze T, Berteaux D, Larouche P, Sears R (2007) Influence of thermal fronts on habitat selection by four rorqual whale species in the Gulf of St. Lawrence. Mar Ecol Prog Ser 335:207–216

Dorman DC, Brenneman KA, McElveen AM, Lynch SE, Roberts KC, Wong BA (2002) Olfactory transport: a direct route of delivery of inhaled manganese phosphate to the rat brain. J Toxicol Environ Health A 65(20):1493–1511

Dorman DC, Struve MF, Wong BA, Dye JA, Robertson ID (2006) Correlation of brain magnetic resonance imaging changes with pallidal manganese concentrations in rhesus monkeys following subchronic manganese inhalation. Toxicol Sci 92(1):219–227

Egler SG, Rodrigues-Filho S, Villas-Boas RC, Beinhoff C (2006) Evaluation of mercury pollution in cultivated and wild plants from two small communities of the Tapajos gold mining reserve, Para State, Brazil. Sci Total Environ 368(1):424–433

Eisler R (2004) Mercury hazards from gold mining to humans, plants, and animals. Rev Environ Contam Toxicol 181:139–198

Ericksen JA, Gustin MS, Schorran DE, Johnson DW, Lindberg SE, Coleman JS (2003) Accumulation of atmospheric mercury in forest foliage. Atmos Environ 37(12):1613–1622

Erikson KM, Dorman DC, Lash LH, Aschner M (2008) Duration of airborne-manganese exposure in rhesus monkeys is associated with brain regional changes in biomarkers of neurotoxicity. Neurotoxicology 29(3):377–385

Eto K (2000) Minamata disease. Neuropathology 20(suppl):S14–S19

Evans HL, Garman RH, Weiss B (1977) Methylmercury: exposure duration and regional distribution as determinants of neurotoxicity in nonhuman primates. Toxicol Appl Pharmacol 41(1):15–33

Fire SE, Flewelling LJ, Wang Z, Naar J, Henry MS, Pierce RH, Wells RS (2008) Florida red tide and brevetoxins: association and exposure in live resident bottlenose dolphins (*Tursiops truncatus*) in the eastern Gulf of Mexico, USA. Mar Mamm Sci 24:831–844

Gahan JB, Travis BV, Morton PA, Lindquist AW (1945) DDT as a residual-type treatment to control *Anopheles quadrimaculatus*. J Econ Entomol 38:251–253

Gannon DP, Berens EJ, Camilleri SA, Gannon JG, Brueggen MK, Barleycorn A, Palubok V, Kirkpatrick GJ, Wells RS (2009) Effects of *Karenia brevis* harmful algal blooms on nearshore fish communities in southwest Florida. Mar Ecol Prog Ser 378:171–186

Gardner SW, Kendall DR, Odom RR, Windom HL, Stephens JA (1978) The distribution of methylmercury in a contaminated salt marsh ecosystem. Environ Pollut 15:243–251

Geraci JR (1989) Clinical investigation of the 1987-88 mass mortality of bottlenose dolphins along the U.S. central and south Atlantic coast. Final Report to National Marine Fisheries Service, U.S. Navy, Office of Naval Research, and Marine Mammal Commission

Grandjean P, Weihe P, White RF, Debes F, Araki S, Yokoyama K, Murata K, Sorensen N, Dahl R, Jorgensen PJ (1997) Cognitive deficit in 7-year-old children with prenatal exposure to methylmercury. Neurotoxicol Teratol 19(6):417–428

Guilarte TR, McGlothan JL, Degaonkar M, Chen MK, Barker PB, Syversen T, Schneider JS (2006) Evidence for cortical dysfunction and widespread manganese accumulation in the nonhuman primate brain following chronic manganese exposure: a ^{1}H-MRS and MRI study. Toxicol Sci 94(2):351–358

Guo YL, Chen YC, Yu ML, Hsu CC (1994) Early development of Yu-Cheng children born seven to twelve years after the Taiwan PCB outbreak. Chemosphere 29(9–11):2395–2404

Haase RF, McCaffrey RJ, Santiago-Rivera AL, Morse GS, Tarbell A (2009) Evidence of an age-related threshold effect of polychlorinated biphenyls (PCBs) on neuropsychological functioning in a Native American population. Environ Res 109(1):73–85

Hall AJ, McConnell BJ, Rowles TK, Aguilar A, Borrell A, Schwacke L, Reijnders PJH, Wells RS (2006) An individual based model framework to assess the population consequences of polychlorinated biphenyl exposure in bottlenose dolphins. Environ Health Perspect 114(1):60–64

Herman DP, Burrows DG, Wade PR, Durban JW, Matkin CO, LeDuc RG, Barrett-Lennard LG, Krahn MM (2005) Feeding ecology of eastern North Pacific killer whales *Orcinus orca* from fatty acid, stable isotope, and organochlorine analyses of blubber biopsies. Mar Ecol Prog Ser 302:275–291

Hernandez-Aguilar RA, Moore J, Pickering TR (2007) Savannah chimpanzees use tools to harvest the underground storage organs of plants. Proc Natl Acad Sci USA 104(49):19210–19213

Higham JP, Warren Y, Adanu J, Umaru BN, MacLarnon AM, Sommer V, Ross C (2009) Living on the edge: life-history of olive baboons at Gashaka-Gumti National Park, Nigeria. Am J Primatol 71(4):293–304

Hockings KJ, Anderson JR, Matsuzawa T (2009) Use of wild and cultivated foods by chimpanzees at Bossou, Republic of Guinea: feeding dynamics in a human-influenced environment. Am J Primatol 71(8):636–646

Honda K, Fujise Y, Tatsukawa R (1986) Age-related accumulation of heavy metals in bone of the striped dolphin, *Stenella coeruleoalba*. Mar Environ Res 20:143–160

Irvine AB, Scott MD, Wells RS, Kaufmann JH (1981) Movements and activities of the Atlantic bottlenose dolphin, *Tursiops truncatus*, near Sarasota, Florida. Fish Bull US 79:671–688

Jacobson JL, Jacobson SW (1997) Evidence for PCBs as neurodevelopmental toxicants in humans. Neurotoxicology 18(2):415–424

Janmaat KRL, Olupot W, Chancellor RL, Arlet ME, Waser PM (2009) Long-term site fidelity and individual home range shifts in *Lophocebus albigena*. Int J Primatol 30(3):443–466

Jedrychowski W, Perera F, Jankowski J, Mrozek-Budzyn D, Mroz E, Flak E, Edwards S, Skarupa A, Lisowska-Miszczyk I (2009) Gender specific differences in neurodevelopmental effects of prenatal exposure to very low-lead levels: the prospective cohort study in three-year olds. Early Hum Dev 85(8):503–510

Jolly A, Pride E (1999) Troop histories and range inertia of *Lemur catta* at Berenty, Madagascar: a 33-year perspective. Int J Primatol 20:359–373

Kishimba MA, Henry L, Mwevura H, Mmochi AJ, Mihale M, Hellar H (2004) The status of pesticide pollution in Tanzania. Talanta 64(1):48–53

Kuehl DW, Haebler R (1995) Organochlorine, organobromine, metal, and selenium residues in bottlenose dolphins (*Tursiops truncatus*) collected during an unusual mortality event in the Gulf of Mexico. Arch Environ Contam Toxicol 28:494–499

Lahaye V, Bustamante P, Dabin W, Van Canneyt O, Dhermain F, Cesarini C, Pierce GJ, Caurant F (2006) New insights from age determination on toxic element accumulation in striped and bottlenose dolphins from Atlantic and Mediterranean waters. Mar Pollut Bull 52:1219–1230

Lahaye V, Bustamante P, Law RJ, Learmonth JA, Santos MB, Boon JP, Rogan E, Dabin W, Addink MJ, Lopez A, Zuur AF, Pierce GJ, Caurant F (2007) Biological and ecological factors related to trace element levels in harbour porpoises (*Phocoena phocoena*) from European waters. Mar Environ Res 64:247–266

Lahvis GP, Well RS, Kuehl DW, Stewart JL, Rhinnehart HL, Via CS (1995) Decreased lymphocyte responses in free-ranging bottlenose dolphins (*Tursiops truncatus*) are associated with increased concentrations of PCBs and DDT in peripheral blood. Environ Health Perspect 103:62–72

Lavery TJ, Butterfield N, Kemper CM, Reid RJ, Sanderson K (2008) Metals and selenium in the liver and bone of three dolphin species from South Australia, 1988–2004. Sci Total Environ 390:77–85

Lavery TJ, Kemper CM, Sanderson K, Schultz CG, Mitchell JG, Seuront L (2009) Heavy metal toxicity of kidney and bone tissues in South Australian adult bottlenose dolphins (*Tursiops aduncus*). Mar Environ Res 67:1–7

Law RJ (1996) Metals in marine mammals. In: Beyer WN, Heinz GH, Redmon-Norwood AW (eds) Environmental contaminants in wildlife: interpreting tissue concentrations. CRC, Boca Raton

Law RJ, Fileman CF, Hopkins AD, Baker JR, Harwood J, Jackson DB, Kennedy S, Martin AR, Morris RJ (1991) Concentrations of trace metals in the livers of marine mammals (seals, porpoises and dolphins) from waters around British Isles. Mar Pollut Bull 22:183–191

Law RJ, Jones BR, Baker JR, Kennedy S, Milne R, Morris RJ (1992) Trace metals in the livers of marine mammals from the Welsh coast and the Irish Sea. Mar Pollut Bull 24:296–304

Learmonth JA (2006) Life history and fatty acid analysis of harbour porpoises (*Phocoena phocoena*) from Scottish waters. Ph.D. Thesis, University of Aberdeen

Lin KC, Guo NW, Tsai PC, Yang CY, Guo YLL (2008) Neurocognitive changes among elderly exposed to PCBs/PCDFs in Taiwan. Environ Health Perspect 116(2):184–189

Lockyer C, Desportes G, Hansen K, Labberté S, Siebert S (2003) Monitoring growth and energy utilisation of the harbour porpoise (*Phocoena phocoena*) in human care. In: Haug T, Desportes G, Víkingsson GA, Witting L (eds) Harbour porpoises in the North Atlantic Marine Mammal Commission, vol 5. NAMMCO Scientific Publications, Tromsø

Long M, Reid RJ, Kemper CM (1997) Cadmium accumulation and toxicity in the bottlenose dolphin, the common dolphin, and some dolphin prey species in South Australia. Austr Mammal 20:25–33

Lopez-Espinosa MJ, Granada A, Carreno J, Salvatierra M, Olea-Serrano F, Olea N (2007) Organochlorine pesticides in placentas from Southern Spain and some related factors. Placenta 28(7):631–638

Lusseau D (2003) Effects of tour boats on the behavior of bottlenose dolphins: using Markov chains to model anthropogenic impacts. Conserv Biol 17:1785–1793

Mackey EA, Oflaz RD, Epstein MS, Buehler B, Porter BJ, Rowles T (2003) Elemental composition of liver and kidney tissues of roughtoothed dolphins (*Steno bredanensis*). Arch Environ Contam Toxicol 22:523–532

Manirakiza P, Covaci A, Nizigiymana L, Ntakimazi G, Schepens P (2002) Persistent chlorinated pesticides and polychlorinated biphenyls in selected fish species from Lake Tanganyika, Burundi, Africa. Environ Pollut 117(3):447–455

Mann J, Connor RC, Barre LM, Heithaus MR (2000) Female reproductive success in bottlenose dolphins (*Tursiops* sp.): life history, habitat, provisioning, and group-size effects. Behav Ecol 11:210–219

Martineau D, Beland P, Desjardins C, Lagace A (1987) Levels of organochlorine chemicals in tissues of beluga whales (*Delphinapterus leucas*) from the St. Lawrence Estuary, Quebec, Canada. Arch Environ Contam Toxicol 16:137–147

McSweeney DJ, Baird RW, Mahaffy SD (2007) Site fidelity, associations, and movements of Cuvier's (*Ziphius cavirostris*) and Blainville's (*Mesoplodon densirostris*) beaked whales off the island of Hawai'i. Mar Mamm Sci 23(3):666–687

Mertl-Millhollen AS (2000) Tradition in *Lemur catta* behavior at Berenty Reserve, Madagascar. Int J Primatol 21(2):287–297

Metcalfe C, Koenig B, Metcalfe T, Paterson G, Sears R (2004) Intra- and inter-species differences in persistent organic contaminants in the blubber of blue whales and humpback whales from the Gulf of St. Lawrence, Canada. Mar Environ Res 57:245–260

Meynier L, Pusineri C, Spitz J, Santos MB, Pierce GJ, Ridoux V (2008) Intraspecific dietary variation in the short-beaked common dolphin *Delphinus delphis* in the Bay of Biscay: importance of fat fish. Mar Ecol Prog Ser 354:277–287

Munns WR (2006) Assessing risks to wildlife populations from multiple stressors: overview of the problem and research needs. Ecol Soc 11(1)

Murray CM, Gilby IC, Mane S, Pusey AE (2008) Adult male chimpanzees inherit maternal ranging patterns. Curr Biol 18:20–24

Natoli A, Birkun A, Aguilar A, Lopez A, Hoelzel AR (2005) Habitat structure and the dispersal of male and female bottlenose dolphins (*Tursiops truncatus*). Proc R Soc B 272:1217–1226

Nigro M, Leonzio C (1996) Intracellular storage of mercury and selenium in different marine vertebrates. Mar Ecol Prog Ser 135:137–143

Nishida T (1990) The chimpanzees of the Mahale mountains: sexual and life history strategies. University of Tokyo Press, Tokyo

Nowacek SM, Wells RS, Solow AR (2001) Short-term effects of boat traffic on bottlenose dolphins, *Tursiops truncatus*, in Sarasota Bay, Florida. Mar Mamm Sci 17:673–688

Noyes PD, McElwee MK, Miller HD, Clark BW, Van Tiem LA, Walcott KC, Erwin KN, Levin ED (2009) The toxicology of climate change: environmental contaminants in a warming world. Environ Int 35(6):971–986

O'Shea TJ (1999) Environmental contaminants and marine mammals. In: Reynolds JE III, Rommel SA (eds) Biology of marine mammals. Smithsonian Institution Press, Washington, DC

O'Shea TJ, Aguilar A (2001) Cetacea and Sirenia. In: Shore RF, Rattner BA (eds) Ecotoxicology of wild mammals. Wiley, New York

Ogawa H, Idani G, Moore J, Pintea L, Hernandez-Aguilar A (2007) Sleeping parties and nest distribution of chimpanzees in the savanna woodland, Ugalla, Tanzania. Int J Primatol 28:1397–1412

Oliveira FA (1997) General overview of vector control in relation to the insecticide pollution in Brazil. In: International workshop on organic micropollutants in the environment IBCCF-URFJ. Note no. 1097. AB-Dlo, Ministry of Agriculture, The Netherlands

Owen ECG, Hofmann S, Wells RS (2002) Ranging and social association patterns of paired and unpaired adult male bottlenose dolphins, *Tursiops truncatus*, in Sarasota, Florida, provide no evidence for alternative male strategies. Can J Zool 80:2072–2089

Passos CJ, Mergler D, Fillion M, Lemire M, Mertens F, Guimaraes JR, Philibert A (2007) Epidemiologic confirmation that fruit consumption influences mercury exposure in riparian communities in the Brazilian Amazon. Environ Res 105(2):183–193

Rattner BA (2009) History of wildlife toxicology. Ecotoxicology 18(7):773–783

Ribas-Fito N, Cardo E, Sala M, Eulalia de Muga M, Mazon C, Verdu A, Kogevinas M, Grimalt JO, Sunyer J (2003) Breastfeeding, exposure to organochlorine compounds, and neurodevelopment in infants. Pediatrics 111(5 Pt 1):e580–e585

Röllin H, Mathee A, Levin J, Theodorou P, Wewers F (2005) Blood manganese concentrations among first-grade schoolchildren in two South African cities. Environ Res 97(1):93–99

Röllin HB, Rudge CV, Thomassen Y, Mathee A, Odland JO (2009) Levels of toxic and essential metals in maternal and umbilical cord blood from selected areas of South Africa: results of a pilot study. J Environ Monit 11(3):618–627

Rowe CR (2008) "The calamity of so long life:" life histories, contaminants, and potential emerging threats to long-lived vertebrates. Bioscience 58(7):623–630

Samuel AM, Worthy GAJ (2004) Variability in fatty acid composition of bottlenose dolphin (*Tursiops truncatus*) blubber as a function of body site, season, and reproductive state. Can J Zool 82:1933–1942

Schneider JS, Decamp E, Koser AJ, Fritz S, Gonczi H, Syversen T, Guilarte TR (2006) Effects of chronic manganese exposure on cognitive and motor functioning in non-human primates. Brain Res 1118(1):222–231

Schoeffner DJ, Thorgeirsson UP (2000) Susceptibility of nonhuman primates to carcinogens of human relevance. In Vivo (Athens, Greece) 14(1):149–156

Schwacke LH, Voit EO, Hansen LJ, Wells RS, Mitchum GB, Hohn AA, Fair PA (2002) Probabilistic risk assessment of reproductive effects of polychlorinated biphenyls on bottlenose dolphins (*Tursiops truncatus*) from the Southeast United States coast. Environ Toxicol Chem 21(12):2752–2764

Scott MD, Wells RS, Irvine AB (1990) A long-term study of bottlenose dolphins on the west coast of Florida. In: Leatherwood S, Reeves RR (eds) The bottlenose dolphin. Academic Press, San Diego

Sha JC, Gumert MD, Lee BP, Jones-Engel L, Chan S, Fuentes A (2009) Macaque–human interactions and the societal perceptions of macaques in Singapore. Am J Primatol 71(10):825–839

Simonich SL, Hites RA (1995) Global distribution of persistent organochlorine compounds. Science 269(5232):1851–1854

Sing KA, Hryhorczuk D, Saffirio G, Sinks T, Paschal DC, Sorensen J, Chen EH (2003) Organic mercury levels among the Yanomama of the Brazilian Amazon Basin. Ambio 32(7):434–439

Smith A (1991) Chlorinated hydrocarbon insecticides. In: Hayes WJ, Laws ER (eds) Handbook of pesticide toxicology. Academic Press, San Diego

Stockin KA, Law RJ, Duignan PJ, Jones GW, Porter L, Mirimin L, Meynier L, Orams MB (2007) Trace elements, PCBs and organochlorine pesticides in New Zealand common dolphins (*Delphinus* sp.). Sci Total Environ 387:333–345

Suutari A, Ruokojarvi P, Hallikainen A, Kiviranta H, Laaksonen S (2009) Polychlorinated dibenzo-p-dioxins, dibenzofurans, and polychlorinated biphenyls in semi-domesticated reindeer (*Rangifer tarandus tarandus*) and wild moose (*Alces alces*) meat in Finland. Chemosphere 75(5):617–622

Takayama S, Sieber SM, Dalgard DW, Thorgeirsson UP, Adamson RH (1999) Effects of long-term oral administration of DDT on nonhuman primates. J Cancer Res Clin Oncol 125(3–4):219–225

Tanabe S (2002) Contamination and toxic effects of persistent endocrine disrupters in marine mammals and birds. Mar Pollut Bull 45:69–77

Tanabe S, Iwata H, Tatsukawa R (1994) Global contamination by persistent organochlorines and their ecotoxicological impact on marine mammals. Sci Total Environ 154:163–177

Thorgeirsson UP, Dalgard DW, Reeves J, Adamson RH (1994) Tumor incidence in a chemical carcinogenesis study of nonhuman primates. Regul Toxicol Pharmacol 19(2):130–151

Turusov V, Rakitsky V, Tomatis L (2002) Dichlorodiphenyltrichloroethane (DDT): ubiquity, persistence, and risks. Environ Health Perspect 110(2):125–128

Vahter M (1999) Methylation of inorganic arsenic in different mammalian species and population groups. Sci Prog 82:69–88

Van Straaten P (2000) Mercury contamination associated with small-scale gold mining in Tanzania and Zimbabwe. Sci Total Environ 259(1–3):105–113

Vandelannoote A, Robberecht H, Deelstra H, Vyumvuhore F, Bitetera L, Ollevier F (1996) The impact of the River Ntahangwa, the most polluted Burundian affluent of Lake Tanganyika, on the water quality of the lake. Hydrobiologia 328(2):161–171

Voldner EC, Yi-Fan L (1995) Global usage of selected persistent organochlorines. Sci Total Environ 160-161:201–210

Webb M, Cain K (1982) Commentary: functions of metallothionein. Biochem Pharmacol 31:137–142

Weiss JM, Bauer O, Bluthgen A, Ludwig AK, Vollersen E, Kaisi M, Al-Hasani S, Diedrich K, Ludwig M (2006) Distribution of persistent organochlorine contaminants in infertile patients from Tanzania and Germany. J Assist Reprod Genet 23(9-10):393–399

Wells RS (1991) The role of long-term study in understanding the social structure of a bottlenose dolphin community. In: Pryor K, Norris KS (eds) Dolphin societies: discoveries and puzzles. University of California Press, Berkeley

Wells RS (2003) Dolphin social complexity: lessons from long-term study and life history. In: de Waal FBM, Tyack PL (eds) Animal social complexity: intelligence, culture, and individualized societies. Harvard University Press, Cambridge

Wells RS, Scott MD (1990) Estimating bottlenose dolphin population parameters from individual identification and capture-release techniques. In: Hammond PS, Mizroch SA, Donovan GP (Eds) Individual recognition of cetaceans: use of photo-identification and other techniques to estimate population parameters. Report of the International Whaling Commission, Special Issue 12, Cambridge

Wells RS, Scott MD (1997) Seasonal incidence of boat strikes on bottlenose dolphins near Sarasota, Florida. Mar Mamm Sci 13:475–480

Wells RS, Scott MD (1999) Bottlenose dolphin *Tursiops truncatus* (Montagu, 1821). In: Ridgway SH, Harrison R (eds) Handbook of marine mammals, vol 6. The second book of Dolphins and Porpoises, Academic, San Diego

Wells RS, Scott MD (2009) Common bottlenose dolphin (*Tursiops truncatus*). In: Perrin WF, Würsig B, Thewissen JGM (eds) Encyclopedia of marine mammals, 2nd edn. Elsevier, San Diego

Wells RS, Scott MD, Irvine AB (1987) The social structure of free-ranging bottlenose dolphins. In: Genoways H (ed) Current mammalogy, vol 1. Plenum, New York

Wells RS, Hofmann S, Moors TL (1998) Entanglement and mortality of bottlenose dolphins (*Tursiops truncatus*) in recreational fishing gear in Florida. Fish Bull 96(3):647–650

Wells RS, Rhinehart HL, Cunningham P, Whaley J, Baran M, Koberna C, Costa DP (1999) Long distance offshore movements of bottlenose dolphins. Mar Mamm Sci 15:1098–1114

Wells RS, Tornero V, Borrell A, Aguilar A, Rowles TK, Rhinehart HL, Hofmann S, Jarman WM, Hohn AA, Sweeney JC (2005) Integrating life-history and reproductive success data to examine potential relationships with organochlorine compounds for bottlenose dolphins (*Tursiops truncatus*) in Sarasota Bay, Florida. Sci Total Environ 349:106–119

Wells RS, Allen JB, Hofmann S, Bassos-Hull K, Fauquier DA, Barros NB, DeLynn RE, Sutton G, Socha V, Scott MD (2008) Consequences of injuries on survival and reproduction of common bottlenose dolphins (*Tursiops truncatus*) along the west coast of Florida. Mar Mamm Sci 24:774–794

Whitehead H (2003) Sperm whales: social evolution in the oceans. University of Chicago Press, Chicago

WHO (1989) DDT and its derivatives. Environmental health criteria 83. World Health Organization, Geneva

Wren CD (1986) A review of metal accumulation and toxicity in wild mammals. I. Mercury. Environ Res 40(1):210–244

Zolman ES (2002) Residence patterns of bottlenose dolphins (*Tursiops truncatus*) in the Stono River estuary, Charleston County, South Carolina, U.S.A. Mar Mamm Sci 18:1879–1892

Zühlke U, Weinbauer MG (2003) The common marmoset (*Caliithrix jacchus*) as a model in toxicology. Toxicol Pathol 31(1):123–127

Part IV
Selected Topics in Comparative Behavior

Chapter 17
Observing and Quantifying Cetacean Behavior in the Wild: Current Problems, Limitations, and Future Directions

Janet Mann and Bernd Würsig

A subgroup of dusky dolphins "boisterously" leaping. Without behavioral context, it is difficult to know whether these leaping animals represent a mating group, with often several males chasing a female in probable estrus; or whether it is a feeding group, with dolphins leaping to rapidly and simultaneously access a school or shoal of small fish just below the surface. (Off Kaikoura, New Zealand, summer 2011–2012, by Anke Kügler)

J. Mann (✉)
Reiss Science Building, Department of Biology, Georgetown University, Washington, DC 20057, USA
e-mail: mannj2@georgetown.edu

B. Würsig
Ocean and Coastal Sciences Building, Department of Marine Biology, Texas A&M University, 200 Seawolf Pkwy., Galveston, TX 77553, USA

J. Yamagiwa and L. Karczmarski (eds.), *Primates and Cetaceans: Field Research and Conservation of Complex Mammalian Societies*, Primatology Monographs, DOI 10.1007/978-4-431-54523-1_17, © Springer Japan 2014

Abstract Behavioral research and analysis is prone to both error and bias, particularly in the early stages of a discipline, in part because it is widely (and erroneously) believed that "behavior" is rather simple and can be easily described or quantified. However, since the 1970s for terrestrial animals, and since the late 1990s for marine mammals, systematic protocols of data gathering and ever more sophisticated modeling and multivariate statistical techniques have been described, largely to reduce problems of bias and pseudoreplication. With modern observational protocols, often enhanced by sophisticated multivariable data-gathering tools, the future for more accurate assessments, and therefore interpretations, of the sophisticated social behaviors of wild cetaceans seems assured.

Keywords Ad libitum • Animal behavior • Behavioral sampling • Data tags • Events • Fission–fusion • Focal animal following • Point sampling • Quantitative methods • Sampling errors • Scan sampling • States

17.1 Introduction

Mapping cetacean behavior is critical to evolutionary approaches and conservation management. How can we understand the basic biology, life history, and evolution of a species, *and* address critical conservation questions, without at least some rudimentary appreciation of their ranging, foraging, social, and parental behavior? Although many people are fascinated with animal behavior, evident by the number and popularity of nature shows, a common misconception of amateur and even senior scientists is the assumption that studying behavior is easy. The premise is that we are all observers of behavior, at least within our own species, so compared to gene sequencing, neuroscience, or biochemistry, mere "behavior" is something with which we are intimately familiar, regardless of training. Historically, such overconfidence plagued field studies of animal behavior until the 1970s, and descriptive studies often overinterpreted behaviors that happened to be noticed. Following Jeanne Altmann's publication on sampling techniques for behavioral studies (Altmann 1974), which distinguished between ad libitum ("ad lib") and more quantitative methods, many observers of terrestrial species and systems were more careful and explicit in both defining behaviors with an ethogram and finding appropriate sampling methods to approximate frequency and duration. A similar general shift was later introduced to cetacean researchers (Mann 1999). To date, a large number of papers on cetacean behavior fail to estimate either frequency or duration, except for a limited range of behaviors (e.g., diving intervals), possibly because focal sampling methods require individual recognition of animals, usually from natural marks (Würsig and Würsig 1977). Given the task of observing animals that are difficult to identify, fast moving, wide ranging, leave no scats or tracks, and spend most of their lives out of the sight of surface-dwelling observers, it is no surprise that few studies present basic activity budget data.

17.2 Challenges and Solutions of Behavioral Data Gathering

The challenges confronting students of marine mammal behavioral descriptions are to reduce observer and sampling biases and expand and refine sampling and analytical techniques that yield useful information. Ethologists studying birds, burrowing animals, and forest species have similar difficulties of investigating cryptic species, and their subjects do not have to show themselves: at least cetaceans need to come to the surface at regular intervals to breathe! This need allows for visually tracking individuals or groups, but has several limitations. First, many behaviors, especially foraging, occur at depth, and second, surfacing intervals are strongly influenced by the behaviors themselves. Thus, although it might be important to record surface and dive times, behavior sampling must also account for subsurface periods. Similarly, nocturnal periods are ignored for most cetacean studies, although it is widely recognized that cetaceans are *cathemeral*, that is, active day and night.

When behavioral information is gathered by eye from surface vessels or shore, the limitations of the viewing platform demand careful interpretation of the data. For example, in Shark Bay, Australia, socializing by Indo-Pacific bottlenose dolphins (*Tursiops aduncus*) typically involves prolonged periods at or near the surface where continuous sampling or point sampling is possible. Deep-water foraging involves long dives and short intervals at the surface. During socializing, surface and subsurface behaviors are similar. During foraging, the dolphin sometimes rests at the surface and forages during dives. If sampling records were limited to surface observations, foraging activity budgets would be grossly underestimated. To systematically capture the stream of behavior, the observer must make inferences about what is occurring subsurface, but could indicate which behaviors are directly observed (at or near surface) or based on diving behavior. The validity of the inferences depends on other "confirming" observations, such as fish catches, acoustic behavior, or matching surface with subsurface behavior (Vaughn et al. 2009). For example, if 3-min point sampling intervals are used to quantify delphinid behavior, then the samples might be marked as surface or subsurface (e.g., social-surface, social-subsurface, travel-subsurface, forage-subsurface). In Shark Bay, bottlenose dolphin dives average about 1 min in deep water, and nearly continuous observation is possible in shallower water, enabling us to quantify activity budgets and track behavior at or beneath the surface (Gibson and Mann 2008; Mann et al. 2008).

As was pointed out by Mann (1999), it is important that types of behaviors are broadly but accurately categorized. *Events* (brief behaviors such as surface displays, or dive types) are usually not timed and can be readily converted into rates, either rate per unit time (e.g., an individual dives 13 times per hour) or as a rate during another behavior. For example, dolphin A dives 22 times per hour of foraging and 13 times per hour of resting. *States* are typically longer behaviors that are either timed (e.g., onset and offset of foraging bouts) or estimated using quantitative measures such as scan or point sampling (Altmann 1974). Behavioral events such as fish catches, dive types, and particulars of interactions help confirm the behavioral state. Events are easily missed unless they regularly occur at the surface, but states should not be.

To interpret event rates fairly, researchers need to be careful to avoid observer bias (systemic, nonrandom sampling errors). Such biases might be implicit or explicit but are especially likely for animals that conduct much of their behavior subsurface or otherwise out of sight and have therefore plagued many cetacean studies (Mann 1999). For example, calves might catch small fish that are less visible to observers than adult fish catches. Thus, one might underestimate the rate at which calves catch fish relative to juveniles or adults. Or, if juvenile mating behaviors were raucous and tended to occur at the surface, but adult matings were subsurface and less obvious, then comparing mating rates between different age-sex classes would be futile. This bias might be exaggerated further if there were other interactions by age and sex (e.g., season). If, however, individual traits did not affect the likelihood of observing a behavior, then one could determine relative differences in mating behavior by age, sex, or season. How one interprets events largely depends on the sampling protocol (group or individual), how visible or obvious the behavior is, intrinsic biases to observability, and the sampling method (i.e., did the observer record all events of one type or just a subset for an individual or group?).

Even ad lib event sampling can add to a dataset and, as Altmann (1974) pointed out, these samples can be used for sociometric analysis, especially if direction is important. For example, in many delphinid societies, fission–fusion is a central feature. Often one group or individual clearly is the "joiner" and others are "joined." Similarly, a subgroup or individual may leave a group and the others are left. Such directionality might be extremely informative and can be used for social network analyses (i.e., in-degree or out-degree; see Stanton and Mann 2013). In Indo-Pacific bottlenose dolphins, adult males might join females, and females might often leave males, but females almost never join up with males. This type of information can reveal much about male–female relationships. An observer might not always be able to record who leaves and who joins during individual or group follows, but so long as ad lib join–leave events are not biased (e.g., the observer is biased by recording leaves only if they are females or joins only when they are males), then directionality can be quantitatively analyzed.

Given the difficulty in following most cetacean species, observers must first select an appropriate sampling protocol that captures the behavior(s) of enough individuals to be representative. There is a tradeoff between the number of individuals sampled and how often or intensively the same individuals are sampled. Typically, researchers use surveys (transect or opportunistic) to increase the number of individuals sampled in the population, or they follow groups (group-follow) or individuals (focal animal sampling) and collect more detailed and repeated measures on the same individuals (see Mann 1999). Surveys are useful for keeping track of individual life histories, ranging, and significant events. Group follows tend to be most useful when individuals cannot be tracked or the main research question focuses on group behaviors, as is particularly likely when animals stay in relatively stable groups (e.g., pilot whales, false killer whales, killer whales, sperm whales). Under these conditions, scan sampling can be a good relative measure of behavior. If fissions and fusions are common, then it is critical to have clear protocols to guide the observer on the "group" with which to stay. Otherwise, the observer might have a tendency to always stay with the larger or more interesting group, or not even notice if an individual or several animals left. Biases are likely to emerge if one always stays with the larger group, for example, so techniques for capturing the diversity of group behavior would be needed. Recently, we found that surveys of mothers and calves grossly underestimated the amount of time they spend separated relative to focal follows (Gibson and Mann 2009). The probable explanation is that observers are more likely to approach and sample adults, and if the calf joined at some point, observers might infer that the calf had been there all along.

Surveys are good for sampling a large number of individuals and across different time periods. Although each sampling point for an individual might be considered independent on a given day, surveys can inflate association patterns, that is, the "gambit of the group" where all individuals in a group are considered associated even though there might be strong preferences within the group (Whitehead 2008). One way to reduce this bias is to use weighted data (e.g., half-weight coefficients) and only consider those above a certain threshold to be associated (Franks et al. 2010). Regardless, sample size (sighting record per individual) has an immense impact on the validity of such estimations. We recently used bottlenose dolphin surveys over a 22-year period to determine social preferences between tool-using (with marine basket sponges) and non-tool-using dolphins. We took several precautions to reduce bias. We used weighted coefficients (affinity indices) with a very large sample size (average of 75 surveys per individual), and if individuals were not alive at the same time, the data for that dyad were coded as missing. Because we wanted to control for factors such as sex, maternal kinship, and range overlap, we used a multiple regression quadratic assignment procedure (Dekker et al. 2007). This permutation method allowed us to discriminate between the multiple factors that are likely to influence association by incorporating multiple matrices into one analysis while accounting for the structural autocorrelation that is inherent to social networks (Mann et al. 2012). Our analysis showed a clear pattern where sponger (tool-using) dolphins preferentially associated with each other over non-spongers (Mann et al. 2012). Such methods are likely to gain popularity as long-term datasets grow in size and complexity.

Ranging estimates are best achieved with systematic (e.g., transect) survey sampling, but are also plagued by inadequate sample sizes and pseudo-replication when groups are moderately stable. Fixed kernel densities (Gaussian distribution) are commonly used (Worton 1989; Seaman and Powell 1996), but a new adaptive local convex hull method outperforms traditional kernel density (KD) methods (*a*-LoCoH; Getz and Wilmers 2004; Getz et al. 2007). Urian et al. (2009) found that more than 100 points were needed to capture home range estimates using traditional methods, but few studies achieve this. For Shark Bay dolphins, we found that beyond 50 points, KD home range sizes did not change in a systematic way, but *a*-LoCoH home ranges did. Thus, to examine relative home range sizes (e.g., to compare males and females), we selected a random subset of 50 points for each animal (Patterson 2012). This method is recommended because any differences between groups cannot then be attributed to differences in sample size. Although it is tempting to use all one's data, randomized subsampling is preferable when variation in sample size biases the analysis. Long-term information on ranges of known animals and their mothers can provide important insights to bisexual philopatry, mother–son avoidance, and the role of fission–fusion societies (Tsai and Mann 2013).

For detailed behavioral information, individual focal follows are optimal because the observer is less likely to make sampling errors while observing the stream of behavior of one animal. However, such follows depend on individual identification or at least being able to identify the same animal throughout the follow. For example, a calf might not be "identifiable" by photo-identification but can be followed because it is distinctive enough from others in the group. Follows can also be quite short (e.g., 5 min) if longer follows are too difficult. Short sequential follows of all individuals in a group can provide information similar to scan sampling and are sometimes easier if behaviors are difficult to identify. When aggregations are very large (sometimes hundreds or even thousands), systematic sampling can still occur, but might involve sampling smaller clusters within the larger group or scan-sampling every tenth dolphin in view. The important point is to establish clear protocols that minimize bias regardless of sampling conditions.

Central to all these issues is establishing protocols that yield adequate sample sizes for drawing inferences about the population or group of interest. To reduce sampling error (variation from one sample to another, usually the standard error of the estimate, or the coefficient of variation, which expresses the standard error as a percentage of the estimate) and bias (e.g., selection bias, measurement bias, statistical bias), care must be taken to repeat samples, avoid pitfalls in the selection of subjects and measurement of behavior, and finally, apply appropriate statistical techniques. On the last point, which has received little attention here, pseudoreplication is a particular problem with many animal behavior studies, and cetacean studies in particular (Milinski 1997).

17.3 Technological Advances in Studying Behavior

In this overview, we have concentrated on the kind of behavioral information that can be gleaned from watching animals quite close up, as from a small boat near an individual or group. We acknowledge that the mere presence and (usually) noise of the boat engine can cause some degree of disruption of the "normal" behavioral repertoire of the watched animals. With careful boat approaches by experienced operators, such disruption is usually minimal (Bejder et al. 2006). For some species of cetaceans, observations can also be made from shore, with binoculars, still cameras with long lenses, digital video cameras, and theodolite tracking (the latter for more accurate positional information), and with the advantage that no disruption is made; however, shore-based observations provide a less intimate view of the animals or group (Würsig et al. 1991; Lundquist et al. 2012). Observations can also be made from circling aircraft, but this technique has been used mainly for large whales that can be identified from above, although some successful group-structure data have been gained on delphinids in clear waters of the open Pacific Ocean. This technique can also disturb animals, and care must be taken while circling (from an altitude of at least 450 m) to stay outside of the "cone of sound" of airplane noise underwater and to ensure that the plane's shadow does not fall onto the animals being observed.

Nowacek and colleagues (Nowacek 2002; Maresh et al. 2004) used a helium-filled aerostat (blimp) fitted with a videocamera and tethered to a boat, with two hydrophones, to record detailed foraging sequences in bottlenose dolphins. These innovative methods greatly expand our view into the world of smaller cetaceans. We predict that the recent rapid development of remote-controlled ("drone") mini-copters equipped with high-resolution cameras, operated from shore or vessel, will become a modern staple of group formation and behavioral research (NOAA 2012).

"Observations" of behavior do not need to be only by eye, but can involve acoustic studies of the especially soniferous delphinids, and various techniques of developed or developing electronic monitors (on the animals) or remote sensing devices. An up-to-date example is the use of so-called DTags (for "data tags") that can be placed onto a cetacean by suction cup or small barb attachment. Such tags have provided valuable insights. For example, exciting new information has become available with such tags for short-finned pilot whales (*Globicephala macrorhynchus*) that were discovered to echolocate for prey many hundreds of meters below them and then rapidly plunge-dive to depth to attack (Aguilar Soto et al. 2008). The DTag to accomplish this was outfitted with high-frequency echolocation detection and storage capability, as well as a depth sensor, triaxial accelerometer, and magnetometer for pitch, roll, and heading information in three-dimensional space. Because such devices fall off the animal and float, they can be recovered for data retrieval, and used again. A future application could be on multiple animals of a group, so that better social data can be obtained for animals at depth. At any rate, one device on one animal can already be thought of as an extension of a "focal follow" beyond that possible by visual assessment alone.

Other technological advances have greatly expanded research potential for marine mammals. However, most of the devices placed on animals have to date been used largely on pinnipeds and larger whales, and their development for smaller delphinids is just becoming practical as the result of miniaturization. The tried-and-true technique is conventional radio-tracking, but larger-distance satellite tracking is becoming practical for even small delphinids. Video chips, memory storage, and battery systems are becoming ever smaller, so that it is now feasible to use video recorders for underwater swimming, foraging, and social behavior information of even smaller cetaceans, although such devices have been used for more than 10 years on pinnipeds (Williams et al. 2004) and large whales. Anything put onto an animal can also be built to gather environmental data and can therefore enhance not only our understanding of behavior but of ecology as well. An overview of modern data acquisition systems (and their promise) is provided by Read (2009).

As can be the case for observations from surface vessels, a device put onto an animal, even for short times or by what seems a benign attachment technique (such as a suction cup system for delphinids), can be bothersome to the animal and change its behavior. One technique of gathering dive and foraging information that has been used for spinner (*Stenella longirostris*) and dusky (*Lagenorhynchus obscurus*) dolphins uses a sophisticated multi-beam "fish-finder" sonar array and mathematical algorithms to reconstruct information on potential prey and the depths and kind of dives of dolphins (Benoit-Bird et al. 2004, 2009). However, the high-frequency sonar itself might cause some behavioral change, and at any rate, one needs to be directly over the diving animals of interest for such information. Another technique not yet fully explored for delphinids is three-dimensional array passive listening of their own acoustics, for positional data during different behavioral states (Schotten et al. 2004), and for determination of which dolphin is vocalizing when incorporated with a video camera (Schotten et al. 2005).

17.4 Conclusions

Observation and quantification of behavior can proceed in many different ways, with and without enhancement by modern data acquisition techniques. Direct and remotely sensed data can also be augmented by, for example, scat samples or protein analyses relating to diet, and genetic sampling to examine relatedness and mating patterns. Such samples can be gleaned directly from behind or on the swimming animals, or from short-term captures in special situations (Wells et al. 1999). A practical set of research directions will probably involve more integration of variable data-gathering platforms, so that, for example, a group of cetaceans with several DTags or other electronic devices can be watched by eye from shore (in special situations), from a surface vessel, or via a remote-controlled mini-copter at the same time that detailed biological and oceanographic data are gathered. The future of behavioral observations of cetaceans is bright, and we only caution that behavioral patterns be well defined and data gathering be as representative as possible.

References

Aguilar Soto N, Johnson MP, Madsen PT, Díaz F, Domínguez I, Brito A, Tyack P (2008) Cheetahs of the sea: deep foraging sprints in short-finned pilot whales off Tenerife (Canary Islands). J Anim Ecol 77:936–947

Altmann J (1974) Observational study of behavior: sampling methods. Behaviour 49:227–267

Bejder L, Samuels A, Whitehead H, Gales N, Mann J, Connor R, Heithaus M, Watson-Capps J, Flaherty C (2006) Shift in habitat use by bottlenose dolphins (*Tursiops* sp.) exposed to long-term anthropogenic disturbance. Conserv Biol 20:1791–1798

Benoit-Bird KJ, Würsig B, McFadden CJ (2004) Dusky dolphin (*Lagenorhynchus obscurus*) foraging in two different habitats: active acoustic detection of dolphins and their prey. Mar Mamm Sci 20:215–231

Benoit-Bird KJ, Dahood AD, Würsig B (2009) Using active acoustics to compare lunar effects on predator-prey behavior in two marine mammal species. Mar Ecol Prog Ser 395:119–135

Dekker D, Krackhardt D, Snijders TAB (2007) Sensitivity of MRQAP tests to collinearity and autocorrelation conditions. Psychometrika 72:563–581

Franks DW, Ruxton GD, James R (2010) Sampling animal association networks with the gambit of the group. Behav Ecol Sociobiol 64:493–503. doi:10.1007/s00265-009-0865-8

Getz WM, Wilmers CC (2004) A local nearest-neighbor convex-hull construction of home ranges and utilization distributions. Ecography 4:489–505

Getz WM, Fortmann-Roe S, Cross PC, Lyons AJ, Ryan SJ, Wilmers CC (2007) LoCoH: nonparameteric kernel methods for constructing home ranges and utilization distributions. PLoS One 2:e207

Gibson QA, Mann J (2008) Early social development in wild bottlenose dolphins: sex differences, individual variation, and maternal influence. Anim Behav 76:375–387

Gibson QA, Mann J (2009) Do sampling method and sample size affect basic measures of dolphin sociality? Mar Mamm Sci 25:187–198

Lundquist D, Gemmell NJ, Würsig B (2012) Behavioural responses of dusky dolphin (*Lagenorhynchus obscurus*) groups to tour vessels off Kaikoura, New Zealand. PLoS One 7(7):e41969

Mann J (1999) Behavioral sampling methods for cetaceans: a review and critique. Mar Mamm Sci 15:102–122

Mann J, Sargeant BL, Watson-Capps J, Gibson Q, Heithaus MR, Connor RC, Patterson E (2008) Why do dolphins carry sponges? PLoS One 3(12):e3868

Mann J, Stanton MA, Patterson EM, Bienenstock EJ, Singh LO (2012) Social networks reveal cultural behaviour in tool using dolphins. Nat Commun 3:980. doi:10.1038/ncomms1983

Maresh JL, Fish FE, Nowacek DP, Nowacek SM, Wells RS (2004) High performance turning capabilities during foraging by bottlenose dolphins (*Tursiops truncatus*). Mar Mamm Sci 20:498–509

Milinski M (1997) How to avoid seven deadly sins in the study of behavior. Adv Stud Behav 26:159–180

NOAA Fisheries (2012) http://www.nmfs.noaa.gov/stories/2012/10/_10_03_12wayne_perryman.html

Nowacek DP (2002) Sequential foraging behaviour of bottlenose dolphins, *Tursiops truncatus*, in Sarasota Bay, FL. Behaviour 139:1125–1145

Patterson EM (2012) Ecological and life history factors influence habitat and tool use in wild bottlenose dolphins (*Tursiops* sp.). Ph.D. Thesis, Department of Biology, Georgetown University, Washington, DC

Read AJ (2009) Telemetry. In: Perrin WF, Würsig B, Thewissen JGM (eds) The encyclopedia of marine mammals, 2nd edn. Academic/Elsevier, San Diego

Schotten M, Au WWL, Lammers MO, Aubauer R (2004) Echolocation recordings and localization of wild spinner dolphins (*Stenella longirostris*) and pantropical spotted dolphins (*S. attenuata*) using a four-hydrophone array. In: Thomas JA, Moss CF, Vater M (eds) Echolocation in bats and dolphins. University of Chicago Press, Chicago, pp 393–400

Schotten M, Lammers MO, Sexton K, Au WWL (2005) Application of a diver-operated 4-channel acoustic-video recording device to study wild dolphin echolocation and communication. J Acoust Soc Am 117:2552

Seaman DE, Powell RA (1996) An evaluation of the accuracy of kernel density estimators for home range analysis. Ecology 77:2075–2085

Stanton MA, Mann J (2013) Social network analysis: applications to primate and cetacean societies. In: Yamagiwa J, Karczmarski L, Takeda M (eds) Primates and cetaceans. Springer, Tokyo

Tsai J-YJ, Mann J (2013) Dispersal, philopatry, and the role of fission–fusion dynamics in bottlenose dolphins. Mar Mamm Sci 29:261–279

Urian KW, Hofmann S, Wells RS, Read AJ (2009) Fine-scale population structure of bottlenose dolphins (*Tursiops truncatus*) in Tampa Bay, Florida. Mar Mamm Sci 25:619–638

Vaughn RB, Würsig B, Packard J (2009) Dolphin prey herding: prey ball mobility relative to dolphin group and prey ball sizes, multispecies aggregations, and feeding duration. Mar Mamm Sci 26:213–225

Wells RS, Boness DJ, Rathbun GB (1999) Behavior. In: Reynolds JE III, Rommel SA (eds) Biology of marine mammals. Smithsonian Institution Press, Washington, DC

Whitehead H (2008) Analyzing animal societies: quantitative methods for vertebrate social analysis. University of Chicago Press, Chicago, IL USA

Williams TM, Fuiman LA, Horning M, Davis RW (2004) The cost of foraging by a marine predator, the Weddell seal, *Leptonychotes weddellii*, pricing by the stroke. J Exp Biol 207:973–982

Worton BJ (1989) Kernel methods for estimating the utilization distribution in home range studies. Ecology 70:164–168

Würsig B, Würsig M (1977) The photographic determination of group size, composition, and stability of coastal porpoises (*Tursiops truncatus*). Science 198:75–756

Würsig B, Cipriano F, Würsig M (1991) Dolphin movement patterns: information from radio and theodolite tracking studies. In: Pryor K, Norris KS (eds) Dolphin societies: discoveries and puzzles. University of California Press, Berkeley

Chapter 18
Social Network Analysis: Applications to Primate and Cetacean Societies

Margaret A. Stanton and Janet Mann

M.A. Stanton (✉)
Department of Biology, Georgetown University, Washington, DC, USA

Department of Anthropology, The George Washington University, Washington, DC, USA
e-mail: mastanton@gwu.edu

J. Mann
Department of Biology, Georgetown University, Washington, DC, USA

Department of Psychology, Georgetown University, Washington, DC, USA

J. Yamagiwa and L. Karczmarski (eds.), *Primates and Cetaceans: Field Research and Conservation of Complex Mammalian Societies*, Primatology Monographs,
DOI 10.1007/978-4-431-54523-1_18, © Springer Japan 2014

Abstract To better address questions concerning animal sociality, animal behaviorists and behavioral ecologists are increasingly turning to the suite of analytical techniques known as social network analysis (SNA). SNA allows for the quantification of multi-actor interactions, thereby providing a more realistic representation of social patterns and relationships. Here, we provide a brief introduction to SNA, consider some of the challenges in studying sociality, and discuss the application of SNA to studies of animal societies, with a focus on primates and cetaceans. Additionally, we present techniques for network comparison and dynamic network analysis developed in the social sciences with exciting potential applications to the study of animal behavior.

Keywords Animal societies • Dolphins • Metrics • Primates • Quantitative methods • Social network analysis

18.1 Introduction

A powerful quantitative tool with which to address the causes and consequences of sociality is social network analysis (SNA). Indeed, social network theory has potential in any discipline that requires the description of complex systems, including physics, psychology, sociology, ethology, neuroscience, cell and molecular biology, ecology, mathematics, military intelligence, and computer science (Wasserman and Faust 1994; Freeman 2004). A social network is defined as actors (or nodes, points, vertices) linked by relationships (edges, links, ties), and the visual representation of these nodes and edges is referred to as a graph (Fig. 18.1). This type of analysis was popularized in the 1970s after Stanley Milgram (1967) examined the social distance between individuals in the United States (U.S.), the results of which are commonly referred to as "Six Degrees of Separation." Later, Watts and Strogatz (1998) formalized Milgram's idea in their description of small-world phenomena, where tightly knit subgroups of individuals are closely connected to each other, but with at least one member maintaining a connection to a separate subgroup.

With recent advances in computing power, SNA has gained momentum in the field of animal behavior (Krause et al. 2007; Wey et al. 2008). Traditional studies of social relationships and structure focus on dyadic interactions, whereas network analysis applies graph theory to quantify multi-actor interactions, thereby providing more realistic representations of the complex societies typically observed in primates and cetaceans. Additionally, by providing more direct measurements of social relationships, rather than proxies such as group size, SNA allows for more in-depth investigations into complex sociality. By quantifying multi-actor interactions, SNA accounts for some of the unavoidable data dependency, which is problematic for traditional statistical analyses. Take, for example, an investigation into the relationship between dominance rank and relationship quality in a savanna baboon troop. With traditional methods, female rank and relationship quality are treated as independent, when in fact they are not. A female who grooms one female cannot simultaneously be grooming another. Rank is determined by who is above and below, so is by definition not

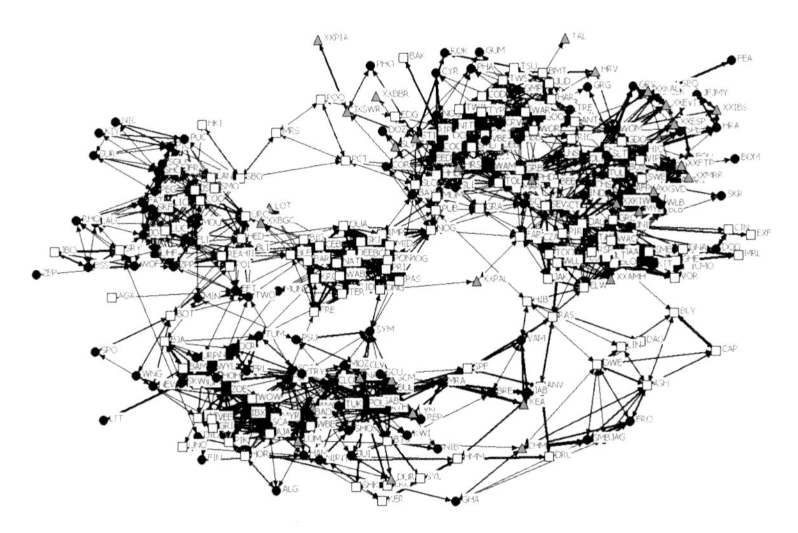

Fig. 18.1 Social network of Shark Bay adult and juvenile dolphins constructed from survey data from 1999 to 2007. Edges are weighted by half-weight coefficient and only those greater than the average (0.13) are included. *Circles*, females; *squares*, males; *triangles*, unknown

independent. SNA treats the relational nature of data as part of the analysis. A network's edges can be directed or undirected, weighted or unweighted. Undirected edges indicate that the relationship is symmetrical, as in the case of a mutual friendship. However, if one individual identifies another as a friend and the sentiment is not reciprocated, the relationship is directed. Additionally, an unweighted edge indicates the presence of a relationship, while a weighted edge can indicate the presence and the strength of a relationship. Edges can also be either positive or negative, as might be quantified when individuals preferentially approach or avoid each other (Wasserman and Faust 1994; Croft et al. 2008; Wey et al. 2008). Network theory is also useful and unique in that it is capable of analysis on multiple levels by characterizing individuals, their subgroups of neighbors, and the network as a whole. Some basic social network metrics including measures of centrality (a node's connection to the rest of the network) and clustering (the tightness of subgroups or cliques) are described in Box 1, but it is most important to note at this juncture that distinct social network metrics provide different information about the same individual, subgroup, or network and that this information is not necessarily accessible using more conventional methods.

18.2 Considerations and Caveats

Despite the usefulness of SNA and the increasing frequency with which these techniques are applied, a number of considerations and caveats are warranted before the initiation of a network study (see James et al. 2009; Croft et al. 2011 for review of

SNA potential pitfalls). One important consideration is how to define a relationship. According to Hinde's (1976) classic framework for the study of social structure, a relationship is defined as successive interactions between individuals. In the study of animal behavior, however, interactions are often difficult to observe and quantify; therefore, relationships are often assessed in terms of association defined by shared group membership with the assumption that associating individuals have the potential to interact. Whitehead and Dufault (1999) refer to this assumption as the "gambit of the group," and for the purposes of SNA, researchers should be aware that networks built from group-defined association data may appear highly clustered by this sampling method. These associations are also typically measured in terms of an association index, such as the half-weight index, that accounts for sampling effort (Cairns and Schwager 1987). Additionally, there is no universally accepted method for determining whether a biologically meaningful relationship exists and should be included in a social network for analysis or whether the observed association is random, erroneous, or biased by sampling method or effort. A common response to this uncertainty to date has been to filter and dichotomize data, thereby only including edges or nodes above a certain value when constructing the network. However, the thresholds at which networks are filtered and dichotomized are arbitrary, and the resulting binary networks are likely oversimplified (Franks et al. 2010). Recent methodological developments are providing researchers with exciting new variations on centrality and clustering metrics for analyzing weighted networks, which are exceedingly useful for investigating animal social networks (Box 1) (Newman 2004; Lusseau et al. 2008).

Weighting edges by association should be considered whenever possible, and algorithm development is ongoing to facilitate further substantive interpretation of weighted graphs. Following construction of the network and calculation of network metrics, weighted or unweighted, the subsequent statistical analysis of network data must proceed with caution. For reasons of their very nature, network data

Box 1 Social network analysis terms. Definitions from Wey et al. (2008), p. 334, and Whitehead (2008), pp. 172–175

Individual measures

Centrality	A measure of an individual's structural importance based on its network position
Degree centrality	Centrality based on the number of direct edges connected to a node
Betweenness centrality	Centrality based on the number of shortest paths between every pair of other nodes in the network that pass through the node of interest
Reach	A measure of indirect connectedness that is defined as the number of nodes two or fewer steps away

(continued)

Box 1 (continued)

Affinity	Average degree of a node's neighbors; a node with high affinity is connected to other nodes of high degree
Intermediate measures	
Clustering coefficient	The density of a node's local network; the number of observed edges between a node's neighbors is divided by the number of possible edges between them
Cliquishness	How much the network is divided into subgroups; a clique is a set of nodes that are all directly connected to each other
Group measures	
Average path length	The average of all path lengths, or number of edges, between all pairs of nodes in the network
Density	The number of observed edges divided by the number of possible edges in the network
Diameter	The longest path length in the network
Weighted measures	
Strength	A measure of weighted degree that is the sum of the weights of the edges connected to a node
Betweenness	Sum of the inverses of the weights on each edge that equals the shortest path lengths that pass through the node
Eigenvector	The corresponding element of the first eigenvector of an association matrix; accounts for both the number and weights of all directly connected edges, as well as indirect connections
Reach	Overall strength of a node's neighbors
Affinity	Average weighted strength of a node's neighbors
Clustering coefficient	A measure of cliquishness or how well connected neighbors are to each other considering the weight on all three edges of each triangle linking the nodes

violate the assumption of independence of traditional statistical methods. Therefore, when analyzing network data the scope of inference is generally constrained, and the statistical significance of network metrics is typically assessed by carefully chosen randomized techniques or models that account for network autocorrelation (Croft et al. 2011). This particular caveat has important implications for network comparison and dynamic network analysis.

18.3 Network Comparison and Dynamic Network Analysis

Two networks can be compared with permutation tests (Manly 2007), determining whether specific network metrics means differ more than random expectations. However, care must be taken to ensure that the networks under comparison are similar in size and density or that measures have been normalized based on the maximum value for a node in that network because most network metrics vary with the number of nodes and edges. An additional methodology for comparing networks regardless of differences in size, or even species, was suggested by Faust and Skvoretz (2002). This technique characterizes networks in terms of their structural properties and measures the similarity of networks based on the parameter estimates for models (exponential random graph models, or "ergms") that predict the probability of network ties. Dynamic social data present additional methodological obstacles, and the techniques for analyzing such data using network theory are still in development or untested on real-life data. Intuitively, social networks are dynamic with relationships forming and fading over time; however, the vast majority of research has focused on static networks that are unable to capture information about changes in the network or the mechanisms related to observed dynamics. In the social sciences, Snijders (1996) and his colleagues (Steglich et al. 2010) developed some dynamic models used in the analysis of dynamic friendship networks. These dynamic models can identify what is likely driving change in social networks over time and could be particularly useful for studying animal social development. Research into dynamic network models is ongoing, and the availability of applied longitudinal datasets will facilitate the creation of exciting additional methodologies.

18.4 Analysis of Primate and Dolphin Social Networks

In the field of animal behavior, SNA is employed, generally on static graphs, to describe complex social structure and to provide insight into studies of cooperation, disease, and information transfer, the different roles of individuals in groups, and the consequences of anthropogenic disturbance on animal societies (see Krause et al. 2007; Wey et al. 2008 for excellent reviews). The specific applications of SNA to animal behavior are too numerous to enumerate in further detail here; however, it is worth noting some of the early and commonly cited animal SNA conducted on wild populations of bottlenose dolphins (*Tursiops* sp.) (Lusseau 2003; Lusseau and Newman 2004; Lusseau et al. 2006; Lusseau 2007). Bottlenose dolphins and some other odontocetes (toothed whales) are attractive candidates for SNA because of their dynamic and complex fission–fusion society. For example, Lusseau (2003) described the network of a relatively small population ($N=64$) of bottlenose dolphins in Doubtful Sound, New Zealand and investigated the theoretical removal of random individuals compared to specific individuals with a large number of associates. The dolphin network appeared robust to removal of random individuals whereas the removal of individuals with high degree increased the network diameter, defined as the average shortest path length between any two nodes, by 20–30 %.

In a separate study, Lusseau et al. (2006) described the network of bottlenose dolphins ($N = 124$) in the inner Moray Firth in eastern Scotland. This study addressed the possible relationship between social structure and geographic preference by assigning dolphins to one of two categories, either (1) always sighted in inner Moray Firth or (2) sighted in inner Moray Firth and elsewhere, and determining whether dolphins in these two categories constitute cliques in the network. The researchers conclude that composition of the two communities identified in the social network matched well with the categories of geographic preference. More recent work in delphinids has continued to address spatial, as well as temporal and ecological, correlates of social network structure to better understand factors influencing social processes (Cantor et al. 2012; Foster et al. 2012).

We have recently published several detailed studies on bottlenose dolphin maternal and calf social networks (Stanton et al. 2011), the relationship between early calf networks and survival during the juvenile period (Stanton and Mann 2012), sex differences in social network metrics (Mann et al. 2012), and how SNA can help identify "culture" in dolphins (Mann et al. 2012). Most of the insights gained from these studies are attributable to the application of social network methods. For example, the likelihood of survival of male dolphins beyond weaning was positively related to eigenvector centrality as calves (Stanton and Mann 2012). This metric is an excellent measure of an individual's importance in the network, and because this method accounts for both direct and indirect ties, we would not have detected this pattern without a SNA approach.

In primates, network theory has been applied to the analysis of grooming interaction networks as well as association networks (Flack et al. 2006; Lehmann and Boesch 2009). An interesting SNA study investigated the roles and structural positions of captive pigtailed macaques (*Macaca nemestrina*) in their social networks. In this case, dominant males perform a policing function by impartially intervening in conflicts between other members of the group. Both the simulated and empirical removal of just a few of these policing individuals altered the macaque social network, in some cases significantly decreasing in the mean degree and increasing clustering coefficients, which the authors conclude destabilizes the group (Flack et al. 2006). Such investigations can be utilized as a means of predicting anthropogenic effects on free-ranging animal social networks. Although many wildlife conservation management plans assume all animals are equal, research such as the macaque study just described, as well as an analysis of a killer whale (*Orcinus orca*) social network, indicates that certain individuals have a disproportionally large impact on their networks and should be differentially accounted for in conservation plans (Williams and Lusseau 2006).

Historically, some primate studies used the term social network when referring to associating individuals and grooming interactions, but do not actually apply network theory when analyzing data. However, as the utility of SNA becomes increasingly apparent, network theory is being applied to datasets from primate field sites to identify differences in association patterns between both individuals and age-sex classes, describe association trends over time, and provide new approaches and perspectives for measuring dominance and other hierarchical structures (Lehmann and Ross 2011; Henzi et al. 2009; Ramos-Fernández et al. 2009; Shizuka and McDonald 2012). In recent years, studies have begun comparing social networks across closely related primate species. For example, in a study of four macaque species that vary in degree of social tolerance, SNA metrics revealed novel dimensions of these otherwise well-characterized societies (Sueur et al. 2011).

18.5 Additional Social Network Applications

The use of SNA in animal behavior is by no means confined to research in cetaceans and primates. Indeed, the use of these methods now ranges from studies of insects (Fewell 2003), rodents (Wey and Blumstein 2010), ungulates (Sundaresan et al. 2007), and social carnivores (Smith et al. 2010). An exceedingly useful social network technique is quadratic assignment procedure (QAP) regression, which allows for the regression of explanatory matrices on a response sociomatrix representing associations or interactions (Krackhardt 1988; Dekker et al. 2007). QAP regression first calculates coefficients by performing an ordinary least squares (OLS) regression, then randomly permutes the response matrix and reruns the OLS regression x number of times to obtain a matrix-specific distribution of coefficients against which the observed matrix coefficients can be compared and statistical significance evaluated. This permutation-based approach avoids the inflation of type I errors caused by the correlational nature of network data (Krackhardt 1988). As with traditional multiple regression, the multiple regression quadratic assignment procedure (MRQAP) allows for the inclusion of multiple factors that may account for variation in a sociomatrix, including factors that are not necessarily of interest that need to be controlled for. Animal behavior researchers are beginning to recognize the usefulness of this analysis (Croft et al. 2011) and have thus far employed QAP regression to investigate factors influencing network structure in yellow-bellied marmots (*Marmota flaviventris*) (Wey and Blumstein 2010) and ring-tailed coatis (*Nasua nasua*) (Hirsch et al. 2012). We have also recently applied this method to our own investigation of the social function of tool use in Shark Bay bottlenose dolphins (*Tursiops* sp.) (Mann et al. 2012). In that study, the MRQAP was used to control for sex as well as geographic distance and maternal relatedness between individuals while investigating whether similarity based on the use of marine sponges as tools is a significant predictor of association and indicative of culture. This was the first study to examine whether tool use or foraging similarity influences social preference. Such an examination is not possible with traditional non-network approaches.

18.6 Conclusion

As indicated here, we have applied SNA in our own work on bottlenose dolphins. Chapter 6 in this volume provides example measures from the networks of two calves, based on different measures of sociality including networks created from association and petting/grooming interactions. As networks may differ based on the types of data used to define relationships between individuals, one exciting new direction for the field involves combining different behaviors (e.g., grooming, aggression, and proximity) into one multidimensional object (Barrett et al. 2012). As evidenced by the pioneering animal network studies described here, SNA of cetacean and primate populations is exceedingly applicable to free-ranging primate and cetacean populations, and the potential inquiries are plentiful.

References

Barrett L, Henzi PS, Lusseau D (2012) Taking sociality seriously: the structure of multi-dimensional social networks as a source of information for individuals. Philos Trans R Soc B 367:2108–2118

Cairns SJ, Schwager SJ (1987) A comparison of association indices. Anim Behav 35:1454–1469

Cantor M, Wedekin LL, Guimarães PR, Daura-Jorge FG, Rossi-Santos MR, Simões-Lopes PC (2012) Disentangling social networks from spatiotemporal dynamics: the temporal structure of a dolphin society. Anim Behav 84:641–651

Croft DP, James R, Krause J (2008) Exploring animal social networks. Princeton University Press, Princeton

Croft DP, Madden JR, Franks DW, James R (2011) Hypothesis testing in animal social networks. Trends Ecol Evol 26:502–507

Dekker D, Krackhardt D, Snijders TAB (2007) Sensitivity of MRQAP tests to collinearity and autocorrelation conditions. Psychometrika 72:563–581

Faust K, Skvoretz J (2002) Comparing networks across space and time, size and species. Sociol Methodol 32:267–299

Fewell JH (2003) Social insect networks. Science 301:1867–1870

Flack JC, Girvan M, de Waal FBM, Krakauer DC (2006) Policing stabilizes construction of social niches in primates. Nature (Lond) 439:426–429

Foster EA, Franks DW, Morrell LJ, Balcomb KC, Parsons KM, van Grinneken A, Croft DP (2012) Social network correlates of food availability in an endangered population of killer whales, *Orcinus orca*. Anim Behav 83:731–736

Franks DW, Ruxton GD, James R (2010) Sampling animal association networks with the gambit of the group. Behav Ecol. Sociobiol 64: 493–503

Freeman LC (2004) The development of social network analysis. Empirical Press, Vancouver

Henzi SP, Lusseau D, Weingrill T, Schaik CP, Barrett L (2009) Cyclicity in the structure of female baboon social networks. Behav Ecol Sociobiol 63:1015–1021

Hinde RA (1976) Interactions, relationships and social structure. Man 11:1–17

Hirsch BT, Stanton MA, Maldonado J (2012) Kinship shapes affiliative social networks but not aggression in ring-tailed coatis. PLoS One 7:e37301

James R, Croft DP, Krause J (2009) Potential banana skins in animal social network analysis. Behav Ecol Sociobiol 63:989–997

Krackhardt D (1988) Predicting with networks: nonparametric multiple regression analysis of dyadic data. Soc Netw 10:359–381

Krause J, Croft DP, James R (2007) Social network theory in the behavioural sciences: potential applications. Behav Ecol Sociobiol 62:15–27

Lehmann J, Boesch C (2009) Sociality of the dispersing sex: the nature of social bonds in West African female chimpanzees, *Pan troglodytes*. Anim Behav 77:377–387

Lehmann J, Ross C (2011) Baboon (*Papio anubis*) social complexity: a network approach. Am J Primatol 73:775–789

Lusseau D (2003) The emergent properties of a dolphin social network. Proc R Soc Lond Ser B Suppl 270:S186–S188

Lusseau D (2007) Why are male social relationships complex in the Doubtful Sound bottlenose dolphin population? PLoS One 2:e348

Lusseau D, Newman MEJ (2004) Identifying the role that animals play in their social networks. Proc R Soc Lond Ser B Suppl 271:S477–S481

Lusseau D, Wilson B, Hammond PS, Grellier K, Durban JW, Parsons KM, Barton TR, Thompson PM (2006) Quantifying the influence of sociality on population structure in bottlenose dolphins. J Anim Ecol 75:14–24

Lusseau D, Whitehead H, Gero S (2008) Incorporating uncertainty into the study of animal social networks. Anim Behav 75:1809–1815

Manly BFJ (2007) Randomization, bootstrap and Monte Carlo methods in biology, 3rd edn. Chapman & Hall, Boca Raton

Mann J, Stanton MA, Patterson EM, Bienenstock EJ, Singh LO (2012) Social networks reveal cultural behaviour in tool using dolphins. Nat Commun 3:980. doi:10.1038/ncomms1983

Milgram S (1967) The small world problem. Psychol Today 2:60–67

Newman MEJ (2004) Analysis of weighted networks. Phys Rev E 70:1–9

Ramos-Fernández G, Boyer D, Aureli F, Vick LG (2009) Association networks in spider monkeys (*Ateles geoffroyi*). Behav Ecol Sociobiol 63:999–1013

Shizuka D, McDonald DB (2012) A social network perspective on measurements of dominance hierarchies. Anim Behav 83:925–934. doi:10.1016/j.anbehav.2012.01.011

Smith JE, Van Horn RC, Powning KS, Cole AR, Graham KE et al (2010) Evolutionary forces favoring intragroup coalitions among spotted hyenas and other animals. Behav Ecol 21:284–303

Stanton MA, Gibson QA, Mann J (2011) When mum's away: A study of mother and calf ego networks during separations in wild bottlenose dolphins (Tursiops sp.). Anim Behav 82: 405–412

Stanton MA, Mann, J (2012) Early social networks predict survival in wild bottlenose dolphins. PLoS One 7:e47808

Snijders TAB (1996) Stochastic actor-oriented models for network. J Math Sociol 21:149–172

Steglich C, Snijders TAB, Pearson M (2010) Dynamic networks and behavior: separating selection from influence. Sociol Methodol 4:329–393

Sueur C, Petit O, De Marco A, Jacobs AT, Watanabe K, Thierry B (2011) A comparative network analysis of social style in macaques. Anim Behav 82:845–852

Sundaresan SR, Fischhoff IR, Dushoff J, Rubenstein DI (2007) Network metrics reveal differences in social organization between two fission–fusion species, Grevy's zebra and onager. Oecologia (Berl) 151:140–149

Wasserman S, Faust K (1994) Social network analysis: methods and applications. Cambridge University Press, Cambridge

Watts DJ, Strogatz SH (1998) Collective dynamics of "small-world" networks. Nature (Lond) 393:440–442

Wey TW, Blumstein DT (2010) Social cohesion in yellow-bellied marmots is established through age and kin structuring. Anim Behav 79:1343–1352

Wey TW, Blumstein DT, Shen W, Jordán F (2008) Social network analysis of animal behaviour: a promising tool for the study of sociality. Anim Behav 75:333–344

Whitehead H, Dufault S (1999) Techniques for analyzing vertebrate social structure using identified individuals: review and recommendations. Adv Stud Behav 28:33–74

Whitehead (2008) Analyzing animal societies: Quantitative methods for vertebrate social analysis. University of Chicago Press, Chicago

Williams R, Lusseau D (2006) A killer whale social network is vulnerable to targeted removals. Biol Lett 2:497–500

Chapter 19
Social Touch in Apes and Dolphins

Michio Nakamura and Mai Sakai

The online version of this chapter (doi: 10.1007/978-4-431-54523-1_19) contains supplementary material, which is available to authorized users

M. Nakamura (✉) • M. Sakai
JSPS Research Fellow, Wildlife Research Center, Kyoto University,
2-24, Tanaka-Sekiden-Cho, Sakyo, Kyoto 606-8203, Japan
e-mail: nakamura@wrc.kyoto-u.ac.jp

J. Yamagiwa and L. Karczmarski (eds.), *Primates and Cetaceans: Field Research and Conservation of Complex Mammalian Societies*, Primatology Monographs, DOI 10.1007/978-4-431-54523-1_19, © Springer Japan 2014

Abstract Social touch, or physical contact among two or more individuals in a nonaggressive context, seems to play important roles among both primates and cetaceans. However, with exception of social grooming among primates, it has rarely been studied in detail. Thus, in this chapter we review the descriptions of social touch in great apes and dolphins from the literature and from our own observations. After reviewing the social grooming among various mammalian taxa, we considered various types of social touch in apes and dolphins in more detail by dividing them into following seven categories: (1) social touch between mother and infant; (2) touch in play; (3) tactile gestures; (4) social grooming; (5) touch in greeting, reassurance, and appeasement; (6) touch to/with genital areas; and (7) simple body contact. Information from scattered descriptions in the literature suggests that social touch is widespread in apes and dolphins, yet frequencies may vary greatly among species. Although there has been no single theory to explain these diverse types of social touch, we briefly review theories that might be of relevance to explaining social touch.

Keywords Dolphins • Flipper rubbing • Great apes • Odontocetes • Physical contact • Social grooming • Social interaction • Tactile communication

19.1 Introduction

Primates and cetaceans form complex and diverse societies (primates: Itani 1977; Smuts et al. 1987; cetaceans: Mann et al. 2000). A society is neither a visible entity nor is it a simple accumulation of individual behaviors but emerges through many types of social interactions among individuals. Thus, it is important to begin by examining social interactions to study different types of societies. Social interactions occur through a number of modalities, and in this chapter, we particularly focus on "social touch"—touch between two or more individuals. This focus is because physical contact, except for grooming behavior, has rarely been studied, even though it seems to play important social roles among both primates and cetaceans.

In addition to the tactile modality, mammals use visual, auditory, kinesthetic, and olfactory modalities in the course of their social interaction. Vocal communication has been extensively studied both in primates (Gouzoules and Gouzoules 2007) and in cetaceans (Tyack 2000, 2003) and compared to human language (primates: Itani 1963; Marler 1965; Arcadi 1996; cetaceans: Yurk et al. 2002). Studies on visual communication in primates have focused on facial displays (van Hooff 1967), sociosexual signals (Wickler 1967), visual gestures (Tomasello et al. 1994), etc. It has also been suggested that visual communication, in addition to verbal communication, plays an important role in human greeting and salutation (Kendon and

Ferbar 1973/1996). Cetaceans, except for the several species that live in rivers, have a well-developed sense of vision (Tyack 2000). Visual communication in cetaceans, especially postures and movements in aggression and sexual contexts, has been reported (Tavolga and Essapian 1957; Caldwell and Caldwell 1972; reviewed by Madsen and Herman 1980; Herman and Tavolga 1980; Samuels and Gifford 1997). For example, mouth-opening and jaw-clapping (dolphins produce a sound by quickly snapping their jaws) have been interpreted as threatening. Many delphinids also have various skin pigmentations and characteristics. For example, killer whales (*Orcinus orca*) and Commerson's dolphins (*Cephalorhynchus commersonii*) have white and black pigments, spotted dolphins (*Stenella attenuata* and *Stenella frontalis*) have spots, and Risso's dolphins (*Grampus griseus*) have scratches on the body (Hartman et al. 2008). These skin characteristics may serve to identify species, individuals, sex and age classes, and reproductive state (Madsen and Herman 1980; Herman and Tavolga 1980) or may serve as camouflage or disruptive coloration against predators (Tyack 2000). Cetaceans may also use body postures, body movements, bubbling and splashing for visual communication but these have not yet been studied in detail.

Olfactory signals are thought to be important, especially in prosimian primates (Scordato and Drea 2007). Cetaceans are considered to lack the sense of olfaction (Breathnach 1960; Morgane and Jacobs 1972; Kishida et al. 2007) and, compared to humans, they have a limited or perhaps different sense of taste (Tyack 2000).

Few studies on tactile senses have been conducted. In humans, the importance of touch to form intimate relationships between mothers and their offspring and between sexual partners has been discussed (Montagu 1971; Morris 1971/1993), and touching among mammals, especially among primates, has also been previously reviewed. Darian-Smith (1982) reviewed touching among primates in relationship to neural processing but did not consider the interactional aspects. Hertenstein et al. (2006) reviewed tactile communication in humans, nonhuman primates, and rodents; however, in the case of nonhuman primates, they exclusively studied grooming behaviors and did not include other forms of tactile communication. Thus far, only a few reviews of tactile communication have been conducted in cetaceans (Herman and Tavolga 1980; Tyack 2000).

19.2 Scope of This Chapter

We define *social touch* as physical contact of any part of the body of one individual with a part of the body of another individual in any way. However, physical contacts that harm the partner's body or that make the partner flee (i.e., aggression) are not addressed in this chapter. Defined this way, social touch may be used in various contexts in different species, as we see in later sections. Consequently, there seems to be no single theory to explain all the forms of social touch. In addition, because most forms of social touch have not been studied in detail, we first need to accumulate knowledge of what is actually occurring in nature. In this chapter, we first try to descriptively show the diversity and similarities of social touch, focusing especially on how different species use it in different contexts. Among various types of social touch, social grooming may be exceptional, in that many detailed studies have been conducted, and accordingly, several explanatory theories have been proposed. Although these theories may not have been intended to cover social touch in general, in the final section, we examine whether these theories can be generalized to social touch.

In the following section, we first briefly review social grooming among mammals. We then review various types of social touch in great apes and dolphins (relatively small odontocetes), sometimes referring to several species outside these taxa when relevant. Because quantitative comparisons of social touch are difficult at this point, the comparisons herein are primarily based on behavioral patterns. As descriptions of most types of social touch in great apes and dolphins are scattered into various behavioral domains, we divided them into seven categories for the reader's convenience, but these categories are not always mutually exclusive.

19.3 Social Grooming in Mammals

Social grooming is a type of social interaction involving social touch that is observed in various animal species (reviewed by Spruijt et al. 1992), including arthropods and birds (where it is called "preening"). Because of this ubiquity, social touch and social grooming are often regarded as synonymous; as already mentioned, Hertenstein et al. (2006) focused solely on social grooming when reviewing tactile communication. However, social grooming or preening is only one type of social touch.

In many mammalian species, social grooming consists of cleaning the surface of the body with the mouth or tongue. Even excluding maternal licking of neonates to clean their fur and the licking of genital areas to stimulate excretion, social grooming is widely observed in various mammalian taxa (Diprotodonta: Blumstein and Daniel 2003; Artiodactyla: Hall 1983; Perissodactyla: Kimura 2000; Chiroptera: Wilkinson 1986; Carnivora: van den Bos 1998; Rodentia: Stopka and Graciasova 2001; Primates: Sparks 1967).

In addition to primates, social grooming has been studied extensively in ungulates. Mooring et al. (2004) observed 60 species of ungulates in zoos and reported social grooming in 19 species belonging to Cervidae and Bovidae. Although they did not report grooming in the genus *Equus*, these animals are known to frequently groom each other (Crowell-Davis et al. 1986; Kimura 1998, 2000; Rho et al. 2007). Therefore, the figure of 19 species may be an underestimate.

Behavioral patterns of ungulate grooming are not uniform. For example, impalas (*Aepyceros melampus*) groom by making upward movements with their lower incisors or tongue (Hart and Hart 1992), while horses (*Equus caballus*) groom with their upper incisors (Crowell-Davis et al. 1986). Red deer (*Cervus elaphus*: Hall 1983) and impalas (Hart and Hart 1992) always groom unidirectionally, whereas horses sometimes groom each other simultaneously (Crowell-Davis et al. 1986).

Among other mammals, social grooming is most frequent among primates (see reviews by Sparks 1967; Goosen 1987). Consequently, many studies have been conducted on social grooming among various species of primates from different perspectives such as hygienic function (Hutchins and Barash 1976), reciprocity (Muroyama 1991), exchange with support in fights by higher-ranking individuals (Seyfarth and Cheney 1984), exchange with allomothering (Muroyama 1994), correlation with group size (Dunbar 1991), and tension reduction (Schino et al. 1988). In general, platyrrhine primates groom less frequently than catarrhine primates, but it has been suggested that other forms of social touch, such as embracing, are used for similar functions in similar situations (Slater et al. 2007).

Compared to the standard grooming practices in mammals, grooming in primates is unique because most simian species groom with their hands. Lemurs (prosimians) use a toothcomb (Rosenberger and Strasser 1985) for grooming, and thus their grooming habits resemble mammalian standards. However, simian primates also use their mouths for grooming. For example, patas monkeys (*Erythrocebus patas*) frequently groom with their mouths (Starkey et al. 1989), and chimpanzees (*Pan troglodytes*) often touch their lips to the hair or skin of the groomee. Social grooming is usually performed within species, but interspecies grooming sometimes occurs among primates (Abordo et al. 1975; Freeland 1981; Ihobe 1990; John and Reynolds 1997).

19.4 Social Touch in Great Apes and Dolphins

19.4.1 Social Touch Between Mother and Infant

19.4.1.1 Apes

Mother–infant contact is critical for all primate species. An infant will not survive without contact from his or her mother or an adequate substitute. A famous example is Harlow's (1958) experiment involving the isolation of an infant rhesus monkey (*Macaca mulatta*) from its mother. The infant who was separated from its mother clung more often to a substitute mother made of cloth that did not provide milk than to a substitute mother made of wire that did provide milk. It may be ethically inappropriate to conduct such experiments on apes, but in sanctuaries, orphaned infant apes need to be constantly hugged by human caretakers, and when this is not possible, the infants often hug each other. These observations suggest that great ape infants also need to constantly touch other individuals (usually their own mothers).

In fact, except for some nidicolous primates such as galagos and tarsiers, most primate infants maintain continuous contact with their mother by clinging to their mother's belly for a long time after birth. Great apes are no exception, and infants cling to their mother for at least a few months after birth. For the first several days, the mother supports the infant with her hands and thighs (van Lawick-Goodall 1967; van Schaik 2004). After several days, the infant can cling to its mother's belly on its own by grasping the hair. As the motor skills of the infants develop, the time they spend on their mother's back gradually increases [chimpanzee: van Lawick-Goodall 1967; gorilla (*Gorilla* spp.): Fossey 1979; bonobo (*Pan paniscus*): Kano 1986]. Around this stage, infants begin to occasionally wander away from their mother during feeding or resting, but they immediately return to their mother's belly when they feel uneasy; this lasts until they are 3–4 years of age. A chimpanzee infant first breaks contact from its mother at 16 weeks of age; it is carried on its mother's back from 7–8 months to 3–6 years of age (van Lawick-Goodall 1967; Hiraiwa-Hasegawa 1990). A bonobo infant may ride on its mother's back at the age of 7–8 months but is usually carried on the mother's belly. In fact, in an emergency, even a 4-year-old infant may be carried on its mother's belly (Kano 1986). According to van Noordwijk and van Schaik (2005), independence of locomotion is achieved at about 3–4 years of age in gorillas, chimpanzees, and orangutans (*Pongo* spp.). However, some may depend on their mothers, even though they are completely capable of moving on their own. For example, a male chimpanzee at Gombe was sometimes carried on his mother's back until he was 8 years of age (Goodall 1990).

Although other female great apes carry their infant on their belly and then on their back, orangutan mothers carry their infant on their flank and do not change the position to the back (MacKinnon 1974), which may be related to the arboreal nature of female orangutans that almost never move on the ground.

Among the great apes, gorillas may be the earliest to gain independence from their mothers. Infant mountain gorillas (*Gorilla beringei*) maintain almost continuous physical contact with their mothers until they are about 5 months of age; mother–infant contact is around 50 % by 13 months of age and decreases to 30 % at 36 months (Fossey 1979). Wild western gorilla (*Gorilla gorilla*) infants remain in contact with their mothers for a slightly longer duration; they maintain physical contact for 100 % (median) of the time until they are 9–12 months of age (Nowell and Fletcher 2007). Gorillas are usually weaned at 3–4 years of age (Watts and Pusey 1993), whereas chimpanzees are weaned later: 3 years of age at the earliest but usually at about 5 years (Pusey 1983). Bonobos are also weaned at about 5 years of age (Kuroda 1991). Orangutans, as already seen, achieve independence of locomotion at about the same age as other great apes, but they are not weaned until they are about 7 years of age. They maintain nipple contact and sleep in the same bed with their mothers (van Noordwijk and van Schaik 2005), which may be related to the orangutan having the longest interbirth interval among great apes (Galdikas and Wood 1990). Generally, the timing of weaning and the end of bed sharing (social touch during sleep) coincide.

19.4.1.2 Dolphins

Social touch between mothers and calves occurs in many species in the wild [bottlenose dolphins (*Tursiops* sp.): Mann and Smuts 1998, 1999; Mann and Watson-Capps 2005; Gibson and Mann 2008; Indo-Pacific bottlenose dolphins (*Tursiops aduncus*): Sakai et al. 2006a; Atlantic spotted dolphins (*Stenella frontalis*): Dudzinski 1998; belugas (*Delphinapterus leucas*): Krasnova et al. 2006; Heaviside's dolphins (*Cephalorhynchus heavisidii*) and dusky dolphins (*Lagenorhynchus obscurus*): Sakai, personal observation] and in captivity [bottlenose dolphins: Cockcroft and Ross 1990; belugas (*Delphinapterus leucas*): Morisaka, personal communication; killer whales, finless porpoises (*Neophocaena phocaenoides*), Pacific white-sided dolphins (*Lagenorhynchus obliquidens*), and Commerson's dolphins: Sakai et al. 2013]. Social touch between mothers and calves seems to be fundamental as it is observed widely across a variety of social structures (e.g., group or solitary living, high or low gregariousness).

Some mothers push their newborns up to the surface just after the delivery to support breathing [spinner dolphins (*Stenella longirostris*): Johnson and Norris 1994; bottlenose dolphins, finless porpoises, and Commerson's dolphins: Sakai et al. 2013; belugas: Morisaka, personal communication]. Beluga mothers let their calves ride on their backs (Krasnova et al. 2006). In many delphinids, swimming in echelon position (the calves swim alongside their mother, roughly parallel, and at the level of the mother's flank above the midline, but are no farther away than 30 cm: Mann and Smuts 1999) and in infant position (infants swim with their mothers such that the melon or head lightly touches the mother's abdomen) are typical (bottlenose dolphins: Mann and Smuts 1999; Atlantic spotted dolphins: Miles and Herzing 2003; Indo-Pacific bottlenose dolphins: Sakai et al. 2010; Pacific white-sided dolphins and Commerson's dolphins: Sakai, personal observation; killer whales: Asper et al. 1988). It has been suggested that when calves swim in the echelon position, they receive hydrodynamic benefits and mothers shoulder the cost (bottlenose dolphins: Weihs 2004; Noren 2008; Noren et al. 2008). Maintaining proximity may be as important in dolphins as it is in apes. However, dolphins cannot hold their infant using hands, so mothers might keep and feel their infant hydrodynamically, especially when visibility is poor.

Mammary bump, wherein the calves bump the mammary glands of their mother, usually using their head while swimming below her, is observed in some odontocetes (Atlantic spotted dolphins: Miles and Herzing 2003; bottlenose dolphins: Morisaka et al. 2005; Commerson's dolphins: Sakai et al. 2013; Indo-Pacific bottlenose dolphins and Heaviside's dolphins: Sakai, personal observation ; belugas: Morisaka, personal communication). This behavior tends to occur before nursing. Flipper rubbing (see Sect. 19.4.4.2) between mothers and calves is also observed in several species (Atlantic spotted dolphins: Dudzinski 1998; bottlenose dolphins: Mann and Smuts 1999; Indo-Pacific bottlenose dolphins: Sakai et al. 2006a; Pacific white-sided dolphins and Commerson's dolphins: Sakai et al. 2013, and mothers usually rub their calves more often than the reverse. Mother and infant also make simple contact with flipper to body or body to body. An infant swimming in echelon position may also perform flipper rubbing or body-to-body rubbing with other females (bottlenose dolphins: Mann and Smuts 1999; Indo-Pacific bottlenose dolphins: Sakai, personal observation). See Sect. 19.4.6.2 about touch to/with genital area in mother-and-calf pairs of dolphins.

19.4.2 Touch in Play

19.4.2.1 Apes

Most social play also involves touch. Great apes begin to play during infancy, first with their mother and then with others. Juveniles prefer to play with same-age peers (gorillas: Fossey 1979; chimpanzees: Hayaki 1985). Orangutan infants play with their mothers until they are weaned, and then they frequently play with other juveniles when they have the opportunity (van Noordwijk and van Schaik 2005). One unique characteristic of great apes is that not only immature individuals but also adults play with each other (chimpanzees: Nishida 1981; gorillas: Yamagiwa 1987; bonobos: Enomoto 1990; orangutans: Maple 1982) or with immature individuals.

Patterns of social touch in play include *aeroplane*, *finger wrestling*, *gentle touching*, *slapping*, *biting*, *kicking with the heel*, *tickling*, *stamping*, *wrestling*, *jumping on*, *holding*, and *patting* (for definitions, see Nishida et al. 1999; Palagi 2006). The overall patterns are similar among apes, but some differences exist: for example, *tickling* is not common among wild gorillas (Fossey 1979), and *play slapping*, *tickling*, *gentle grabbing*, and *aeroplane* during play between mature and immature conspecifics are more frequent in chimpanzees than in bonobos (Palagi 2006). Immature gorillas play among themselves by using silverbacks as playing fields: they climb up and slide down the silverback and pull his hair (Schaller 1965).

Except during consortship, wild orangutan adult males and immature individuals seldom stay close to each other; however, in captivity, adult males also actively play with immature individuals, and their play patterns almost resemble those of chimpanzees and gorillas (Zucker et al. 1978). A comparative study showed that chimpanzees more often run or hit, whereas orangutans more often hold tightly with their hands and bite (Maple 1982).

19.4.2.2 Dolphins

Social play involving physical contact (e.g., pushing with the rostrum) starts 2 weeks after birth in bottlenose dolphins, and newborns plays mainly with other young animals (Tizzi et al. 2001). Sex play with physical contact (*mounting*, *goosing*, *push-ups*, and *petting* involving genital contact) is reported among bottlenose dolphin calves (Mann 2006).

Although a few instances of social play without social touch have been observed in bottlenose dolphins (Mann and Smuts 1999) and Indo-Pacific bottlenose dolphins (Sakai, personal observation), social play with body contact seems to occur less frequently in dolphins than in great apes.

19.4.3 *Tactile Gestures*

19.4.3.1 Apes

In great apes, touch has also been studied from the perspective of gestural communication. Although it is uncommon to include social touch among human gestures, perhaps because hand and other body movements are studied in close relationship to speech, in the studies of great ape gestures, social touch is often included. According to Morris (1994), who listed 653 human gestures from many cultures, only approximately 5 % (32 gestures) involve physical touch. In contrast, 39.4 % (15/38) of chimpanzee gestures are tactile (Call and Tomasello 2007), as are 40 % (8/20) in bonobos (Pika 2007), 33.3 % (11/33) in gorillas (Pika et al. 2003), and 48.3 % (14/29) in orangutans (Liebal 2007). This difference between humans and great apes may be largely because of the different focus of those studies: Although Morris (ibid.) focused only on greeting and symbolic gestures in various cultures, studies on gestures in great apes have not been as limited. For example, Pika et al. (2003) included *embrace*, *grab*, *hand on*, *long touch* (>5 s), *pull*, *punch*, *prod*, *push*, *slap*, *touch* (<5 s), and *grab-push-pull* as tactile gestures, many of which are not usually considered in human gestural studies. In gorillas, these tactile gestures occur in more contexts than do visual and auditory gestures (Pika et al. 2003). Tanner and Byrne (1999), who also studied gorilla gestures in captivity, included only two tactile gestures: *tactile close* and *tap other.* In wild gorillas, touch is often used for inviting play or homosexual behavior (Yamagiwa 1987).

19.4.3.2 Dolphins

It is possible that dolphins use their various postures as gestures in visual communication (Madsen and Herman 1980), in addition to their pigmentation. It is also possible that some tactile social behaviors such as *flipper rubbing*, *goosing*, *rubbing*, *petting*, *bonding*, or *contact swimming* serve as gestures, depending on the definition. However, no detailed studies have been conducted on tactile gestures in dolphins.

19.4.4 *Social Grooming*

19.4.4.1 Apes

The frequency of social grooming varies greatly among great apes. According to a review by Lehmann et al. (2007), the proportion of time spent in social grooming is 0 % in orangutans, 0.09 % in western gorillas, 1 % in eastern gorillas, 5.7 % in bonobos, 8.27 % in western chimpanzees, and 11.67 % in eastern chimpanzees. Considerable differences exist between species, but these rates should be considered as approximate figures. Moreover, we have to be cautious about this simple quantitative comparison because many factors seem to influence these figures, such as differences in observational conditions, degrees of habituation, and degrees of dispersal. For example, in a captive colony, bonobos spent 13.6 % of their time on social grooming (Franz 1999).

Nevertheless, we can at least say that chimpanzees and bonobos engage in social grooming more often than gorillas and orangutans. Chimpanzees occasionally spend 1–2 h continuously grooming each other (Goodall 1965; Nishida 1970), although differences may exist among individuals and among populations. Bonobos can also spend more than 2 h grooming each other (Kano 1980; Kuroda 1980).

Gorillas spend less time grooming compared with time spent in proximity, and most grooming is conducted between mother and offspring (Schaller 1965; Harcourt 1979). Gorilla males seldom groom each other when in multimale bisexual groups (once for 10 min during 420.6 h of observation) but do so more often in all-male groups (91 times for 26.23 h during 407.4 h of observation) (Robbins 1996).

Because wild orangutans seldom spend time with each other, the opportunity for social grooming is rare, but they do sometimes groom each other when together. During consortship, males groom females (MacKinnon 1974). In captivity, social grooming was the second most frequent social interaction in an adult orangutan group (Edwards and Snowdon 1980). In a comparative study, grooming rates did not differ between groups of juvenile chimpanzees and orangutans in captivity (Nadler and Braggio 1974).

The quality of grooming is not uniform among species; thus, it may not be appropriate to judge grooming solely on the basis of its duration. For example, two chimpanzees often groom each other simultaneously, and multiple individuals can be connected in a grooming chain (Nakamura 2003) (Fig. 19.1). However, such mutual grooming and polyadic grooming is not common for other great apes (bonobos: Kano 1998). Furthermore, it is known that locality-specific patterns of grooming exist in wild chimpanzees, such as the *grooming hand-clasp* (McGrew and Tutin 1978; Nakamura 2002) and *social scratch* (Nakamura et al. 2000; Nishida et al. 2004). Furthermore, facial grooming may be of special significance as well (Nishida and Hosaka 1996). In captivity, bonobos groom the face of other individuals more often than chimpanzees do (de Waal 1988). Although many primates, such as macaques, groom mostly with both hands, chimpanzees often groom with one hand, and removal movements are less frequent in such one-hand grooming (Zamma 2011). In captive chimpanzees, grooming with one hand accounted for 76.64 % of the total grooming; 48.60 % of this was accompanied with the use of the mouth (Hopkins et al. 2007). Such use of the mouth is also frequent in wild chimpanzees (Nakamura, personal observation). Although ectoparasites such as lice are removed during grooming (Zamma 2002), a hygienic function seems to be less involved in some instances of one-handed grooming.

In bonobos, there is no grooming after copulation (Kano 1986), but short grooming often takes place after copulation in chimpanzees (Goodall 1986). Generally, humans are excluded in discussions on grooming, but humans do groom socially. Usually,

Fig. 19.1 Grooming chain of three chimpanzee females

it takes the form of lice removal (San Bushmen: Sugawara 1984; Efe pygmies: Bailey and Aunger 1990), but grooming is often performed between intimate couples (Americans: Nelson and Geher 2007) even without hygienic function.

19.4.4.2 Dolphins

Flipper rubbing (Fig. 19.2; Online Resource) in dolphins may be an affiliative behavior similar to social grooming in primates (Norris 1991; Dudzinski 1998; Mann and Smuts 1998, 1999; Sakai et al. 2006a). Flipper rubbing is defined as a frictional physical contact in which one dolphin contacts another dolphin with its pectoral fin (flipper). Body movements during flipper rubbing differ between species; bottlenose dolphins and Indo-Pacific bottlenose dolphins move their flipper from side to side (directed along the cross-body axis) actively and often repeatedly. Pacific white-sided dolphins usually do not move their pectoral fin repeatedly, and rubbing lasts only a few seconds. Commerson's dolphins have saw-toothed serrations on the leading edge of pectoral fins and tend to use this part for flipper rubbing. They often move their flipper back and forth (parallel to the body axis) and swim alongside the partner with the flipper on the partner's body, which may yield a frictional effect.

The majority of flipper rubbing by wild Indo-Pacific bottlenose dolphins is performed dyadically, and the participation of three or more dolphins is very rare (Sakai et al. 2006a). Dolphins often switch roles: the rubbing individual is subsequently rubbed by the partner. The face is rubbed significantly more frequently than

Fig. 19.2 Flipper-to-body rubbing in Indo-Pacific bottlenose dolphins (A movie of this behavior is available as an online material at doi: 10.1007/978-4-431-54523-1_19.) A dolphin rubbed the genital area of the other dolphin by moving its left pectoral fin

would be expected by its share of the total body surface. The animal being rubbed tends to have initiated the interaction, while the animal who is rubbing the partner tends to terminate it. The animal being rubbed often assume various postures (e.g., side-up, upside-down), while the rubbing animal usually remain horizontal. These observations suggest that the former receives some benefit during flipper rubbing. Dolphins around Mikura Island, Tokyo, Japan, tend to rub with the left flipper (Sakai et al. 2006b). Similar left-side bias exists in adult male Commerson's dolphins, as 92 % of vibrating touches with a flipper to females were by left flippers (Johnson and Moewe 1999), which may be related to the fact that they tend to have their serrations on their left flippers (Goodall et al. 1988).

Adult and subadult Indo-Pacific bottlenose dolphins tend to perform flipper rubbing with individuals of the same age and sex, whereas calves do so almost exclusively with their mothers (Sakai et al. 2006a). The fact that dolphins choose their rubbing partners suggests that this behavior fulfils some social function. Because impalas did not choose preferential grooming partners, Hart and Hart (1992) concluded that grooming in impalas served only hygienic functions. The fact that mothers rub their calves more often than the reverse also suggests that this is a type of caring behavior that benefits the recipients (calves). During heterosexual flipper rubbing, males rub females more often than vice versa, suggesting that the males impart some immediate benefits to the females, although the precise nature of these benefits remains unclear. One possible benefit is that the frictional contact during flipper rubbing facilitates hygiene of the body surface. Flipper rubbing seems to effectively remove old skin from the body surface. Many small, whitish fragments of old skin often dissipate like smoke from the part being rubbed in the wild and in captivity (Sakai, personal observation). However, we have never seen the dolphins rubbing parasites off the body surface during flipper rubbing, although we often observe soft-bodied barnacles (*Xenobalanus* spp.) and remoras (*Echeneis* spp.) attached to the body surface of the wild dolphins. Another, but not mutually exclusive, possibility is that dolphins engage in flipper rubbing to have body contact that is simply pleasurable. When training bottlenose dolphins, stroking the body can be an effective reward (Defran and Milberg 1973; Herman and Tavolga 1980), and many species in captivity solicit stroking from their handlers (Defran and Pryor 1980). With respect to captive bottlenose dolphins, Tavolga and Essapian (1957) described flipper rubbing as a precopulatory behavior, whereas Weaver (2003) and Tamaki et al. (2006) suggested that it helps appease or repair relationships after agonistic behavior. Tamaki et al. (2006) also made preliminary suggestions for the functions of third-party flipper rubbing, including tension reduction by the third party and displacement as a result of aggressive interactions.

In questionnaire surveys of odontocete behavior in captivity (Defran and Pryor 1980; Nakahara and Takemura 1997), flipper rubbing has been reported to be performed frequently by finless porpoises, Commerson's dolphins, common dolphins (*Delphinus delphis*), spinner dolphins, rough-toothed dolphins (*Steno bredanensis*), Indo-Pacific bottlenose dolphins, killer whales, and false killer whales (*Pseudorca crassidens*); occasionally by Amazon River dolphins (*Inia geoffrensis*), belugas, harbor porpoises (*Phocoena phocoena*), Pacific white-sided dolphins (*Lagenorhynchus obliquidens*), bottlenose dolphins, short-finned pilot whales (*Globicephala macrorhynchus*); and rarely by Risso's dolphins and pilot whales (*Globicephala* sp.). Dusky dolphins and Heaviside's dolphins also perform flipper rubbing (Sakai, personal observation).

A few comparative studies of flipper rubbing suggest that there are several similarities in the characteristics of flipper rubbing among Indo-Pacific bottlenose dolphins, Atlantic spotted dolphins, and bottlenose dolphins (Dudzinski et al. 2009, 2010).

19.4.5 Touch in Greeting, Reassurance, and Appeasement

19.4.5.1 Apes

Chimpanzees frequently touch each other in the context of encounters, social excitement, and anxiety: the types of touch include *touch face*, *hand in mouth*, *hold hand*, *put hand on the back*, *mount*, *embrace* (Fig. 19.3), *embrace-half*, *open mouth kiss*, *touch with lower lip*, and *put mouth to the body* (Nishida et al. 1999). Interestingly, touch in such contexts is not very frequent among bonobos, except among immature individuals (Mori 1983; Kano 1986; de Waal 1988), although this species shares many behavioral patterns with chimpanzees. For example, kissing is frequent among adult chimpanzees in the context of greeting and social tension but is observed only in juvenile bonobos and not adult bonobos (Kano 1986, 1998; de Waal 1988). Mori (1983) argued that such a lack of appeasement behaviors in bonobo adults is related to the lack of ritual charging displays in bonobo males.

Fig. 19.3 Embrace between adult male chimpanzees

It seems that gorillas also seldom touch in such contexts. In a wild, all-male mountain gorilla group, the males embraced or mounted after nonaggressive interactions but not after aggressive interactions (Yamagiwa 1987); thus, such social touch did not serve to reduce tension. In wild female gorillas, postconflict embraces generally increased tolerance between the individuals, but not always (sometimes fighting ensued after the embrace) (Watts 1995). It may be that eye contact rather than physical contact serves as a tension-reducing function in gorillas (Yamagiwa 1992).

In wild orangutans, almost no greeting behaviors occur on the meeting of two subgroups (MacKinnon 1974; Galdikas 1979); but in captivity, behavioral patterns such as *hand fondling*, *touch*, *extend hand*, *grab*, *mouth*, and *kiss* have been reported (Edwards and Snowdon 1980). Many of these behaviors are common in chimpanzees.

19.4.5.2 Dolphins

When two groups of southern resident killer whales meet, they participate in social behavior termed the *greeting ceremony* (Osborne 1986). As already mentioned, flipper rubbing might serve an appeasement function in captive bottlenose dolphins (Weaver 2003; Tamaki et al. 2006). *Contact swimming*, in which one dolphin rests its flipper against the flank of another dolphin behind the latter's flipper and below or just posterior to the dorsal fin, has possible direct benefits, including stress reduction in female Indian Ocean bottlenose dolphins (Connor et al. 2006). It is also observed in captive bottlenose dolphins (Caldwell and Caldwell 1972), wild Indo-Pacific bottlenose dolphins (Sakai, personal observation), and wild Atlantic spotted dolphins (Dudzinski 1998).

19.4.6 Touch to or with Genital Areas

19.4.6.1 Apes

Genito-genital touch is most obvious in bonobos; in the context of social tension, they often make genital contact with each other, represented by *genito-genital rubbing* between females (Kano 1980; Kuroda 1980; Kitamura 1989). Mounting and rump–rump contact between males are also frequent, but in the strictest sense, they do not constitute genito-genital contact. Between males, genito-genital contact (i.e., *penis fencing*) is reported in bonobos but is rare (Kano 1998). Touching between genital or perigenital areas, as occurs during male–female copulation, male–male mounting, and rump–rump contact, may serve to reduce social tension in bonobos; thus, "bonobos use sex at moments when chimpanzees kiss and embrace" (de Waal 1988).

As bonobos seem to generalize "sex" in various social contexts and combinations, it is puzzling that bonobo infants almost never copulate with their mothers (Kano 1986; Kuroda 1991). In contrast, chimpanzee infants quite often copulate with their mothers (Nishida 1981). Captive orangutans also participate in maternal mounting (Maple 1982), wherein a mother rubs her genitals against the genitals of her infant.

In chimpanzees, except for the usual copulation between a male and a female, genito-genital touching is rare (but see Zamma and Fujita 2004). Instead, genitals are typically touched with the hands, such as holding the testes in the hand and shaking the penis with the hand (Nishida et al. 1999), often in the context of appeasement and reassurance. Conversely, bonobos do not hold their testes in such contexts (Kano 1998). Chimpanzee males often touch the female genitals with their hands, and females touch each other's genitals simultaneously in one group (Nakamura and Nishida 2006).

In an all-male group of wild mountain gorillas, homosexual mountings often occurred, but they did not seem to be related to tension reduction (Yamagiwa 1987). Touching of the genitals with the hands was also reported as a homosexual behavior in all-male gorilla groups (Robbins 1996). In a female-only group of captive gorillas, the females also participated in homosexual behaviors (Fischer and Nadler 1978). Genital touching occurs between male and female captive gorillas (Hess 1973).

Although social touch in general is rare in orangutans, there have been several observations of genital contact, including a female that invited an adult male for sexual play by touching the nonerect penis, a male that thrust into the back of a female, a subadult male and a subadult female that played together (bit, tickled, and grabbed each other) after inspection of the genitals by the male (MacKinnon 1979), and a young female that invited an adult male for copulation several times by touching the penis of the male (Schüermann 1982).

Genital inspection occurs in most species of great apes. In these behaviors, after touching the genitals with their fingers, the apes often smell or lick the fingers.

Fig. 19.4 An adult male chimpanzee checking the genitalia of an adult female

In chimpanzees, males usually inspect the genitals of adult females (Nishida 1997) (Fig. 19.4). A similar pattern is also observed in captive orangutans (Edwards and Snowdon 1980). Gorilla mothers in captivity perform genital inspection of their infant offspring, which is thought to promote urination and evacuation, and continues until the infants are about 3 years of age (Hess 1973). Borneo orangutans (*Pongo pygmaeus*) once copulated ventro-ventrally after playful genital inspection (MacKinnon 1979). Subadult male orangutans sometimes touch female genitals with their fingers or mouths (Galdikas 1979).

Touching the genitals with the mouth is rare in chimpanzees, with some exceptions. For example, mothers sometimes mouth the penis of their own offspring to calm them down at the time of weaning (Nishida 1981). In orangutans, males typically touch the genitals of the females before copulation (Galdikas 1979), and subadult males also touch genitals with their mouths. Nulliparous female orangutans also lick or mouth the male's penis during copulation (Schüermann 1982). Orangutan mothers in captivity mouth their newborn infants' penises (Maple 1982). Captive gorillas perform genital–oral contact (Hess 1973), and captive bonobos also participate in oral sex during play among immature individuals (de Waal 1988).

Many mammals usually copulate dorsoventrally; however, orangutans usually copulate ventro-ventrally (Galdikas 1979), with the female often lying supine for about 25–40 min (up to 1 h) (Schüermann 1982; Suzuki 2003). In a male gorilla group, 16 of 98 homosexual mountings were ventro-ventral (Yamagiwa 1987). Bonobos also often copulate ventro-ventrally, and genito-genital rubbing between females is, in most cases, ventro-ventral. In chimpanzees, almost all copulation is dorsoventral.

19.4.6.2 Dolphins

In general, the bottlenose dolphin is well known as a "sexual" animal (see Mann 2006). Tavolga and Essapian (1957) reported genital contact as a part of precopulatory behavior in captive bottlenose dolphins. In *stroking*, one dolphin strokes the genital area of the partner with the tip of a fluke. In *nuzzling*, one animal applies its closed snout at the genital area of the partner and then moves the snout around. In *mouthing*, a female takes a male's genital region between the teeth. Mann (2006) defined four "sociosexual" behaviors—*mounting*, *goosing*, *push-ups*, and *sociosexual petting*—during genital contact in Indian Ocean bottlenose dolphins in Shark Bay. Mounting is manifested in dorsoventral, lateral-ventral, and ventro-ventral forms, of which the dorsoventral is the most common (Mann 2006). Goosing occurs when a dolphin brings his or her beak into contact with the genital area of the recipient. Push-ups involve one dolphin pushing up the genital area of another with his or her head, usually to clear the water. Sociosexual petting is defined by flipper-to-genital contact, when one dolphin either strokes the genital area of another with his or her flipper or inserts the flipper into the genital slit of another. Connor et al. (2000) observed reciprocal mounting, in which participants exchanged roles. Up to eight individuals participated in sociosexual bouts, and these often involved partner exchanges (synchronous sociosexual behavior) (Mann 2006). Male calves engage in sociosexual behavior (sex play) more frequently than female calves and adults do. When mothers engaged in sociosexual behavior, their calf was almost always the partner, and 100 % of the male calves mounted their mothers. Male calves also mounted their grandmother, maternal sister, and brother. Genital petting was the least common of the sociosexual behaviors. Nearly half of female–female bouts involved sociosexual petting. In contrast, none of the male–male bouts involved sociosexual petting.

Indo-Pacific bottlenose dolphins near Mikura Island also perform sociosexual behaviors (mainly mounting: Fig. 19.5). Two to fourteen participants often exchanged the roles of performer and recipient (Jiroumaru, personal communication). Subadult males tend to exhibit sociosexual behaviors more frequently than other age-sex classes. Calves also mount their mothers. Mother and calf pairs (nine instances), a subadult female pair (two instances), an adult female pair, an adult male pair, a subadult male pair, and a subadult heterosexual pair (one instance in each pair) were involved in sociosexual petting (Sakai, personal observation).

Instances in other species include spinner dolphins that conducted goosing in captivity (Johnson and Norris 1994). Two captive beluga females engaged in sociosexual behavior (Morisaka, personal communication). Captive beluga males also engaged in sociosexual behavior; male-on-female pelvic thrusting varied significantly across months, with a clear peak in March, although male-on-male pelvic thrusting did not differ across months (Glabicky et al. 2010).

Xian et al. (2010) studied the development of the sociosexual behaviors of a captive male Yangtze finless porpoise calf. The behavior appeared at 1 month postpartum and the mother was his preferred partner, but other adults of both sexes were also involved. Males of this species also participate in sociosexual behavior among

Fig. 19.5 Sociosexual behavior in Indo-Pacific bottlenose dolphins. Several subadult male dolphins erected their penises and mounted the other dolphin

themselves (Wu et al. 2010). Subadult males did so more frequently than adult males; thus, this behavior was not quantitatively correlated with testicular volume. Captive male finless porpoises in Japan tried to insert their penises into the blowholes of other males (Sakai, personal observation). Three Heaviside's dolphins chased and splashed each other, two of whom displayed an erection (Sakai, personal observation); this suggested that they also engage in sociosexual behavior. A captive male Commerson's dolphin calf was first observed with an erection 29 days postpartum, and the calf attempted to mount his mother at 43 days postpartum (Sakai, personal observation). The mother rubbed her genital area against the flipper of her male calf (Sakai, personal observation).

According to Nakahara and Takemura (1997), attempts to have same-sex intercourse are frequent in Commerson's dolphins and Indo-Pacific bottlenose dolphins; occasional in belugas, Pacific white-sided dolphins, finless porpoises, killer whales, and bottlenose dolphins; and rare in Risso's dolphins, short-finned pilot whales, false killer whales, and harbor porpoises.

19.4.7 Simple Body Contact

19.4.7.1 Apes

In addition to social touches in obvious contexts and functions, there are cases where individuals simply make physical contact with each other. Such simple body contacts seem to be more frequent among gorillas who do not often groom each other and thus may be compensating for the latter with the former: female often put their heads on the back of a silverback male or lean toward him when resting (Schaller 1965; Watts 1992). In male gorilla groups, males spent 218 h (accumulated) of the 407.4-h observation period in physical contact with each other (Robbins 1996).

19.4.7.2 Dolphins

Many captive odontocetes seek out body contact [*Stenella* sp., common dolphins, Pacific white-sided dolphins, killer whales, Risso's dolphins, *Globicephala* sp., false killer whales, tucuxis (*Sotalia fluviatilis*), Dall's porpoises (*Phocoenoides dalli*), belugas, Amazon River dolphins: Caldwell and Caldwell 1972]. Finless porpoises, harbor porpoises, Commerson's dolphins, Pacific white-sided dolphins, common dolphins, spinner dolphins, rough-toothed dolphins, Indo-Pacific bottlenose dolphins, killer whales, and false killer whales frequently use their flippers for touching, whereas Amazon River dolphins, bottlenose dolphins, Risso's dolphins, and short-finned and long-finned pilot whales do so occasionally, and beluga whales rarely (Defran and Pryor 1980; Nakahara and Takemura 1997). There are some reports of simple touch in wild dolphins (Atlantic spotted dolphins and Indo-Pacific bottlenose dolphins: Paulos et al. 2008; Dudzinski et al. 2009; Aoki et al. 2013).

Several species of dolphins rub their bodies against the bodies of other individuals (bottlenose dolphins: Holobinko and Waring 2010; Atlantic spotted dolphins: Dudzinski 1998; Indian Ocean bottlenose dolphins: Connor et al. 2000; Indo-Pacific bottlenose dolphins, captive finless porpoises, and Commerson's dolphins: Sakai, personal observation). Although this behavior seems to be a type of affiliative behavior resembling flipper rubbing, as yet no research has clarified its functions.

Finless porpoises have a low ridge covered in thick, denticulated skin on their dorsal side, and they use the denticulation for body-to-body rubbing. Such external morphology may be a specialization for tactile behaviors. If so, tactile behavior is important in such species.

Captive killer whales engage in mouth-to-mouth touch (Sakai, personal observation). Some species use the rostrum for contacting, nudging, or rubbing the body part of another in a non-agonistic interaction (bottlenose dolphins: Holobinko and Waring 2010; Hawaiian spinner dolphins: Johnson and Norris 1994; Indo-Pacific bottlenose dolphins: Sakai, personal observation).

19.5 Diversity and Similarities

Information obtained from scattered descriptions in the literature shows that social touch seems to play an important role in social interactions and that both forms and frequencies vary largely, even among apes and dolphins.

One similarity between primates and dolphins may be that they touch others with forelimbs more often than with their mouths, although the morphology of the forelimbs differs greatly between these taxa. Nonhuman primates have acquired dexterous forelimbs along with opposable thumbs, but still use their forearms for locomotion. Dolphins do not always use their flippers for locomotion, in contrast to most quadruped mammals. The different characteristics of the forelimbs may enable both taxa to use their forelimbs for social touch more often than other mammalian species. Because of adaptation to the aquatic environment, dolphins cannot use the flippers as dexterously as primates use their hands, which may be one reason that fewer types of social touch have been observed in dolphins. As mentioned earlier, many other mammals use their mouths for touching others, but because dolphins cannot move their necks and some species have beaks, they do not use their mouths for this purpose, except in dominant, aggressive, or sexual contexts.

Another similarity may be the frequent touching of genital areas. It is possible that in both taxa, touching the genitals may have acquired some social functions other than reproduction. It seems that they also show preferential touching of the face, suggesting their relative interests in others' faces.

Forms of social touch are not necessarily similar between species that are phylogenetically close. For example, Amazon River dolphins (phylogenetically, an older species among odontocetes) stroke and touch, but pilot whales or belugas rarely do so. Chimpanzees and bonobos are the most closely related among the great apes, but the function of genital contact in these species differs considerably. This relative independence from phylogeny implies that social touch is used flexibly. Thus, simple comparisons of the frequencies of superficially similar behaviors may not always be useful for understanding these interactions; rather, we need to consider the forms, participants, and contexts related to social touching.

Social structures such as the degree of cohesiveness or whether the animals are living in the wild or in captivity may also influence the significance of social touch. Further quantitative comparisons in relation to phylogeny, social structures, and environmental factors may reveal the significance of the roles that social touch serves in these complex societies.

19.6 Theories in Relationship to Social Touch and Their Applicability

As we mentioned in the Introduction, there has been no single theory to explain these diverse types of social touch, because social touch covers various domains of behavior and may serve several different social functions. In this section, we briefly review theories that might be relevant in the explanation of social touch and see their applicability.

Zahavi (1977) suggested that aggression and other types of stressful behavior involved in the bonding mechanism may function as a test of the strength of the bond (test of bonding theory). This theory covers seemingly costly behaviors to be used in the affiliative contexts by the cost of the recipient (or both participants) enabling the performer(s) to test the reliability of the information. Some forms of social touch, such as *biting* and *slapping* in social play, are similar to motions seen in aggression (Bateson 1955/1972); thus, they might cause some extent of pain to the recipient. These behaviors may be explained to function as a test for the strength of social bonds (Zahavi 1977). However, it should be noted that even within social play, other types of social touch, such as *aeroplane*, *finger wrestling*, or *tickling*, do not seem to have corresponding motions in aggression nor do they seem painful to the recipient.

We can think of some more direct benefits to the performer(s) in some types of social touch. Most physical touch between mother and infant can be understood in the context of maternal investment (Trivers 1972), in which mothers pay the cost to increase the benefit to their offspring; this may be a simple explanation for some types of social touch such as for nutrition, thermoregulation, protection, support for transportation, and in the case of dolphins, support for breathing. In such cases of social touch, how and when a mother touches her offspring seem to be directly connected to the benefit of the offspring. Conversely, in other contexts, such as in mating, there seems to be little meaning to discuss why social touch has to take place. Because mammals inevitably fertilize internally, they cannot produce the next generation without touching the mate. Although it may seem straightforward to think that social touch in mating contexts is basically a result of the process of copulation itself, this does not exclude the possibility that some forms of physical touch in a mating context play other social functions than just reproduction.

It may be useful to study several theories that explain the function of social grooming to extend them to social touch in general. In most mammalian grooming behaviors, there is a primary hygienic function of removing ectoparasites or debris from the body surface (Hart and Hart 1992; Tanaka and Takefushi 1993). It is also known that, at least in primate grooming, there are hedonistic benefits to the recipient (Keverne et al. 1989). Therefore, even without an ultimate benefit of reduced parasites, grooming can occur with the proximate benefit that the recipient feels good if he/she is groomed. Usually it is thought that the groomee enjoys such a benefit, and the groomer pays some cost, at least by means of time and energy required for grooming. Although such costs and benefits are often only presumed but rarely measured, such assumptions have led researchers to propose theories about this behavior. For example, grooming between non-kin is most often discussed in relationship to reciprocal altruism (Trivers 1971); grooming (= service) can be exchanged for support (Seyfarth 1977), foods (de Waal 1997), or grooming itself (Hart and Hart 1992). The first two possibilities may require the participants to have some extent of cognitive abilities to remember for some time the partner(s) and the amount(s) of service he/she made and received; however, if the participants exchange grooming immediately, such as in impala allogrooming, they do not need to have such cognitive abilities because they can just follow a simple tit-for-tat model (Hart and Hart 1992). This model was originally formulated in the repeated Prisoner's Dilemma game (Axelrod and Hamilton 1981), where two individuals reciprocate the same or similar services in a repeated sequence of bouts. Because the tit-for-tat model is limited to interactions between two individuals, Connor (1995) proposed the parceling model in which individuals divide service into parcels so that cheating becomes unprofitable. Biological market theory (Noë and Hammerstein 1995) summarized various kinds of social exchanges, where grooming is traded for itself or in exchange for other social commodities (Barrett et al. 1999). To apply such theories to other types of social touch, first, the many types of social touch that are mutually performed should be excluded as they do not fit reciprocal altruism models that deal with behaviors beneficial to one party and costly to the other. Therefore, some asymmetrical types of social touch, such as flipper rubbing, may be explained within a reciprocal altruism model, provided that touch gives a certain benefit to the recipient, which needs to be examined in the future.

A different, but not necessarily exclusive, approach to grooming is that it is used to generate tolerance for the coexistence of the participants. An easily understood example is reconciliation, in which affiliative behaviors, including grooming, increase after an aggressive interaction between the participants (Sade 1972; de Waal and van Roosmalen 1979). Another example is tension reduction (Schino et al. 1988), in which grooming decreases the likelihood of future aggression. Such a social function may be useful for maintaining a cohesive social group, which matches with the knowledge that grooming frequency increases with group size, at least in catarrhine primates (Dunbar 1991). Dunbar (1993) further extended this to the evolution of language, because both grooming and conversation serve as bonding mechanisms. This bonding theory may be applicable to other forms of social touch, in that it generally increases mutual tolerance or reduces social stress after conflicts or when conflicts are likely to occur. If physical touch decreases the tension of the partner (or both), not only grooming but also any type of social touch can be used for this bonding purpose. Logically, the behavior does not even have to be touch; any type of behavior (e.g., calling each other) can be used for social bonding, but only if the behavior reduces the tension of the participants. Even though the social bonding theory is broadly applicable, it does not explain why a certain type of behavior, such as social touch, actually has such an effect. Thus, to explain the evolution of social touch with this theory, we first need to confirm that touch generally reduces the tension of the participants; then, we need to test whether tactile modality is suitable for such a function, compared to other sensory modalities.

Finally, we would like to point out that studies supporting the aforementioned theories sometimes disregard qualitative differences between behaviors that are given the same name. For example, although we can easily compare the extent of grooming in different species or age-sex classes, it is not guaranteed that the same amount of grooming has the same survival value in different species, different sexes, or even different individuals (see also Nakamura 2003). We also need to accumulate minute and careful descriptions of social interactions and the dynamism of animal societies.

Acknowledgments M.N. thanks the Tanzania Commission for Science and Technology, the Tanzania Wildlife Research Institute, the Tanzania National Parks, the Mahale Mountains National Park, and the Mahale Wildlife Research Centre for permission to conduct the field research at Mahale. M.S. thanks the people of the Mikura Islands and the staff of Kamogawa Sea World, Toba Aquarium, and Osaka Aquarium Kaiyukan. M.N.'s field study was financially supported by grants from Japanese MEXT (#16255007, #19255008 to T. Nishida, #16770186 to M.N.) and that of M.S. by grants from the Circle for the Promotion of Science and Engineering, the Inui Memorial Trust for Research on Animal Science, and JSPS.

References

Abordo EJ, Mittermeier RA, Lee J, Mason P (1975) Social grooming between squirrel monkeys and uakaris in a seminatural environment. Primates 16:217–221

Aoki K, Sakai M, Miller PJO, Visser F, Sato K (2013) Body contact and synchronous diving in long-finned pilot whales. Behav Process 99:12–20

Arcadi AC (1996) Phrase structure of wild chimpanzee pant hoots: patterns of production and interpopulation variability. Am J Primatol 39:159–178

Asper ED, Young WG, Walsh MT (1988) Observations on the birth and development of a captive born killer whale. Int Zoo Yearb 27:295–304

Axelrod R, Hamilton WD (1981) The evolution of cooperation. Science 211:1390–1396

Bailey RC, Aunger R (1990) Humans as primates: the social relationships of Efe pygmy men in comparative perspective. Int J Primatol 11:127–146

Barrett L, Henzi SP, Weingrill T, Lycett JE, Hill RA (1999) Market forces predict grooming reciprocity in female baboons. Proc R Soc Lond B Biol Sci 266:665–670

Bateson G (1955/1972) A theory of play and fantasy: a report of theoretical aspects of the project for study of the role of paradoxes of abstraction in communication. Reprinted in: Steps to an ecology of mind. University of Chicago Press, Chicago, pp 177–193

Blumstein DT, Daniel JC (2003) Red kangaroos (*Macropus rufus*) receive an antipredator benefit from aggregation. Acta Ethol 5:95–99

Breathnach AS (1960) The cetacean central nervous system. Biol Rev Camb Philos Soc 35:187–230

Caldwell DK, Caldwell MC (1972) Senses and communication. In: Ridgway SH (ed) Mammals of the sea. Thomas, Springfield, pp 406–502

Call J, Tomasello M (2007) The gestural repertoire of chimpanzees (*Pan troglodytes*). In: Call J, Tomasello M (eds) Gestural communication of apes and monkeys. Erlbaum Associates, Hillsdale, pp 17–39

Cockcroft VG, Ross GJB (1990) Observations on the early development of a captive bottlenose dolphin calf. In: Leatherwood S, Reeves RR (eds) The bottlenose dolphin. Academic, San Diego, pp 461–485

Connor RC (1995) Impala allogrooming and the parceling model of reciprocity. Anim Behav 49:528–530

Connor R, Wells RS, Mann J, Read AJ (2000) The bottlenose dolphin: social relationships in a fission–fusion society. In: Mann J, Connor RC, Tyack PL, Whitehead H (eds) Cetacean societies. University of Chicago Press, Chicago, pp 91–126

Connor R, Mann J, Watson-Capps J (2006) A sex-specific affiliative contact behavior in Indian Ocean bottlenose dolphins, *Tursiops* sp. Ethology 112:631–638

Crowell-Davis SL, Houpt KA, Carinu CM (1986) Mutual grooming and nearest-neighbor relationships among foals of *Equus caballus*. Appl Anim Behav Sci 15:113–123

Darian-Smith I (1982) Touch in primates. Annu Rev Psychol 33:155–194

de Waal FBM (1988) The communicative repertoire of captive bonobos (*Pan paniscus*), compared to that of chimpanzees. Behaviour 106:183–251

de Waal FBM (1997) The chimpanzee's service economy: food for grooming. Evol Hum Behav 18:375–386

de Waal FBM, van Roosmalen A (1979) Reconciliation and consolation among chimpanzees. Behav Ecol Sociobiol 5:55–66

Defran RH, Milberg L (1973) Tactile reinforcement in the bottlenosed dolphin. In: Proceedings of 10th annual conference of sonar and diving mammals, Stanford Research Institute, Palo Alto

Defran RH, Pryor K (1980) The behavior and training of cetaceans in captivity. In: Herman LM (ed) Cetacean behavior. Wiley, New York, pp 319–362

Dudzinski KM (1998) Contact behavior and signal exchange in Atlantic spotted dolphins (*Stenella frontalis*). Aquat Mamm 24:129–142

Dudzinski KM, Gregg JD, Ribicc CA, Kuczaj SA (2009) A comparison of pectoral fin contact between two different wild dolphin populations. Behav Process 80:182–190

Dudzinski KM, Gregg JD, Paulos RD, Kuczaj SA (2010) A comparison of pectoral fin contact behaviour for three distinct dolphin populations. Behav Process 84:559–567

Dunbar RIM (1991) Functional significance of social grooming in primates. Folia Primatol (Basel) 57:121–131

Dunbar RIM (1993) Coevolution of neocortical size, group size and language in humans. Behav Brain Sci 16:681–735

Edwards SD, Snowdon CT (1980) Social behavior of captive, group-living orang-utans. Int J Primatol 1:39–62

Enomoto T (1990) Social play and sexual behavior of the bonobo (*Pan paniscus*) with special reference to flexibility. Primates 31:469–480

Fischer RB, Nadler RD (1978) Affiliative, playful, and homosexual interactions of adult female lowland gorillas. Primates 19:657–664

Fossey D (1979) Development of the mountain gorilla (*Gorilla gorilla beringei*): the first thirty-six months. In: Hamburg DA, McCown ER (eds) The great apes. Benjamin/Cummings, Menlo Park, pp 139–184

Franz C (1999) Allogrooming behavior and grooming site preferences in captive bonobos (*Pan paniscus*): association with female dominance. Int J Primatol 20:525–546

Freeland WJ (1981) Functional aspects of primate grooming. Ohio J Sci 81:173–177

Galdikas BMF (1979) Orangutan adaptation at Tanjung Putting Reserve: mating and ecology. In: Hamburg DA, McCown ER (eds) The great apes. Benjamin/Cummings, Menlo Park, pp 195–233

Galdikas BMF, Wood JW (1990) Birth spacing patterns in humans and apes. Am J Phys Anthropol 83:185–191

Gibson QA, Mann J (2008) Early social development in wild bottlenose dolphins: sex differences, individual variation and maternal influence. Anim Behav 76:375–387

Glabicky N, DuBrava A, Noonan N (2010) Social-sexual behavior seasonality in captive beluga whales (*Delphinapterus leucas*). Polar Biol 33:1145–1147

Goodall J (1965) Chimpanzees of the Gombe Stream Reserve. In: De Vore I, De Vore I (eds) Primate behavior. Holt, Rinehart & Winston, New York, pp 425–473

Goodall J (1986) The chimpanzees of Gombe: patterns of behavior. Harvard University Press, Cambridge

Goodall J (1990) Through a window: thirty years with the chimpanzees of Gombe. Weidenfeld & Nicolson, London

Goodall RNP, Galeazzi AR, Sobral AP (1988) Flipper serration in *Cephalorhynchus commersonii*. In: Brownell R Jr, Donovan GP (eds) Biology of the genus *Cephalorhynchus*. Reports of the International Whaling Commission, Special Issue 9, pp 161–171

Goosen C (1987) Social grooming in primates. In: Mitchel G, Erwin J (eds) Comparative primate biology, vol 2B. Liss, New York, pp 107–131

Gouzoules H, Gouzoules S (2007) The conundrum of communication. In: Campbell CJ, Fuentes A, MacKinnon KC, Panger M, Bearder SK (eds) Primates in perspective. Oxford University Press, New York, pp 621–635

Hall MJ (1983) Social organisation in an enclosed group of red deer (*Cervus elaphus* L.) on Rhum: II. Social grooming, mounting behaviour, spatial organisation and their relationships to dominance rank. Z Tierpsychol 61:273–292

Harcourt AH (1979) Social relationships among adult female mountain gorillas. Anim Behav 27:251–264

Harlow HF (1958) The nature of love. Am Psychol 13:673–685

Hart BL, Hart LA (1992) Reciprocal allogrooming in impala, *Aepyceros melampus*. Anim Behav 44:1073–1083
Hartman KL, Visser F, Hendriks AJE (2008) Social structure of Risso's dolphins (*Grampus griseus*) at the Azores: a stratified community based on highly associated social units. Can J Zool 86:294–306
Hayaki H (1985) Social play of juvenile and adolescent chimpanzees in the Mahale Mountains National Park, Tanzania. Primates 26:343–360
Herman LM, Tavolga WN (1980) The communication systems of cetaceans. In: Herman LM (ed) Cetacean behavior. Wiley, New York, pp 149–209
Hertenstein MJ, Verkamp JM, Kerestes AM, Holmes RM (2006) The communicative functions of touch in humans, nonhuman primates, and rats: a review and synthesis of the empirical research. Genet Soc Gen Psychol Monogr 132:5–94
Hess JP (1973) Some observations on the sexual behaviour of captive lowland gorillas, *Gorilla g. gorilla* (Savage and Wyman). In: Michael RP, Crook JH (eds) Comparative ecology and behaviour of primates. Academic, London, pp 507–581
Hiraiwa-Hasegawa M (1990) Maternal investment before weaning. In: Nishida T (ed) The chimpanzees of the Mahale Mountains. University of Tokyo Press, Tokyo, pp 257–266
Holobinko A, Waring GH (2010) Conflict and reconciliation behavior trends of the bottlenose dolphin (*Tursiops truncatus*). Zoo Biol 29:567–585
Hopkins WD, Russell JL, Remkus M, Freeman H, Schapiro SJ (2007) Handedness and grooming in *Pan troglodytes*: comparative analysis between findings in captive and wild individuals. Int J Primatol 28:1315–1326
Hutchins M, Barash DP (1976) Grooming in primates: Imprecation for its utilitarian function. Primates 17:145–150
Ihobe H (1990) Interspecific interactions between wild pygmy chimpanzees (*Pan paniscus*) and red colobus (*Colobus badius*). Primates 31:109–112
Itani J (1963) Vocal communication of the wild Japanese monkey. Primates 4:11–66
Itani J (1977) Evolution of primate social structure. J Hum Evol 6:235–243
John T, Reynolds V (1997) Budongo Forest chimpanzee grooms a red-tailed monkey. Pan Afr News 4:6
Johnson CM, Moewe K (1999) Pectoral fin preference during contact in Commerson's dolphins (*Cephalorhynchus commersonii*). Aquat Mamm 25:73–77
Johnson CM, Norris KS (1994) Social behavior. In: Norris KS, Würsig B, Wells RS, Würsig M (eds) The Hawaiian spinner dolphin. University of California Press, Berkeley, pp 243–286
Kano T (1980) Social behavior of wild pygmy chimpanzees (*Pan paniscus*) of Wamba: a preliminary report. J Hum Evol 9:243–260
Kano T (1986) The last ape. Dôbutsu-sha, Tokyo (in Japanese)
Kano T (1998) A preliminary glossary of bonobo behaviors at Wamba. In: Nishida T (ed) Comparative study of the behavior of the genus *Pan* by compiling video ethogram. Nisshindo, Kyoto, pp 39–81 (in Japanese)
Kendon A, Ferbar A (1973/1996) A description of some human greetings. In: Sugawara K, Nomura M (eds) The body as a communication. Taishûkan-shoten, Tokyo, pp 136–188 (translation into Japanese)
Keverne EB, Martensz ND, Tuite B (1989) Beta-endorphin concentrations in cerebrospinal fluid of monkeys are influenced by grooming relationships. Psychoneuroendocrinology 14:155–161
Kimura R (1998) Mutual grooming and preferred associate relationships in a band of free-ranging horses. Appl Anim Behav Sci 59:265–276
Kimura R (2000) Relationships of the type of social organization to scent-marking and mutual-grooming behaviour in Grevy's (*Equus grevyi*) and Grant's zebras (*Equus burchelli bohmi*). J Equine Sci 11:91–98
Kishida T, Kubota S, Shirayama Y, Fukami H (2007) The olfactory receptor gene repertoires in secondary-adapted marine vertebrates: evidence for reduction of the functional proportions in cetaceans. Biol Lett 3:428–430
Kitamura K (1989) Genito-genital contacts in the pygmy chimpanzee (*Pan paniscus*). Afr Study Monogr 10:49–67

Krasnova VV, Bel'kovich VM, Chernetsky AD (2006) Mother–infant spatial relations in wild beluga (*Delphinapterus leucas*) during postnatal development under natural conditions. Biol Bull 33:53–58

Kuroda S (1980) Social behavior of the pygmy chimpanzees. Primates 21:181–197

Kuroda S (1991) Pygmy chimpanzee. Chikuma-shobô, Tokyo (in Japanese)

Lehmann J, Korstjens AH, Dunbar RIM (2007) Group size, grooming and social cohesion in primates. Anim Behav 74:1617–1629

Liebal K (2007) Gestures in orangutans (*Pongo pygmaeus*). In: Call J, Tomasello M (eds) The gestural communication of apes and monkeys. Erlbaum Associates, Hillsdale, pp 69–98

MacKinnon J (1974) The behaviour and ecology of wild orangutans (*Pongo pygmaeus*). Anim Behav 22:3–74

MacKinnon J (1979) Reproductive behavior in wild orangutan populations. In: Hamburg DA, McCown ER (eds) The great apes. Benjamin/Cummings, Menlo Park, pp 257–273

Madsen CJ, Herman LM (1980) Social and ecological correlates of cetacean vision and visual appearance. In: Herman LM (ed) Cetacean behavior. Wiley, New York, pp 101–147

Mann J (2006) Establishing trust: socio-sexual behaviour and the development of male–male bonds among Indian Ocean bottlenose dolphins. In: Volker S, Paul L (eds) Homosexual behaviour in animals. Cambridge University Press, New York, pp 107–130

Mann J, Smuts BB (1998) Natal attraction: allomaternal care and mother–infant separations in wild bottlenose dolphins. Anim Behav 55:1097–1113

Mann J, Smuts BB (1999) Behavioral development in wild bottlenose dolphin newborns (*Tursiops* sp.). Behaviour 136:529–566

Mann J, Watson-Capps JJ (2005) Surviving at sea: ecological and behavioural predictors of calf mortality in Indian Ocean bottlenose dolphins, *Tursiops* sp. Anim Behav 69:899–909

Mann J, Connor RC, Tyack PL, Whitehead H (2000) Cetacean societies. University of Chicago Press, Chicago

Maple TL (1982) Orangutan behavior and its management in captivity. In: de Boer LEM (ed) The orangutan. W. Junk, The Hague, pp 257–268

Marler P (1965) Communication in monkeys and apes. In: De Vore I (ed) Primate behavior. Holt, Rinehart & Winston, New York, pp 544–584

McGrew WC, Tutin CEG (1978) Evidence for a social custom in wild chimpanzees? Man 13:234–251

Miles JA, Herzing DL (2003) Underwater analysis of the behavioural development of free-ranging Atlantic spotted dolphin (*Stenella frontalis*) calves (birth to 4 years of age). Aquat Mamm 29:363–377

Montagu A (1971) Touching. Harper & Row, New York

Mooring MS, Blumstein DT, Stoner CJ (2004) The evolution of parasite-defense grooming in ungulates. Biol J Linn Soc 81:17–37

Morgane PJ, Jacobs MS (1972) Comparative anatomy of the cetacean nervous system. In: Harrison RJ (ed) Functional anatomy of marine mammals, vol 1. Academic Press, London, pp 118–244

Mori A (1983) Comparison of the communicative vocalizations and behavior of group ranging in eastern gorillas, chimpanzees and pygmy chimpanzees. Primates 24:486–500

Morisaka T, Shinohara M, Taki M (2005) Underwater sounds produced by neonatal bottlenose dolphins (*Tursiops truncatus*): II. Potential function. Aquat Mamm 31:258–265

Morris D (1971/1993) Intimate behaviour. Heibon-sha, Tokyo (Japanese edition)

Morris D (1994/1999) Bodytalk. Sansei-dô, Tokyo (Japanese edition)

Muroyama Y (1991) Mutual reciprocity of grooming in female Japanese macaques (*Macaca fuscata*). Behaviour 119:161–170

Muroyama Y (1994) Exchange of grooming for allomothering in female patas monkeys. Behaviour 128:103–119

Nadler RD, Braggio JT (1974) Sex and species differences in captive-reared juvenile chimpanzees and orangutans. J Hum Evol 3:541–550

Nakahara F, Takemura A (1997) A survey on the behavior of captive odontocetes in Japan. Aquat Mamm 23:135–143

Nakamura M (2002) Grooming-hand-clasp in Mahale M group chimpanzees: implication for culture in social behaviors. In: Boesch C, Hohmann G, Marchant LF (eds) Behavioural diversity in chimpanzees and bonobos. Cambridge University Press, Cambridge, pp 71–83
Nakamura M (2003) 'Gatherings' of social grooming among wild chimpanzees: implications for evolution of sociality. J Hum Evol 44:59–71
Nakamura M, Nishida T (2006) Subtle behavioral variation in wild chimpanzees, with special reference to Imanishi's concept of *kaluchua*. Primates 47:35–42
Nakamura M, McGrew WC, Marchant LF, Nishida T (2000) Social scratch: another custom in wild chimpanzees? Primates 41:237–248
Nelson H, Geher G (2007) Mutual grooming in human dyadic relationships: an ethological perspective. Curr Psychol 26:121–140
Nishida T (1970) Social behavior and relationships among chimpanzees of Mahali Mountains. Primates 11:47–87
Nishida T (1981) Observation records of wild chimpanzees. Chûôkôron-sha, Tokyo (in Japanese)
Nishida T (1997) Sexual behavior of adult male chimpanzees of the Mahale Mountains National Park, Tanzania. Primates 38:379–398
Nishida T, Hosaka K (1996) Coalition strategies among adult male chimpanzees of the Mahale Mountains, Tanzania. In: McGrew WC, Marchant LF, Nishida T (eds) Great ape societies. Cambridge University Press, Cambridge, pp 114–134
Nishida T, Kano T, Goodall J, McGrew WC, Nakamura M (1999) Ethogram and ethnography of Mahale chimpanzees. Anthropol Sci 107:141–188
Nishida T, Mitani JC, Watts DP (2004) Variable grooming behaviours in wild chimpanzees. Folia Primatol (Basel) 75:31–36
Noë R, Hammerstein P (1995) Biological markets. Trends Ecol Evol 10:336–339
Noren SR (2008) Infant-carrying behaviour in dolphins: costly parental care in an aquatic environment. Funct Ecol 22:284–288
Noren SR, Biedenbach G, Redfern JV, Edwards EF (2008) Hitching a ride: the formation locomotion strategy of dolphin calves. Funct Ecol 22:278–283
Norris KS (1991) Dolphin days: the life and times of the spinner dolphin. Norton, New York
Nowell A, Fletcher AW (2007) Development of independence from the mother in *Gorilla gorilla gorilla*. Int J Primatol 28:441–455
Osborne RW (1986) A behavioral budget of Puget sound killer whales. In: Kirkevold BC, Lockard JS (eds) Behavioral biology of killer whales. Liss, New York, pp 211–249
Palagi E (2006) Social play in bonobos (*Pan paniscus*) and chimpanzees (*Pan troglodytes*): implications for natural social systems and interindividual relationships. Am J Phys Anthropol 129:418–426
Paulos RD, Dudzinski KM, Kuczaj SA (2008) The role of touch in select social interactions of Atlantic spotted dolphin (*Stenella frontalis*) and Indo-Pacific bottlenose dolphin (*Tursiops aduncus*). J Ethol 26:153–164
Pika S (2007) Gestures in subadult bonobos (*Pan paniscus*). In: Call J, Tomasello M (eds) The gestural communication of apes and monkeys. Erlbaum Associates, Hillsdale, pp 41–67
Pika S, Liebal K, Tomasello M (2003) Gestural communication in young gorillas (*Gorilla gorilla*): gestural repertoire, learning, and use. Am J Primatol 60:95–111
Pusey AE (1983) Mother-offspring relationships in chimpanzees after weaning. Anim Behav 31:363–377
Rho JR, Srygley RB, Choe JC (2007) Sex preferences in Jeju pony foals (*Equus caballus*) for mutual grooming and play-fighting behaviors. Zool Sci 24:769–773
Robbins MM (1996) Male–male interactions in heterosexual and all-male wild mountain gorilla groups. Ethology 102:942–965
Rosenberger AL, Strasser E (1985) Toothcomb origins: support for the grooming hypothesis. Primates 26:73–84
Sade DS (1972) Sociometrics of *Macaca mulatta*. I. Linkages and cliques of grooming matrices. Folia Primatol (Basel) 18:196–223
Sakai M, Hishii T, Takeda S, Kohshima S (2006a) Flipper rubbing behaviors in wild bottlenose dolphins (*Tursiops aduncus*). Mar Mamm Sci 22:966–978

Sakai M, Hishii T, Takeda S, Kohshima S (2006b) Laterality of flipper rubbing behaviour in wild bottlenose dolphins (*Tursiops aduncus*): caused by asymmetry of eye use? Behav Brain Res 170:204–210

Sakai M, Morisaka T, Iwasaki M, Yoshida Y, Wakabayashi I, Seko A, Kasamatsu M, Kohshima S (2013) Mother–calf interactions and social behavior development in Commerson's dolphins (Cephalorhynchus commersonii). J Ethol 31:305–313

Sakai M, Morisaka T, Kogi K, Hishii T, Kohshima S (2010) Fine-scale analysis of synchronous breathing in wild Indo-Pacific bottlenose dolphins (*Tursiops aduncus*). Behav Process 83:48–53

Samuels A, Gifford T (1997) A quantitative assessment of dominance relations among bottlenose dolphins. Mar Mamm Sci 13:70–99

Schaller GB (1965) The behavior of the mountain gorilla. In: DeVore I (ed) Primate behavior: field studies of monkeys and apes. Holt, Rinehart & Winston, New York, pp 324–367

Schino G, Aureli F, Troisi A (1988) Equivalence between measures of allogrooming: an empirical comparison in three species of macaques. Folia Primatol (Basel) 51:214–219

Schüermann C (1982) Mating behaviour of wild orangutans. In: de Boer LEM (ed) The orangutan. W. Junk, The Hague, pp 269–284

Scordato ES, Drea CM (2007) Scents and sensibility: information content of olfactory signals in the ringtailed lemur, *Lemur catta*. Anim Behav 73:301–314

Seyfarth RM (1977) A model of social grooming among adult female monkeys. J Theor Biol 65:671–698

Seyfarth RM, Cheney DL (1984) Grooming, alliances and reciprocal altruism in vervet monkeys. Nature (Lond) 308:541–542

Slater KY, Schaffner CM, Aureli F (2007) Embraces for infant handling in spider monkeys: evidence for a biological market? Anim Behav 74:455–461

Smuts BB, Cheney DL, Seyfarth RM, Wrangham RW, Struhsaker TT (eds) (1987) Primate societies. University of Chicago Press, Chicago

Sparks J (1967) Allogrooming in primates: a review. In: Morris D (ed) Primate ethology. Morrison & Gibb, London, pp 148–175

Spruijt BM, van Hooff JA, Gispen WH (1992) Ethology and neurobiology of grooming behavior. Physiol Rev 72:825–852

Starkey J, Loy J, Novak M, Goy R (1989) Comments on oral grooming by patas monkeys, including a comparison with rhesus macaques. Am J Primatol 18:327–331

Stopka P, Graciasova R (2001) Conditional allogrooming in the herb-field mouse. Behav Ecol 12:584–589

Sugawara K (1984) Spatial proximity and bodily contact among the Central Kalahari San. Afr Study Monogr 3:1–43

Suzuki A (2003) Wonderful societies of orangutans. Iwanami-Shoten, Tokyo (in Japanese)

Tamaki N, Morisaka T, Taki M (2006) Does body contact contribute towards repairing relationships? The association between flipper-rubbing and aggressive behavior in captive bottlenose dolphins. Behav Processes 73:209–215

Tanaka I, Takefushi H (1993) Elimination of external parasite (lice) is the primary function of grooming in free-ranging Japanese macaques. Anthropol Sci 101:187–193

Tanner JE, Byrne RW (1999) The development of spontaneous gestural communication in a group of zoo-living lowland gorillas. In: Parker ST, Mitchell RW, Miles HL (eds) The mentalities of gorillas and orangutans. Cambridge University Press, Cambridge, pp 211–239

Tavolga MC, Essapian FS (1957) The behavior of the bottle-nosed dolphin (*Tursiops truncatus*): mating, pregnancy, parturition and mother–infant behavior. Zoologica 42(1):11–31

Tizzi R, Castellano A, Pace DS (2001) The development of play behavior in a bottlenose dolphin calf (*Tursiops truncatus*). Eur Res Cetaceans 13:152–157

Tomasello M, Call J, Nagell K, Olguin R, Carpenter M (1994) The learning and use of gestural signals by young chimpanzees: a trans-generational study. Primates 35:137–154

Trivers RL (1971) The evolution of reciprocal altruism. Q Rev Biol 46:35–57

Trivers RL (1972) Parental investment and sexual selection. In: Campbell B (ed) Sexual selection and the descent of man, 1871–1971. Aldine, Chicago, pp 136–179

Tyack PL (2000) Functional aspects of cetacean communication. In: Mann J, Connor RC, Tyack PL, Whitehead H (eds) Cetacean societies. University of Chicago Press, Chicago, pp 270–307
Tyack PL (2003) Dolphins communicate about individual-specific social relationships. In: de Waal FBM, Tyack PL (eds) Animal social complexity. Harvard University Press, London, pp 342–361
van den Bos R (1998) The function of allogrooming in domestic cats (*Felis silvestris catus*): a study in a group of cats living in confinement. J Ethol 16:1–13
van Hooff JARAM (1967) The facial displays of the catarrhine monkeys and apes. In: Morris D (ed) Primate ethology. Weidenfeld & Nicolson, London, pp 7–68
van Lawick-Goodall J (1967) Mother–offspring relationships in free-ranging chimpanzees. In: Morris D (ed) Primate ethology. Weidenfeld & Nicolson, London, pp 287–346
van Noordwijk MA, van Schaik CP (2005) Development of ecological competence in Sumatran orangutans. Am J Phys Anthropol 127:79–94
van Schaik C (2004) Among orangutans. Belknap, Cambridge
Watts DP (1992) Social relationships of immigrant and resident female mountain gorillas. I. Male–female relationships. Am J Primatol 28:159–181
Watts DP (1995) Post-conflict social events in wild mountain gorillas (Mammalia, Hominoidea). I. Social interactions between opponents. Ethology 100:139–157
Watts DP, Pusey AE (1993) Behavior of juvenile and adolescent great apes. In: Pereira ME, Fairbanks LA (eds) Juvenile primates. Oxford University Press, Oxford, pp 148–167
Weaver A (2003) Conflict and reconciliation in captive bottlenose dolphins, *Tursiops truncatus*. Mar Mamm Sci 19:836–846
Weihs D (2004) The hydrodynamics of dolphin drafting. J Biol 3:8
Wickler W (1967) Socio-sexual signals and their intra-specific imitation among primates. In: Morris D (ed) Primate ethology. Weidenfeld & Nicolson, London, pp 69–147
Wilkinson GS (1986) Social grooming in the common vampire bat, *Desmodus rotundus*. Anim Behav 34:1880–1889
Wu H, Hao Y, Yu X, Xian Y, Zhao Q, Chen D, Kuang X, Kou Z, Feng K, Gong W, Wang D (2010) Variation in sexual behaviors in a group of captive male Yangtze finless porpoises (*Neophocaena phocaenoides asiaeorientalis*): motivated by physiological changes? Theriogenology 74:1467–1475
Xian Y, Wang K, Dong L, Hao Y, Wang D (2010) Some observations on the sociosexual behavior of a captive male Yangtze finless porpoise calf (*Neophocaena phocaenoides asiaeorientalis*). Mar Freshw Behav Physiol 43:221–225
Yamagiwa J (1987) Intra- and inter-group interactions of an all-male group of Virunga mountain gorillas (*Gorilla gorilla beringei*). Primates 28:1–30
Yamagiwa J (1992) Functional analysis of social staring behavior in an all-male group of mountain gorillas. Primates 33:523–544
Yurk H, Barrett-Lennard L, Ford JKB, Matkin CO (2002) Cultural transmission within maternal lineages: vocal clans in resident killer whales in southern Alaska. Anim Behav 63:1103–1119
Zahavi A (1977) The testing of bond. Anim Behav 25:246–247
Zamma K (2002) Leaf-grooming by a wild chimpanzee in Mahale. Primates 43:87–90
Zamma K (2011) Frequency of removal movements during social versus self-grooming among wild chimpanzees. Primates 52:323–328
Zamma K, Fujita S (2004) Genito-genital rubbing among the chimpanzees of Mahale and Bossou. Pan Afr News 11:5–8
Zucker EL, Mitchell G, Maple T (1978) Adult male–offspring play interactions within a captive group of orangutans (*Pongo pygmaeus*). Primates 19:379–384

Chapter 20
Non-conceptive Sexual Interactions in Monkeys, Apes, and Dolphins

Takeshi Furuichi, Richard Connor, and Chie Hashimoto

T. Furuichi (✉) • C. Hashimoto
Primate Research Institute, Kyoto University, 41-2 Kanrin,
Inuyama, Aichi 484-8506, Japan
e-mail: furuichi@pri.kyoto-u.ac.jp

R. Connor
Biology Department, UMASS-Dartmouth, North Dartmouth, MA 02747, USA

J. Yamagiwa and L. Karczmarski (eds.), *Primates and Cetaceans: Field Research and Conservation of Complex Mammalian Societies*, Primatology Monographs,
DOI 10.1007/978-4-431-54523-1_20, © Springer Japan 2014

Abstract Primates and dolphins exhibit comparable examples of all categories of non-conceptive sexual behaviors, including sexual interactions involving immature individuals, those involving individuals of the same sex, and copulation during the non-conceptive period. Although mammals of other taxa also perform non-conceptive sexual behaviors, the fact that there are so many reports of non-conceptive sexual interactions among higher primates and dolphins suggest a link between the nonreproductive use of sexual behaviors and high intelligence. This link might be because the greater role of learning in sexual behavior expands the possibility for sex to be incorporated into a variety of non-conceptive functions. Non-conceptive sexual behaviors seem to reflect or be influenced by important social factors, including affiliative relations and alliance between individuals of the same or different sex, high social status of females, within-group or between-group tension resolution, mate selection, and infanticide prevention. Animals may employ non-conceptive sexual behaviors to control various important aspects of their relationships with others which they cannot control with other social behaviors, which suggests that instances of non-conceptive sexual behaviors may serve as keys to understanding important aspects of the social relationships or social structure of the species.

Keywords Bonobo • Bottlenose dolphin • Estrus • Homosexual • Non-conceptive • Sexual behavior

20.1 Introduction

An interesting area of convergence between primates and cetaceans is the extent to which sexual behavior occurs in nonreproductive contexts. An overview of these interactions and their possible roles may lead to a better understanding of the evolution of sexual behaviors in mammals, with special reference to the relationship between the use of sexual behaviors and large brains.

Non-conceptive sexual behavior in primates occurs in a variety of contexts and may satisfy proximate functions such as self-satisfaction and greeting, and ultimate functions including the formation or maintenance of relationships, strengthening or confusing paternity, and female control of female–female and male–male competition (Dixson 1998; Sommer and Vasey 2006). Sexual behavior is clearly non-conceptive when it occurs between individuals whose age and sex present no possibility of conception, such as male–male, female–female, adult–immature, or immature–immature interactions. Especially, homosexual behaviors have been observed in many primate species, and researchers have identified and examined the proximate and ultimate roles of such behaviors (Vasey 1995; Dixson 1998; Sommer and Vasey 2006). Even between mature males and females, sexual behaviors may sometimes be used for nonreproductive purposes without involvement of obvious sexual arousal (Furuichi 1987; Dixson 1998). Copulation involving sexual arousal and ejaculation may also be performed for nonreproductive purposes, especially when performed during non-conceptive periods such as during pregnancy and post-partum amenorrhea (chimpanzee (*Pan troglodytes schweinfurthii*): Tutin and McGinnis 1981; bonobo (*Pan paniscus*): Furuichi 1987; Kano 1992; gorilla (*Gorilla beringei beringei*): Harcourt and Stewart 2007; Japanese macaque (*Macaca fuscata*): Takahata 1982; Fujita et al. 2001; Fujita et al. 2004; rhesus macaque (*Macaca mulatta*); capuchin (*Cebus* spp.): Fragaszy et al. 2004). Furthermore, for copulation performed during conceptive periods, we may need to consider nonreproductive explanations, such as infanticide avoidance by paternity confusion, if large numbers of copulations are performed with very low probability of conception (Matsumoto-Oda 1999; Wrangham 2002; Hashimoto and Furuichi 2006a, 2006b).

Non-conceptive sexual behavior has been observed in a range of cetaceans (reviewed in Bagemihl 1999), from river dolphins (*Inia geoffrensis*) to killer whales (*Orcinus orca*) to grey whales (*Eschrichtius robustus*), but here we derive our comparisons almost entirely from observations of bottlenose dolphins (*Tursiops* spp.) in Shark Bay, Western Australia, where the social and mating-system contexts of non-conceptive sexual behavior are better known than for other cetacean populations (Connor et al. 2000).

Life history parameters of the bottlenose dolphins in Shark Bay, including age of weaning, age of first reproduction, interbirth interval, and the duration of postpartum amenorrhea, are similar to values reported in chimpanzees (Connor and Volmer 2009). As we see here, non-conceptive sexual behavior in bottlenose dolphins is used, as it is in primates, in both social (e.g., dominance, bond formation) and reproductive strategies (e.g., confusion of paternity). The mating system of the Shark Bay bottlenose dolphins is characterized by a diffuse mating season in which alliances of two or three males consort with individual females for multiple periods that are thought to correspond to multiple estrous cycles and ovulations (Connor et al. 1996). The dolphin consortships are often initiated and maintained by aggressive herding (Connor et al. 1992a, 1992b, 1996, 2000; Connor and Volmer 2009), which is costly to females (Watson-Capps 2005). These costs, viewed in the context of multiple estrus cycles, led Connor et al. (1996) to predict that infanticide played an important role in the evolution of the bottlenose dolphin mating system. Evidence for infanticide was discovered subsequently in at least two other populations (Patterson et al. 1998; Dunn et al. 2002) but has yet to be confirmed in Shark Bay. Thus paternity confusion, a common strategy females use to reduce the risk of infanticide, stands as an important explanation for some forms of non-conceptive sexual behavior in bottlenose dolphins, as it does in some primates.

Here we focus our discussion mainly on chimpanzees, bonobos, Japanese macaques rhesus macaques, and bottlenose dolphins because their sexual behavior is well studied in this area under natural conditions.

In this chapter, we include for discussion any interactions involving genital contact or genital manipulation in sexual behaviors, irrespective of the involvement of sexual arousal. In fact, it is very difficult to observe sexual arousal of participants by facial, vocal, and physical expressions. Although penile erection can be regarded as a sign of sexual arousal in males, it is sometimes difficult to confirm penile erection during field observations. Furthermore, penile erection is sometimes observed in apparently nonreproductive interactions such as agonistic interactions, display, and play. Sexual arousal of females may sometimes be expressed by estrous calls, approaches, and presenting behavior, but copulation between adult males and females is not always preceded by these behaviors. It seems that sexual arousal is not an essential criterion for the use of sexual behaviors for nonreproductive purposes.

20.2 Sexual Interactions Between Non-conceptive Participants

20.2.1 Interactions Involving Immature Individuals

Although copulation-like behaviors are rather widely seen in immature mammals, bonobos are unusual among primates in that they start exhibiting such behaviors before weaning (Fig. 20.1) (Kano 1992; Furuichi et al. 1998). Although bonobos usually are nursed until they are 3 to 4 years old, bonobo infants start showing sexual behaviors as early as 1 year of age. Three types of sexual behaviors are observed in immature bonobos, and these types reveal sex-based developmental differences (Hashimoto and Furuichi 1994; Hashimoto 1997).

The first type, performed more frequently by males, is sexual behavior during play. While hugging in the ventro-ventral position during play, participants make mutual genital contact. This kind of sexual behavior is also found in chimpanzees (Tutin and McGinnis 1981; Plooij 1984; Goodall 1986) and gorillas (Harcourt et al. 1981; Nadler 1986). Immature bonobos also have sexual interactions while playing with adult males. For example, adult males sometimes hold an infant on their lap in the ventro-ventral position while sitting, and shake the body of the infant with a foot so that their genitals rub with each other. In most cases, the adult male does not have an erection, so this type of sexual behavior seems to be performed without sexual arousal (Hashimoto 1997).

The second category includes copulation-like sexual behaviors, which are observed between immature males and mature females, but not between immature females and mature males. Adult females usually do not resist when infant or juvenile males with erections perform penile insertions. This behavior increases in frequency with age, peaking when males are young juveniles, then declines as attempts at penile insertion by old juvenile and young adolescent males are met with resistance by females and aggression by adult males. Thus, this tolerated copulation-like behavior is a unique sexual behavior found in infant and juvenile males.

Copulation-like sexual behavior is sometimes seen between immature males and their mothers, but the context of such interactions seems to be somewhat different from those between immature males and mature females other than the mother. Although immature males usually take the lead of the interaction in the latter case, mothers may also initiate sexual behavior with their male offspring, using sex to soothe the temper

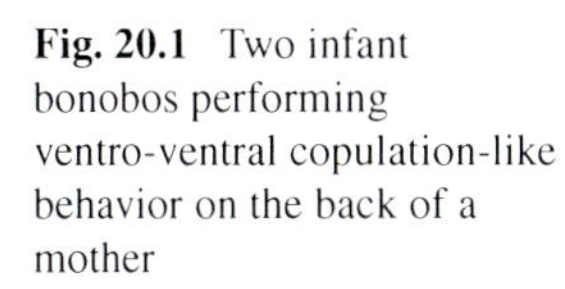

Fig. 20.1 Two infant bonobos performing ventro-ventral copulation-like behavior on the back of a mother

of frustrated or excited infants. The frequency of the mother–offspring copulation-like sexual behavior decreases with the increasing age of the offspring and almost completely vanishes before they reach adolescence. Copulation-like sexual behavior between immature males and their mothers is also observed in chimpanzees, where the ages of participating males, as well as the context of the interactions, are similar to bonobos (Hashimoto, unpublished data).

The third category is sexual behaviors that are used to control social relationships. This behavior is absent among infants. However, immature males start engaging in sexual behaviors with adult males with increasing frequency as they approach adulthood. Most of these behaviors are mounting or rump–rump contact that is usually performed among adult males (see following). They occur during agonistic interactions or when party members are excited at the beginning of a feeding session. In contrast, female juveniles are rarely involved in this kind of sexual interaction with adult males or females. It seems that immature females do not commit to social relationships in the group until they emigrate from their natal group and enter a new one as adolescents. Upon entering a new group, immigrant adolescent females frequently perform sexual behaviors with one another and with senior adult females. Thus, the development of sexual behaviors during the immature period in this male-philopatric species seems to reflect sex differences in social status and life history. Another type of sexual behavior in this category, unique to bonobos, occurs when adult females hold their infant ventro-ventrally while standing quadruped and rub their genitals against the genitals of the infant. As is the case for genito-genital rubbing among adult females (see following), this seems to be a kind of tension-reducing behavior by mothers that follows aggression received from other bonobos.

Bottlenose dolphins might be called "aquatic bonobos" when it comes to the frequency and variety of non-conceptive sexual behavior, and in the category of immature sex it is clear that they exceed bonobos by a considerable margin. Although immature bonobos may exhibit sexual behavior by the time they are 1 year old (Furuichi et al. 1998; Kano 1992), male bottlenose dolphins only 2 days old have been observed engaging in sexual behavior with their mothers. Mann (2006) examined sexual behavior among infant Indian Ocean bottlenose dolphins in Shark Bay (average weaning age = 4 years). Sexual behavior included mounting, probing the genital slit with the rostrum (= "goosing"), pressing the head into another dolphin's genital area to push them up, and contact between the pectoral fin and genital area. Approximately half the observations were same-sex interactions. Homosexual interactions were observed much more often among male than female infants, and those among males tended to involve pairs and trios whereas female infant homosexual behavior was limited to pairs. Male infants were also observed to engage in sexual behavior with their mothers more often than female infants (Mann 2006). The rate that male dolphin calves engage in sexual behavior is extraordinarily high: some 40 times that of adult female bonobos, who have a reputation for frequent sexual behavior. Even female dolphin infants exceed the bonobo rate by a considerable margin (Mann 2006).

Juvenile sexual behavior in Shark Bay sometimes mimics adult consortship behavior as two dolphins will temporarily "herd" another, with mounting and goosing. These bouts are distinguished from true adult consortship behavior by three characteristics: (1) role switches, as the identity of the "herded" dolphin changes, (2) the "herded" dolphin is often male, and (3) the herding behavior ends when the social bout ends and the dolphins return to nonsocial activities such as resting, traveling or foraging (Connor et al. 2000). Such observations support practice, dominance, and male–male bonding functions for juvenile sexual behavior.

20.2.2 Interactions Involving Mature Participants of the Same Sex

Sexual behavior between mature individuals of the same sex is widely seen in many kinds of mammals (Vasey 1995; Sommer and Vasey 2006). Although the mounting behaviors between males are well recognized, mainly as ritualized dominance interactions, sexual behaviors between females are also observed in many primate species, and a variety of functions have been proposed for them (Vasey 1995). Female

Japanese macaques and rhesus macaques show mounting behaviors similar to male–female copulation (Kapsalis and Johnson 2006; Vasey 2006). Females will mount estrous females and even show thrusting movements. Kapsalis and Johnson (2006) suggested that rhesus females perform the sexual behaviors to establish new affiliative relationships and alliances, because such behavior was frequently observed following the loss of alliance partners through artificial trapping or mortality. On the other hand, Vasey (2006) performed quantitative analyses on the probable functions of female–female mounting in Japanese macaques, including alliance formation, dominance demonstration, acquisition of alloparental care, acquisition of opposite-sex mates, reconciliation, and regulation of social tension, but none of these functions was statistically supported. The author speculated that females perform homosexual behavior in pursuit of the proximate benefit of pleasure, and that such behavior is a by-product of female–male mounting that females employ to prompt sexually disinterested or sluggish males to copulate with them.

Bonobos exhibit various other types of same-sex sexual behaviors. Among these, genito-genital rubbing behavior, in which two females hug each other ventro-ventrally and rub their genitals repeatedly in a rapid lateral motion of their hip, has received considerable attention with respect to its social roles. Similar to chimpanzees, bonobos form male-philopatric groups. Although males stay in their natal group throughout their life, females usually transfer between groups in early adolescence (Kano 1982, 1992; Furuichi 1989; Gerloff et al. 1999; Hashimoto et al. 2008). Therefore, most females found in a group are thought to be unrelated, except for the case in which some related females immigrate from the same group. Compared to males, female chimpanzees do not often participate in large mixed-sex parties or have social interactions with each other. This tendency may be partly explained by the lack of kin-relations among females and by forging constraints as they range and feed alone or in small parties (Wrangham 1979; Pusey and Packer 1987; Furuichi 2006). However, bonobo females are found in mixed parties even more frequently than males and have various social interactions with each other, including grooming, food sharing, cofeeding, and genito-genital rubbing (Kuroda 1980; White 1988; Furuichi 1989; Kano 1992; Mulavwa et al. 2008).

Females usually perform genito-genital rubbing in two contexts. First, they perform genito-genital rubbing in tense situations when they are excited upon arriving at feeding sites, when they hear vocalizations of different groups, and when they are involved in agonistic interactions. Genito-genital rubbing in these contexts seems to regulate tension, because after genito-genital rubbing females usually continue feeding or resting in a more relaxed state (de Waal 1987; Furuichi 1989; Kano 1992; Hohmann and Fruth 2000). Second, genito-genital rubbing promotes the establishment or maintenance of affiliative social relationships. When adolescent females immigrate into a new group, they have no relatives or other close female associates. In such circumstances, immigrant females tend to choose a specific senior female with whom to associate, and solicit her for various social interactions, including genito-genital rubbing, food sharing, and cofeeding. When these immigrants find the senior female feeding in a certain position, they approach and solicit a bout of genito-genital rubbing, after which they beg for food or cofeed beside her, even when food is abundant elsewhere. When two or more adolescent females immigrate into a group around the same time, they tend to associate closely and frequently perform genito-genital rubbing with each other (Furuichi 1989; Idani 1991).

Male–male sexual interactions are common in bonobos, including mounting and rump–rump contact (two males bring their rumps together and hit their genitals against each other repeatedly in a quick forward-and-back motion) (Kuroda 1980; de Waal 1987; Kano 1992). In contrast to sexual behaviors between adult females, those between adult males are almost exclusively performed in tense situations. Sexual behaviors may appease an excited male showing display behaviors or may be used to reconcile immediately after agonistic interactions (Furuichi and Ihobe 1994). Interestingly, bonobo males rarely show behaviors such as hugging, touching, embracing, kissing, or pant-grunting that are well developed in chimpanzees for greeting, conflict resolution, reconciliation, or reassurance. In these contexts, male bonobos usually employ sexual behaviors. Also, although chimpanzee males use pant-grunts and other behaviors to express or confirm dominance relationships, bonobo males rarely show such behaviors. The typical rump–rump contact is a symmetrical behavior, and even when they use mounting, males alternate roles as if they are avoiding the expression of a dominance relationship. Thus the use of sexual behaviors by bonobo males seems to reflect social relationships that are egalitarian compared to chimpanzees.

The male–male sexual behaviors in bonobos do not necessarily involve sexual arousal. The participants do not usually show penile erection, and even when they do, we cannot know whether it is caused by sexual arousal or excitement. In fact, male bonobos often show penile erection in various situations that may not involve sexual arousal, including during display, agonistic interactions, play, and when they find preferred foods. One of the authors, Furuichi, observed only one case of male–male mounting that involved ejaculation during his 28 months of field studies. By contrast, male–male mounting in gorillas is usually performed in a manner similar to male–female copulation in terms of the behavioral pattern and vocalizations, and ejaculation was confirmed in 2 of 97 cases, which hints at a higher frequency than bonobos (Yamagiwa 2006).

Homosexual behavior in adult bottlenose dolphins is much more commonly observed in males than females (Fig. 20.2). The same sexual behaviors employed by infants, mounting and goosing, are used frequently by adults. The review by Connor et al. (2000) concluded that sexual behavior, including mounting with erections, is used in both affiliative and agonistic contexts. At one extreme, one adult alliance herded a maturing male pair for more than an hour, even using aggressive vocal signals typical of herding (Connor and Smolker 1996). The mounting and goosing were conducted in an energetic, almost violent manner. On another occasion mounting between two allied males was clearly nonaggressive, occurring in a slow relaxed manner. Occasionally, older larger males permit smaller male calves and juveniles to mount them as well (Connor et al. 2000).

To examine the sexual interactions of male dolphins, one of the authors, Connor, extracted observations of the two most unambiguous and easily detected types of sexual behaviors, mounting and goosing, from 155 focal follows (552 hours) on

Fig. 20.2 Mounting behavior between adult male bottlenose dolphins

alliance-forming males. Two of the alliances were categorized as young (one trio of two immature and one mature male and one pair of maturing males), three alliances were mature (one pair and two trios), and two alliances (a pair and a trio) were considered "old" mature based on extensive ventral speckling (Connor et al. 2000). Other males were occasionally present during follows.

Sexual behavior (goosing or mounting) was observed in 54 of the 155 follows. There were interesting differences in the sexual behavior of older and younger males as well as in heterosexual versus homosexual interactions. Following Mann (2006), any sexual behavior that followed another within 5 min was considered to be part of the same bout. Considering first the recipients of sexual behavior, an individual was scored as a recipient only once per alliance per follow, regardless of the number of mounts or gooses he or she received during the follow. Young males targeted males (23) more than females (15) for sexual interactions whereas mature and old mature males targeted females (24) more than males (14) ($\chi^2=4.266$, $p=0.039$). Eighteen females were recipients of sexual behavior from mature males but the 15 bouts by young males were distributed among only 7 females. Of the 13 males targeted by young males, 9 were also in the young (juvenile or maturing) category while 4 were mature males. Of the 7 males targeted by mature males, 5 were young and 2 were mature.

Some bouts included only goosing or mounting while other bouts included both. Scoring a bout based on the first sexual behavior in the bout, the 15 males (9 young and 6 mature) targeted for sex by males were more often mounted (22) than goosed (15), while the 24 female targets were more often goosed (30) than mounted (17) ($\chi^2=4.515$, $p=0.037$). The higher proportion of mounting to goosing in male–male compared to male–female interactions may seem surprising, but our surprise likely reflects the fact that we understand the normal reproductive function of mounting, but not goosing. If, for example, goosing, conducted in the appropriate manner, serves to enhance female receptivity, then the high proportion of goosing in male–female interactions makes sense. Regardless, a key result is that homosexual interactions constitute a significant proportion of male dolphin sexual interactions for both young and mature individuals.

20.3 Copulation Involving Mature Males and Females During the Non-conceptive Period

20.3.1 Adolescent Infertility

There is a significant gap between first estrus and first birth in Japanese macaques. For example, on Yakushima Island where wild Japanese macaques have been observed without artificial provisioning for more than 30 years, females tend to show first estrus in the mating season at the age of 3 to 4 years. However, they seldom give birth in the next breeding season when they are 4 years old, but usually give birth at the age of 5 or 6 years (Takahata et al. 1998). There are no detailed reports on adolescent infertility in other macaque species, which may be partly because of difficulties in detecting a short gap between first estrus and first conception in nonseasonal-breeding species living in warmer environments. In Japanese macaques, which mate seasonally in autumn, the delay of conception by a few months results in a delay of first delivery by a year.

Adolescent infertility is also reported for female chimpanzees. In Mahale Mountains in Tanzania, females usually show first estrus during late adolescence at the age of 10 years (128 months), and they emigrate from their natal groups at the age of 11.27 years. The mean number of months elapsing from immigration into a new group and first birth was 32 months (Nishida et al. 1990, 2003). Thus, females have a period of adolescent infertility of about 4 years if we assume that females immigrate into a new group immediately after emigrating from their natal group. More detailed, but atypical, data obtained from four females who did not transfer groups until their first birth showed that females experienced adolescent infertility for 1 year 2 months to 4 years 1 month (Nishida et al. 2003).

It is more difficult to estimate the duration of adolescent infertility in bonobos because there has been no observed case in which females stayed in their natal group until the first birth. Female bonobos do not usually start estrus or copulation until they leave their natal group at the age of 6 to 10 years. Females who temporarily joined the study groups at the estimated age of 7 to 9 years did not show estrus, but females who joined the study group at the estimated age of 10 years performed copulation from the beginning and gave first birth at the age of 13 to 15 years. Thus, although there may be some error in the age estimates, there seems to be a period of adolescent infertility of 2 years or more (Hashimoto et al. 2008; Furuichi et al. 2012).

The function of adolescent infertility in Japanese macaques is unknown. Because they live in female-philopatric troops where females have stable social relationships with related females, there is no urgent need for females to form social relationships with males via copulation. The length of adolescent infertility in Japanese macaques is rather short and variable because of nutritional conditions. In fact, in an artificially provisioned group in Arashiyama, 3.9 % of females gave birth at the age of 4 years during the breeding season following the mating season in which they showed first estrus (Koyama et al. 1992), while the age of first birth was delayed under poor nutritional conditions (Watanabe et al. 1992). By contrast, the long adolescent infertility in chimpanzees and bonobos may confer a benefit on adolescent females, because copulation with males in the new group to which they immigrate may help establish stable social bonds and reduce the risk of infanticide. Although immigrant adolescent female bonobos do not attract much sexual attention from males, they show prolonged estrus, approach males, try to have various interactions, and copulate with a higher frequency than older adult females (Furuichi 1989, 1992; Idani 1991; Kano 1992; Furuichi and Hashimoto 2004). Adolescent female chimpanzees also show irregular and long-lasting sexual swelling, but the frequency of copulation is not as high as that of adult females (Goodall 1986; Pusey 1990; Hashimoto unpublished data).

In Shark Bay, some pre-parturient females may be consorted by adult males during more than one season before they conceive. However, this has not been quantitatively distinguished from the same phenomenon in females that have previously given birth, as they may also be consorted for two or more consecutive seasons before giving birth again.

20.3.2 Estrus During Non-conceptive Periods of Adult Females

Estrus and copulation during pregnancy are seen in many primate species including capuchins (Fragaszy, et al. 2004), Japanese macaques (Takahata 1982; Fujita et al. 2004), chimpanzees(Tutin and McGinnis 1981), bonobos (Furuichi 1987; Kano 1992), and gorillas (Harcourt and Stewart 2007). Non-conceptive periods of adult females occur during postpartum amenorrhea, pregnancy and during parts of the estrous cycle that are far from ovulation.

In capuchins, copulations occur frequently outside the females' periovulatory phase, including during pregnancy and postpartum amenorrhea. Copulations occur frequently in socially tense situations, during play, and during group formation in captivity (Dixson 1998; Fragaszy et al. 2004). Thus, capuchins seem to use copulation for social purposes such as forming affiliative relationships or resolving tensions.

Although the period differs from site to site because of climate differences, Japanese macaques have a mating season of 4–5 months around autumn, and females show cyclic estrus during this period (Takahata 1980). It is quite interesting, however, that many females conceived during the first estrous period but still continued showing cyclic estrus and copulated after conception (Takahata 1982; Fujita et al. 2004). In an extreme report for a wild group in Kinkazan Island, as many as 80 % of females conceived during the first estrous cycle but still continued showing cyclic estrus (Fujita et al. 2004).

There are two main hypotheses for the role of estrus during pregnancy in Japanese macaques. Females sometimes form specific relationships with certain males through repeated copulation during the mating season, and such a relationship may provide benefits such as support during agonistic interactions and competition over food (Takahata 1982). In this case, females may benefit from repeated copulations after conception. Another possibility concerns intergroup competition. In Japanese macaques, males seeking better mating opportunities transfer among female-philopatric troops (Suzuki, et al. 1998). When two troops meet at the boundary area of their home ranges, females confront each other in "frontlines," and males actively fight for the troops to which they belong at that moment (Saito et al. 1998). Thus, troops having more males may enjoy a competitive advantage. Males tend to appear around troops that have more females in estrus during the mating season, and some of these males continue to reside in the troop afterward (Furuichi 1985; Sprague 1989). Therefore, repeated estrus by females during pregnancy may aid in the recruitment of males.

Female chimpanzees and bonobos show cyclic estrus (Furuichi 1987; Nishida, et al. 1990; Wallis 1997). Even after conception, female chimpanzees may have a further two estrous cycles (Tutin and McGinnis 1981). Female bonobos also continue showing cyclic estrus until 1 month before giving birth, although the cycles are not as regular as before conception (Furuichi 1987; Kano 1992).

Female bonobos also show estrus during postpartum amenorrhea. Although female chimpanzees do not exhibit estrus until the weaning of offspring at about 4 years of age (Nishida et al. 1990; Wallis 1997), female bonobos start showing estrus about 1 year after giving birth while nursing their offspring (Kano 1992; Furuichi and Hashimoto 2002). The interbirth interval for bonobos with a surviving offspring is 4.8 years. If we deduct 1 year of postpartum amenorrhea and 7.6 months of gestation, the period for female bonobos to show estrus before conception is as long as 3 years (Furuichi et al. 1998; Furuichi and Hashimoto 2002). Why do female bonobos resume estrus so early during lactation when there is no possibility of conception?

Furuichi and Hashimoto (2002) showed that the variation in interbirth interval is significantly larger for bonobos than for chimpanzees. With this finding, they suggested that the restriction for the timing of conception is less for bonobos than for chimpanzees. The cost of travel between distant food patches prohibits female chimpanzees from having two dependent offspring at the same time. Therefore, female chimpanzees need to wait for the weaning and independence of one offspring before having the next. However, in bonobo habitats key food patches are larger and many small food patches exist among those large food patches (White and Wrangham 1988; Wrangham 2000; Furuichi, unpublished data), so the daily range required for obtaining adequate nutrition is reduced (Furuichi et al. 2008; Furuichi 2009). Therefore, female bonobos, traveling more slowly, may not suffer

Fig. 20.3 A female bonobo carrying two infants. Such females have been observed both in a provisioned group (this photograph) and unprovisioned groups

the debilitating costs that two dependent offspring would impose on chimpanzees. In fact, female bonobos sometimes give birth before their previous offspring achieves independence, and will carry one infant on her back and the other on her chest (Fig. 20.3). This difference may partly explain why female bonobos can resume estrus earlier, with a possibility, although not high, of a short interbirth interval.

The estrous period in bonobos, because it occurs during gestation and postpartum amenorrhea, is considerably prolonged. Considering various reproductive parameters, Furuichi and Hashimoto (2002) calculated that female chimpanzees show estrus for only about 5 % of the adult life, even if we include estrus during the early stage of gestation. Thus, only 1 of 20 females show estrus at a time, and estrous sex ratio (or operational sex ratio), which is the proportion of adult males to a female showing estrus at one time, is as great as 20 if there are the same number of males and females in a group. Because the actual number of males is smaller than that of females, probably because of mortality from severe sexual competition, the estrous sex ratio was lower than 20, but still as high as 4.2 for chimpanzees in Mahale, and 12.3 for Gombe (Furuichi and Hashimoto 2002). Such a high estrous sex ratio may have produced severe sexual competition among males that, in turn, imposes significant costs on females. In contrast, owing to a much longer estrous period during gestation and a long estrous period in postpartum amenorrhea, female bonobos show estrus for as much as 27 % of their adult life. Therefore, the estrous sex ratio was as low as 2.8 for bonobos at Wamba, even though there were similar numbers of males and females in a group.

Male sexual competition is reduced in bonobos compared to chimpanzees. In bonobos at Wamba, the frequency of male–male aggression over estrous females is very low. Because of the high percentage of the females in estrus, there are a number of estrous females in a group of bonobos at a given time, which reduces the ability of high-ranking males to monopolize females and allows for female mate choice (Furuichi 1997; Furuichi and Hashimoto 2002). Compared with chimpanzees, bonobos form more stable mixed-sex parties. Females join such parties more frequently than do males; female social status is almost equivalent to that of males, females have priority of access to food, and females can control group movements and ranging. All these advantages for females might be related to the reduced male–male competition and female mate choice (Kano 1982, 1992; White 1988; Furuichi 1989, 1997, 2011; Furuichi et al. 2008; Mulavwa et al. 2008; Stevens et al. 2008).

The Shark Bay bottlenose dolphins have a diffuse mating season that extends from the austral spring though early summer. However, females that conceive during the mating season (September–December) were often herded during the late winter period (July–August), before the mating season. Although this pre-mating season behavior may simply reflect additional ovulations that fail to produce pregnancy (bottlenose dolphins are known to have two to seven ovulations during the year they conceive), it may also reflect anovulatory cycles, a phenomenon also known from captive studies (Connor et al. 1996). The evolutionary reason for anovulatory cycles in bottlenose dolphins may be the same as that for multiple ovulatory cycles: confusion of paternity associated with reducing the risk of infanticide (Connor et al. 1996).

Consortships with pregnant female bottlenose dolphins have been documented in Shark Bay, but it is not clear how common this phenomenon is (Connor et al. 1996). Also, we must be cautious about interpreting such behavior in terms of female tactics because some consortship behavior may relate more to male–male alliance bonds than male and female mating tactics. For several years three mature males were among a group of dolphins that were provisioned with dead fish each day (Connor and Smolker 1995). Relationships among these three males were highly unstable with respect to the formation of consortships with females (Connor and Smolker 1996), and they consorted with more females in a nonfertile reproductive state (e.g., females with newborn calves) than was typical of nonprovisioned males in the area (Connor et al. 1996). Non-conceptive herding by the provisioned males suggests that some cases of herding have more to do with maintaining fragile male–male bonds than conception, but the behavior may have simply been a maladaptive effect of the provisioning itself. However, a male-bonding function is supported by observations following the formation of a new alliance between two nonprovisioned adult males in 1994; each of the first 8 days they were observed in the new alliance they herded a different female, including one that was unlikely to be receptive as she had a 1.5-year-old calf (cycling typically resumes when a calf is 2.5 years old), was not herded again that year by other males, and did not conceive until the following year (Connor and Mann 2006).

20.4 Excessive Number of Copulations with Low Probability of Conception

Although it is difficult to know the probability of conception of wild animals, female chimpanzees present a good example of copulation with a low probability of conception. Female chimpanzees usually resume cycling after weaning an infant, show estrus for about 12 days before monthly ovulation, and typically exhibit five to nine cycles before conception (Hasegawa and Hiraiwa-Hasegawa 1983; Nishida et al. 1990; Wallis 1997). They show high proceptivity during estrus, especially during the periovulatory periods when the frequency of copulation is very high (Tutin 1979; Wrangham 2002; Hashimoto and Furuichi 2006a, 2006b). In an extreme case in the Kalinzu Forest, Uganda, when high-ranking males failed to herd an estrous female, the female copulated more than 60 times with 12 males in a day (Furuichi, personal observation). Although female bonobos show estrus during a larger proportion of their adult life than do chimpanzees, the frequency of copulation during the estrous period is much higher for female chimpanzees than for female bonobos (Furuichi and Hashimoto 2002). It is estimated that female chimpanzees sometimes copulate several hundred or even more than a thousand times from the initiation of estrus to conception (Matsumoto-Oda 1999; Wrangham 2002; Hashimoto and Furuichi 2006a, 2006b; Watts 2007). If the possibility of conceiving during a given copulation is extremely low, we may need to consider such copulation as a kind of non-conceptive sexual behavior and to seek to understand the benefit for females. Why do female chimpanzees need to perform such a large number of copulations? Why do not female chimpanzees conceive on the first ovulation even if they copulate numerous times with many males during the periovulatory period?

To date, a number of hypotheses have been proposed for the role of such an excessive number of copulations. Females may choose a desirable father of the offspring through sperm competition (best male hypothesis), females may copulate with many males to form or maintain familiar relationships with males (many males hypothesis), females may use such a high frequency of copulation as a social passport, or females may copulate with many males to avoid infanticide by confusing paternity (Boesch and Boesch-Achermann 2000; Furuichi and Hashimoto 2002; Wrangham 2002).

To distinguish among these hypotheses, it would be useful to have a better understanding of the proximate factors involved, including the hormonal and genetic mechanisms underlying the low probability of conception. Studies on human females showed that there is a graded continuum from fully fecund ovarian cycles through follicular and luteal suppression, anovulation, oligomenorrhea, to amenorrhea, and that ovulation is strongly impacted by nutritional condition (Ellison et al. 1993). If females show fecund ovarian cycles from the beginning but do not achieve successful conception for 5 to 9 months, we may need to consider whether there exists a threshold for the combination of genotypes, such as complementary major histocompatibility complex (MHC) types. In this case the "best" male is not an absolute standard but is relative to the female's genotype. On the other hand, if females show estrus without fecund ovarian cycles, we may need to consider possible social factors that favored the evolution of a female pseudo-estrus.

During the year they conceive, female bottlenose dolphins in Shark Bay are consorted by male alliances for varying periods of time that may span several months or longer (Connor et al. 1996). Such extended attractive periods may correspond to the multiple ovulations and, as we noted, even anovulatory cycles. Evidence that consortships are coerced is observed in about half the cases (Connor et al. 1996; Watson-Capps 2005) and, given the difficulty of observing the brief episodes of consortship aggression, the actual percentage is likely much higher (see Connor and Volmer 2009). The costs of enduring aggressively maintained consortships (see Watson-Capps 2005) focuses our attention on the nature of the benefits to multiple cycling that would outweigh such costs. Connor et al. (1996) suggested that paternity confusion to reduce infanticide risk was the likely answer, and infanticide was soon discovered in European and North American populations (Patterson et al. 1998; Dunn et al. 2002). To a large extent dolphin infanticide is "cryptic," because the lethal wounds from blows are internal and an infant victim that strands may not exhibit obvious injury.

20.5 Conclusion

When we enumerate non-conceptive sexual interactions in nonhuman primates and dolphins, we realize that such behavior covers a range of phenomena with an equally impressive variety of possible explanations. It was surprising that primates and dolphins exhibit comparable examples of all categories of non-conceptive sexual behaviors (Table 20.1). There are many more reports on non-conceptive sexual behaviors in other mammal species (Dixson 1998; Sommer and Vasey 2006). Furthermore, because it is difficult to know the periods of pregnancy, postpartum amenorrhea, and cases of low probability of conception of wild animal populations, many more instances of non-conceptive copulation will be found when we carry out long-term detailed observation of animals. Nevertheless, the fact that there are so many reports of non-conceptive sexual interactions among primates and dolphins may tell us something important about the evolution of the nonreproductive use of sexual behaviors and high intelligence. An important role for learning in sexual behavior, for at least some primates, is clear. Some higher primates have difficulties in performing copulation if they are raised in isolation. The implication that learning plays a role in sexual behavior likely expands the possibility for sex to be incorporated into a greater variety of nonreproductive functions.

Each example of non-conceptive sexual behavior that we discussed in this chapter seems to reflect or be influenced by important social factors. Copulation during the postconception periods in Japanese macaques may provide important roles for female social status or intergroup conflict. Homosexual behaviors among female macaques seem to contribute to the formation of affiliative relationships under

Table 20.1 Non-conceptive sexual behaviors and proposed functions

	Nonhuman primates	Bottlenose dolphins
1. Sexual interactions between non-conceptive participants		
(a) Involving immature individuals	*Sexual behavior performed during play* – Observed among immatures in chimpanzees, bonobos, gorillas, etc. – More frequently performed by males (bonobos) – Also observed between immatures and adult males	*Sexual behavior with nonidentified purposes* – Including mounting, probing the genital slit with the rostrum, pressing the head into another's genital area, and contact between the pectoral fin and genital area – Extremely frequent as compared with those in bonobos, and more frequent for males – Performed both between same-sex and opposite-sex individuals – Those among males sometimes involve more than two individuals – Male infants engage in sexual behavior with their mothers more often than female infants
	Copulation-like behaviors – Observed between immature males and non-kin adult females, but not between immature females and adult males, in bonobos and chimpanzees – Adult females show great tolerance – Decreases in older juvenile years or early adolescence because of increased intolerance of adult males and females	
	Copulation-like behavior with mother – Observed in chimpanzees and bonobos – Immature males usually take the lead with sexual arousal – Mothers sometimes take the lead for soothing the temper of frustrated or excited infants – Almost completely disappears before sexual maturity	*Sexual behavior that mimics adult consortship behavior* – Two immatures herd another with mounting and rostro-genital contact (goosing) – Distinguished from those by adults by (1) role switches, (2) males are often herded, (3) herding ends when social bout ends – For practice of herding and male–male bonding
	Sexual behavior for the control of social relationships – Observed in many primate species (mounting) and in bonobos (rump–rump contact) – Mostly observed between immature males with increasing frequency as they approach adulthood (bonobos) – Immature females rarely perform this behavior, but begin performing it frequently (genito-genital rubbing) after immigration to other groups (bonobos) – Mothers sometime perform sexual behavior with their infant to control their own tension (bonobos)	

(continued)

Table 20.1 (continued)

	Nonhuman primates	Bottlenose dolphins
(b) Involving mature participants of the same sex	*Japanese macaque* – Females show mounting behavior similar to male–female copulation – Quantitative analysis did not support any functions, including alliance formation, dominance demonstration, acquisition of alloparental care, acquisition of opposite-sex mates, reconciliation, and regulation of social tension: just for fun? – By-product of female–male mounting that females employ to solicit sexually disinterested or sluggish males *Rhesus macaque* – Females show mounting behavior similar to male–female copulation. – Establishment of new affiliative relationships and alliances, following the loss of alliance patterns through artificial trapping or mortality *Bonobo* – Female–female sexual interactions for formation or maintenance of new affiliative relationships, soothing tension, and resolution of agonistic interaction – Male–male sexual interactions for soothing tension and resolution of agonistic interactions – In contrast to male–male mounting in other mammals, dominance is rarely expressed in the male–male sexual interactions	– Same suite of sexual behaviors employed by calves are used frequently by adults – Much more commonly observed among males – Mounting is used in both affiliative and agonistic contexts – For male-bonding and dominance
2. Copulation involving mature males and females during non-conceptive period		
(a) During adolescent infertility	*Japanese macaque* – First estrus at 3 or 4 years old but usually give birth at the age of 5 or 6 years (Yakushima) – Variable and short period of infertility and unspecified benefit *Chimpanzee and bonobo* – Chimpanzees in Mahale: First estrus at 10 years old and emigrate from natal group at 11.27 years old. First delivery in 32 months from immigration into a new group; thus more than 3 years of infertility	– Males herd prepartum females for more than one season before conception, but same may also occur for adult females

	– Bonobos at Wamba: First estrus occurs at immigration into a new group at around 10 years old and first delivery occurs at 13–15 years old; thus more than 2 years of infertility – Adolescent females sometimes show noncyclic continuous estrus – Helps immigrant females establish stable social bonds and reduces risk of infanticide	
(b) During non-conceptive periods of adult females	*Capuchin* – Frequent copulation outside the periovulatory phase including during pregnancy and postpartum amenorrhea – Occurring frequently in socially tense situation, during play, and during group formation in captivity *Japanese macaque* – Cyclic estrus during 4–5 months of mating season, but many females conceive during the first estrus – Formation of affiliative relationships with troop males – Recruitment of non-troop males *Chimpanzee* – Two estrous cycles after conception, probably because of estrogen secreted from the placenta *Bonobo* – Continuous estrus until 1 month before giving birth – Estrus during postpartum amenorrhea – Paternity confusion – Controls intermale aggression by reducing estrous (operational) sex ratio – Leads to high social status of females	– Males sometimes are attracted to females for periods exceeding the duration of rising estrogen levels – Males occasionally herd females in late pregnancy, during first 2 weeks after parturition, and the day after the loss of a nursing infant – Consortships with nonfertile females is observed more frequently in captivity – Herding nonfertile females for a male-bonding function
3. Excessive number of copulations with low probability of conception	*Chimpanzee* – Very high frequency of copulation with many males during the estrous periods – Hundreds or even more than a thousand copulations during 6–9 months from resumption of estrus to successful conception – Various hypotheses including many male hypothesis, best male hypothesis, paternity confusion hypothesis, and high genetic threshold hypothesis	– Females are consorted by multiple male alliances for several months or longer, including multiple ovulations and anovulatory periods – Coerced consortships are observed in half of cases – Paternity confusion (infanticide observed in some populations)

some unstable conditions. Capuchins perform copulation during the non-conceptive periods to help form affiliative relations or to resolve tensions. The difference in the sexual behaviors between immature males and females in bonobos may reflect the life history of each sex. Prolonged estrus of female bonobos during postpartum amenorrhea and gestation may be related to the high social status of females in their male-philopatric groups. The excessive number of copulations with a low probability of conception of female chimpanzees may be understood as playing roles in mate selection, formation of a familiar relationship with males, and infanticide prevention, and the homosexual behaviors and cooperative herding of females in non-conceptive period in bottlenose dolphins may contribute to male–male bonding. Thus, animals may employ sexual behaviors to control various important aspects of their relationships with others that they cannot control with other ordinary social behaviors such as grooming, following, cofeeding, fighting, or displaying dominance or subordinance. This study suggests to us that studies of non-conceptive sexual behavior may inform us about key aspects of social relationships and social structure in other species.

Acknowledgments Studies on chimpanzees and bonobos by Furuichi and Hashimoto were mainly funded by the Japan Society for the Promotion of Science (JSPS) Grants-in-aid for Scientific Research, JSPS Core-to-Core program, JSPS International Training Program, JSPS Asia-Africa Science Platform Program, JSPS Institutional Program for Young Researcher Overseas Visits, Japan Ministry of the Environment (JME) Global Environment Research Fund, JME Environment Research and Technology Development Fund, the National Geographic Fund for Research and Exploration, and Toyota Foundation. We thank Drs. Takayoshi Kano, Toshisada Nishida, Juichi Yamagiwa, Tetsuro Matsuzawa and other members of Primate Research Institute and Laboratory of Human Evolution of Kyoto University, and Drs. Mwanza Ndunda, Mbangi Norbert Mulavwa and other stuff of Ministry of Scientific Research of D.R. Congo for their continued support for our studies. The data on dolphins analyzed by Connor were funded by an NSF Dissertation Improvement Grant and grants from The National Geographic Society and a Fulbright Fellowship to Australia.

References

Bagemihl B (1999) Biological exuberance: animal homosexuality and natural diversity. St. Martins Press, New York

Boesch C, Boesch-Achermann H (2000) The chimpanzees of the Tai forest: behavioural ecology and evolution. Oxford University Press, New York

Connor RC, Smolker RA (1995) Seasonal changes in the stability of male–male bonds in Indian Ocean bottlenose dolphins (*Tursiops* sp.). Aquat Mamm 21:213–216

Connor RC, Smolker RA (1996) “Pop” goes the dolphin: a vocalization male bottlenose dolphins produce during consortships. Behaviour 133:643–662

Connor RC, Volmer NL (2009) Sexual coercion in dolphin consortships: a comparison with chimpanzees. In: Muller MN, Wrangham RW (eds) Sexual coercion in primates and humans: an evolutionary perspective on male aggression against females. Harvard University Press, Cambridge, pp 218–243

Connor RC, Smolker RA, Richards AF (1992a) Two levels of alliance formation among male bottlenose dolphins (*Tursiops* sp.). Proc Natl Acad Sci USA 89:987–990

Connor RC, Smolker RA, Richards AF (1992b) Dolphin alliances and coalitions. In: Harcourt AH, de Waal FBM (eds) Coalitions and alliances in animals and humans. Oxford University Press, Oxford

Connor RC, Richards AF, Smolker RA, Mann J (1996) Patterns of female attractiveness in Indian Ocean bottlenose dolphins. Behaviour 133:37–69

Connor RC, Wells R, Mann J, Read A (2000) The bottlenose dolphin: social relationships in a fission–fusion society. In: Mann J, Connor R, Tyack P, Whitehead H (eds) Cetacean societies: field studies of whales and dolphins. University of Chicago Press, Chicago, pp 91–126

de Waal FBM (1987) Tension regulation and nonreproductive functions of sex in captive bonobos (*Pan paniscus*). Natl Geogr Res 3:318–335

Dixson AF (1998) Primate sexuality: comparative studies of the prosimians, monkeys, apes, and human beings. Oxford University Press, Oxford

Dunn DG, Barco SG, Pabst DA, McLellan WA (2002) Evidence for infanticide in bottlenose dolphins of the western north Atlantic. J Wildl Dis 38:505–510

Ellison PT, Panter-Brick C, Lipson SF, O'Rourke MT (1993) The ecological context of human ovarian function. Human Reproduction 8:2248–2258

Fragaszy DM, Visalberghi E, Fedigan LM (2004) The complete capuchin. Cambridge University Press, Cambridge

Fujita S, Mitsunaga F, Sugiura H, Shimizu K (2001) Measurement of urinary and fecal steroid metabolites during the ovarian cycle in captive and wild Japanese macaques, *Macaca fuscata*. Am J Primatol 53:167–176

Fujita G, Sugiura H, Mitsunaga F, Shimizu K (2004) Hormone profiles and reproductive characteristics in wild female Japanese macaques (*Macaca fuscata*). Am J Primatol 64:367–375

Furuichi T (1985) Inter-male associations in a wild Japanese macaque troop on Yakushima Island, Japan. Primates 26:219–237

Furuichi T (1987) Sexual swelling receptivity and grouping of wild pygmy chimpanzee females at Wamba Zaire. Primates 28:309–318

Furuichi T (1989) Social interactions and the life history of female *Pan paniscus* in Wamba Zaire. Int J Primatol 10:173–197

Furuichi T (1997) Agonistic interactions and matrifocal dominance rank of wild bonobos (*Pan paniscus*) at Wamba. Int J Primatol 18:855–875

Furuichi T (2006) Evolution of the social structure of hominoids: reconsideration of food distribution and the estrus sex ratio. In: Ishida H, Tuttle R, Pickford M, Ogihara N, Nakatsukasa M (eds) Human origins and environmental backgrounds. Springer, New York, pp 235–248

Furuichi T (2009) Factors underlying party size differences between chimpanzees and bonobos: a review and hypotheses for future study. Primates 50:197–209

Furuichi T (2011) Female contribution to the peaceful nature of bonobo society. Evol Anthropol 20:131–142

Furuichi T, Idani G, Ihobe H, Hashimoto C, Tashiro Y, Sakamaki T, Mulavwa MN, Yangozene K, Kuroda S (2012) Long-term study on wild bonobos at Wamba, Luo Scientific Reserve, D. R. Congo: towards the understanding of female life history in a male-philopatric species. In: Kappler PM, Watts DP (eds) Long term field study of primates. Springer-Verlag, Berlin, Heidelberg, pp 413–433

Furuichi T, Hashimoto C (2002) Why female bonobos have a lower copulation rate during estrus than chimpanzees. In: Boesch C, Hohmann G, Marchant LF (eds) Behavioural diversity in chimpanzees and bonobos. Cambridge University Press, New York, pp 156–167

Furuichi T, Hashimoto C (2004) Sex differences in copulation attempts in wild bonobos at Wamba. Primates 45:59–62

Furuichi T, Ihobe H (1994) Variation in male relationships in bonobos and chimpanzees. Behaviour 130:211–228

Furuichi T, Idani G, Ihobe H, Kuroda S, Kitamura K, Mori A, Enomoto T, Okayasu N, Hashimoto C, Kano T (1998) Population dynamics of wild bonobos (*Pan paniscus*) at Wamba. Int J Primatol 19:1029–1043

Furuichi T, Mulavwa M, Yangozene K, Yamba-Yamba M, Motema-Salo B, Idani G, Ihobe H, Hashimoto C, Tashiro Y, Mwanza N (2008) Relationships among fruit abundance, ranging rate, and party size and composition of bonobos at Wamba. In: Furuichi T, Thompson J (eds) The bonobos: behavior, ecology, and conservation. Springer, New York, pp 135–149

Gerloff U, Hartung B, Fruth B, Hohmann G, Tautz D (1999) Intracommunity relationships, dispersal pattern and paternity success in a wild living community of bonobos (*Pan paniscus*) determined from DNA analysis of faecal samples. Proc R Soc Lond B 266:1189–1195

Goodall J (1986) The chimpanzees of Gombe: patterns of behavior. Harvard University Press, Cambridge

Harcourt AH, Stewart KJ (2007) Gorilla society: conflict, compromise, and cooperation between sexes. Chicago University Press, Chicago

Harcourt AH, Stewart KJ, Fossey D (1981) Gorilla reproduction in the wild. In: Graham CE (ed) Reproductive biology of the great apes: comparative and biomedical perspectives. Academic, New York, pp 265–279

Hasegawa T, Hiraiwa-Hasegawa M (1983) Opportunistic and restrictive mating among wild chimpanzees in the Mahala Mountains, Tanzania. J Ethol 1:75–85

Hashimoto C (1997) Context and development of sexual behavior of wild bonobos (*Pan paniscus*) at Wamba, Zaire. Int J Primatol 18:1–21

Hashimoto C, Furuichi T (1994) Social role and development of noncopulatory sexual behavior of wild bonobos. In: Wrangham R, McGrew WC, De Waal FBM, Heltne PG (eds) Chimpanzee cultures. Harvard University Press, Cambridge, London, pp 155–168

Hashimoto C, Furuichi T (2006a) Comparison of behavioral sequence of copulation between chimpanzees and bonobos. Primates 47:51–55

Hashimoto C, Furuichi T (2006b) Frequent copulations by females and high promiscuity in chimpanzees in the Kalinzu forest, Uganda. In: Newton-Fisher NE, Notman H, Paterson JD, Reynolds V (eds) Primates of western Uganda. Springer, New York, pp 247–257

Hashimoto C, Tashiro Y, Hibino E, Mulavwa M, Yangozene K, Furuichi T, Idani G, Takenaka O (2008) Longitudinal structure of a unit-group of bonobos: male philopatry and possible fusion of unit-groups. In: Furuichi T, Thompson J (eds) The bonobos: behavior, ecology, and conservation. Springer, New York, pp 107–119

Hohmann G, Fruth B (2000) Use and function of genital contacts among female bonobos. Anim Behav 59:107–120

Idani G (1991) Social relationships between immigrant and resident bonobo (*Pan paniscus*) females at Wamba. Folia Primatol (Basel) 57:83–95

Kano T (1982) The social group of pygmy chimpanzees (*Pan paniscus*) of Wamba. Primates 23:171–188

Kano T (1992) The last ape: pygmy chimpanzee behavior and ecology. Stanford University Press, Palo Alto

Kapsalis E, Johnson RL (2006) Getting to know you: female–female consortships in free-ranging rhesus monkeys. In: Sommer V, Vasey PL (eds) Homosexual behaviour in animals: an evolutionary perspective. Cambridge University Press, Cambridge, pp 220–237

Koyama N, Takahata Y, Huffman MA, Norikoshi K, Suzuki H (1992) Reproductive parameters of female Japanese macaques: thirty years data from the Arashiyama troops, Japan. Primates 33:33–47

Kuroda S (1980) Social behavior of the pygmy chimpanzees. Primates 21:181–197

Mann J (2006) Establishing trust: socio-sexual behaviour and the development of male–male bonds among Indian Ocean bottlenose dolphins. In: Sommer V, Vasey PL (eds) Homosexual behaviour in animals: an evolutionary perspective. Cambridge University Press, New York, pp 107–130

Matsumoto-Oda A (1999) Female choice in the opportunistic mating of wild chimpanzees (*Pan troglodytes schweinfurthii*) at Mahale. Behav Ecol Sociobiol 46:258–266

Mulavwa M, Furuichi T, Yangozene K, Yamba-Yamba M, Motema-Salo B, Idani G, Ihobe H, Hashimoto C, Tashiro Y, Mwanza N (2008) Seasonal changes in fruit production and party size of bonobos at Wamba. In: Furuichi T, Thompson J (eds) The bonobos: behavior, ecology, and conservation. Springer, New York, pp 121–134

Nadler R (1986) Sex-related behavior of immature wild mountain gorillas. Dev Psychobiol 19(2):125–137

Nishida T, Takasaki H, Takahata Y (1990) Demography and reproductive profiles. In: Nishida T (ed) The chimpanzees of the Mahale Mountains: sexual and life history strategies. University of Tokyo Press, Tokyo, pp 63–97

Nishida T, Corp N, Hamai M, Hasegawa T, Hiraiwa-Hasegawa M, Hosaka K, Hunt KD, Itoh N, Kawanaka K, Matsumoto-Oda A, Mitani JC, Nakamura M, Norikoshi K, Sakamaki T, Turner L, Uehara S, Zamma K (2003) Demography, female life history, and reproductive profiles among the chimpanzees of Mahale. Am J Primatol 59:99–121

Patterson IAP, Reid RJ, Wilson B, Grellier K, Ross HM, Thompson PM (1998) Evidence for infanticide in bottlenose dolphins: an explanation for violent interactions with harbour porpoises? Proc R Soc Lond B 265:1–4

Plooij FX (1984) The behavioral development of free-living chimpanzee babies and infants. Ablex, Norwood

Pusey A (1990) Behavioral-changes at adolescence in chimpanzees. *Behaviour* 115:203–246

Pusey AE, Packer C (1987) Dispersal and philopatry. In: Smuts BB, Cheney DL, Seyfarth RM, Wrangham RW, Struhsaker TT (eds) Primate societies. University of Chicago Press, Chicago, pp 250–266

Saito C, Sato S, Suzuki S, Sugiura H, Agetsuma N, Takahata Y, Sasaki C, Takahashi H, Tanaka T, Yamagiwa J (1998) Aggressive intergroup encounters in two populations of Japanese macaques (*Macaca fuscata*). Primates 39:303–312

Sommer V, Vasey PL (2006) Homosexual behavior in animals. Cambridge University Press, Cambridge

Sprague DS (1989) Male intertroop movement during the mating season among the Japanese macaques of Yakushima Island, Japan. A dissertation presented to the Faculty of the Graduate School of Yale University

Stevens JMG, Vervaecke H, Van Elsacker L (2008) The bonobo's adaptive potential: social relations under captive conditions. In: Furuichi T, Thompson J (eds) The bonobos: behavior, ecology, and conservation. Springer, New York, pp 19–38

Suzuki S, Hill DA, Sprague DS (1998) Intertroop transfer and dominance rank structure of nonnatal male Japanese macaques in Yakushima, Japan. Int J Primatol 19(4):703–722

Takahata Y (1980) The reproductive biology of a free-ranging troop of Japanese monkeys. Primates 21(3):303–329

Takahata Y (1982) The socio-sexual behavior of Japanese monkeys. Z Tierpsychol 59:89–108

Takahata Y, Suzuki S, Agetsuma N, Okayasu N, Sugiura H, Takahashi H, Yamagiwa J, Izawa K, Furuichi T, Hill DA, Maruhashi T, Saito C, Sato S, Sprague DS (1998) Reproduction of wild Japanese macaque females of Yakushima and Kinkazan Islands: a preliminary report. Primates 39:339–349

Tutin CEG (1979) Mating patterns and reproductive strategies in a community of wild chimpanzees (*Pan troglodytes schweinfurthii*). Behav Ecol Sociobiol 6:29–38

Tutin CEG, McGinnis PR (1981) Chimpanzee reproduction in the wild. In: Graham CE (ed) Reproductive biology of the great apes: comparative and biomedical perspectives. Academic Press, New York, pp 239–264

Vasey PL (1995) Homosexual behavior in primates: a review of evidence and theory. Int J Primatol 16:173–204

Vasey PL (2006) The pursuit of pleasure: an evolutionary history of female homosexual behaviour in Japanese macaques. In: Sommer V, Vasey PL (eds) Homosexual behaviour in animals. Cambridge University Press, Cambridge, pp 191–219

Wallis J (1997) A survey of reproductive parameters in the free-ranging chimpanzees of Gombe National Park. J Reprod Fertil 109:297–307

Watanabe K, Mori A (1992) Characteristic features of reproduction of Koshima monkeys, *Mucaca fuscata fuscata*: a summary of thirty-four years of observation. Primates 33:1–32
Watson-Capps J (2005) Female mating behavior in the context of sexual coercion and female ranging behavior of bottlenose dolphins in Shark Bay, Western Australia. Ph.D. thesis, Georgetown University, Washington, DC
Watts DP (2007) Effects of male group size, parity, and cycle stage on female chimpanzee copulation rates at Ngogo, Kibale National Park, Uganda. Primates 48:222–231
White FJ (1988) Party composition and dynamics in *Pan paniscus*. Int J Primatol 9:179–193
White FJ, Wrangham RW (1988) Feeding competition and patch size in the chimpanzee species *Pan paniscus* and *Pan troglodytes*. Behaviour 105:148–164
Wrangham RW (1979) Sex differences in chimpanzee dispersion. In: Hamburg DA, McCown ER (eds) The great apes. Benjamin/Cummings, Menlo Park, pp 481–489
Wrangham RW (2000) Why are male chimpanzees more gregarious than mothers? A scramble competition hypothesis. In: Kappeler PM (ed) Primate males: causes and consequences of variation in group composition. Cambridge University Press, Cambridge, pp 248–258
Wrangham R (2002) The cost of sexual attraction: is there a trade-off in female *Pan* between sex appeal and received coercion? In: Boesch C, Hohmann G, Marchant LF (eds) Behavioral diversity in chimpanzees and bonobos. Cambridge University Press, New York, pp 204–215
Yamagiwa J (2006) Playful encounters: the development of homosexual behaviour in male mountain gorillas. In: Homosexual behaviour in animals: an evolutionary perspective. Cambridge University Press, New York, pp 273–293

Chapter 21
A Mix of Species: Associations of Heterospecifics Among Primates and Dolphins

Marina Cords and Bernd Würsig

A dusky dolphin (*right*) and common dolphin (*left*) swim closely together in a mixed-species group in Admiralty Bay, New Zealand. (Photograph credit: Chris Pearson)

M. Cords (✉)
Department of Ecology, Evolution and Environmental Biology,
Columbia University, New York, NY 10027, USA
e-mail: mc51@columbia.edu

B. Würsig
Department of Marine Biology, Texas A&M University, Galveston, TX 77553, USA

J. Yamagiwa and L. Karczmarski (eds.), *Primates and Cetaceans: Field Research and Conservation of Complex Mammalian Societies*, Primatology Monographs,
DOI 10.1007/978-4-431-54523-1_21,

Abstract Among mammals, associations of two or more species are likely to involve taxa that are also gregarious intraspecifically, such as primates and delphinids. Although these two groups generally differ in habitat, diet, and the stability of their social units, they share mixed-species association as a conspicuous aspect of their behavior. We compare the general features of such associations in both groups and review the evidence for particular adaptive explanations and proximate mechanisms. On the whole, delphinid associations seem more likely to involve fluid individual membership and hybridization. Random chance seems unlikely to explain many associations in both taxa, but it can be challenging to rule out a shared attraction to environmental features as a driver. Both antipredator and foraging-benefit functions of mixed-species grouping are more directly supported for primates than for dolphins but are plausible adaptive explanations for both groups. Costs of association are better supported in primates, which face feeding competition and increased energetic burden; for dolphins, these costs appear to be minimal, and direct heterospecific social interactions, including harassment, may be more important. Vocal and visual signals may mediate associations, but comparatively little is known about such proximate mechanisms in comparison to adaptive function. Future study of delphinid associations will benefit from some of the approaches taken by primatologists, including the comparison of animals in and out of association, the correlation of association with environmental variables, and the comparison of different communities with different demographic or ecological characteristics.

Keywords Antipredator behavior • Competition • Foraging • Interspecific association • Mixed-species association • Mixed-species school • Multispecies aggregation Null models • Polyspecific association

21.1 Introduction

Assemblages including individuals of more than one species occur in a broad range of vertebrates (Dickman 1992). Flocks of birds in tropical regions or overwintering at high latitudes, clusters of tadpoles in forest ponds (Glos et al. 2007), schools of coral reef fish, and rays resting on the floor of oceanic bays (Semeniuk and Dill 2006) are among the diverse associations of heterospecific animals reported in the literature. Among mammals, associations are especially well known among primates, cetaceans, and savanna ungulates (Stensland et al. 2003), although they occur among other orders as well (e.g., carnivores: Minta et al. 1992; rodents and carnivores: Waterman and Roth 2007), and may also include nonmammalian taxa as partners. Among mammals, the most conspicuous mixed-species assemblages involve taxa that are gregarious intraspecifically as well, so that entire social units are associating with one another (Stensland et al. 2003).

Biologists studying these assemblages have focused on understanding why they occur. Some represent random meetings of animals that live sympatrically at high densities. Others result when animals with similar needs are attracted independently to the same environmental features, such as roosting or feeding sites. Most interesting, biologically, however, are the cases in which heterospecifics are attracted to one another per se. We refer to these cases in which members of one or more species are attracted to heterospecifics as "associations," in distinction to the broader term "assemblage," which applies to any situation in which heterospecifics are together.

Most investigations of associations focus on their adaptive value for the members of one or multiple associating species by identifying environmental factors that make associating beneficial even if there are also certain costs. Changes in the environmental features related to costs and benefits have the potential to explain temporal or spatial variation in the occurrence of associations. Although such functional investigations predominate in the literature, they have been complemented in a few cases by studies of the proximate behavioral mechanisms that bring and keep heterospecific individuals together. In this context, researchers examine communication systems or the spatial positioning of different individuals, for example. Although interesting in their own right, these mechanisms can also shed light on the costs and benefits that accrue to participants in mixed-species groups.

In this chapter, we review research that has been carried out on primates and delphinids and the mixed-species associations in which they participate. Although both taxa form associations with taxonomically distant animals (e.g., primates with birds: Boinski and Scott 1988; Seavy et al. 2001; Hankerson et al. 2006; King and Cowlishaw 2009; and dolphins with fish: Perrin et al. 1973; Scott and Cattanach 1998; Das et al. 2000; with birds: Würsig and Würsig 1980; Au and Pitman 1986; Camphuysen and Webb 1999; Vaughn et al. 2010; or with other mammals such as dugongs: Kiszka 2007; sea lions: Bearzi 2006; fur seals: Vaughn et al. 2007; or large whales: Vaughn et al. 2007; or large whales: Weller et al. 1996, Stockin et al. 2009), we focus on those that occur among primates and among dolphins, respectively.

The two taxa are similar in being highly social but differ in social organization, diet, and habitat (Gowans et al. 2008), factors that should relate to mixed-species association. In particular, primates that associate with other species live in stable groups of their own species, whereas the corresponding smaller delphinids that have been studied to date exhibit more fission–fusion social dynamics. Primates are largely herbivorous animals, typically including a high proportion of plant material (fruits and/or leaves) in their diets, quite different from carnivorous delphinids. Both fruiting trees and schools of fish typically represent spatially clumped resources for which consumers must actively search, but plant foods are fixed in space, making them easier to find and revisit, and setting the stage for contest competition for access to feeding sites. The open ocean habitat of some delphinids that form associations is considerably more vast and featureless than the forested habitat of most primates, presenting additional challenges for locating prey and avoiding predators. Although prey is abundant in the open ocean, it is ephemeral and occurs unpredictably in space and time, necessitating extensive travel by delphinids in search of a meal. Nearshore environments provide more structured habitats for delphinids and a more predictable supply of food, but the abundance is limited relative to open ocean sources. As a result, and also because crypsis is a viable strategy to avoid predation near shore, communities are smaller and more sedentary in such environments, thus resembling primates more closely than do their open-ocean counterparts (Gowans et al. 2008). Exceptions occur in areas where deep-water canyons or other features are close to shore, allowing for abundant mesopelagic prey to rise toward the surface at night, and for large "oceanic-type" schools of dolphins (Würsig and Würsig 2010; Smultea and Bacon 2012).

Despite these differences, both primates and delphinids are among mammals in which mixed-species association has been especially noted (Stensland et al. 2003). For both taxa, we review the general characteristics of these associations and consider available information that explains their occurrence from a functional perspective. Specifically, we review null models, as well as the two major adaptive hypotheses for mixed-species association. The adaptive reasons for mixed-species association in primates usually involve improved or more efficient predator defense, foraging success, or both (Stensland et al. 2003). Cords (2000) and Heymann and Buchanan-Smith (2000) provide comprehensive reviews of the relevant literature for primates, so here we focus on the kinds of observations that support these particular adaptive functions of association in primates. Similar reviews for delphinids are not available, most likely reflecting the fact that little research on delphinids has focused explicitly on this topic: our coverage of findings from delphinids is necessarily more descriptive. Finally, we briefly discuss proximate behavioral mechanisms that contribute to their formation and persistence.

21.2 Overview of Associations in Primate and Dolphins

Mixed-species association in primates is especially well known among African forest guenons (genus *Cercopithecus*: Gautier-Hion 1988; Cords 1990a; Buzzard 2010), between guenons and *Colobus* monkeys (Honer et al. 1997; Chapman and Chapman 2000), and among the South American tamarins (genus *Saguinus*) and Goeldi's monkey (*Callimico*: Heymann and Buchanan-Smith 2000; Rehg 2006), as well as the larger South American monkeys *Cebus* (capuchins) and *Saimiri* (squirrel monkeys: Pinheiro et al. 2011). In many cases, associations in these taxa involve three species or more. Apes, Asian monkeys, and prosimians generally seem to associate less, although there are exceptions (Freed 2006). Although there are reports of a single individual from one species associating with a heterospecific group (Fleury and Gautier-Hion 1997; Tutin 1999; Detwiler et al. 2005), most mixed-species associations of primates involve two or more heterospecific groups moving together, with individuals at least partially spatially intermingled, over various periods. Among tamarins and some guenons, associations with a particular heterospecific group may be essentially permanent, whereas in other cases partner groups may join together and split apart more frequently, typically spending several hours in association as they move together through multiple feeding trees. Transient associations can vary enormously in duration, however, and sometimes last for days before breaking up. In such cases, a group of a given species may associate in series with more than one group of a different species.

Dolphins form mixed-species associations especially often in deep oceanic waters, but also near shore and over continental slopes. In deep waters, an especially well-known association has been observed between pantropical spotted (*Stenella attenuata*) and spinner (*S. longirostris*), dolphins, often in association with yellowfin (*Thunnus albacares*) and skipjack (*Katsuwonus pelamis*) tuna (Au 1991). However, other deep-water associations are also common, especially in and near the tropics. These associations may include up to four species of greatly varying body size, ranging from the small-bodied *Stenella* spp. to much larger Risso's dolphins (*Grampus griseus*) and short-finned pilot whales (*Globicephala macrorhynchus*: Bearzi 2005 provides an overview), as well as Risso's dolphins with a host of other species (Würsig and Würsig 1980; Smultea and Bacon 2012). Common mixed-species associations in nearshore waters include common dolphins (both *D. delphis* and *D. capensis*) with common bottlenose dolphins (*Tursiops truncatus*) and (separately) with dusky dolphins (*Lagenorhynchus obscurus*: Markowitz 2004). In addition, single individuals are known to associate with groups of another species in this environment. One of many examples is an individual long-finned pilot whale (*Globicephala melas*) associating with groups of Atlantic white-sided dolphins (*Lagenorhynchus acutus*) over 6 years (Baraff and Asmutis-Silvia 1998); however, it is likely that such examples simply stand out near shore because of human presence (often, tourism), and because it is dramatic to see (in the present example) a group of smaller, lighter, white-sided dolphins with the large bulbous-headed black pilot whale.

Mixed-species associations in delphinids differ from those in primates in ways that reflect the nature of the social organization of the individual species involved. As already noted, most smaller delphinids that associate with heterospecifics live in fission–fusion societies, in contrast to their primate counterparts that live in stable groups (Gowans et al. 2008). In fission–fusion systems, conspecific individuals change intra- and intergroup affiliations, sometimes on an hourly basis. It is not surprising then that mixed-species associations also tend to be ephemeral in delphinids. For example, it is unlikely that the association of Pacific white-sided dolphins (*Lagenorhynchus obliquidens*) and northern right whale dolphins (*Lissodelphis borealis*) seen in Monterey Bay, California, lasts longer than several hours, or that it includes the same individuals from day to day (Würsig, personal observation). Such changeable individual membership in delphinid associations, along with the absence of semipermanent associations reported in some monkey communities, contrasts especially with primate examples. Another noteworthy aspect of interspecific relationships in delphinids is hybridization: interspecies copulations have often been described for dolphins and other odontocetes, and photographic and genetic evidence of hybrid births is becoming quite common [for example, between Atlantic spotted (*Stenella frontalis*) and common bottlenose dolphins: Herzing et al. 2003, and short-beaked common (*D. delphis*) and dusky dolphins: Reyes 1996, and Würsig, personal observation; see also Bérubé 2009]. Although hybridization also occurs among a few primate species that associate in heterospecific groups, it appears to be rare in most cases (Detwiler et al. 2005). Quantitative comparisons between the taxa are not possible, however.

Little is currently known about variation in mixed-species grouping as a function of ecological context. Primatologists are just beginning to evaluate how the relationships between species differ across communities, whereas there are no detailed comparisons of this sort for delphinids. In most cases, primatologists are limited to comparing only two communities that contain a given species pair (Cords 1990a; Holenweg et al. 1996; Chapman and Chapman 2000; Porter and Garber 2007). Both the frequency of association and the effects of association on the behavior of those involved may differ considerably from one community to another. With just two or a few data points, one can usually identify some plausible explanations for why the differences occur, but many more data are needed to use intercommunity comparisons as an effective tool to identify the factors that drive and limit associations. At present, we can simply say that the ecological context in which any two associates find themselves matters substantially. Population (or group) densities of subject species, their partners, and others with whom they may interact (such as competitors) create various demographic scenarios that are likely to influence the probability of association. For example, some researchers have suggested, both for primates and for delphinids, that associations with heterospecifics may occur especially often when conspecifics are rare or unavailable as social partners (Fleury and Gautier-Hion 1997; Frantzis and Herzing 2002). The situation of "lone sociable" dolphins interacting with a school of heterospecifics may be an example of this social attraction, at least at times. In addition, demographic factors may interact with variation in resource abundance and predator pressure, to influence diets, group size, and ranging patterns, which in turn may determine the value and hence the frequency of association with other species (Cords 1990a).

21.3 Null Models

Primatologists have led the way in devising methods to distinguish assemblages that result from mutual attraction from those that are random or which reflect independent interactions with the same environmental features. Waser (1982, 1984), considering the assemblages of mangabeys with other sympatric monkeys, devised null models based on the physics of gas molecules. These models predict the frequency and duration of associations between groups of different species and were used in subsequent empirical studies to determine whether associations occurred more often than expected by chance (Cords 1987; Whitesides 1989; Holenweg et al. 1996; Porter and Garber 2007). Hutchinson and Waser (2007) updated and corrected some unsatisfactory or erroneous aspects of the initial models, including the problem of statistical comparisons of observed and expected encounter rates which, first-generation models did not address and the issue of nonrandom movement patterns which earlier models ignored despite their occurrence in nature. Although Hutchinson and Waser commented explicitly that updated predictions of encounter durations will not alter conclusions about the nonrandom nature of mixed-species association in African monkeys based on that parameter, conclusions based on encounter rates or proportion of time spent in mixed-species associations may merit reevaluation. Simulation models provide an alternative approach for generating null expectations of association time. Whitesides (1989) pioneered this approach in his studies of associations of forest monkeys in Sierra Leone, some of which he found to participate in associations in excess of a random expectation. Simulations of animal movements and encounters based on random walks were also developed by Hutchinson and Waser (2007) but have not yet been applied to mixed-species associations.

Researchers have not tested the null hypothesis of random association in dolphins using such predictive models because many of the input parameters needed to derive a null expectation (travel speeds, group diameters, and group densities) are poorly known or not known at all. Furthermore, no models to date capture the dynamic group sizes characteristic of fission–fusion societies, such as those of many of the smaller delphinids. Nevertheless, it is likely that the large-school association of spinner and pantropical spotted dolphins in the eastern tropical Pacific can occur for many days at a time, and possibly for weeks to months (Norris and Dohl 1980), and dusky and short-beaked common dolphins have been reidentified together for up to several weeks in summer in New Zealand (Würsig, unpublished data). Off the Atlantic coast of Costa Rica, common bottlenose dolphins often associate with Guyana (formerly termed tucuxi) dolphins (*Sotalia guianensis*), and known individuals of each species repeatedly socialize together (often aggressively, as discussed later), day after day (Acevedo-Gutiérrez et al. 2005; May-Collado 2010). We emphasize that when members of two or more species of dolphins are traveling together in the expanse of the ocean for even more than a few minutes, chance seems an unlikely explanation. These cases are quite unlike the generally short-term association of, for example, short-beaked common, Atlantic spotted, and common bottlenose dolphins when feeding on an aggregated food supply off the Azores (Clua and Grosvalet 2001).

The null expectations generated by gas or simulation models are limited because they have not incorporated environmental features, such as feeding or resting sites, that might independently attract heterospecific groups and thus bring them together. Such ecological attractors could greatly increase the expected frequency or duration of associations, even when heterospecifics are not drawn to each other directly, and thus null models incorporating such factors would provide a more conservative benchmark against which to compare observed values. Primates are well known to share certain resources, such as large fruiting trees in their forest habitats. Waser and Case (1981) derived a model to explain competitive relationships between monkey species based on the chances of each species finding a common feeding site and the chances of one group supplanting another there. The model can be used to predict encounter rates between species. Although Skorupa (1983) later challenged application of this model to the Kibale Forest monkey data to which it was originally applied, suggesting that variable land tenure systems need to be accounted for, this model remains one of the only attempts to generate quantitative predictions about how often heterospecifics should congregate at a common resource without being attracted to one another per se.

Again, however, no such model has been developed for or applied to associations of delphinids. It seems clear, however, that, some delphinid associations are simply aggregations around shared food. Such aggregation has been reasonably well described, for example, for short-beaked common dolphins and common bottlenose dolphins near the Azores (Clua and Grosvalet 2001). The common dolphins appear to herd prey fish into tight balls near the surface, and bottlenose dolphins at times show up and at least temporarily displace the common dolphins to feed on the ball of fish.

Another argument against more complex null hypotheses incorporating attractive resources comes from the behavior of associating individuals: if a particular resource brings them together, then associations should occur only or mainly in the presence of that resource. Such an evaluation may sometimes be straightforward, but we emphasize three further points. First, some animals, including many forest monkeys, eat from a wide variety of plants, which they prefer to different degrees. In such cases, it may not be straightforward for an observer to differentiate the particularly attractive food resources that drive ranging decisions from those eaten opportunistically when an animal happens to pass by, and yet this distinction is important for ruling out resource-based associations by this method. However, this concern may be minimal in cases where associations continue for many hours, as when heterospecific monkeys pass through a series of feeding trees together: an identical long route through an array of potential feeding sites is hard to reconcile with movement decisions that are independent of the heterospecific group (Holenweg et al. 1996; Cords 1987). Second, there may be cases in which a common resource does bring heterospecific animals together, so association results from mutual attraction to a resource, and yet their ensuing behavioral interactions may be far from random. Although it is neither a primate nor a delphinid example, the coordination between badgers and coyotes hunting ground squirrels is a case in point (Minta et al. 1992). Such examples illustrate that it is important to examine behavioral interactions, and not merely the rates of association or the amount of time that heterospecifics spend together (Stensland et al. 2003). Third, in some cases, heterospecifics may occur together only when exploiting a common resource, and yet this pattern results *not* from independent attraction to the resource, but rather from a special motivation to exploit particular resources jointly because it is beneficial. For example, dusky and short-beaked common dolphins seem to coordinate herding of prey (Markowitz 2004), and it is likely that members of both species benefit from more efficient herding (as in Würsig and Würsig 1980, for one species) by increasing their chance of a bigger meal. In this case, attraction to a common resource is coupled with obvious benefits related to joint exploitation.

21.4 Adaptive Explanations: Predator Defense

Protection against predators may be enhanced in mixed-species groups through various behavioral mechanisms including improved early warning (because additional eyes and ears, possibly tuned to different types or locations of predators, are available), greater ability to evade detection by the predator (selfish herd effect), heightened confusion of a predator's attack, reduced probability of being the one victim, and more effective active defense. Not all these mechanisms need apply in any particular case (Heymann and Buchanan-Smith 2000). Observations that associating primate species share common predators and respond to one another's alarm calls (Heymann and Buchanan-Smith 2000; Eckardt and Zuberbühler 2004) generally support this hypothesis but are not critical tests. Direct comparisons of predator success on single- versus mixed-species groups are seldom possible because predation is so rarely observed. Thus, the most compelling evidence for a predator-defense function comes from behavioral observations that correlate risk with association status.

Where risk is variable over time, it is possible to see if it correlates with the formation of mixed-species associations. Several studies of primates have taken this approach. For example, Noë and Bshary (1997) reported that red colobus (*Piliocolobus badius*) monkeys in the Ivory Coast increase association rates with highly vigilant Diana monkeys (*Cercopithecus diana*) during the time of year when chimpanzees (which prey on red colobus) hunt most often. In addition, playbacks of chimpanzee calls, mimicking the presence of these predators, induced associations between the monkeys, or prolonged the duration of those in progress. Similarly, Chapman and Chapman (2000) report that red colobus (*Piliocolobus tephrosceles*) in East Africa form more associations in areas with higher chimpanzee density, and at times when their groups contain a relatively large number of vulnerable infants. Others have examined vigilance levels as a function of association with heterospecifics. Blue and redtail monkeys (*Cercopithecus mitis* and *C. ascanius*) reduce individual vigilance levels when associated with each other (Cords 1990b), as one would expect if they feel safer in mixed-species groups. Diana monkeys reduce individual vigilance levels when associated with *Cercopithecus campbelli* (Wolters and Zuberbühler 2003). Similarly, studies of both wild and captive tamarins show that individuals reduce vigilance levels in mixed- versus single-species groups, although the total number of vigilant individuals increases (Heymann and Buchanan-Smith 2000; but see Garber and Bicca-Marquez 2001). In tamarins, there are species differences in forms of vigilance (Peres 1993; Buchanan-Smith and Hardie 1997): red-bellied tamarins glance upward more often than the saddlebacks with whom they often associate. Captive saddlebacks in single-species groups look up more often than they do when associated with heterospecifics, as if compensating for the greater danger of aerial predators in these circumstances.

Mixed-species associations of dolphins may also benefit by greater protection through early warning or at least a greater dilution effect of more animals present when encountered by, for example, a single large predatory shark. However, there is less information on these points than exists from the more detailed and longer-term studies of primates, and it is primarily the behavior of animals in association that suggests the antipredator advantages of mixed-species association for delphinids. For example, when large delphinids such as pilot whales travel with small ones such as *Stenella* spp., it is possible that because the stronger (and lower frequency) echolocation clicks of the large animals travel substantially further than those of the smaller ones, a greater detection benefit accrues to the *Stenella* spp. as they pay attention to the potential fright/flee movements of their heterospecific associates. Unfortunately, there are no known data to bear on this idea. Although it is a case of association with a non-dolphin species, we have seen individuals of mixed schools of Eastern Tropical Pacific spotted and spinner dolphins being alerted to attack by a large shark from behind and below (where dolphins do not have eyes, or effective echolocation) by the rapid outward movement, away from the shark, of the tuna that were traveling below. Thus, the dolphins—both species—were alerted to the presence of danger by association with tuna. Spinner dolphins feed and rest at different times than do spotted dolphins, with different states of alertness in the two species at any one time likely improving vigilance in mixed-species groups throughout the day. Norris and Dohl (1980) and Norris et al. (1994) postulated that enhanced predator detection is traded off between species, and that this is an important benefit driving associations of heterospecifics in especially dangerous "shark waters" of the deep tropical ocean. The recent report of Kiszka et al. (2011) bears on this idea. They found that spinner and pantropical spotted dolphins associated more often when traveling than when foraging, and that they used deeper water when associated than when alone; accordingly, they suggested that spinner dolphins associate with spotted dolphins for safety against surprise attacks by sharks while the spinners are transiting from one rest area to the next. Similarly, island- or atoll-associated spinner and spotted dolphins tend to occur as single species, but travel in multispecies associations when far from shore, even though their diets are similar near and far from shore (Würsig et al. 1994).

We have no information on vigilance differences of dolphins in and out of mixed-species associations. However, Markowitz (2004) showed that dusky dolphins off Kaikoura, New Zealand, and the nearshore-living Hector's dolphins (*Cephalorhynchus hectori*) of that same area sometimes form multispecies nursery groups, especially in summer when killer whales (*Orcinus orca*) are most prevalent, and it is possible that this is an antipredator vigilance response. At the same time, dusky dolphins are also more often joined by generally smaller groups of short-beaked common dolphins in summer months. However, because the common dolphins are at the edge of their range and occur off Kaikoura less often in non-summer periods, it is not certain whether this latter affiliation is indeed in direct response to danger from killer whales (Markowitz 2004), although this has been suggested (Srinivasan and Markowitz 2010).

Another form of predator defense, namely mobbing, may be facilitated when heterospecifics associate. Among primates, mobbing individuals approach and sometimes threaten a predator while giving alarm calls: most likely this behavior reduces the success rate of the predator, for whom any element of surprise is ruined. For small monkeys, however, mobbing may be risky. Although sample sizes are tiny, Heymann and Buchanan-Smith (2000) suggest that mobbing in tamarins may be more likely when they are in mixed- versus single-species groups. Others have concluded that mobbing—including attacks of the predator—by heterospecifics is an attractive service to those less inclined to engage in this risky behavior, perhaps because they are smaller bodied and more vulnerable (Struhsaker 1981; Cords

1987; Gautier-Hion and Tutin 1988). Eckardt and Zuberbühler (2004) argued that ready and conspicuous adult male mobbing of eagles by putty-nosed monkeys (*Cercopithecus nictitans*) made it worthwhile for socially dominant Diana monkeys to tolerate their competitors in mixed-species associations.

Apparent mobbing in dolphins has been mentioned but not well described, as the approach by the generally smaller dolphins to their potential predators, killer whales or sharks, is generally not as easily observed as monkeys interacting with their predators in daylight. Nevertheless, Markowitz (2004) and the second author have on several occasions seen dusky dolphins rapidly approaching killer whales in apparent mobbing behavior, and Jefferson et al. (1991) postulated that similar behavior may occur often between prey and predator.

In sum, evidence for antipredator advantages of mixed-species association is quite strong in primates, where researchers have been able to correlate association with risk and have witnessed asymmetries in mobbing behavior that suggest that some associates are receiving a guard-like service from others. Some observations of delphinids are consistent with risk-sensitive association patterns, but the role of mobbing behavior in predator deterrence, and the influence of mixed-species association on mobbing behavior, remain largely unknown at present.

21.5 Adaptive Explanations: Improved Foraging

Improved predator defense in mixed-species groups may enable more advantageous foraging. For example, both red colobus and Diana monkeys are more likely to feed at or near ground level when associating with largely terrestrial sooty mangabeys (*Cercocebus atys*), which are more likely to sound the alarm to terrestrial predator models than their more arboreal partners (McGraw and Bshary 2002). Diana monkeys show similar increases in feeding at lower forest levels when associating with *Cercopithecus campbelli* (Wolters and Zuberbühler 2003), which frequents lower forest strata more often than the dianas. Similarly, mustached and red-bellied tamarins (*Saguinus mystax* and *S. labiatus*) are more likely to feed at ground level when associating with saddle-back tamarins (*S. fuscicollis*) than when on their own, both in the wild and in captivity (Heymann and Buchanan-Smith 2000). Saddlebacks generally spend more time in vegetation nearer the ground than their associates. Shared and improved vigilance in these mixed-species groups is well documented (see earlier). Several other studies have documented an increase in foraging niche width or foraging area as a function of association, and present at least some evidence that improved predator avoidance allows the niche expansion (Gautier-Hion et al. 1983; Cords 1987; Porter 2001; Porter and Garber 2007). Many of these examples highlight a particular advantage for *mixed*-species groups: given species differences in the forest strata most often used, there are particular advantages to associating with heterospecifics that are more alert to predators in certain microhabitats. In primates, those microhabitats are often height dependent.

Other foraging advantages of primate mixed-species association are known. One species may make food available to another in a very direct way, as when mustached tamarins flush insects downward where saddleback tamarins subsequently capture them (Peres 1992a) or when grey-cheeked mangabeys (*Lophocebus albigena*) expose the pulp of large hard-shelled fruits which redtails (*Cercopithecus ascanius*) can then eat (Struhsaker 1981). Similarly, bark-gouging marmosets (*Callithrix emiliae*) may make gum available to associated mustached tamarins (Lopes and Ferrari 1994). More indirectly, one species may guide another to advantageous feeding sites. For example, Terborgh (1983) and Cords (1987) argued that members of a species with a larger home range (squirrel monkeys, *Saimiri sciureus*, and redtails, respectively) benefited from associating with groups of a sympatric species with smaller ranges (capuchins, *Cebus apella* and *C. albifrons*, and blue monkeys, respectively) by avoiding areas recently depleted by the latter. Their arguments were somewhat indirect, however, based primarily on ranging patterns rather than direct assessments of food discovery rate. In addition, the frequency of association correlated with dietary overlap, as expected if finding shared food sources is a primary advantage of associating. Podolsky (1990), measuring feeding time on different kinds of resources, added the observation that the greater swath of a traveling mixed-species group ensured a greater likelihood of discovery of certain types of food sources, namely smaller fruiting trees (5–15 m crown diameter), relative to what would occur in a single-species group.

We have no direct information on improved foraging in delphinid mixed-species associations. It seems likely that two vigilant species, especially if they feed at different times or in a different manner, as in the spinner-spotted dolphin system in the Eastern Tropical Pacific (Norris and Dohl 1980), could each gain a foraging advantage from the presence of the other. Advantages in finding food may also result if one associating species has longer-distance echolocation capabilities and the other species is able to take advantage of this (an unsubstantiated postulation). Dusky dolphins herd schooling fishes in shallow nearshore waters of Argentina and certain bays in New Zealand. In Argentina, Risso's dolphins at times associate with dusky dolphins at the periphery of their school, and it has been surmised that the Risso's are taking advantage of larger prey that aggregate where dusky dolphins have herded fish (Würsig and Würsig 1980). As already mentioned, short-beaked common dolphins in New Zealand were observed engaging in apparent coordinated fish-herding behavior with dusky dolphins (Markowitz 2004), presumably enhancing their own foraging success, and potentially contributing to that of the dusky dolphins. No agonistic or competitive interactions were evident between the two species, suggesting that common dolphins were at least tolerated by the more numerous dusky dolphins. Quérouil et al. (2008) postulate that delphinid mixed-species associations off the Azores may more often be for enhanced foraging than for predator avoidance or other social reasons.

As in primates, it is certainly possible that the larger swath taken up by a mixed-species dolphin group covers more area than a single-species group traveling alone, thus increasing discovery rate of certain prey, especially those that aggregate or school. Furthermore, one species adept at herding prey (such as dusky and bottlenose dolphins) could have another dolphin species take advantage of this situation, and feed directly on the herded prey (but this could lead to potentially destructive competition), or feed on prey that have been attracted to the primary herded prey, as suggested earlier. Other than dolphins, larger fish, small sharks, diving and hovering birds, sea lions, and fur seals take advantage of the prey herded by the dusky dolphins (Würsig et al. 2007; Vaughn et al. 2008), but we know of no clear situation where two or more dolphin species travel together while only one does the work. One potential for such social parasitism may exist with common bottlenose dolphins taking advantage of herding efforts by short-beaked common and pan-tropical spotted dolphins (Clua and Grosvalet 2001). Further research focusing on foraging advantages in delphinid mixed-species asscociations will help evaluate these plausible but speculative ideas about food-finding advantages.

Finally, we must include joint territorial defense as a foraging advantage that may accrue to certain primates in mixed-species groups. In tamarins, heterospecific groups may form essentially permanent associations, and neighboring pairs defend territorial boundaries against one another (Garber 1988; Peres 1992b). Garber (1988) correlated success in these encounters with group size, but interpretation of these results as indicating an advantage of mixed-species association is not entirely clear because group size in his sample depended on the number of adults of only one species, which happened to be more active in range defense (Heymann and Buchanan-Smith 2000). It remains possible, however, that the less active partner species benefits from the range defense of the more active one: Peres (1992b) showed that saddleback tamarins have higher insect foraging success rates in the center of the range which is successfully defended by the partner mustached tamarins, whereas more peripheral areas that were less successfully defended reduced foraging rates for the saddlebacks. Outside the genus *Saguinus*, this kind of joint territorial defense is unknown. Furthermore, the importance of joint territorial defense is not supported by all studies of callitrichines (Rehg 2006).

In dolphins, there appears to be little active defense of space, except for some specific nearshore examples. Hawaiian spinner dolphins that have entered a bay for daytime rest may actively exclude other spinners from entering that same bay (Norris et al. 1994), and similar site defense may occur across species, with bottlenose dolphins of northern Scotland, for example, at times being highly aggressive against harbor porpoises (*Phocoena phocoena*), and thereby probably restricting harbor porpoise range (Ross and Wilson 1996). A similar situation may be occurring off Costa Rica, with bottlenose dolphins possibly "harassing" the smaller Guyana dolphins, but here the situation is further complicated by the facts that the same individuals co-occur, at least for some time, there is interspecies sexual activity, and some evidence for hybridization (Acevedo-Gutiérrez et al. 2005; May-Collado 2010). We can conceive of mixed-species associations that form highly integrated (although ephemeral; see earlier) societies as perhaps better keeping an undesired third species out of the society (say, spotted and spinner dolphins coordinating against false killer whales, *Pseudorca crassidens*, known to prey occasionally on these smaller species; Perryman and Foster 1980), but there are no actual data to bear on the subject.

In sum, data from primates show that association with heterospecifics can increase access to feeding areas and particular foods in beneficial ways. For delphinids, there are no comparable direct data, but aspects of the natural history of associating species suggest possible foraging advantages that merit further study. Research strategies similar to those used for primates—in which predation risk and feeding behavior are related to association frequency—are likely to provide useful information for delphinids in the future.

We close our discussion on adaptive benefits with a reminder to future researchers that the particular benefits that apply may vary considerably from case to case. Studies of primates have shown that researchers convinced that associations are nonrandom often look for evidence for many possible adaptive benefits, but find only certain benefits supported in their study system. Furthermore, even in one setting, the advantages—and their relative importance—for any one participating species may differ relative to other participating species. Last, as already noted, the same species may associate for different reasons in different ecological settings. Overall, the ability to associate with other species thus seems to offer its practitioners a flexible way to solve a variety of problems.

21.6 Costs and Limits to Association

Mixed-species associations may also be costly, and costs, as benefits, may not be experienced similarly by the participants. Although in some cases costs may be minimal (Porter et al. 2007), in others they may act to limit the occurrence of associations, despite the benefits that could accrue (Chapman and Chapman 2000; Rehg 2006).

In primates, feeding competition is the cost most often considered, and direct observations of contests, as well as documentation of reduced feeding rates in association, provide evidence that it is important in some cases (Cords 1987; Heymann and Buchanan-Smith 2000; Rehg 2006). Heymann (1997) has emphasized the importance of body weight ratios in explaining why some callitrichines do not associate: similar body weights dictate similar feeding ecology, with too much competition to allow stable associations (Heymann and Buchanan-Smith 2000). However, to the extent that members of different species diverge in their dietary requirements when shared food is in short supply, and share food only when it is relatively abundant, feeding competition is not a necessary consequence of association with other species. In addition, some primate species, although socially subordinate to their partners, have found coping mechanisms: they may rush to a feeding site ahead of their dominant associates, to have some uninterrupted feeding time, or may linger at the site longer (Cords 1987; Bicca-Marquez and Garber 2003). Some of the documented cases of foraging niche expansion may also reflect ways of coping with the threat of contest competition by broadening the array of foods consumed.

Another cost for primates is the energetic burden of extra travel required when a group increases in size and therefore depletes local feeding areas more rapidly. Many reports of primate mixed-species association reveal that one or more associating partners increase their rate of movement in association (Cords 2000; Chapman and Chapman 2000).

It is possible that mixed-species association in dolphins is also related to the costs of feeding competition, although there is no direct evidence. Association may be more common in the open ocean than along shore because dolphins in the open ocean tend to travel great distances while feeding both on resources near the surface (such as by pantropical spotted dolphins) and on those associated with the deep scattering layer that rises toward the surface at night (such as by spinner dolphins). Thus, open ocean dolphins of different species can remain together and yet diverge in their feeding niches (Scott and Cattanach 1998); furthermore, they are so much on the move—perhaps to evade predators—that they are unlikely to deplete their resources in any one place. Their association is therefore likely to be one whose main function is minimizing predation although, as previously mentioned, there may be an element of differential food finding involved as well. Near shore, however, dolphins often have quite restricted ranges, as in particular bays or inlets, and it is more likely that local food resources would be depleted more quickly, making association more costly. Alternatively, however, reduced association near shore may reflect reduced benefits, rather than or in addition to increased costs, in this habitat: nearshore dolphins have the variability of the shore "to hide" from predators, and by having one wall of the shoreline and another wall of the bottom close beneath them, they may minimize the element of predator surprise. These protective walls do not exist in the open ocean, where another species may be sought out for partial protection instead. We stress the speculative nature of these considerations, however, given the lack of direct evidence.

Although traveling on land tends to be expensive for vertebrates (for a chimpanzee example, see Sockol et al. 2007), dolphins expend relatively little energy when they travel (Williams 1999). Moving greater distances to minimize depletion of local food supplies, if it occurs, is probably not an appreciable extra expense of mixed-species associations in dolphins, in contrast to the situation with primates. One exception is likely to hold for mothers and their young offspring, as Noren (2008) has shown that mothers travel more slowly when accompanied by an infant. Mothers and infants, who tend to avoid large groups that are sexually active or "boisterous," may be less inclined to move within a highly social mixed-species group (Pearson 2011).

For dolphins, interspecific harassment seems to reach intensities not reported in primates, and its occurrence may at times limit associations. In the clear waters of the Bahamas, Herzing and Johnson (1997) and Herzing et al. (2003) described larger offshore common bottlenose dolphin males attempting to force copulations on smaller Atlantic spotted dolphins (*Stenella frontalis*) and male spotted dolphins forming temporary coalitions to ward off and "mob" the larger aggressors. A somewhat similar but perhaps more often aggressive set of interactions has been described for spinner and pantropical spotted dolphins of Hawai'i (Psarakos et al. 2003). There, copulations often seemed to be consensual, but spinner dolphins almost always reacted aggressively toward spotted dolphins when the latter first approached. Psarakos et al. (2003) speculated that although appearing to cooperate, individuals of a dominant species are actually being aggressive toward those of inferior status, and the copulations are not in fact consensual (as has been postulated for primates; Strier 2006). It is unclear for bottlenose and Guyana dolphins whether the larger bottlenose are indeed the dominant aggressor (Acevedo-Gutiérrez et al. 2005). Nevertheless, copulations by like-sized species such as short-beaked common dolphins and dusky dolphins off New Zealand appear to be truly consensual. In both these species, both males and females mate with multiple partners, and sexual competition within each species probably occurs mainly at the sperm volume level. The noncompetitive nature of between-species copulations may be a mere by-product of the polygynandrous mating systems (see also Kenagy and Trombulak 1986 and Brownell and Ralls 1986) together with the proximity of heterospecific mating partners.

In sum, although the notion that costs may limit mixed-species associations seems applicable to both primates and delphinids, the case appears to be stronger for primates. Contest and scramble competition—the latter reflected in increased travel rates in association—are well documented, whereas such competition appears relatively unimportant in open-ocean delphinids, where food is less limiting and travel costs are low. In nearshore environments, competition could be a more important factor, but there is no direct evidence. Interspecific harassment in delphinids has been reported and may be a form of contest competition; however, it seems to take forms, such as forced interspecific copulations, never reported in primates.

21.7 Mechanisms That Start or Perpetuate Associations

Surprisingly little is known about the behavioral mechanisms underlying mixed-species group formation and persistence. Among primates, studies of callitrichines suggest quite strongly that vocalizations are important in attracting heterospecific groups to one another (Cords 2000). Tamarins make antiphonal loud contact calls and respond to those of other species. Windfelder's (2001) playback experiments showed clearly that a heterospecific's call was sufficient to attract association partners.

Guenon males also produce loud calls, and Zuberbühler's work has shown that members of one species attend to the calls of another: for example, they respond to another species' alarm call as if the predator were present (Wolters and Zuberbühler 2003; Eckardt and Zuberbühler 2004). To date, however, there is no evidence that these male calls are related to mixed-species association, even though males of different species may counter-call in long bouts of vocalizations (Cords 1987). It has been noted that other group members vocalize at higher rates during associations, a difference that was significant for Campbell's monkeys when together with Diana monkeys (Wolters and Zuberbühler 2003); this increased rate of vocalization may have an interspecific function, although other possibilities remain.

Many dolphins produce calls that sound quite similar to those of humans, but with species-specific characteristics. Overall, however, little progress has been made in deciphering the details of dolphin communication in any one species, let alone in mixed-species associations. It is very likely that members of different species can at least determine each other's state of arousal, such as whether fearful or aggressive, just as human researchers have learned to interpret some of these basic vocal messages (Dudzinski et al. 2009). Visual communication is also possible among dolphin species. Postural stances can be shared across species; for example, an S-shaped adult body posture that seems to come from a tightening of mid-body musculature signifies aggression (in at least common bottlenose and spinner dolphins), as does a particular type of jaw clap, tooth raking and biting, and abrupt bubble blowing (Psarakos et al. 2003). Frantzis and Herzing (2002) report synchronized swimming among three dolphin species in the Gulf of Corinth as coordinating associations. Affiliative behavior may be indicated by presenting the belly to another animal, whether of the same or different species (e.g., Atlantic spotted and common bottlenose dolphins; Herzing and Johnson 1997). Bottlenose dolphins of Isla del Coco, Costa Rica increase whistling when feeding compared to when not feeding, and this increase may serve to call others to the feeding activity, increasing cooperative strategies to contain the prey and thereby presumably allowing each individual to feed in a more efficient fashion (Acevedo-Gutiérrez and Stienessen 2004). Bottlenose dolphins of Moray Firth, Scotland, make a low-frequency bray call that is food related, possibly serving incidentally to inform others as well (Janik 2000).

Adjustments to movements are a less conspicuous way of forming and perpetuating mixed-species associations. In primates, they have been described mainly anecdotally. Cords (2000) reviewed this literature, and documented adjustments of travel rates, trajectories, habitat and microhabitat use, and scheduling of major activities. Such adjustments may be critical in maintaining associations, because heterospecifics may move at different rates, and have different priorities related to where and when they move. In dolphins, we suspect that one determinant of the end of an association is when one species travels too fast or dives too deeply for another to follow. One species may also travel outside the realm of another's preferred habitat, as when dusky dolphins move away from the nearshore murky water environment preferred by their occasional associate, the Hector's dolphin (Markowitz 2004).

Understanding mechanisms that lead to the formation and persistence of mixed-species associations can help decipher the relative benefits of participation. One would expect the animals that have the most to gain to be most responsible for initiating or perpetuating associations. Indeed, on this basis, Teelen (2007) argued that associations between redtail and red colobus monkeys in Uganda serve to protect redtails from eagles, rather than red colobus from chimpanzees: the redtails consistently initiated, maintained, and ended the encounters, but it was red colobus who suffered most from chimpanzee predation. Similarly, Cords (1987) found redtail monkeys more responsible than blue monkeys in initiating and terminating associations, and thus corroborated more indirect evidence about the ways in which associations benefited both species, but the redtails especially. One needs to be somewhat careful in interpreting all behavioral adjustments as indicating an interest in (or benefit from) association: it is also possible that some adjustments are forced upon participants who are merely coping with heterospecifics and unable to control their presence. In dolphins, for example, it often occurs that a few members of one species interact with many members of another, as in Kaikoura, New Zealand when a dozen or so short-beaked common dolphins associate with several hundred dusky dolphins (Würsig et al. 2007). In these cases, the benefits to members of the larger group are probably minimal, as are the costs, whereas the benefits for the less numerous "joiners" might be substantial. The joiners appear to be tolerated, and may provide elements of social stimulation even if they provide little or nothing in the way of enhanced predator detection, predator defense, or foraging success. Approaches and departures from a heterospecific association are probably more informative than changes in activity and diet for this reason, for both primates and dolphins.

21.8 Conclusion

Clearly our comparative review must be viewed as preliminary, given the limited study directed to the question of between-species relations, and particularly association, in delphinids. Nonetheless, many aspects of these associations seem similar in dolphins and the better-studied primates. Perhaps the most noteworthy contrasts suggested by the information available concern the roles of feeding competition and travel costs as constraints on the formation and persistence of associations. Feeding competition appears to be an important factor in explaining variation among primates, both between species and between communities (Cords 1990a), while there is no evidence that it is important in dolphin associations. In fact, cooperative herding of mobile prey may actually improve individual foraging success in dolphins. Although between-species aggression does occur in delphinids, it is not yet clear if it inhibits association despite foraging or antipredator benefits. Travel costs, higher for terrestrial mammals and exacerbated in larger groupings for primates, are also less likely to limit associations of delphinids as they do in primates (Pearson 2011). Future study of dolphin associations will benefit from some of the approaches taken by primatologists, including the comparison of animals in and out of association, the correlation of association with environmental variables, and the comparison of different communities with different demographic or ecological characteristics. Given the logistics of studying highly mobile marine mammals with very flexible social groupings, it will likely not be easy to fill in these gaps, but we encourage fellow researchers to try!

21.9 Acknowledgments

We thank Heidi Pearson, Timothy Markowitz, Robin Vaughn, Juichi Yamagiwa, and three anonymous reviewers for making suggestions that improved the manuscript, and Chris Pearson for the photograph.

References

Acevedo-Gutiérrez A, Stienessen SC (2004) Bottlenose dolphins (*Tursiops truncatus*) increase number of whistles when feeding. Aquat Mamm 30:357–362

Acevedo-Gutiérrez A, DiBerardinis A, Larkin S, Larkin K, Forestell P (2005) Social interactions between tucuxis and bottlenose dolphins in Gandoca-Manazanillo, Costa Rica. LAJAM 4:49–54

Au DW (1991) Polyspecific nature of tuna schools: shark, dolphin, and seabird associates. Fish Bull 89:342–354

Au DW, Pitman RL (1986) Seabird interactions with dolphins and tuna in the eastern tropical Pacific. Condor 88:304–317

Baraff LS, Asmutis-Silvia RA (1998) Long-term association of an individual long-finned pilot whale and Atlantic white-sided dolphin. Mar Mamm Sci 14:155–161

Bearzi M (2005) Dolphin sympatric ecology. Mar Ecol Res 1:165–175

Bearzi M (2006) California sea lions use dolphins to locate food. J Mammal 87:606–617

Bérubé M (2009) Hybridism. In: Perrin WF, Würsig B, Thewissen JGM, Perrin WF, Würsig B, Thewissen JGM (eds) Encyclopedia of marine mammals, 2nd edn. Elsevier/Academic, Amsterdam, pp 588–592

Bicca-Marquez JC, Garber PA (2003) Experimental field study of the relative costs and benefits to wild tamarins (*Saguinus imperator* and *S. fuscicollis*) of exploiting contestable food patches as single- and mixed-species troops. Am J Primatol 60:139–153

Boinski S, Scott PE (1988) Association of birds with monkeys in Costa Rica. Biotropica 20:136–143

Brownell RL Jr, Ralls K (1986) Potential for sperm competition in baleen whales. Rep Int Whaling Comm 8:97–112, Special issue

Buchanan-Smith HM, Hardie SM (1997) Tamarin mixed-species groups: an evaluation of a combined captive and field approach. Folia Primatol (Basel) 68:272–286

Buzzard PJ (2010) Polyspecific association of *Cercopithecus campbelli* and *C. petaurista* with *C. diana*: what are the costs and benefits? Primates 51:307–314

Camphuysen K, Webb A (1999) Multi-species feeding associations in North Sea seabirds: jointly exploiting a patchy environment. Ardea 87:177–198

Chapman CA, Chapman LJ (2000) Interdemic variation in mixed-species association patterns: common diurnal primates in Kibale National Park, Uganda. Behav Ecol Sociobiol 47:129–139

Clua É, Grosvalet F (2001) Mixed-species feeding aggregation of dolphins, large tunas and seabirds in the Azores. Aquat Living Resour 14:11–18

Cords M (1987) Mixed-species association of *Cercopithecus* monkeys in the Kakamega Forest, Kenya. Univ Calif Publ Zool 117:1–109

Cords M (1990a) Mixed-species association of East African guenons: general patterns or specific examples? Am J Primatol 21:101–114

Cords M (1990b) Vigilance and mixed species association in some East African forest guenons. Behav Ecol Sociobiol 26:297–300

Cords M (2000) Mixed-species association and group movement. In: Boinski S, Garber P (eds) On the move: how and why animals travel in groups. University of Chicago Press, Chicago, pp 73–99

Das K, LePoint G, Loizeau V, DeBacker V, Dauby P, Bouquegneau JM (2000) Tuna and dolphin associations in the North Atlantic: evidence of different ecological niches from stable isotope and heavy metal measurements. Mar Pollut Bull 40:102–109

Detwiler KM, Burrell AS, Jolly CJ (2005) Conservation implications of hybridization in African cercopithecine monkeys. Int J Primatol 26:661–684

Dickman CR (1992) Commensal and mutualistic interactions among terrestrial vertebrates. Trends Ecol Evol 7:194–197

Dudzinski KM, Thomas JA, Gregg JD (2009) Communication. In: Perrin WF, Würsig B, Thewissen JGM (eds) Encyclopedia of marine mammals, 2nd edn. Elsevier/Academic, Amsterdam, pp 260–269

Eckardt W, Zuberbühler K (2004) Cooperation and competition in two forest monkeys. Behav Ecol 14:400–411

Fleury MC, Gautier-Hion A (1997) Better to live with allogenerics than to live alone? The case of single male *Cercopithecus pogonias* in troops of *Colobus satanas*. Int J Primatol 18:967–974

Frantzis A, Herzing DL (2002) Mixed-species associations of striped dolphins (*Stenella coeruleo-alba*), short-beaked common dolphins (*Delphinus delphis*), and Risso's dolphins (*Grampus griseus*) in the Gulf of Corinth (Greece, Mediterranean Sea). Aquat Mamm 28:188–197

Freed BZ (2006) Polyspecific associations of crowned lemurs and Sanford's lemurs in Madagascar. In: Gould L, Sauther ML (eds) Lemurs: ecology and adaptation. Springer, New York, pp 111–131

Garber PA (1988) Diet, foraging patterns and resource defense in a mixed species troop of *Saguinus mystax* and *S. fuscicollis* in Amazonian Peru. Behaviour 105:18–34

Garber PA, Bicca-Marquez JC (2001) Evidence of predator sensitive foraging and traveling in single-and mixed-species tamarin troops. In: Miller L (ed) Eat or be eaten: predator-sensitive foraging among primates. Cambridge University Press, Cambridge, pp 138–153

Gautier-Hion A (1988) Polyspecific associations among forest guenons: ecological, behavioural and evolutionary aspects. In: Gautier-Hion A, Bourliere F, Gautier JP, Kingdon J (eds) A primate radiation: evolutionary biology of the African guenons. Cambridge University Press, Cambridge, pp 452–476

Gautier-Hion A, Quris R, Gautier JP (1983) Monospecific vs. polyspecific life: a comparative study of foraging and anti-predatory tactics in a community of *Cercopithecus* monkeys. Behav Ecol Sociobiol 12:325–335

Gautier-Hion A, Tutin CE (1988) Simultaneous attack by adult males of a polyspecific troop of monkeys against a crowned eagle. Folia primatol. 51:149–151

Glos J, Dausmann KH, Linsenmair KE (2007) Mixed-species social aggregations in Madagascan tadpoles: determinants and species composition. J Nat Hist 41:1965–1977

Gowans S, Würsig B, Karczmarski L (2008) The social structure and strategies of delphinids: predictions based on an ecological framework. Adv Mar Biol 53:195–294

Hankerson SJ, Dietz JM, Raboy BE (2006) Associations between golden-headed lion tamarins and the bird community in the Atlantic Forest of southern Bahia. Int J Primatol 27:487–495

Herzing DL, Johnson CM (1997) Interspecific interactions between Atlantic spotted dolphins (*Stenella frontalis*) and bottlenose dolphins (*Tursiops truncatus*) in the Bahamas, 1985–1995. Aquat Mamm 23:85–99

Herzing DL, Moewe K, Brunnick BJ (2003) Interspecies interactions between Atlantic spotted dolphins, *Stenella frontalis*, and bottlenose dolphins, *Tursiops truncatus*, on Great Bahama Bank, Bahamas. Aquat Mamm 29:335–341

Heymann EW (1997) The relationship between body size and mixed-species troops of tamarins (*Saguinus* spp.). Folia Primatol (Basel) 68:287–295

Heymann EW, Buchanan-Smith HM (2000) The behavioural ecology of mixed-species troops of callitrichine primates. Biol Rev 75:169–190

Holenweg AK, Noë R, Schabel M (1996) Waser's gas model applied to associations between red colobus and Diana monkeys. Folia Primatol (Basel) 67:125–136

Honer OP, Leumann L, Noe R (1997) Dyadic associations of red colobus and Diana monkey groups in the Tai National Park, Ivory Coast. Primates 38:281–291

Hutchinson JMC, Waser PM (2007) Use, misuse and extension of "ideal gas" models of animal encounter. Biol Rev 82:335–359

Janik VM (2000) Food-related bray calls in wild bottlenose dolphins (*Tursiops truncatus*). Proc R Soc Lond 267:923–927

Jefferson TA, Stacey PM, Baird RW (1991) A review of killer whale interactions with other marine mammals: predation to co-existence. Mamm Rev 21:151–180

Kenagy GJ, Trombulak SC (1986) Size and function of mammalian testes in relation to body size. J Mammal 67:1–22

King AJ, Cowlishaw G (2009) Foraging opportunities drive interspecific associations between rock kestrels and desert baboons. J Zool 277:111–118

Kiszka JJ (2007) Atypical associations between dugongs (*Dugong dugon*) and dolphins in a tropical lagoon. J Mar Biol Assoc UK 87:101–104

Kiszka J, Perrin WF, Pusineri C, Ridoux V (2011) What drives island-associated tropical dolphins to form mixed-species associations in the southwest Indian Ocean? J Mammal 92:1105–1111

Lopes M, Ferrari S (1994) Foraging behavior of a tamarin group (*Saguinus fuscicollis wedeli*) and interactions with marmosets (*Callithrix emiliae*). Int J Primatol 15:373–387

Markowitz T (2004) Social organization of the New Zealand dusky dolphin. Ph.D. Dissertation, Wildlife and Fisheries Sciences, Texas A&M University, College Station

May-Collado L (2010) Changes in whistle structure of two dolphin species during interspecific associations. Ethology 116:1065–1074

McGraw WS, Bshary R (2002) Association of terrestrial mangabeys (*Cercocebus atys*) with arboreal monkeys: experimental evidence for the effects of reduced ground predator pressure on habitat use. Int J Primatol 23:311–325

Minta SC, Minta KA, Lott DF (1992) Hunting associations between badgers (*Taxidea taxus*) and coyotes (*Canis latrans*). J Mammal 73:814–820

Noë R, Bshary R (1997) The formation of red colobus–Diana monkey associations under predation pressure from chimpanzees. Proc R Soc Lond B 264:253–259

Noren SR (2008) Infant carrying behaviour in dolphins: costly parental care in an aquatic environment. Funct Ecol 22:284–288

Norris KS, Dohl TD (1980) The structure and functions of cetacean schools. In: Herman LM (ed) Cetacean behavior: mechanisms and functions. Wiley, New York, pp 211–261

Norris KS, Würsig B, Wells RS, Würsig M (1994) The Hawaiian spinner dolphin. University of California Press, Berkeley, p 408

Pearson H (2011) Sociability of female bottlenose dolphins (*Tursiops* spp.) and chimpanzees (*Pan troglodytes*): understanding evolutionary pathways toward social convergence. Evol Anthropol 20:85–95

Peres CA (1992a) Prey-capture benefits in a mixed-species group of Amazonian tamarins, *Saguinus fuscicollis* and *S. mystax*. Behav Ecol Sociobiol 31:339–347

Peres CA (1992b) Consequences of joint territoriality in a mixed-species group of tamarin monkeys. Behaviour 123:220–246

Peres CA (1993) Anti-predator benefits in a mixed-species group of Amazonian tamarins. Folia Primatol (Basel) 61:61–76

Perrin WF, Warner RR, Fiscus CH, Holts DB (1973) Stomach contents of porpoise, *Stenella* spp., and yellowfin tuna, *Thunnus albacares*, in mixed-school aggregations. Fish Bull 71:1077–1092

Perryman WL, Foster TC (1980) Preliminary report on predation by small whales, mainly the false killer whale, *Pseudorca crassidens*, on dolphins (*Stenella* spp. and *Delphinus delphis*) in the eastern tropical Pacific. Administrative report LJ-80-05. National Marine Fisheries Service, La Jolla

Pinheiro T, Ferrari SF, Lopes MA (2011) Polyspecific associations between squirrel monkeys (*Saimiri sciureus*) and other primates in eastern Amazonia. Am J Primatol 73:1145–1151

Podolsky RM (1990) Effects of mixed-species association on resource use by *Saimiri sciureus* and *Cebus apella*. Am J Primatol 21:147–158

Porter LM (2001) Benefits of polyspecific association for the Goeldi's monkey (*Callimico goeldi*). Am J Primatol 54:143–158

Porter LM, Garber PA (2007) Niche expansion of a cryptic primate, *Callimico goeldii*, while in mixed species troops. Am J Primatol 69:1340–1353

Porter LM, Sterr SM, Garber PA (2007) Habitat use and ranging behavior of *Callimico goeldi*. Int J Primatol 28:1035–1058

Psarakos S, Herzing DL, Marten K (2003) Mixed-species associations between pantropical spotted dolphins (*Stenella attenuata*) and Hawaiian spinner dolphins (*Stenella longirostris*) in Oahu, Hawaii. Aquat Mamm 29:390–395

Quérouil S, Silva MA, Cascão I, Magalhães S, Seabra MI, Machete MA, Santos RS (2008) Why do dolphins from mixed-species associations in the Azores? Ethology 114:1183–1194

Rehg JA (2006) Seasonal variation in polyspecific associations among *Callimico goeldii*, *Saguinus labiatus*, and *S. fuscicollis* in Acre, Brazil. Int J Primatol 27:1399–1428

Reyes JC (1996) A possible case of hybridism in wild dolphins. Mar Mamm Sci 12:301–307

Ross HM, Wilson B (1996) Violent interactions between bottlenose dolphins and harbour porpoises. Proc R Soc Lond 263:283–286

Scott MD, Cattanach KL (1998) Diel patterns in aggregations of pelagic dolphins and tuna in the eastern Pacific. Mar Mamm Sci 14:401–422

Seavy NE, Apodaca CK, Balcomb SR (2001) Associations of crested guineafowl *Guttera pucherani* and monkeys in Kibale National Park, Uganda. Ibis 143:310–312

Semeniuk CAD, Dill LM (2006) Anti-predator benefits of mixed-species groups of cowtail stingrays (*Pastinachus sephen*) and whiprays (*Himantura uarnak*) at rest. Ethology 112:33–43

Skorupa JP (1983) Monkeys and matrices: a second look. Oecologia (Berl) 57:391–396

Smultea MA, Bacon CE (2012) A comprehensive report of aerial marine mammal monitoring in the Southern California range complex: 2008–2012. EV5 Environmental. Naval Facilities Engineering Command Southwest (NAVFAC SW), San Diego

Sockol MD, Raichlen DA, Ponzer H (2007) Chimpanzee locomotor energies and the origin of human bipedalism. Proc Natl Acad Sci USA 104:12265–12269

Srinivasan M, Markowitz TM (2010) Predator threats and dusky dolphin survival strategies. In: Würsig B, Würsig M (eds) The dusky dolphin: master acrobat off different shores. Elsevier/Academic Press, Amsterdam, pp 133–150

Stensland E, Angerbjorn A, Berggren P (2003) Mixed species groups in mammals. Mamm Rev 33:205–223

Stockin KA, Binedell V, Wiseman N, Brunton DH, Orams MB (2009) Behavior of free-ranging common dolphins (*Delphinus* sp.) in the Hauraki Gulf, New Zealand. Mar Mamm Sci 25:283–301

Strier KB (2006) Primate behavioral ecology, 3rd edn. Allyn & Bacon, Boston

Struhsaker TTS (1981) Polyspecific associations among tropical rain-forest primates. Z Tierpsychol 57:268–304

Teelen S (2007) Influence of chimpanzee predation on associations between red colobus and red-tailed monkeys at Ngogo, Kibale National Park, Uganda. Int J Primatol 28:593–606

Terborgh J (1983) Five New World primates. Freeman, New York

Tutin CEG (1999) Fragmented living: behavioral ecology of primates in a forest fragment in the Lopé Reserve, Gabon. Primates 40:249–265

Vaughn RL, Shelton DE, Timm LL, Watson LA, Würsig B (2007) Dusky dolphin (*Lagenorhynchus obscurus*) feeding tactics and multi-species associations. N Z J Mar Freshw Res 41:391–400

Vaughn RL, Würsig B, Shelton DS, Timms LL, Watson LA (2008) Dusky dolphins influence prey accessibility for seabirds in Admiralty Bay, New Zealand. J Mammal 89:1051–1058

Vaughn RL, Würsig B, Packard J (2010) Dolphin prey herding: prey ball mobility relative to dolphin group and prey ball size, multispecies associates, and feeding duration. Mar Mamm Sci 26:213–225

Waser PM (1982) Primate polyspecific associations: do they occur by chance? Anim Behav 30:1–8

Waser PM (1984) "Chance" and mixed-species associations. Behav Ecol Sociobiol 15:197–202

Waser PM, Case TJ (1981) Monkeys and matrices: on the coexistence of "omnivorous" forest primates. Oecologia (Berl) 49:102–108

Waterman JM, Roth JD (2007) Interspecific associations of Cape ground squirrels with two mongoose species: benefit or cost? Behav Ecol Sociobiol 61:1675–1683

Weller DW, Würsig B, Whitehead H, Norris JC, Lynn SK, Davis RW, Clauss N, Brown P (1996) Observations of an interaction between sperm whales and short-finned pilot whales in the Gulf of Mexico. Mar Mamm Sci 12:588–594

Whitesides GH (1989) Interspecific associations of Diana monkeys, *Cercopithecus diana*, in Sierra Leone, West Africa: biological significance or chance? Anim Behav 37:760–776

Williams TM (1999) The evolution of cost efficient swimming in marine mammals: limits to energetic optimization. Philos Trans R Soc Lond 354:193–220

Windfelder T (2001) Interspecific communication in mixed-species groups of tamarins: evidence from playback experiments. Anim Behav 61:1193–1201

Wolters S, Zuberbühler K (2003) Mixed-species associations of Diana and Campbelli's monkeys: the costs and benefits of a forest phenomenon. Behaviour 140:371–385

Würsig B, Würsig M (1980) Behavior and ecology of the dusky dolphin, *Lagenorhynchus obscurus*, in the South Atlantic. Fish Bull 77:871–890

Würsig B, Würsig M (2010) The dusky dolphin: master acrobat off different shores. Elsevier/Academic, Amsterdam

Würsig B, Wells RS, Norris KS (1994) Food and feeding. In: Norris KS, Würsig B, Wells RS, Würsig M (eds) The Hawaiian spinner dolphin. University of California Press, Berkeley, pp 216–231

Würsig B, Duprey N, Weir J (2007) Dusky dolphins (*Lagenorhynchus obscurus*) in New Zealand waters: present knowledge and research goals. DOC Research and Development Series 270. Department of Conservation, Wellington

Index

J. Yamagiwa and L. Karczmarski (eds.), *Primates and Cetaceans: Field Research and Conservation of Complex Mammalian Societies*, Primatology Monographs, DOI 10.1007/978-4-431-54523-1, © Springer Japan 2014

Printed by Publishers' Graphics LLC